TRANSACTIONS

OF THE

AMERICAN PHILOSOPHICAL SOCIETY

HELD AT PHILADELPHIA

FOR PROMOTING USEFUL KNOWLEDGE

NEW SERIES—VOLUME XXIV

PHILADELPHIA:

Published by the Society

1935

CONTENTS OF VOLUME XXIV

TRANSACTIONS

OF THE

AMERICAN PHILOSOPHICAL SOCIETY

HELD AT PHILADELPHIA

FOR PROMOTING USEFUL KNOWLEDGE

NEW SERIES—VOLUME XXIV

PART II.   JUNE, 1935

A Commentary on Loureiro's "Flora Cochinchinensis"

E. D. Merrill

PHILADELPHIA:

THE AMERICAN PHILOSOPHICAL SOCIETY

104 South Fifth Street

1935

LANCASTER PRESS, INC., LANCASTER, PA.

A COMMENTARY ON LOUREIRO'S "FLORA COCHINCHINENSIS"

By E. D. MERRILL

There are certain types of botanical publications that have caused much trouble to systematists who have attempted to monograph various groups of plants. The difficulties in dealing with such works are due to various causes, chiefly inaccurate, inadequate, or indefinite original descriptions of genera and species included in them and, as far as later investigators are concerned, absence or inaccessibility of authentically named specimens for purposes of study and comparison. Where later systematists have had access to botanical material examined by earlier authors in the preparation of their publications, the problems as a rule have been easily solved. In works like Blanco's "Flora de Filipinas" where that author preserved no botanical material, and Loureiro's "Flora Cochinchinensis" where such botanical material as was prepared by him has partly been destroyed, the problem of determining the status of numerous genera and species described, in relation to those proposed and described by other authors, becomes a distinctly complicated one. In the case of those publications based on regions where the floras, for all practical purposes, are now thoroughly well known, it has usually been possible to determine the status of most of the forms described by the early authors; but when older publications were based on material from regions as yet imperfectly explored, such as the Philippines and Indo-China, the problem is decidedly more difficult.

The author of a monographic treatment of any natural group, such as a family, a genus, or a subgenus, normally attempts to account for all species and binomials proposed in such groups, while the author of a flora of a definite region normally attempts to account for all species that have been credited by other authors to the area he is attempting to cover. All systematists realize that some monographic treatises and various published floras fall far short of this indicated goal. This is because some authors, either because of personal inclination or belief, on account of various difficulties encountered, or because of the absence or inaccessibility of authentic material, or the unavailability of certain published papers, follow the lines of least resistance, treating fully those species well known to them, treating sketchily those imperfectly known, and leaving a residue of obscure ones in such categories as "species incertae sedis," "species dubiae," "species exclusae" and "inextricabiles," and not infrequently ignoring binomials entirely which, for one reason or another, are more or less obscure.

One regrettable result of the publication of such works as those of Blanco and Loureiro is that systematic botanical literature is overburdened with a large number of binomials either proposed by the original authors or those based by later systematists on the original imperfect descriptions that most modern authors have not been able to place to their full satisfaction. Loureiro's binomials and those proposed by his successors, but based on his original descriptions, bulk large in many lists of unknown or imperfectly understood genera and species. It has therefore seemed desirable to make a rather intensive study of all of his descriptions with a view to placement of his binomials, as far as they can be placed in the light of our present knowledge, in relation to those proposed by other authors, and at the same time to attempt to account for the very numerous binomials proposed by later authors but based on Loureiro's descriptions.

CERTAIN PRE-LINNAEAN PUBLICATIONS

Among the pre-Linnaean publications on the Indo-Malaysian flora, whose illustrations and descriptions typify a large number of binomials proposed by Linnaeus and his successors, may be mentioned Rheede's "Hortus Malabaricus," eleven volumes, folio (1678–1703); Rumphius' "Herbarium Amboinense," six volumes, and "Auctuarium," folio (1741–55); Burman's "Thesaurus Zeylanicus" (1737); and Linnaeus' "Flora Zeylanica" (1747). In the case of the last two works much of the botanical material on which they were based is still extant, but of the first two it is manifest that no botanical material was prepared or, if specimens were prepared, this was done with no idea of their preservation; hence all binomials based on the work of Rheede and Rumphius must be interpreted solely on the basis of the generalized non-technical descriptions and rather crude illustrations given in the original publications.

Burman's material on which the "Thesaurus Zeylanicus" was based is, at least in part, preserved in the Delessert Herbarium, now at the Jardin Botanique at Geneva, Switzerland, and that on which the "Flora Zeylanica" was based is now preserved in the herbarium of the British Museum, Natural History, London. The extant "Thesaurus Zeylanicus" material does not appear to have been critically studied by any modern botanist, but Trimen [1] has done this for Hermann's Ceylon collections on which the "Flora Zeylanica" was based. Trimen's work is invaluable as an aid in determining the exact status of many of the Linnaean binomials typified by the specimens named and briefly described in the "Flora Zeylanica" (1747) and even more briefly characterized in the "Species Plantarum" (1753). Trimen, however, neither in his original paper nor in his succeeding work, "Handbook of the Flora of Ceylon" (1894–1900), interpreted the Linnaean binomials in accordance with the principles of priority but usually adopted names in current use instead of making the new combinations necessitated by the rule of priority.

Several attempts have been made to interpret the species described by Rheede [2] and by Rumphius [2] in view of their importance in relation to the exact status of numerous bi-

[1] **Trimen, H.** Hermann's Ceylon herbarium and Linnaeus' ' Flora Zeylanica.' Journ. Linn. Soc. Bot. 24: 129–155. 1887.

[2] **Burman, J.** Index universalis in sex tomos et auctuarium herbarii Amboinensis Cl. Georgii Everhardi Rumphii. Herb. Amb. Auct. [1–20]. 1755.

Burman, J. Index alter in omnes tomos herbarii Amboinensis cl. G. Everhardi Rumphii, quem de novo recensuit, auxit et emendavit Joannes Burmannus [1–22]. 1769.

This was issued in connection with Burman's index to Rheede's Hortus Malabaricus and is sometimes bound in the last volume of that work.

Anonymous. Register op het Ambons Kruid-Boek von G. E. Rumphius. 1–16. 1764.

Copy in the library of the British Museum, Natural History. See Britton, J. Journ. Bot. 56: 363–364. 1918.

Linnaeus, C. Herbarium Amboinense, quod consens. experient. Facult. Medicae in Regia Academia Upsalensi, sub praesidio viri nobilissimi atque experientissimi, Dn. Doct. Caroli Linnaei . . . publico examini submittit, Alumnus Regius Olavus Stickman. i–iv, 1–28. 1754.

Republished with slight alterations under the title: Herbarium Amboinense, sub praesidio D. D. Car. Linnaei, proposuit Olavus Stickman. Amoen. Acad. 4: 112–143. 1759.

nomials, based wholly or partly on data given in these pre-Linnaean works.   The problems involved are by no means simple, and none of the publications based on these two fundamental botanical works are entirely satisfactory.   What is particularly needed in reference to the unsolved problems of Rheede and Rumphius are more intensive explorations of the classical localities in India and in the Moluccas, the field work to be carried on over considerable periods of time with special reference not only to the descriptive data given by these pre-Linnaean authors but also with very special reference to the native names cited, habitats, and indicated economic uses of the plants they described.   Until this is done and these data correlated with binomial nomenclature, we shall continue to guess at what numerous binomials typified by references to these pre-Linnaean works are supposed to represent, but with more comprehensive collections, together with full field data from the classical localities, we can in most cases approach the position of exactness.

CERTAIN POST-LINNAEAN PUBLICATIONS

Among the post-Linnaean publications on the Indo-Malaysian flora that have caused numerous difficulties to later systematists are Burman's "Flora Indica" (1768); Loureiro's "Flora Cochinchinensis" (1790, Willdenow's edition, 1793); and Blanco's "Flora de Filipinas" (1837, ed. 2, 1845).

Some years ago [3] I made an attempt to determine the status of Burman's new species as far as this could be done from a study of the short descriptions and the illustrations; but this work needs correction and amplification through an examination of Burman's extant types at Geneva.   Unfortunately the Burman herbarium was not retained as a special collection, the specimens being scattered through the large general herbarium, so that it is frequently difficult to locate specific types, and some of these are apparently no longer available.   Various specialists, in monographing genera or families, have examined many of Burman's specimens and have prepared amplified descriptions based, at least in part, on the original material.

Blanco, as noted above, preserved no botanical material, describing his species from time to time over a period of many years as he had the opportunity of examining fresh specimens.   His work is notably uneven and naturally contains numerous errors, both of

Buchanan-Hamilton, F.   Commentary on the Herbarium Amboinense. Liber Primus.   Mem. Wern. Soc. 5: 307–383. 1826.

Buchanan-Hamilton, F.   A commentary on the second book of the Herbarium Amboinense.   Op. cit. 6: 268–333. 1832.

Henschel, A. G. E. T.   Clavis Herbarii Amboinensis: in his Vita G. E. Rumphii. 139–202. 1833.

Hasskarl, J. K.   Neuer Schlüssel zu Rumph's Herbarium Amboinense.   Abh. Naturf. Gesellsch. Halle 9: 145–389. 1866.   Reprint 1–247. 1866.

Merrill, E. D.   An interpretation of Rumphius's Herbarium Amboinense.   Bur. Sci. Publ. 9: 1–595, map, 1917.

Burman, J.   Flora Malabarica, sive index in omnes tomos horti malabarici, quem juxta normam a botanicis hujus aevi receptam conscripsit, et ordine alphabetico digessit. 1–10. 1769.

Dennstedt, A. W.   Schlüssel zum Hortus Indicus Malabaricus. 1–40. 1818.

Buchanan-Hamilton, F.   Commentary on the Hortus Malabaricus. 1–410. 1822.

[Dillwyn, L. W.]   A review of the references to the Hortus Malabaricus of Henry van Rheede van Draakenstein. i–viii, 1–69. 1839.

[3] Merrill, E. D.   A review of the new species of plants proposed by N. L. Burman in his Flora Indica. Philip. Journ. Sci. 19: 329–388. 1921.

observation and of interpretation.   In 1905 I made a preliminary study of Blanco's species [4] for use chiefly as a guide for what needed to be done, and thirteen years later published a more extensive work [5] in which each Blancoan species was more or less critically considered.   This publication was supplemented by sixteen sets of duplicate botanical specimens, each set containing 1060 specimens, which were distributed to the larger herbaria in Europe, America, and Asia, the specimens being selected to represent Blanco's species as I then understood them.   Through an intensive knowledge of the Philippine flora, particularly of those regions familiar to Blanco, and through a study of each description together with a careful consideration of other data, native names, localities, habitats, times of flowering, and economic uses, supplemented by special trips to special localities to search for individual species, it became possible definitely to place all but one of Blanco's twenty-three new genera, and all but about fifty of the 1136 species and varieties described by him, including the 636 that he described as new.   In my studies of such species based on Rumphius and on those proposed by Burman, Blanco, Llanos [6] and others, I was actuated by a desire to determine as far as possible the status of the numerous species described by these authors, and the numerous new binomials proposed by later authors but based wholly on these usually inexact, often incomplete, and in other ways unsatisfactory early descriptions, and to correlate the species with those described by other authors under other names.

In 1919 I completed a preliminary study of Loureiro's species, following the same principles that had guided me in the other studies mentioned above.   The result was i–xxxvii + 1–693 pages of typescript which was prepared in sextuplicate.   To stimulate further work on the numerous unsolved problems, copies were sent to the British Museum, Natural History, London; the Museum d'histoire naturelle, Paris; the Institut scientifique, Saigon, Indo-China; the United States Department of Agriculture, Washington; and to the Canton Christian College (now Lingnan University), Canton, China.   This manuscript stimulated the preparation of several important papers [7] based on Loureiro's extant types in the herbarium of the British Museum which have solved numerous problems in relation to Loureiro's genera and species, problems that could scarcely otherwise have been solved because of Loureiro's faulty descriptions; yet while I, and doubtless some other botanists, look on these contributions to stability in nomenclature as distinctly important, others will sympathize with Gagnepain [8] whose review of Moore's papers merely states: "L'auteur, après Elmer D. Merrill, a essayé s'appuyant sur la collection de Londres, de donner aux plantes de Loureiro une synonimie certaine.   Malgré de très louables efforts, il n'a pas toujours réussi."   Moore incidentally placed definitely about twenty-five genera described by Loureiro, many of which previous authors had failed to interpret (even to the extent of determining to what families they belonged), supplied the information that rendered it possible to place two other misunderstood and unplaced genera, and definitely settled the status of about fifty species, previously just as doubtful as the genera above mentioned.

---

[4] **Merrill, E. D.**   A review of the identifications of the species described in Blanco's Flora de Filipinas. Govt. Lab. Publ. [Philip.] **27**: 1–132. 1905.

[5] **Merrill, E. D.**   Species Blancoanae.   A critical revision of the Philippine species of plants described by Blanco and by Llanos.   Bur. Sci. Publ. **12**: 1–423. 1918.

[6] The Philippine species of Llanos are considered in Species Blancoanae.

[7] See special bibliography, p. 23.

[8] **Gagnepain, F.**   Bull. Soc. Bot. France **73**: 752–753. 1926.

He also indicated certain valid Indo-China species that had been overlooked by the authors of the "Flore générale de l'Indo-Chine," of which actual types are extant.

THE BEARING OF THE INTERNATIONAL CODE ON THE PRESENT PROBLEM

As long as botanists were content to follow conventional usage in adopting binomials without regard to the historical aspects of each individual case and without regard to priority—and some botanists still do this—the question of the exact identity of a doubtful species proposed by any early author was perhaps of little importance. With the rapidly increasing tendency to adopt the principle of priority, modified by the lists of *nomina generica conservanda* approved by the International Botanical Congresses held at Vienna (1905), Brussels (1910), and Cambridge (1930), the exact status of each unit, whether genus or species, proposed by early authors and long considered as imperfectly known or of doubtful status, becomes distinctly important in connection with the question of stability in nomenclature. If we are to follow the principles of priority in selecting names of described species, we cannot hope even to approach the desired stability until the exact status of a high percentage of all doubtful species proposed by early authors shall be determined. It is evident that the desired end cannot be attained by even the most critical revision of any one or two of the early botanical works which contain the descriptions of numerous new genera and species, the status of many of which are uncertain. The solutions of numerous problems presented by the publications of Rheede, Rumphius, Burman, Blanco, Loureiro, and others, demand an intensive and sympathetic study of their works, some more particularly in reference to the known extant botanical collections on which they were based, others with intensive field work in the classical localities whence the several authors secured their material. In the latter cases the field work should be combined with an intensive study of each individual description and illustration and of all additional data given by each author in comparison with extensive collections of botanical material.

While the preliminary examination of Loureiro's work in 1919 clearly indicated that the proper interpretation of his numerous genera and species presented an exceedingly difficult series of problems, yet it was believed that an application of the methods followed in the study of the Rumphian and Blancoan problems would yield productive results, and that it was not only possible but highly probable that a high percentage of Loureiro's numerous doubtful genera and species could definitely be placed. From the standpoints of priority and of stability in nomenclature, a critical study of Loureiro's "Flora Cochinchinensis" seemed to me to be the most important need in reference to all of the post-Linnaean publications appertaining to the flora of the Indo-Malaysian region.

The International Code of Botanical Nomenclature recognizes the principle of priority, limited only by the approved lists of *nomina generica conservanda*, and most botanists now follow this Code in principle. We can therefore no longer ignore the imperfectly described species of various early authors. To each individual case the historical method of research should be applied, and where the actual types are no longer extant, all possible means should be employed for the purpose of locating definitely the various more or less doubtful species and determining their status in reference to those described by other authors. Dr. H. Handel-Mazzetti [9] has urged that in order not to upset established and accepted nomenclature more than is necessary, great caution should be exercised in adopting Loureiro's

[9] Rept. Proc. Fifth Internat. Bot. Congr. Cambridge, 536. 1931.

names.   All taxonomists will admit the correctness of this position; yet when the definite status of any of Loureiro's numerous doubtful species can be determined, the logical course to follow is to accept the name in accordance with the rules of procedure in such cases, provided it is valid and has priority.   Without some definitely established system for determining the proper binomial for each individual species and for adjusting the synonymy in each individual case, it would be useless to attempt an interpretation of the numerous binomials based on the work of such early authors as Rheede, Rumphius and the older Burman, or the species described under the binomial system by the younger Burman, Loureiro, and Blanco.   If we assume one attitude toward Linnaean species, are we justified in adopting a totally different attitude toward Loureiroan, Blancoan, or de Candollean species, or even those proposed by our own contemporaries?   The task, if performed at all, should be done in accordance with the principles of priority.   Accordingly, in my previous publications of this type I have not hesitated to adopt older, valid, specific names, even when they replaced long-established and well-known binomials in current use and do not hesitate to make new combinations in the present work where such a course is indicated.

It is realized that this procedure is markedly different from that of most of the contributors to Lecomte's "Flore générale de l'Indo-Chine" (1906–35+) which covers the area of the old Kingdom of Cochinchina.   In this modern work one would logically expect to find, if not admitted as a species, at least mentioned as synonyms or in notes, all or most of Loureiro's species that were based on Cochinchina material, as well as the even more numerous binomials proposed by later authors but based on Loureiro's descriptions of Cochinchina plants.   Yet within the families treated in the published parts of this important contribution to our knowledge of the flora of tropical Asia up to the end of 1934, I note nearly 400 binomials, either those of Loureiro or those based on Loureiro's descriptions of Indo-China plants, that are not mentioned; and some of these omissions even include genera and species recognized by all botanists as valid and of which the actual types are extant.

Some botanists do not fully approve of the type of bibliographical and botanical work characterized by my attempts to clarify the binomials based on Rumphius' "Herbarium Amboinense," and more definitely to place the species proposed by Blanco, Llanos, Burman f., and Loureiro, in so far as this work involves changes in accepted nomenclature.   On the other hand, the value of such work is appreciated by others, in spite of the necessary changes of names involved.   While I personally consider this type of work important, I can scarcely expect that my contemporaries and successors will agree with me in all of my interpretations, or will accept all the proposed changes in nomenclature.   The facts are presented as clearly as possible and the data are thus available to those who care to check my conclusions.   It gives me distinct personal satisfaction to clarify the status of an overlooked, forgotten, or obscure genus or species, and my own reaction is that this type of work fully justifies the time and effort devoted to it.

With due mental reservations as to the type of taxonomic work prosecuted by that remarkable but erratic genius, C. F. Rafinesque, which is rather generally ignored by most taxonomists, I cannot refrain from quoting the first paragraph from the preamble to the fourth and last part of his "Flora Telluriana" (1838):

"In the process of this work I have met with many interruptions and disappointments. It is neither easy nor agreeable to stem the current of botanical errors and blunders, and whoever swims against the streams of scientific prejudice may reckon on difficulties.   I have met such in all my attempts to increase and correct knowledge; but I persevere neverthe-less, and write for posterity rather than the actual Schools. I feel that my weary labors are not now appreciated except by a few, but am confident that in 50 years hence they will be more valued. Of this I have received already some assurances, when young and skilful Botanists have partly approved and adopted my views."

Rafinesque was at least optimistic regarding the value of his work to posterity, but he would probably be gravely disappointed at the current lack of appreciation of his "weary labors" even one hundred years after his results were published.

LOUREIRO AND HIS WORK

João de Loureiro, S.J., was born in Lisbon, Portugal, in 1710. From Macao, where he had resided for four years, having previously spent three years in Goa, he was sent on a special mission to Cochinchina in 1742. Finding that missionary activities as such were not in favor there, he entered the service of the King of Cochinchina as mathematician and naturalist.

Loureiro remained in Cochinchina for nearly thirty-six years, with the exception of one short interval (1750–52) when he made a trip to India, being forced to leave Cochin-china because of a violent outbreak of persecution. While in Cochinchina his chief place of residence was its capital city, Hue; hence botanical material from the immediate vicinity of Hue is of distinctly great importance in connection with any attempt to interpret Lou-reiro's Indo-China species, for Hue is the classical locality for several hundred species. In December 1777 Loureiro proceeded to Bengal, Pondichery, Macao, and Canton, and at Canton for the next three years he continued his botanical activities, leaving China in March 1781 on his return to Portugal. Bad weather prevented the ship on which he was a passenger from rounding the Cape of Good Hope, forcing its return to Mozambique, whence he finally departed for Lisbon early in 1782. This interruption gave him an opportunity of making certain botanical collections in Mozambique, Zanzibar, and in tropical East Africa. During the remainder of his life he apparently remained in Lisbon, where he died on October 18, 1791.

Loureiro's chief publication, the "Flora Cochinchinensis," [10] was completed in 1788 and published by the Academy of Science in Lisbon in 1790, two quarto volumes of 744 pages, the pagination continuous. The first part includes the introduction and pages 1 to 354, the second part pages 355 to 744, with one page of errata. The appearance of this work, containing as it did original descriptions of no less than 185 new genera and nearly 1300 species of which about 630 were described as new, created enough of a sensation in European botanical circles that three years later Willdenow [11] issued a second edition of it in Berlin, with the addition of some brief notes. This edition is merely a republication of the original work with a very few minor changes and corrections, no important changes in nomenclature, with some not very important footnotes giving certain reductions, and some suggested alliances. Some of the mutilated type used to indicate certain phonetic values

---

[10] **Loureiro, J. de.** Flora Cochinchinensis: sistens plantas in regno Cochinchina nascentes. Quibus accedunt aliae observatae in Sinensi imperio, Africa orientali, Indiaeque locis variis. Omnes dispositae secun-dum systema sexuale Linnaeanum. i–xx, 1–744 [errata 1]. 1790.

[11] **Loureiro, J. de.** Flora Cochinchinensis . . . denuo in Germania edita cum notis Caroli Ludovici Willdenow. i–xxiv, 1–882. 1793.

in local names, and some of the special diacritical marks used in the first edition, are eliminated.   Like the first edition, it appeared in two volumes with continuous pagination.

While Loureiro's chief publication is this important botanical work, a number of other papers were prepared by him, mostly published after his death in 1791.   These are as follows:

Loureiro, J.   Memoria sobre o algodaõ; sua cultura, e fabrica.   Mem. Econ. Acad. Sci. Lisb. 1: 32–40. 1789.

――――― Da trasplantaçaõ das arvores mais uteis de paizes remotos.   Mem. Econ. Acad. Sci. Lisb. 1: 152–163. 1789.

――――― Da incerteza que ha acerca da origem da Gomma Myrrha.   Dá-se noticia de hum arbusto, que tem as mesmas qualidades, e virtudes.   Mem. Acad. Sci. Lisb. 1: 379–387. 1797.

――――― Memoria sobre a natureza, e verdadeira origem do Páo de Aguila.   Mem. Acad. Sci. Lisb. 1: 402–415. 1797.

――――― Memoria sobre huma especie de petrificaçaõ animal.   Mem. Acad. Sci. Lisb. 2: 47–55. 1799.

――――― Exame phisico, e historico.   Se ha, ou tem havido no mundo diversas especies de homens?   Mem. Acad. Sci. Lisb. 2: 56–81. 1799.

――――― Descripçaõ botanica das cúbebas medicinaes.   Mem. Acad. Sci. Lisb. 2: 82–87. 1799.

――――― Consideraçaõ phisica, e botanica da planta Aerides, que nasce, e se alimenta no Ar.   Mem. Acad. Sci. Lisb. 2: 88–98. 1799.

――――― [Two letters to P. Eckart]: In Hoffler, P.   Historia Cochinchinae. 1803.

――――― Observationes astronomicae a P. Joanne de Loureiro. Soc. Jesu, in regno Cochinchinae habitae in urbe Sinoae Regis sede.   Mem. Acad. Sci. Lisb. 3(2): 1–6. 1814.

That Loureiro was a man of remarkable attainments is evidenced not only by his published papers, but also by the diverse unpublished manuscripts bequeathed by him to the Academy of Science at Lisbon.   Some idea of the extent of these can be gained from Gomes' [12] statement that these consisted of twelve large octavo volumes written on Chinese paper in Chinese characters supposed to consist of a history of Anam; two volumes of drawings representing minerals, plants, and animals; two large volumes containing three hundred ninety-seven colored drawings of plants with their local and Latin names; a "flora iconographica" of Cochinchina written in Anamese; and an Anamese-Portuguese dictionary.

Loureiro, after his return to Portugal, submitted a manuscript entitled "Nova Genera Plantarum" to Sir Joseph Banks in London for publication, but was urged by the recipient to reconsider it in connection with publications of other authors.   The English botanists apparently realized that many of the new genera proposed had already been published under other names and were undoubtedly influenced in their recommendations to Loureiro by definite knowledge based on an actual examination of some of his specimens, then in London, that this was the case.

Loureiro had no formal training as a botanist but became interested in the study of plants chiefly from his personal interest in the possibility of using native drugs in the place

―――――

[12] **Gomes, B. A.**   Elogio historico do Padre João de Loureiro.   Mem. Acad. Sci. Lisb. Cl. Pol. Mor. Bel.-Let. n.s. 4(1): 5–6. 1868.

of those known to and used by Europeans, for of course it was practically impossible for him to secure them in Indo-China.   He gained his knowledge of European literature appertaining to drug plants from a copy of one of the numerous Spanish editions of Dioscorides' "Materia Medica" by A. de Laguna.   His interest in materia medica naturally attracted his attention to botany, and he gained his first knowledge of the Linnaean system of classification from copies of Linnaeus' "Genera Plantarum," "Systema Naturae," and "Philosophica Botanica"—works which he secured through Thomas Riddell, captain of an English ship at Canton.   Captain Riddell also placed Loureiro in correspondence with Sir Joseph Banks in London, which led to his sending an important collection of specimens to London in 1779.   In the actual preparation of his "Flora Cochinchinensis," which was finished in Lisbon in October, 1788, he consulted numerous other botanical publications, a bibliographic list of ninety-six titles being appended to the introduction to his work.   The volumes most consulted appear to have been Reichardt's edition of Linnaeus' "Systema Plantarum" (1779–80), Rumphius' "Herbarium Amboinense," and Rheede's "Hortus Malabaricus."   Gomes' quotation ascribed to Schreber,[13] repeated by an anonymous writer in Broteria 5: 103. 1906: "Mirandum est sane virum omnibus libris destitutum tam erudite de plantis potuisse judicare," perhaps gives Loureiro somewhat more credit than he himself would claim, because he was not wholly without botanical books.

The Scope of Loureiro's and Blanco's Floras

In dealing with such works as Blanco's "Flora de Filipinas" and Loureiro's "Flora Cochinchinensis," one is always at a loss to explain the basis of selection, for the reason that so many common and conspicuous species that obviously must have been familiar to them are not considered.   Neither includes many orchids, grasses, sedges, or ferns; yet the regions covered are particularly rich in representatives of these groups.   Conspicuous, common, and economically important trees and shrubs are missing in very large numbers. No high-altitude species are considered by either author.   In the case of Loureiro it is manifest that he was very greatly influenced by the medicinal or reputed medicinal qualities of the plants that he actually described.   He was not particularly interested in plants of little or no economic value, and like Blanco manifestly did not plan his "flora" to be a complete one of the regions covered; and as with Blanco, for reasons of inaccessibility, high-altitude plants were not available to him.   Manifestly but a small percentage of Loureiro's species came from the primary forests, but most of them came from the settled and cultivated areas, thickets, and second-growth forests.   This is evidenced by the small percentage of endemic species and the large percentage of widely distributed ones among those actually described by him.

On the other hand, Loureiro collected and described certain very conspicuous species that appear in no Indo-China collections made between 1780 and 1927, in spite of the great amount of field work accomplished in that country in the past ninety years. Bauhinia coccinea (Lour.) DC. and Clianthus scandens (Lour.) Merr. are conspicuous examples of these—plants with masses of showy flowers that would normally not be overlooked by any collector.   It is suspected that they are of local occurrence in Indo-China and not species of general distribution.   Other species observed and described by Loureiro have not ap-

[13] This quotation is really from Willdenow. Fl. Cochinch. ed. Willd. Praefatio III–IV. 1793.

peared in any modern collection.   It is highly probable that some of the species, still known only from Loureiro's descriptions and still considered to be of doubtful status, have not been rediscovered by modern explorers.   It is manifest that the Hue region in Indo-China is one still worthy of long-continued and intensive botanical exploration, not only in reference to its interesting flora as such, but also regarding still unsolved problems concerning Loureiro's species.

The title "Flora Cochinchinensis" is a somewhat misleading one.   Although the percentage of species from Cochinchina is much higher than from any other country, yet several hundred species were described from China and a considerable number from other parts of the Orient, from the Philippines to India and tropical East Africa.   The sources of his species are: from Cochinchina alone about 697; from China alone about 254; from Cochinchina and China together about 292; from tropical East Africa 29; from Mozambique 9; from Zanzibar 8; from India 5; from the Malay Peninsula 2; and from the Philippines, Madagascar, and Sumatra 1 each.   The problem of interpreting Loureiro's species is thus distinctly more difficult of solution than was the interpretation of Blanco's Philippine species, as the latter author considered only plants from one general and restricted region.

Unfortunately, some authors who have intensively studied the floras of some of the regions mentioned above failed to realize the desirability of attempting to determine the status of Loureiro's species.   The general result is that practically all of the genera and species described by Loureiro in 1790 have been redescribed under other names.   In spite of Loureiro's errors of commission and omission, I can see no reason why his names of 1790 should not be accepted in place of those proposed a hundred or more years later for the same species, as long as the earlier names are valid and as long as the identity of the species concerned can reasonably be determined.   To me the greater error is the comparatively modern redescription of plants that were already named and well characterized in 1790. In such cases is it logical to accept specific names published since the beginning of the present century for species that were sufficiently well characterized by Loureiro more than one hundred years earlier, which bear specific names that are valid within their respective genera, and which are manifestly identical with the more recently described ones?   Here are some illustrative cases:

Tabernaemontana bovina Lour. (*T. tonkinensis* Pitard 1933).

Diospyros lobata Lour. (*D. odoratissima* Lecomte 1928).

Elaeocarpus sylvestris (Lour.) Poir. (*E. decipiens* Hemsl. 1886; *E. glabripetalus* Merr. 1922; *E. kwangtungensis* Hu 1924).

Alchornea rugosa (Lour.) Muell.-Arg. (*A. hainanensis* Pax & Hoffm. 1914).

Glochidion pilosum (Lour.) Merr. (*G. annamense* Beille 1927).

Baccaurea sylvestris Lour. (*B. annamensis* Gagnep. 1927).

Quercus concentrica Lour. (*Q. sabulicola* Hickel & Camus 1921).

Gnetum indicum (Lour.) Merr. (*G. montanum* Marcgr. 1930).

Barringtonia cochinchinensis (Lour.) Merr. (*B. annamica* Gagnep. 1918).

Ardisia loureiriana (G. Don) Merr. (*Rhododendron loureiroianum* G. Don 1834, based on *Azalea punctata* Lour.) (*Ardisia expansa* Pitard 1930).

Desmodium rubrum (Lour.) DC. (*D. carlesii* Schindl. 1920).

Ormosia pinnata (Lour.) Merr. (*O. hainanensis* Gagnep. 1914).

Pueraria montana (Lour.) Merr. (*P. tonkinensis* Gagnep. 1916).

Ficus simplicissima Lour. (*F. palmatiloba* Merr. 1928).

Ochna integerrima (Lour.) Merr. (*Discladium harmandii* Van Tiegh. 1902; *Ochna harmandii* Lecomte 1911).

Hedyotis simplicissima (Lour.) Merr. (*H. subdivaricata* Drake 1922).

Gmelina racemosa (Lour.) Merr. (*G. balansae* Dop 1914).

Lindera myrrha (Lour.) Merr. (*Lindera eberhardtii* Lecomte 1913).

The list could be greatly extended, particularly by adding those species renamed and redescribed by various nineteenth century authors who, like some twentieth century authors, failed to recognize that Loureiro had anticipated them.

Various Opinions of Loureiro's Work

Hooker f.[14] makes the following statement regarding Loureiro's work:

"The 'Flora Cochinchinensis' of Loureiro, though it relates to a country beyond our limits, contains so many forms identical with those of Ava and Malaya, that we shall have frequent occasion to refer to it. Father Loureiro, a native of Portugal, resided for thirty-six years in the kingdom of Cochin-China, whither he proceeded as a missionary, but finding that Europeans were not permitted to reside there without good cause, entered the service of the King as chief mathematician and naturalist.[15] Though he had no acquaintance with the science of botany, the difficulty of procuring European medicines induced him to direct his attention to native drugs; and with a zeal of which we have unfortunately too few instances, he prosecuted his botanical studies, and so successfully, notwithstanding his want of early education, as to produce a work of standard value. The 'Flora Cochinchinensis' was published at Lisbon, in two volumes quarto, in 1790; a second edition, edited by Willdenow, with a few notes, appeared in octavo, at Berlin, in 1793. As was to be expected, in a work devoted to the botany of a previously unexplored tropical region, the 'Flora Cochinchinensis' contained a great amount of novelty; but the absence of plates, and a defective terminology, caused by a want of familiarity with the labours of other botanists, render the descriptions often obscure, so that a number of the genera described by Loureiro have not yet been identified, while others, not being recognized, have been described as new, and renamed by subsequent botanists."

A. de Candolle,[16] in his discussion of "Descriptions énigmatiques de groupes naturels," in concluding that Father Velloso, author of the "Flora Fluminensis" (1825) and "Florae Fluminensis Icones" (1827), is the most culpable of botanical authors in the number of genera proposed that were considered to be of doubtful status in 1880, remarks:

"Le Père Blanco est à peu près au même rang, tandis que le Père Loureiro avait eu au moins le mérite d'envoyer en Europe quelques plantes sèches, au moyen desquelles on peut comprendre une partie de ses descriptions. Il est à regretter que ces révérends ecclésiastiques, et même le Père Plumier, leur prédécesseur, ne se soient pas contentés d'écrire des homélies. Bonnes on les aurait lues, mauvaises on les aurait mises de côté; tandis qu'en histoire naturelle l'existence de certains noms et de certaines planches rend nécessaire de consulter indéfiniment les plus mauvais ouvrages."

[14] **Hooker, J. D. & Thomson, T.** Flora Indica, introductory essay. 46. 1855.

[15] " He styles himself, in his own narrative, ' rebus mathematicis et physicis praefectum.' "

[16] **Candolle, A. de.** La Phytographie, 141. 1880.

Hooker also speaks of Blanco's "Flora de Filipinas" as a botanical curiosity and chides him for "his want of acquaintance with scientific works," saying that "so many well-known plants are treated as new, that we consider it undesirable to devote time to their identification." But I can scarcely subscribe to Hooker's opinion as on the whole Blanco's work was better than that of Loureiro, and a distinctly higher percentage of his genera and species have definitely been correlated with those of other authors than is the case with those of Loureiro. Of the twenty-three new genera described by Blanco but one, *Saola*, remains unplaced, and only nine of his one thousand one hundred thirty-six species and varieties, of which six hundred and thirty-six were proposed as new, still remain wholly doubtful; that is, doubtful even as to their proper families, while but about forty additional species cannot be placed closer than the genera to which they apparently belong. These reductions have been made, it should be remembered, solely from the published record and in the entire absence of any botanical material prepared by Blanco. As contrasted to the present status of Blanco's genera and species, Loureiro stands about as follows: Four genera based by later authors on Loureiro's descriptions quite unknown; about 23 species that cannot be placed in their proper families; and about 140 additional ones that cannot be correlated with those of other authors, of which about 80 can be placed in their proper genera and about 60 are doubtful as to their generic position. Over 1100 can absolutely be placed as to their species. These figures include all species that Loureiro characterized under binomial names, whether described as new or correctly or erroneously ascribed to Linnaeus, and include 55 cellular cryptogams so inadequately described that only two or three are actually identifiable from the descriptions.

LOUREIRO'S HERBARIUM MATERIAL

Loureiro fortunately prepared some herbarium material, and it is known that he took with him to Lisbon, on his return to Portugal in 1782, representatives of some of the species he described in the form of dried specimens. He had sent several hundred other specimens to Europe some years before his departure from the Orient. It is manifest, however, from Gomes' statement that Loureiro gave little attention to the problem of associating his accepted binomials with his herbarium specimens; nor did he do this consistently with the material he sent to Europe before he left the Orient.

The data given by Gomes in his "Elogio historico do Padre João de Loureiro" under the heading "Os herbarios de Loureiro" [17] are of sufficient importance to warrant their reproduction verbatim, particularly because his lists have apparently been overlooked by all botanists since they were published, and are not available in most botanical libraries. Gomes' paper was read on April 30, 1865, but not published until three years later. It is to a certain degree unfortunate that the data were published in the form presented, in that they contain a number of Loureiro's herbarium names, including various binomials that had previously not been published, none of which, except a few new ones proposed by Desvaux, appear in botanical literature to date. The lists, however, throw much light on Loureiro's methods and clearly bring out the fact that he proposed numerous names that for one reason or another he abandoned in favor of others when he actually did publish his findings. Doubtless some of these generic names were those included in his "Nova

[17] **Gomes, B. A.** Mem. Acad. Sci. Lisb. Cl. Pol. Mor. Bel.-Let. n.s. **4**(1): 25–31. 1868.

Genera Plantarum," the manuscript of which was sent to London from Lisbon, but which was not published.

Gomes lists thirty-seven Loureiro specimens as preserved in Lisbon in 1865, and eighty-seven (not eighty-eight) as preserved in the herbarium of the Muséum d'Histoire Naturelle, Paris. The list of Paris specimens was credited by Gomes to A. L. de Jussieu, but was apparently based on the Desvaux manuscript that is attached to the package of Loureiro specimens in the Paris herbarium; it was copied in Paris for Doctor Gomes by Arthur Morelet.

The quoted statement regarding this material reproduced below (pp. 13–19) applies equally to the Loureiro specimens in the herbarium of the British Museum. Probably some botanists would hesitate to accept these Loureiro specimens as actual "types" in the generally accepted sense of that word, particularly where Loureiro himself did not add his published binomial to the actual specimen. However, the specimens are at least those that Loureiro prepared to represent his genera and species; many of them bear the local names cited by him; others bear his published Latin names; and most of them doubtless represent material that he actually used in preparing his descriptions. As far as these specimens agree with his published descriptions, and as far as they can be correlated with the latter, my opinion is that they should be accepted as types. Some years ago Doctor F. Gagnepain kindly supplied me with a manuscript list of the Loureiro specimens in the Paris herbarium. In this list ninety specimens are included, but two mentioned by Gomes are lacking, *Phyteuma* = *P. bipinnata* Lour., and *Faskia divaricata* Lour. = *Nerium scandens* Lour. = *Strophanthus dichotomus* DC. = *S. caudatus* (Burm. f.) Kurz. Gagnepain, however, lists four species that do not appear in the enumeration as published by Gomes. These are:

Campylus sinensis Lour. = *Tinospora tomentosa* Miers, det. Gagnepain.[18]

Volkameria inermis Lour. [*Clerodendrum inerme* Gaertn.].

Polygonum tataricum Lour. [*Polygonum fagopyrum* Linn.].

Ganosma inodora Lour. = *Gymnema inodorum* Decne., det. Decaisne.

In 1931 Dr. A. C. Smith, of the staff of the New York Botanical Garden, prepared for me photographs of all the Loureiro specimens preserved in Paris which have proved to be of material assistance in elucidating his species.

In the lists as given below, Gomes has been followed verbatim. Additions and a few manifest corrections appear in square brackets; minor errors appear as in the original. The names of botanists given in parentheses are those of the individuals who have made or verified the identifications as taken from Gagnepain's list. The original work in correlating these Loureiro specimens with the latter's descriptions was done by A. N. Desvaux.

OS HERBARIOS DE LOUREIRO

[p. 25]

"Das plantas que Loureiro remettêra da Asia existe na Academia das Sciencias de Lisboa um pequeno numero, de que damos a relação. Ha outra porção no Museu de Paris, cuja lista, d'elle alcançada pelos cuidados do sr. Arthur Morelet, egualmente publicamos. Além d'estas plantas devem existir algumas no *British Museum* em Londres, e talvez existi-

---

[18] Loureiro's generic description of *Campylus* does not remotely apply to any menispermaceous plant, yet the species description was apparently based on a *Tinospora*, and the Loureiro specimen is a *Tinospora*.

rão por outras partes.   Das que se conservam em Lisboa, em Paris e em Londres, dá noticia, no Musée botanique de B. Delessert, o auctor d'essa obra e conservador do interessante estabelecimento a que ella se refere, o sr. A. Lasègue, a pag. 323, 348.   Conforme [p. 26] o proprio testemunho d'este auctor os dois fragmentos do herbario de Loureiro teriam existido primitivamente reunidos no Museu de Lisboa, e talvez os acompanhasse então porção maior de plantas de egual procedencia.   Diz o sr. Lasègue que a porção d'este herbario actualmente no Museu de Paris é a menor das duas, existentes ali e em Lisboa; de facto porém não succede hoje assim, por quanto possuimos apenas trinta e sete exemplares d'estas plantas, sendo as de Paris mais de oitenta.   Se com effeito as duas porções de plantas fizeram parte de mesma collecção, como tudo o indica, não é difficil atinar com a origem da separação.   Mas não sirva isso para recordar um facto que deve ser tido unicamente em conta dos accidentes de guerra, e que para nós teve sobeja compensação no modo por que estes preciosos restos, documento da actividade e zêlo scientifico do nosso missionario, teem sido respeitosamente conservados no Museu de Paris, onde acharam quem tanto os soubesse apreciar, e muito os aproveitasse em beneficio da sciencia de todos.   Na lista que damos das plantas de Loureiro existentes em Lisboa, vão os nomes que achámos escriptos com a propria lettra de Loureiro nos papeis que envolvem cada um dos objectos; e como estes nomes não são sempre os da *Flora Cochinchinensis*, ajuntámos os que ahi vem, e lhes correspondem, á vista das descripções confrontados com os caracteres verificados pelo estudo dos exemplares; procedendo assim de modo analogo ao que fôra praticado para a lista das plantas de Paris por Antoine Laurent de Jussieu, na revista que se diz fizera d'essas plantas o distincto botanico francez.

Lista de 37 plantas de Loureiro conservadas no Museu da Academia Real das Sciencias de Lisboa

Nomes escriptos por Loureiro	Nomes determinados pela comparação dos objectos com o texto da Flora Cochinchinense
Amomum arboreum-Sumatriae	*Amomum Arboreum* Lour.
Amomum-Me tlé	*Amomum Globosum* Lour.
Amomum galanga-Cây Rieng	*Amomum Galanga* Lour.
Amomum-Mé tlé bà	*Amomum Hirsutum* Lour.
Abrus precatorius-Daû dó	*Abrus Precatorius* Lour.
Casuarina africana	*Casuarina Africana* Lour.
Caesalpina Sapãn	*Caesalpina Sappan* Lour.
Cephalanthus Dioicus-Deei Trôp	*Cephalanthus Procumbens* Lour.
Cephalanthus Stellatus-Ri-ri bou gaó	*Cephalanthus Stellatus* Lour.
Convolvulus-Bim bim lá dua	*Convolvulus Aggregatus* Lour.
Curcuma longa	*Curcuma Longa* Lour.
Curcuma rotunda	*Curcuma Rotunda* Lour.

[p. 27]

Dimocarpus Longan-Cây Nhon	*Dimocarpus Longan* Lour.—*Euphoria longa-*[*na*] Lamk.
Erythrina-Cay boung	*Erythrina Corallodendrum* Lour.—*Erythrina Indica* Lamk.
Flagellaria catenata-Mây báóc	*Flagellaria Indica* Lour.

Flagellaria repens-Mây báóc bò cây   Flagellaria Repens Lour.—Pothos scandens Spreng.

Flagellaria petraea-Mây dá

Grammicarpus-Dâu Chi   Coronilla Cochinchinensis Lour.

Laurus Caryophyllata-Cay ranh ranh   Laurus Caryophyllus Lour.

Laurus curvifolia-Mieng Sanh Caõ Lá   Laurus Curvifolia Lour.

Laurus cinnamomum   Laurus Cinnamomum Lour.

Laurus myrrha   Laurus Myrrha Lour.—Tetranthera trinervia Spreng.

Melodorum-Bõ giẽ   Melodorum Fruticosum Lour.

Michelia Champava-Hoa Sú   Michelia Champava Lour.

Melastoma-Cây Mua   Melastoma Septemnervia Lour.

Ploca amentacea-Dâi Mâm

Phyllodes placentaria   Phyllodes Placentaria Lour.

Poinciana pulcherrima-Hoa phung   Poinciana Pulcherrima Lour.

Piper-Tieo bõ   Piper Nigrum Lour.

Piperis species-Tlâù   Piper Betle Lour.

Ruhelia-Sài hô   Ruellia Antipora Lour.

Tabernaemontana-Sung tlân bò   Tabernaemontana Bovina Lour.

Tamarindus-Me   Tamarindus Indica Lour.

Uvaria-Mu tru   Uvaria Zeylanica Lour.

Van pi Sinensis   [ex nom. Chin. van pi = Clausena lansium (Lour.) Skeels].

Winterania-Madagascar   Winterania Canella Lour.

Zeydora agrestis-Sanà rùng   Dolichos Montanus Lour.

[p. 28]

Lista de 88 [87] plantas de Loureiro conservadas no Museu do Jardim das Plantas em Paris, com a synonimia e mais indicações de Antoine Laurent de Jussieu

Amomum Zingiber   Amomum Zingiber Lour. et Linn. [Zingiber officinale Rosc. (Gagnepain)].

Keranthera   Curcuma Longa Lour. [Zingiber zerumbet Sm. (Gagnepain)].

Cochlia   Garciana Cochinchinensis Lour.—Philydrum lanuginosum Banks ex de Cand.

Lobus   Salomonia Cantoniensis Lour.—Polygaleae. [Baillon, Gagnepain].

Striga   Striga Lutea Lour.

Cleianth[us] coccineus   Volkameria Angulata Lour. est Clerodendrum paniculatum Linn.

Botrus   Porphyra Dichotoma Lour.—Callicarpae species. (Callicarpa purpurea Juss.).

Oikia   Phyla Chinensis Lour. est Verbena nodiflora Linn. [Lippia nodiflora Rich. (Baillon)].

Cephalanthus monas   Cephalanthus Montanus Lour.

Muringuizingui   Allasia Payos Lour.—Affinit ignota. [Vitex (Baillon)].

Carandás   Carissa Carandás ? Lour. [Carissa africana A. DC. (A. de Candolle)].

Pareira brava	*Botria Africana* Lour.—*Sarmentaleae* seu *Vites*. [*Vitis* (Baillon)].
Dissolen[a]	*Dissolena Verticillata* Lour.—*Vitices.* [*Rauwolfia* Pierre].
Phyteuma	*Phyteuma Bipinnata* Lour.—*Sambucus ebuloides* de Cand.
Faskia divaricata	*Nerium Scandens* Lour.—*Strophanthus dichotomus* de Cand.
Pavetta sinensis	*Pavetta Arenosa* Lour. [*P. indica* ? Pierre; Gagnepain].
Argyreia	*Argyreia Acuta* Lour.—*Convolvulaceae* [Gagnepain]
Thela alba	*Thela Alba* Lour.—*Plumbago Zeylanica* Linn.
Gentiana scandens	*Gentiana Scandens* Lour.—*Paederia foetida* Linn. [Pierre].
Gardenia sinensis	*Gardenia Volubilis* Lour.—*Rubiaceae.* [*Ichnocarpus*, Pierre].
Xylochus	*Xilochus* Lour. inedit. *Antidesma alexiteria* Linn.
Stylidium Bauthas	*Stylidium Chinense* Lour.—Affinit ignota. [*Marlea* (Decaisne) begonifolia (Gagnep.)].
Matricaria	*Matricaria Cantoniensis* Lour. [*Hisutsua cantoniensis* DC. ?].
Perihola-Xich laong	*Rhamnus Lineatus* Linn. Lour. [*Berchemia* (Baillon)].
Heloda	*Hydrolea Inermis* Lour.
Trisanthus	*Trisanthus Cochinchinensis* Lour. est *Hydrocotyle lunata* Lamk.
Tamaris [*Tamarix*] *sinica*	*Tamarix Chinensis* Lour.
Plectronia Chinensis	*Plectronia Chinensis* Lour. [*Panax loureiriana*].
[p. 29]	
Gloriosa luxurians	*Hemerocallis Fulva* Lour. [*H. esculenta* Desv.].
Hemisus	*Acanthus Ilicifolius* Lour. Linn. [*Dilivaria* (Baillon)].
Ezehlsia palma-phat Dien	*Dracaena Ferrea* Linn. Lour.
Xiphidium-tave tien	*Liriope Spicata* Lour.—*Dianella* ?.
Dracaena alliaria	*Ornithogalum Sinense* Lour.—*Scilla* ?.
Spathium	*Spathium Chinense* Lour.—*Aponogeton monostachium* Linn. [*Saururus loureiri* Dcne.].
Ribera	*Lagunea Cochinchinensis* Lour.—*Polygonum lagun*[e]*a.* [Desv.].
Polyg. tinctorium	*Polygonum Tinctorium* Lour. Linn.
Trapela	*Primula Mutabilis* Lour.—*Hortensia.* [*Hydrangea hortensia*].
Xylosma Cochine	*Daphne Cannabina* Lour.—*Daphne* ? [*Wikstroemia cannabina* Dcne.].
Rheum Cantoniense	*Rheum Barbarum* Lour. non Linn.—*Rumex.*
Quinarius Van Pimone	*Quinaria Lansium* Lour.—*Cockia.* [*Cookia punctata* Lam.].
Ophispermum	*Ophispermum Sinense* Lour. Affinit. ignota. [*Aquilaria chinensis* Spreng. (Lecomte)].
Mekistus sinensis	*Quisqualis Indica* Lour. [Gagnepain].
Egkianthus	*Enkianthus Biflora* Lour. [Baillon].
Dumula sinens.	*Limonia Monophylla* Lour.
Libaria	*Aubletia Ramosissima* Lour.—*Zizyphi* species [*Paliurus* ? *aubletia* DC. (Baillon)].
Hedona-Yu-mi	*Hedona Chinensis* Lour.—*Lychnis grandiflora.*
Ngaoc	*Hecatonia Palustris* Lour.—*Ranunculus sceleratus* Linn. [Baillon].
Myrt. Sinensis	*Myrtus Sinensis* Lour.—*Symplocos Sinica* de Cand. [*Symplocos chinensis* (Lour.) Druce].

Crataeg. sinensis-Ngulin mone	Crataegus Rubra Lour. [Raphiolepis rubra Lindl. (Spach)].
Spiraea sinensis-Ngulin mone	Spiraea Cantonensis Lour. [S. lanceolata Poir. (Baillon)].
Thea olearia	Thea Oleosa Lour. [T. sasanqua Pierre (Gagnepain)].
Thea Canton.	Thea Cantonensis Lour. [T. chinensis Linn. (Baillon)].
Mangueiro	Thilachium Africanum Lour.—Thilachium ovalifolium Juss. Herb.—Capparideae.
Dentidia Nankinensis	Dentidia Nankinensis Lour.—Labiatae. [Perilla (Baillon)].
Stachys artemisia	Stachys artemisia Lour.—Leonurus Sibiricus Linn.
Clemat. minor	Clematis Minor Lour. [C. chinensis Retz. (Gagnepain)].
Arthroda	Desmos Chinensis Lour.—Unona discolor Vahl. [Baillon].
Dodecatria	Dodecadia Agrestis Lour.—Grewia [microcos aff.].
Polycaulis	Corchorus Angulatus ? Lour.—Inedit. [C. acutangulus Lam. (Gagnepain)].
Rhizanota Cannabina	Corchorus Capsularis Lour. Linn. [Gagnepain].
Ligustrum	Ligustrum Sinense Lour.
[p. 30]	
Phyllimorphus	Capparis Magna Lour. [Crataeva ? macrocarpa (Gagnepain)].
Lagerstroemia	Lagerstroemia Indica Lour. et Linn.
Viribiri	Martynia Zanguebaria Lour.—Podalium ? [Pretrea zanguebarica Gagnep.].
Canutia	Cornutia Quinata Lour.—Vitex leucoxylon Linn. [= Vitex quinata F. N. Will.].
Ahcantina	[Phoberos chinensis Lour. (Baillon) = Scolopia chinensis Clos (Gagnepain)].
Hebdoma	Septas Repens Lour.—Gratiola Monniera Linn.
Kirphum	Campsis Adrepens Lour.—Bignonia sinensis Lam.
Mutondo	Corypha [Cordyla] Africana Lour. [Calycandra pinnata Guill. & Perr. (Baillon)].
Lipara nigra	Pimela Nigra Lour.—Canarium pimela Kon. [Canarium nigrum Engl. (Guillaumin)].
Sebifera	Sebifera Glutinosa Lour.—Litsea chinensis Lam. [Tetranthera (Pierre), Litsea sebifera (Lecomte)].
Gonus	Gonus Amarissimus Lour.—Brucea amarissima Des. [B. antidysenterica Lam.].
Ricinus apalta [apelta]	Ricinus Apelta Lour. [Mallotus apelta Muell.-Arg. (Mueller-Arg.); Echinus apelta Baill. (Baillon)].
Muthona	Triphaca Africana Lour.—Sterculiaceae ? [(Baillon) = Sterculia triphaca B. Br.].
Tridemis	Tridesmis Tomentosa Lour.—Crotonis spec. [Croton tomentosus Muell.-Arg. (Mueller-Arg.)].
Morella	Morella Rubra Lour.—Affinitas ignota. [Myrica (Baillon)].
Nymphantus	Nymphantus Niruri Lour.—Phyllanthus [urinaria Linn. (Mueller-Arg.)].

Hoan Semg	*Aristotelea Spiralis* Lour.—*Orchidea* [= *Spiranthes aristotelea* (Raeusch.) Merr.].
Polytoma inodora	*Epidendrum Tuberosum* Lour. Linn. ?
Polytoma odorifera	*Aerides Odorata* Lour.
Tropha	[*Pachyrhizus* (Baillon); *Pueraria* (Gagnepain)].
Rhynchosia	*Rhynchosia Volubilis* Lour. [Gagnepain].
Ploca humilis	*Hedysarum Reniforme* Lour. non Linn. [*Lourea obcordata* Desv. (Gagnepain)].
Plagium	*Cytisus Cajan* Lour. Linn. [*Cajanus indicus* Spreng. (Baillon, Gagnepain)].
Derris	*Derris Trifoliata* Lour.—*Leguminosa* [*Derris uliginosa* Roxb. (Gagnepain)].
Kercops	*Polygala Glomerata* Lour. [(Gagnepain)].
Mopex Sinensis	*Urena Polyflora* Lour. [*Helicteres lanceolata* DC. (Gagnepain)].

Esta relação veiu acompanhada com a observação de ter sido escripta pela propria mão de Antoine Laurent de Jussieu, e de existir com as plantas de Loureiro no Museu de Paris a seguinte nota de lettra e auctor differente.

Observations sur 80 et quelques plantes de la Flore de la Cochinchine.

Des circonstances particulières ayant enrichi le muséum de Paris d'un certain nombre de plantes de l'herbier du missionaire portuguais Loureiro, nous avons eu d'autant plus de plaisir à les examiner qu'elles ont fait partie de l'herbier qui a servi à la description des plantes publiées dans la *Flora Cochinchi* [p. 31] *nensis*. On sait que Loureiro n'a pas été assez heureux pour mettre au jour le fruit de ses travaux sur la botanique, tant dans la Cochinchine que dans la Chine et la partie occidentale de l'Afrique, et qu'il est mort à Lisbonne dans le temps qu'il s'occupait à pourvoir au moyen de publier son manuscrit. Il parait d'après ce que nous avons observé sur les 80 et quelques plantes de son herbier, qu'il n'avait pas eu le temps de porter les noms des plantes définitivement adoptés sur son manuscrit; ou bien que, s'en rapportant plus à son manuscrit où les descriptions etaient faites avec soin qu'à une collection qui pouvait être détruite par diverses circonstances, il n'avait pas attaché beaucoup d'importance à étiqueter exactement les échantillons qu'il possédait. Il en résulte que les plantes n'ont point été nommées, ou qu'un très petit nombre d'entre elles portent des noms correspondants à ceux de la Flore. Dans le haut de la feuille sont inscripts seulement la classe et l'ordre de Linné dans lesquels la plants doit être portée; on y trouve encore, quelquefois, un nom générique qui, presque toujours, se trouve changé dans l'ouvrage, le nom spécifique étant cependant demeuré le même; on remarque aussi, chez plusieurs plantes, au-dessus de l'inscription de la classe et de l'ordre, un nom vulgaire, quelquefois orthographié différemment qu'il ne l'est dans l'ouvrage imprimé. C'est avec ce peu d'indications que nous sommes parvenus à retrouver les noms de toutes ces plantes et à acquerir, par là, des idées précieuses sur plusieurs genres que Loureiro avait établi et qui ne peuvent plus exister, ou qui mériteraient d'être examinés. Quelques soient les erreurs que cet auteur a commises, il est à remarquer que les plantes sont, en général, très bien décrites, et qu'il est facile de vérifier son exactitude dès que l'ont peut avoir acquis la certitude de l'identité d'espèce.

Na mesma nota existe em seguida uma discussão a respeito de muitos generos ou espe-cies, como *Salomonia*, *Allasia*, etc.; mas esta parte não nos foi enviada, só veiu d'ella a indicação.

É interessante esta nota pela revelação do modo por que estão as plantas de Loureiro no herbario do Museu de Paris. É este modo exactamente o mesmo que se observa na pequena porção de plantas conservadas no Museu de Lisboa. Sabemos que em uma e outra parte ellas estão como embrulhadas em papel chinez, de certo o mesmo em que as envolveu Loureiro, porque é n'esse papel que existem escriptas com a sua propria lettra as indicações a que se refere a nota do herbario de Paris.

No que se enganou porém o auctor da nota foi em suppôr que a Flora Cochinchinensis não fôra impressa em vida de Loureiro, por cuanto esta impressão verificou-se no anno de 1790, e Loureiro morreu no immediato, em 1791; sendo certo que elle proprio vigiára ainda e superintendêra essa impressão, apesar da edade muito adiantada em que se achava."

The most important extant Loureiro collection of botanical specimens is that in the herbarium of the British Museum, Natural History, in London. In a list supplied by that institution, about 228 species are included as being represented by herbarium specimens or checked as being in the herbarium in its copy of Loureiro's "Flora Cochinchinensis," although some of the checked specimens cannot now be located. The Loureiro material, however, is distributed through the large general herbarium, and while it is probable that some of the specimens may have been destroyed, it is also probable that others will even-tually be located. These Loureiro specimens, like those in Lisbon and Paris, sometimes bear Loureiro's binomial in his handwriting, sometimes an unpublished manuscript name, sometimes a local name, sometimes no technical name or other data except an annotation to the effect that the specimen was received from Loureiro. The correlation with Lou-reiro's descriptions, in the case of the inadequately labeled specimens, was largely done by Dryander. Many of these types have been critically examined by such botanists as R. Brown, Hiern, Seemann, Pierre, Moore, Britten, Rolfe, Kränzlin, Reichenbach f., Bennett, Miers, Haviland, and others, and some I have personally examined. Wherever published records have been found based on an examination of Loureiro's types, these have been indicated either in the bibliographic references or in the discussions appertaining to the species in question.

In 1774 [19] Loureiro sent about sixty specimens from Cochinchina to Europe, a few of these being cited by such authors as Bergius, "Materia Medica" 5. 1778, and by the younger Linnaeus, "Supplementum" 331. 1781. Yet Doctor R. E. Fries, writing in July, 1919, informed me that he could find no Loureiro specimens in Bergius' herbarium. The Loureiro specimen of *Acosta spicata*, illustrated and described by Swartz (Weber & Mohr Beitr. 1: 6. *pl. 2.* 1805) as *Vaccinium orientale* Sw., is not preserved in Swartz's herbarium, and Doctor Samuelsson thinks it probably was based on the specimen in the Banksian her-barium.

In 1779 Loureiro, then in Canton, sent another lot of about 230 specimens to Sir Joseph Banks in London, these being the ones now preserved in the herbarium of the British Museum, Natural History. A few of the extant British Museum specimens, however, are from the herbarium of Robert Brown, and these apparently were received from the Paris herbarium.

[19] Fl. Cochinch. Introd. XI.

Gomes notes Lasègue's [20] statement to the effect that Loureiro's herbarium was preserved in Lisbon in 1845, but throws little or no light on what became of it. He lists thirty-seven specimens as then (1865) preserved in Lisbon, eighty-seven as then preserved in the Museum of Natural History, Paris, and refers to others, the number not indicated, as being preserved in the herbarium of the British Museum. A. de Candolle, [21] writing in 1880, definitely states, on the authority of a letter from Doctor Gomes in Lisbon, that Loureiro's herbarium, being found to be in a bad state of preservation, had been destroyed; this statement probably applies to the remnant of Loureiro's herbarium, i.e., the thirty-seven specimens listed in 1865. The statement received from Paris, repeated by Gomes, merely mentions the fact that the Paris material was secured under particular circumstances and he (Gomes) speaks of it as one of the accidents of war. It is recorded by de Candolle that Geoffroy Saint Hilaire secured eighty-three specimens from Loureiro's herbarium in Lisbon in 1808, which were deposited in the herbarium of the Museum of Natural History, Paris; [22] Gomes lists eighty-seven specimens, but Gagnepain's manuscript list of 1924 gives a total of ninety. This Loureiro material at Paris is preserved as an individual collection, and to the package is attached Desvaux's manuscript enumeration of the species, with numerous reductions indicated by him to which later authors have added others.

Contrasted with Lasègue's statement of 1845 that Loureiro's herbarium, except the small part that was to be found in Paris, was then in Lisbon, is the statement of an anonymous writer in Flora [23] which it seems desirable to quote in full.

"Bekanntlich sind die Systematiker über gar manche von Loureiro in seiner Flora Cochinchianensis [sic] beschriebene Pflanzen nicht im Klaren und nur die Ansicht der Originalexemplare wäre im Stande über dieselben Licht zu verbreiten. Leider scheint alle Hoffnung verloren, die Pflanzen je wieder zu finden. Als nämlich Lissabon durch die Franzosen erobert wurde, lies der Marshall Junot nebst anderen botanischen Schätzen die dort aufbewahrt wurden, auch das Herbar Loureiro's einpacken und nach Paris abgehen; den Empfangschein darüber kann man im Lissaboner Naturaliencabinet sehen. Ob die kostbaren Pakete je an den Ort ihrer Bestimmung gelangt sind, darüber hat man durchaus nichts ermitteln können; im Jardin des plantes will man nichts davon wissen, und es ist allerdings sehr möglich, dass jene botanischen Schätze irgendwie auf der Reise vernichtet wurden."

The above statement is definite and categorical, yet it seems strange that Gomes, writing five years later, should have been ignorant of this and should not have known of Marshal Junot's signed receipt for the material which was stated in 1860 to be then extant in Lisbon. This alleged transaction occurred in 1807. Whatever the fate of Loureiro's herbarium, whether destroyed in Lisbon, whether packed for shipment to Paris and lost or destroyed in transit, the fact remains that of about 1300 species described by him, less than one-fourth are now represented by extant types. The internal evidence is that the

[20] **Lasègue, A.**   Musée botanique de M. Benjamin Delessert. 323, 348. 1845.

[21] **Candolle, A. de.**   La Phytographie. 429. 1880.

[22] French interest in the Loureiro herbarium in Lisbon was possibly due, in part, to the series of papers then being published by A. L. de Jussieu on the relationship of various Loureiroan genera to those described by other authors; see p. 25.

[23] **Anonymous.**   Loureiro's Herbar.   Flora **43**: 207–208. 1860.

entire Loureiro herbarium consisted only of the specimens now preserved in Paris, and the small remnant that was retained in Lisbon, and that Loureiro actually prepared specimens only of those which he sent to Europe in 1774 and 1779, and about 127 specimens that he took with him on his return voyage to Lisbon; the total would be slightly over 400 specimens, of which apparently about 100 have, for one reason or another, been lost or destroyed. This conclusion is based on the facts that of the 227 species listed in the British Museum copy of Loureiro's "Flora Cochinchinensis" all but 2 or 3 are from Indo-China, i.e., those specimens actually sent by Loureiro from Canton in 1779; of the 90 in the Paris herbarium about 80 are from China, 7 from Africa and only 4 from Indo-China; and of the 37 listed by Gomes as being formerly preserved in Lisbon, 33 are from Indo-China, and one each from China, Africa, Sumatra, and Madagascar. The Paris collection then represents for the most part those specimens actually prepared by Loureiro after he arrived in Canton in 1778. The small Lisbon collection probably represents the special herbarium mentioned by Almeida [24] as having been presented to the botanical garden at Ajuda in 1789. There then seems to be no warrant for charging the French with having taken more than the 90 specimens listed above on pages 15 to 18. The difference between the contents of the "Paris" collection and that of the smaller "Lisbon" one, as to origin, would seem to indicate two different collections, the first almost wholly of Chinese plants, the second mostly of Indo-China ones.

Almeida [25] himself states regarding the herbaria at the botanical garden at Ajuda: "O antigo museu da Ajuda, apesar da pobreza em herbários, que lá foi encontrar—desfalcados não só pela invasão francesa, mas principalmente pelo abandono da su conservaçao." On page 234 of the same work in discussing certain old accession records of the botanical garden at Ajuda which he saw in the home of Doctor Alves de Sá, he quotes the following: "Joao Loureiro deu, no ano de 1789, um pequeno herbário de Plantas da Cochinchina, algumas sementes das ditas, utensílios vários e móveis domésticos daquele País." It is highly probable that the Loureiro specimens actually listed by Gomes (page 14) were those in this small herbarium, and that this herbarium was independent of the one presented by him to the Academy of Science, and which was taken to Paris by the French in 1807 or 1808.

In view of the fact that the value of the Loureiro specimens was not appreciated in Lisbon, and that they apparently were destroyed some time before 1874, the fact that an important part of the herbarium was transmitted to Paris in 1807 or 1808, where the value of the material was appreciated, may be looked upon as one of the fortunate accidents of war.

Mr. Achilles Machado, Secretary of the Academy of Science of Lisbon, in answer to my enquiry, courteously supplied the following information under date of April 20, 1932. Mr. Machado secured his information from Prof. José Joaquim de Almeida, who supplied the following data:

"At the time when Bernardino Gomes, the younger, wrote his eulogy on Loureiro in 1865, the Botanical Garden and the Museum of the Academy of Science at Ajuda, about eight kilometers from Lisbon, had been a part of the Polytechnical School since 1839, but the herbarium was not actually transferred to the Polytechnical School until 1874. It is

[24] **Almeida, J. de.** O Dr. Frederico Welwitsch e a sua obra em Angola 1: 207–208 [no date].
[25] **Almeida, J. de.** *l.c.*

certain that Gomes' statements were based on documents rather than on an examination of the material he mentions, as the Conde de Ficalho in charge of the Botanical Section of the Polytechnical School inspected the herbarium of the Museum of the Academy of Science in 1874 and stated that it was then in poor condition, partly because of the material removed by the French, but principally because of the practical abandonment of the collection; that is, the herbarium had not been properly cared for. He mentions Ferreira's Amazon collections but says nothing about those of Loureiro. Thus the thirty-seven Loureiro specimens mentioned by Gomes were not transferred to the Polytechnical School as they had apparently deteriorated so badly at Ajuda as to be valueless, and had apparently been discarded before 1874. The 397 colored drawings of the iconographic flora of Cochinchina written in the native language, bequeathed to the Academy by Loureiro with other manuscripts, are not at the Polytechnical School, and no longer exist at the Academy. I saw some of the drawings at the home of Dr. Alves de Sá, a son of the Visconde de Alves de Sá, nephew of Dr. José de Sá dos Santos Vale, who was Director of the Museum and of the Botanical Garden at Ajuda in 1834–35. . . ."

Mr. de Almeida concludes that the thirty-seven Loureiro specimens listed by Gomes are no longer extant, were not extant in 1874, and that the only specimens that were saved to science from the Loureiro collections of the Museum of the Academy of Science are those that were transferred to Paris in 1807 or 1808.

While in the discussion of the individual species I have generally mentioned those that are represented by extant Loureiro specimens in London and in Paris, I have not thought it worth while to so mention the thirty-seven Lisbon specimens listed by Gomes as they no longer exist.

Loureiro's name looms large in the annals of systematic botany because of the large number of genera and species described by him as new; because of the very numerous unsolved problems as to the status of his genera and species and their relationship with those described by other botanists; and because of the same problems in relation to the larger number of binomials and many new generic names proposed by his successors but based on his original descriptions. In somewhat over eighty binomials the specific name is derived from Loureiro and six generic names have been proposed in his honor. These are:

Lourea Necker (1790). A valid genus of the Leguminosae.

Loureira Raeuschel (1797) based on *Schrebera* Retzius, and a synonym of *Elaeodendron* Jacquin; not based on any Loureiroan genus or species.

Loureira Cavanilles (1799), a synonym of *Jatropha* Linnaeus; not based on any Loureiroan genus or species.

Loureira Meisner (1837) based on *Toluifera cochinchinensis* Lour. and a synonym of *Glycosmis* Correa.

Lourya Baillon (1887), a synonym of *Peliosanthes* Andrews; not based on any Loureiroan genus or species.

Neolourya Rodriguez (1934); not based on any Loureiroan genus or species.

Publications or commentaries on Loureiro's "Flora Cochinchinensis" as to the entire work do not exist, although most of his species have been considered by other authors in the 144 years that have elapsed since the work was published. Most of these post-Loureiroan references are merely repeated or abstracted descriptions and add little to our knowledge of the individual species described; in fact, many of the authors following Loureiro have complicated rather than simplified the situation through proposing 54 new generic names and about 750 new binomials based on Loureiro's often more or less imperfect and incomplete descriptions, rarely supplemented by an examination of an authentic specimen. These names were proposed by such authors as Blume, Cambessides, A. Chevalier, Choisy, Dietrich, G. Don, Hooker & Arnott, A. L. de Jussieu, O. Kuntze, Nees, Persoon, Pierre, Poiret, Raeuschel, Rafinesque, M. Roemer, Roemer & Schultes, Sprengel, Steudel, Vahl and many others. Various authors, such as Baillon, Britten, Bunting, Gagnepain, Koenig, Moore, R. Brown, Seemann, Mueller-Arg., Tandy, A. de Candolle, Swartz, and others, have published more extended descriptions or critical notes based on the extant Loureiro specimens in London and Paris.

Special Publications on Loureiro's Work

To prepare a complete bibliography of all works in which Loureiro's genera and species appear, would be a very heavy task and one that would serve no useful purpose, as references to Loureiro are found in all standard works and monographic treatises dealing with families or genera with which Loureiro was concerned, and in all or most standard works dealing with the floras of British India, Indo-China, China, Japan, Malaysia, the Philippines, Australia, Polynesia, and tropical Africa. References to Loureiro are scattered through the voluminous periodical literature and independent treatises from shortly after 1790 to the present date, while a considerable number of special articles have been published on the identity of certain genera and species proposed by Loureiro, some of these papers being based on an examination of extant types, and others based on a study of Loureiro's descriptions with or without reference to material from the classical localities. A bibliography of this type of literature, including some biographical references, follows.

Special Bibliography, Consisting of Papers Based Largely or in Part on Loureiro's Descriptions and Specimens Together with Certain Biographical References

Anonymous. Loureiro's Herbar. Flora 43: 207–208. 1860. Loureiro's herbarium in Lisbon is said to have been ordered by Marshal Junot to be packed and shipped to Paris [1807], the suggestion being made that it was lost or destroyed in transit. See page 20.

——— Rerum naturalium in Lusitania cultores. P. Joanes de Loureiro e Soc. Jesu. Broteria 5: 98–114. 1906. Pages 98–104 in Latin; pages 105–114 in Portuguese, with a subtitle: P. João de Loureiro da Companhia de Jesus. Includes important biographical data regarding Loureiro and data regarding his herbarium, largely taken from Gomes.

Baillon, H. Sur l'organization et les affinités du Dissolena verticillata Lour. Adansonia 4: 378–382. *pl. 12.* 1864. A complete description and long discussion of this species based on specimens cultivated in Paris.

——— Sur le genre Placus. Bull. Soc. Linn. Paris 1: 282–283. 1881. *Placus* is reduced to *Blumea* on the basis of an examination of the type of *Placus laevis* Lour. in the herbarium of the British Museum.

——— Le Tripinna de Loureiro. Bull. Soc. Linn. Paris 1: 714. 1887. The type specimen in the herbarium of the British Museum = *Vitex* sp.

―――― Sur le Dissolaena verticillata Lour. Bull. Soc. Linn. Paris **1**: 768. 1888. On the basis of an examination of Loureiro's extant type this is shown to be *Rauwolfia verticillata* (Lour.) Baill.

Bonaparte, R. Synonymie des Ptéridophytes décrites par J. de Loureiro 1793 [1790]. Notes Ptérid. **7**: 135–139. 1918. A list with suggested reductions based wholly on Loureiro's descriptions.

Britten, J. Notes on Hoya. Journ. Bot. **36**: 413–418. 1898. Includes a critical note on Loureiro's specimen of *Stapelia cochinchinensis* in the herbarium of the British Museum, confused by R. Brown with *S. chinensis* Lour. (See Traill, below.)

―――― Notes from the National Herbarium. IV. Journ. Bot. **55**: 341–345. 1917. Contains a critical note on the type of *Creodorus odorifer* Lour. in the herbarium of the British Museum = *Chloranthus inconspicuus* Sw.[=*Chloranthus spicatus* (Thunb.) Makino].

―――― Eugenia lucida Banks. Journ. Bot. **58**: 151–152. 1920. Discusses *Opa odorata* Lour. on the basis of Loureiro's type specimen in the herbarium of the British Museum, showing that Seemann (Journ. Bot. **1**: 280. 1863) was in error in identifying *Opa odorata* Lour. as a synonym of *Syzygium lucidum* Gaertn. (See Seemann, B., below.)

Bunting, R. H. The genus Rotula. Journ. Bot. **47**: 269–270. 1909. On the basis of an examination of Loureiro's type specimen in the herbarium of the British Museum, *Rotula* is shown identical with *Rhabdia*, and Loureiro's name *Rotula*, being much the older, is retained.

Burkill, I. H. The lesser yam,—Dioscorea esculenta. Gard. Bull. Straits Settlem. **1**: 396–399. *pl. 7–9.* 1917. *Dioscorea esculenta* Burkill based on *Oncus esculentus* Lour. The illustrations are of the tubers of various races, no. 288 on plate 7 being *khoai tu bua* from Saigon.

Chevalier, A. Liste de quelques espèces de la flore d'Indochine ou du jardin botanique dont la nomenclature est a modifier. Cat. Pl. Jard. Bot. Saigon 63–66. 1919. Includes a number of new binomials based on Loureiro's descriptions. (See page 46.)

―――― & **Poilane, M.** Les Cycas d'Indochine. Rev. Bot. Appl. **4**: 472–474. 1924. Includes a critical note on *Cycas inermis* Lour. with an amplified description based on recently collected material, the species said to be common in southern Anam.

Colmeiro, M. La botánica y los botánicos de la península Hispano-Lusitana. Estudios bibliográficos y biograficos I–XI. 1–216. 1858. Loureiro (Juan) p. 178. Gives brief biographical data regarding Loureiro and some notes regarding his botanical work.

Dammer, U. Callista v. Dendrobium. Gard. Chron. III **47**: 34–35. 1910. Considers Kränzlin's and O'Brien's papers on these two genera, concluding that *Callista amabilis* Lour. is a true *Dendrobium*, on the basis of a critical study of Loureiro's type. (See Kränzlin, O'Brien, and Rolfe.)

Demange, V. Notes sur quelques champignons comestibles, vénéneux ou curieux du Tonkin et de l'Annam. Bull. Écon. Indochine **22**: 592–609. *f. 1–16.* 1919. Supplies clues to the possible identity of a few of Loureiro's species although the paper in no way is based on the latter's work.

Dubard, M. & Eberhardt, P. A propos caoutchoucs de " Tabernaemontana." Description de deux espèces de ce genre. Ann. Sci. Agron. Franç. Étrang. IV **2**: 131–137. *1 f.* 1913. Contains a description and illustration of *Tabernaemontana bovina* Lour.

Gagnepain, F. Discussion de la valeur des caractères des Argyreia acuta Lour., obtusifolia Lour., Championi Benth., et du genre Lettsomia. Bull. Soc. Bot. France **62**:
4–7. 1915. Includes critical notes on Loureiro's extant type of *Argyreia acuta* Lour.
in the herbarium of the Paris Museum and notes on *A. obtusifolia* Lour.

——— Ulmacées et Artocarpacées nouvelles ou litigieuses. Bull. Soc. Bot. France **72**:
804–810. 1926. On pages 806–808 *Trema cannabina* Lour. is extensively discussed and
is considered to be identical with *T. velutina* Blume. Gagnepain did not see Loureiro's
type which is preserved in the herbarium of the British Museum and which is *T. virgata* Blume [*T. amboinensis* (Willd.) Blume, non auctt. plur.].

Gomes, B. A. Elogio historico do Padre João de Loureiro lido na sessão solemne da Aça-
demia Real das Sciencias de Lisboa em 30 de Abril de 1865. Mem. Acad. Sci. Lisboa
Cl. Sci. Mor. Pol. Bel.-Let. n.s. **4**(1): 1–31. 1868. Includes many biographical data
regarding Loureiro, together with published lists of 37 of Loureiro's species preserved
at Lisbon, and 87 preserved in the herbarium of the Muséum d'Histoire Naturelle,
Paris; the Lisbon list was prepared by Gomes, and the Paris list was probably prepared
by A. N. Desvaux, although credited by Gomes to A. L. de Jussieu. These lists are
reproduced verbatim above, pages 14–18.

Hallier, H. Bausteine zu einer Monographie der Convolvulaceen. Bull. Herb. Boiss. **6**:
714–724. 1898. On pages 716–719 *Argyreia arborea* Lour. is critically considered and
reduced to *Cordia myxa* [non Linn.] = *C. dichotoma* Forst. f. on the basis of Loureiro's
description.

Hance, H. F. Note on the Capparis magna, of Loureiro. Journ. Bot. **7**: 41–42. 1869. A
brief note based on a collection made by Swinhoe in Hainan.

——— On the Fagus Castanea of Loureiro's 'Flora Cochinchinensis'; with descriptions
of two new Chinese Corylaceae. Journ. Linn. Soc. Bot. **10**: 199–203. 1869. *Fagus
castanea* Lour. is erroneously interpreted to be the same as *Castanopsis chinensis* Hance
on the basis of Loureiro's description and recently collected material.

——— On the so-called "olives" (Canarii spp.) of southern China. Journ. Bot. **9**: 38–
40. 1871. Contrasts and distinguishes two Loureiroan species, *Canarium album* and
C. pimela, on the basis of material collected in Kwangtung Province.

——— On the genus Fallopia Lour. Journ. Bot. **9**: 239–240. 1871. On the basis of
Loureiro's description and recently collected material this is shown to be the same as
Grewia microcos Linn. [= *Microcos paniculata* Linn.].

Jussieu, A. L. de. Observations sur la famille des plantes Verbenacées. Ann. Mus. Hist.
Nat. [Paris] **7**: 63–77. 1806. Includes *Callicarpa purpurea* Juss. based on *Porphyra
dichotoma* Lour.; *Cornutia* Lour. is reduced to *Vitex* Linn. on the basis of Loureiro's
descriptions.

——— Note sur le genre Physkium de Loureiro. Ann. Mus. Hist. Nat. [Paris] **9**: 402–
404. 1807. *Physkium* is reduced to *Vallisneria* on the basis of Loureiro's description.

——— Note sur quelques genres de la Flore de la Cochinchine de Loureiro, qui ont de
l'affinité avec d'autres genres connus. Ann. Mus. Hist. Nat. [Paris] **11**: 74–76, 150–
152. 1808. Various genera described by Loureiro are discussed and reductions made
or suggested on the basis of Loureiro's descriptions.

——— Suite des observations sur quelques genres de la Flore de Cochinchine de Loureiro,
avec quelques réflexions sur l'Elaeocarpus et les genres qui doivent s'en rapprocher

dans l'ordre naturel.   Ann. Mus. Hist. Nat. [Paris] 11: 231–236, 327–328. 1808; 12: 68–72. 1808.   A continuation of the preceding paper.

——— Suite des observations sur quelques genres de plantes de Loureiro, acompagnées de notes sur ceux qui composent la famille des Anonées.   Ann. Mus. Hist. Nat. [Paris] 16: 338–340. 1810.   A continuation of the preceding papers, including notes on a few species described by Loureiro on the basis of Loureiro's published data.

——— [probably by A. N. Desvaux].   Lista de 88 [87] plantas de Loureiro conservadas no Museu do Jardim das Plantas em Paris, com a synonimia e mais indicaçoes de Antoine Laurent de Jussieu; in Gomes, B. A.   Elogio historico do Padre João de Loureiro. Mem. Acad. Sci. Lisboa, Cl. Sci. Mor. Bel.-Let. n.s. 4(1): 28–31. 1868.   Based on the manuscript list of A. N. Desvaux, preserved with the package of Loureiro specimens in the herbarium of the Muséum d'Histoire Naturelle, Paris.   Reproduced verbatim above, pages 15–18.

Koenig, C.   A few botanical observations.   Ann. Bot. Koenig & Sims 1: 356–368. *pl. 7–8*. 1805.   Includes an amplified description of *Canarium pimela* based on Loureiro's specimen now in the herbarium of the British Museum.

Kränzlin, F.   Callista amabilis Lour.   Gard. Chron. III 46: 354. 1909.   A critical note based on Loureiro's type in the herbarium of the British Museum, considering that *Callista* represents a genus distinct from *Dendrobium*.   (See Dammer, O'Brien, and Rolfe.)

Martelli, U.   Pandanus odoratissimus var. loureirii Gaud.   Univ. Calif. Publ. Bot. 12: 373–374. 1930.   The *nomen nudum*, *Pandanus loureirii* Gaudich., appears in Gaudichaud Bot. Voy. Bonite, associated with *pl. 22, fig. 13*, the figure representing a single drupe.   It was based on material secured by Gaudichaud probably at Tourane, Indo-China, where he collected botanical material.   It is not associated with Loureiro's description of *Pandanus odoratissimus*.   Martelli, not Gaudichaud, is the proper authority for the variety, as he gave the form varietal status in Webbia 4(1): 34. 1913.   Martelli's description of 1930 is based on *Clemens 4343*, from Tourane, near Hue.

Merrill, E. D.   A commentary on Loureiro's Flora Cochinchinensis.   Typescript manuscript i–xxxii. 1–693. 1919.   A preliminary draft of the present consideration of Loureiro's species.   This is the typescript report referred to by S. le M. Moore in his papers on "Identification of Loureiro's specimens in the British Museum Herbarium."   Journ. Bot. 63: 245–256, 281–291. 1925.

——— Additional notes on the Kwangtung flora.   Philip. Journ. Sci. 15: 225–261. 1919. Includes the identifications of a number of previously unplaced Loureiro species based on a study of recently collected Kwangtung specimens, and Loureiro's descriptions.

——— On the application of the generic name Melodorum of Loureiro.   Philip. Journ. Sci. 15: 125–137. 1919.   *Fissistigma* Griff. is accepted for the large number of species erroneously ascribed to *Melodorum*.   The technical generic description of *Melodorum* unquestionably appertains to *Mitrephora* Hooker f. & Thomson; one of the species described is a *Mitrephora*, the other apparently a *Polyalthia*.

——— The identity of the genus Sarcodum Loureiro.   Journ. Bot. 66: 264–265. 1928. This is shown to be the same as *Clianthus*.

——— Factors to be considered in interpreting Loureiro's species.   Journ. N. Y. Bot. Gard. 33: 32–36. 1932.

—— Loureiro and his botanical work. Proc. Amer. Philos. Soc. **72**: 229–239. 1933. A general discussion of Loureiro's work and its significance.

Moore S. le M.   Alabastra diversa.—Part XXIV. Journ. Bot. **52**: 146–151. 1914. Treats the genera *Decadia* Lour. and *Dicalyx* Lour. on basis of Loureiro's types preserved in the herbarium of the British Museum.

—— Identification of Loureiro's specimens in the British Museum Herbarium. Journ. Bot. **63**: 245–256, 281–291. 1925. An important paper in which about 54 of Loureiro's genera and species are discussed on the basis of his extant types in the herbarium of the British Museum.

—— & **Tandy, G.**   Notes on two species of Loureiro's Flora Cochinchinensis. Journ. Bot. **65**: 279–281. 1927. *Bembix* Loureiro is shown to be identical with *Ancistrocladus* Wall., and *Cycas inermis* Lour. is shown to represent a species allied to *C. circinalis* Linn. and *C. rumphii* Miq. on the basis of Loureiro's specimens in the herbarium of the British Museum.

Nelson, J. E.   A little-known botanist. Am. Bot. **25**: 129–133. 1919. Brief and incomplete data regarding Loureiro and his botanical work.

O'Brien, J.   Callista v. Dendrobium. Gard. Chron. III **47**: 19. 1910. Considers Loureiro's genus *Callista* in relation to *Dendrobium*. (See Dammer, Kränzlin, and Rolfe.)

Pierre, L.   Sur deux espèces d'Epicharis produisant les bois dits: Sandal citrin et Sandal rouge. Bull. Soc. Linn. Paris **1**: 289–292. 1881. *Santalum album* Lour. is identified as *Epicharis loureirii* Pierre = *Dysoxylum loureirii* Pierre, on the basis of recent collections and field notes.

—— Sur le genre Stixis Lour. Bull. Soc. Linn. Paris **1**: 652–656. 1887. Ten species are considered, including *S. scandens* Lour., no type extant.

Reichenbach f., H. G.   Galeola Lour. Xen. Orch. **2**: 76–78. 1862. Includes a note on *Galeola nudifolia* Lour., type in the herbarium of the British Museum.

—— Thrixpermum Lour. Xen. Orch. **2**: 120–123. 1865. Includes notes on various Loureiroan types of orchids in the herbarium of the British Museum.

—— Loureiro's Orchideengattungen. Flora **51**: 52–53. 1868. Consists of brief notes on Loureiro's genera and species based on extant types in the herbarium of the British Museum.

Rolfe, R. A.   Donax and Schumannianthus. Journ. Bot. **45**: 242–244. 1907. By a special examination of Loureiro's type of *Donax* in the herbarium of the British Museum, Rolfe clarifies the confusion caused by K. Schumann's misinterpretation of it.

—— Callista amabilis. Orchid Rev. **18**: 99–100, 242. 1910. Considers Kränzlin's and O'Brien's papers on this species and concludes that the living plant identified by Kränzlin as *Callista amabilis* does not represent Loureiro's species. (See Dammer, Kränzlin, and O'Brien.)

Schindler, A. K.   Die Desmodiinen in der botanischen Literatur nach Linné. Fedde Repert. Beih. **49**: 1–371. 1928. Pages 5–6 include Schindler's reductions of Loureiro's species of *Hedysarum*, *Trifolium* and *Grona*, the extant types in the London and Paris herbaria indicated.

Schott, H.   Einiges über Lasia Loureiro's. Bonplandia **5**: 122–129. 1857. Considers *Lasia loureiri* Schott on the basis of Loureiro's published data.

—— Aroideologisches. Bonplandia **7**: 183. 1859. Considers *Arum indicum* Lour. on the basis of Loureiro's published data.

Seemann, B.   Loureiro's Cathetus fasciculata and Camellia drupifera.   Bonplandia **7**: 47–50. 1859.   Critical notes and additional descriptive data based on Loureiro's type of the former in the herbarium of the British Museum and on his description of the latter.

——— On the position of the genera Hydrocotyle, Opa, Commia, and Blastus in the natural system.   Journ. Bot. **1**: 278–282. 1863.   Data are given on *Opa odorata* Lour., *O. metrosideros* Lour. and *Blastus cochinchinensis* Lour. based on Loureiro's types in the herbarium of the British Museum.   (See Britten, J. on *Eugenia lucida* in relation to *Opa odorata* Lour., above.)

Swartz, O.   Acosta spicata Loureiro, eine neue Art von Vaccinium.   Weber & Mohr Beitr. **1**: 4–7. *pl. 2.* 1805.   *Vaccinium orientale* Sw. is proposed as a new name for *Acosta spicata* Lour. and the species illustrated from a Loureiro specimen.   The specimen is not in Swartz's herbarium, the description and illustration being probably based on the specimen in the Banksian herbarium now in the British Museum, Natural History.

Traill, J.   Accounts and descriptions of the several plants belonging to the genus Hoya, which are cultivated in the garden of the Horticultural Society at Chiswick.   Trans. Hort. Soc. London **7**: 16–30. *pl. 1.* 1827.   Gives a note on Loureiro's specimen of *Stapelia* in the herbarium of the British Museum.   (See Britten, above.)

In many of the descriptions of Loureiro's species appearing in standard monographic works, except in the case of those whose status is well understood, nothing is added to the original descriptions which are often more or less imperfect or incomplete, and frequently inaccurate.   In some cases monographers have taken great pains to search for the extant types in Paris and in London, and have published amplified and corrected descriptions based on these original specimens.   In other cases, amplified descriptions have been based on collections made since Loureiro's time, which can, with a fair degree of certainty, be taken to represent the actual species Loureiro had in mind.   In still other cases later authors have interpreted some of Loureiro's descriptions from such data as he gave, and have prepared amplified descriptions of what they took to represent the same species but based on material originating from places remote from the classical localities; and in some cases such descriptions do not apply to the form that Loureiro actually attempted to describe.   In other words, Loureiro's specific names are sometimes currently applied to forms more or less remote from the original type, on the basis of a misinterpretation of the original description.

Loureiro's Genera

Loureiro described 185 new genera.   Of these, up to 1919, fourteen had never been placed and remained as *genera incertae sedis*, some not even having been placed within their proper families.   To this list should be added a considerable number of generic names proposed by later authors but typified by Loureiro's descriptions which have not satisfactorily been placed.   From data published within the past few years by various authors, based on Loureiro's extant types, and supplemented by a study of Loureiro's descriptions, this list of doubtful genera in both categories has been reduced to four, i.e., *Agonon* Raf., *Isgarum* Raf., *Silimanus* Raf. and *Pseudiosma* A. Juss.

   Wherever Loureiro's generic names, although older than those proposed by other authors for the same groups, have been eliminated in favor of more recent names in the approved list of *nomina generica conservanda*, the conserved name has here been adopted and Loureiro's name placed as a synonym.

Loureiro's new genera are listed below in alphabetic sequence with their equivalents as far as known, except that in the case of the generally accepted ones, or those that should be accepted, later synonyms are usually not given; these accepted names are printed in black-faced type.   To the list of currently accepted names I have added *Rotula* Lour. (1790) in place of *Rhabdia* Martius (1827), and *Picria* Lour. (1790) in place of *Curanga* A. L. de Jussieu (1807); while *Desmos* Lour. (1790) is the correct generic name for the so-called species of *Unona* of the Old World.   Numerous species erroneously placed in *Melodorum* Lour. I have previously transferred to *Fissistigma* Griff.[26] and I am now convinced that *Melodorum* as actually described by Loureiro is the same as *Mitrephora* Hook. f. & Th., although one of the two species described is apparently a species of *Polyalthia*.   Otherwise there are no changes made in currently accepted generic nomenclature, except the proposal of a new generic name for the few Asiatic species erroneously placed in *Grona*, as *Grona* Lour. proves to be a synonym of *Desmodium* Desvaux.

Loureiro's New Genera, 1790	Equivalents in Terms of the International Code of Botanical Nomenclature
Abutua	= *Gnetum* Linnaeus (1753)
Acosta	= *Vaccinium* Linnaeus (1753)
Adenodus	= *Elaeocarpus* (Burman) Linnaeus (1753)
Aerides	
Aglaia	
Aidia	= *Fagraea* Thunberg (1782)
Allasia	= *Vitex* (Tournefort) Linnaeus (1753)
Aloexylum	= *Aquilaria* Lamarck (1786)
Anoma	= *Moringa* (Burman) A. L. de Jussieu (1789)
Antherura	= *Psychotria* Linnaeus (1753)
Argyreia	
Aristotelea	= *Spiranthes* L. C. Richard (1818)
Arsis	= *Microcos* Linnaeus (1753)
Astranthus	= *Homalium* Jacquin (1760)
Athruphyllum	= *Rapanea* Aublet (1775)
Aubletia	= *Paliurus* (Tournefort) Miller (1752)
Augia	= *Rhus* (Tournefort) Linnaeus (1753); saltem pro majore parte.
Aulacia	= *Micromelum* Blume (1825)
Axia	= *Boerhaavia* Linnaeus (1753)
Baccaurea	
Balsamaria	= *Calophyllum* Linnaeus (1753)
Barbula	= *Caryopteris* Bunge (1835)
Baryxylum	= *Peltophorum* Walpers (1842) pro majore parte, quoad fl.; *Gymnocladus* Lamarck (1783) p.p., quoad fruct.; ? *Intsia* Thouars (1806), pro minore parte, quoad fruct.

[26] **Merrill, E. D.**   On the application of the generic name Melodorum of Loureiro.   Philip. Journ. Sci. **15**: 125–137. 1919.

Bembix	= *Ancistrocladus* Wallich (1829)
Blastus	
Botria	= *Ampelocissus* Planchon (1887)
Bragantia	= *Apama* Lamarck (1783)
Calispermum	= *Embelia* Burman f. (1768)
Callista	= *Dendrobium* Swartz (1799, 1800)
Calodium	= *Cassytha* Linnaeus (1753)
Campsis	
Campylus [27]	= *Tinospora* Miers (1851)
Cathetus	= *Phyllanthus* Linnaeus (1753)
Centipeda	
Ceraia	= *Dendrobium* Swartz (1799, 1800)
Cerium	= *Lysimachia* (Tournefort) Linnaeus (1753)
Citta	= *Mucuna* Adanson (1763)
Cladodes	= *Alchornea* Swartz (1788)
Coleus	
Columella	= *Cayratia* A. L. de Jussieu (1823) (*Lagenula* Loureiro, 1790; *Pedastis* Rafinesque, 1838; *Causonia* Rafinesque, 1838)
Commia	= *Excoecaria* Linnaeus (1753)
Cordyla	
Craspedum	= *Elaeocarpus* (Burman) Linnaeus (1753)
Creodus	= *Chloranthus* Swartz (1787)
Cubospermum	= *Jussiaea* Linnaeus (1753)
Cyathula	= *Achyranthes* Linnaeus (1753)
Cylindria	= *Linociera* Swartz (1791)
Cyrta	= *Styrax* (Tournefort) Linnaeus (1753)
Dartus	= *Maesa* Forskål (1775)
Dasus	= *Lasianthus* Jack (1823)
Decadia	= *Symplocos* Jacquin (1760)
Dentidia	= *Perilla* Linnaeus (1764)
Derris [28]	
Desmos [29]	
Diaphora	= *Scleria* Bergius (1765)
Diatoma	= *Carallia* Roxburgh (1814)
Dicalyx	= *Symplocos* Jacquin (1760)
Diceros	= *Limnophila* R. Brown (1810)
Dichroa	
Dimocarpus	= *Litchi* Sonnerat (1782)
Diphaca	= *Ormocarpum* Beauvois (1806)
Dissolaena	= *Rauwolfia* (Plumier) Linnaeus (1753)

[27] The single species described, *Campylus sinensis* Lour., is a species of *Tinospora*. The generic description, however, does not apply to any menispermaceous plant.

[28] Conserved name over *Salken* Adans. and *Solori* Adans. 1763, *Deguelia* Aubl. 1775, and *Cylizoma* Neck. 1790.

[29] The proper generic name for the so-called species of *Unona* of the Old World. *Unona* Linnaeus is strictly an American genus.

Dodecadia	= *Pygeum* Gaertner (1788)
Donax	
Drupatris	= *Symplocos* Jacquin (1760)
Dysoda	= *Serissa* Commerson (1789)
Ebenoxylum	= *Maba* J. R. & G. Forster (1776)
Echinus	= *Mallotus* Loureiro (1790)
Echtrus	= *Argemone* (Tournefort) Linnaeus (1753)
Enhydra (*Enydra*)	
Enkianthus	
Eystathes	= *Xanthophyllum* Roxburgh (1814, 1819)
Fallopia	= *Microcos* Linnaeus (1753)
Fibraurea	
Floscopa	
Galeola	
Garciana	= *Philydrum* Banks (1788)
Gela	= *Acronychia* J. R. & G. Forster (1776)
Gemella	= *Allophylus* Linnaeus (1753)
Gonus	= *Brucea* J. F. Miller (1780)
Grammica	= *Cuscuta* (Tournefort) Linnaeus (1753)
Grona	= *Desmodium* Desvaux (1813)
Hecatonia	= *Ranunculus* (Tournefort) Linnaeus (1753)
Hedona	= *Lychnis* (Tournefort) Linnaeus (1753)
Helicia	
Helixanthera [30]	= *Loranthus* Linnaeus (1753)
Heptaca	= *Oncoba* Forskål (1775)
Hexadica	= *Ilex* (Tournefort) Linnaeus (1753)
Hexanthus	= *Litsea* Lamarck (1789)
Homonoia	
Hydrogeton	= *Potamogeton* (Tournefort) Linnaeus (1753)
Knema	
Lagenula	= *Cayratia* A. L. de Jussieu (1833)
Lagunea	= *Polygonum* (Tournefort) Linnaeus (1753)
Lasia	
Lepta	= *Euodia* J. R. & G. Forster (1776)
Limacia	
Liriope	
Mallotus	
Marcanthus	= *Mucuna* Adanson (1763) § *Stizolobium* P. Browne (1756 as a genus)
Mazus	
Medusa	= *Rinorea* Aublet (1775)
Melodorum	= *Mitrephora* Hooker f. & Thomson (1855) [31]

[30] Dr. Danser (Bull. Jard. Bot. Buitenzorg III **11**: 268. 1931) retains this as a genus distinct from *Loranthus*.

[31] The generic *description* clearly appertains to *Mitrephora*; one of the species described, *Melodorum fruticosum*, is apparently a species of *Polyalthia*, the other a *Mitrephora*.

Meteorus	= *Barringtonia* J. R. & G. Forster (1776)
Miltus	= *Gisekia* Linnaeus (1771)
Morella	= *Myrica* Linnaeus (1753)
Muricia	= *Momordica* (Tournefort) Linnaeus (1753)
Nephroia	= *Cocculus* de Candolle (1818)
Neptunia	
Nymphanthus	= *Phyllanthus* Linnaeus (1753)
Octarillum	= *Elaeagnus* (Tournefort) Linnaeus (1753)
Oncinus	= *Melodinus* J. R. & G. Forster (1776)
Oncus	= *Dioscorea* (Plumier) Linnaeus (1753)
Opa	= *Eugenia* (Micheli) Linnaeus (1753) § *Syzygium* Gaertner (1788 as a genus)
Ophelus	= *Adansonia* Linnaeus (1753)
Ophispermum	= *Aquilaria* Lamarck (1786)
Osmanthus	
Oxycarpus	= *Garcinia* Linnaeus (1753)
Oxyceros	= *Randia* Houstoun (1753)
Pedicellia	= *Mischocarpus* Blume (1825)
Pentaloba	= *Rinorea* Aublet (1775)
Phaius	
Phanera	= *Bauhinia* Linnaeus (1753)
Phoberos	= *Scolopia* Schreber (1789)
Phyla	= *Lippia* Houstoun (1753)
Phyllamphora	= *Nepenthes* Linnaeus (1753)
Phyllaurea	= *Codiaeum* (Rumphius) A. L. de Jussieu (1824)
Phyllodes	= *Phrynium* Willdenow (1797)
Physkium	= *Vallisneria* (Micheli) Linnaeus (1753)
Picria	(= *Curanga* A. L. de Jussieu, 1807)
Pimela	= *Canarium* (Rumphius) Linnaeus (1759)
Placus	= *Blumea* de Candolle (1833)
Polia	= *Polycarpaea* Lamarck (1792)
Polychroa	= *Pellionia* Gaudichaud (1826)
Polyozus	= *Canthium* Lamarck (1783)
Polypara	= *Houttuynia* Thunberg (1784)
Polyphema	= *Artocarpus* J. R. & G. Forster (1776)
Porphyra	= *Callicarpa* Linnaeus (1753)
Pselium	= *Pericampylus* Miers (1851)
Pterotum	= ? *Rourea* Aublet (1775)
Pyrgus	= *Ardisia* Swartz (1788)
Pythagorea	= *Homalium* Jacquin (1760)
Quinaria	= *Clausena* Burman f. (1768)
Rapinia	= *Sphenoclea* Gaertner (1788)
Renanthera	
Restiaria	= *Uncaria* Schreber (1789)
Rhaphis	= *Andropogon* Linnaeus (1753) § *Chrysopogon* Trinius (1820 as a genus)

Rhytis	= *Antidesma* (Burman) Linnaeus (1753)
Rotula	(= *Rhabdia* Martius, 1827)
Rynchosia	
Salomonia	
Sarcodum	= *Clianthus* Banks & Solander (1832)
Scutula	= *Memecylon* Linnaeus (1753)
Sebifera	= *Litsea* Lamarck (1789)
Septas	= *Bramia* Lamarck (1785)
Solena	= *Melothria* Linnaeus (1753)
Spathium	= *Saururus* (Plumier) Linnaeus (1753)
Stegosia	= *Rottboellia* Linnaeus f. (1779)
Stemona	
Stephania	
Stigmanthus	= *Randia* Houstoun (1753)
Stigmarota	= *Flacourtia* L'Héritier (1785)
Stixis	
Streblus	
Striga	
Stylidium	= *Alangium* Lamarck (1783)
Tetradium	= *Euodia* J. R. & G. Forster (1776)
Tetrapilus	= *Olea* (Tournefort) Linnaeus (1753)
Thela	= *Plumbago* (Tournefort) Linnaeus (1753)
Thrixspermum	
Thylachium (*Thilachium*)	
Thysanus	= *Cnestis* A. L. de Jussieu (1789)
Tralliana	= *Colubrina* L. C. Richard (1827)
Trema	
Triadica	= *Sapium* (P. Browne) Jacquin (1763)
Tricarium	= *Phyllanthus* Linnaeus (1753)
Triceros	= *Turpinia* Ventenat (1803)
Tridesmis	= *Croton* Linnaeus (1753)
Triphaca	= *Sterculia* Linnaeus (1753)
Triphasia	
Tripinna	= *Vitex* (Tournefort) Linnaeus (1753)
Trisanthus	= *Centella* Linnaeus (1760)
Vanieria	= *Cudrania* Trécul (1847)
Vernicia	= *Aleurites* J. R. & G. Forster (1776)
Zala	= *Pistia* Linnaeus (1753)

An examination of this list brings out certain interesting facts. Of the 185 genera proposed and described by Loureiro as new in 1790, about 115 had already been characterized by other authors under other names previous to 1790. About forty of Loureiro's generic names are valid under all rules of botanical nomenclature. If strict priority be followed as to Loureiro's generic names, where his name was the first one proposed for the group in question, about 64 would be accepted. Thus *Phyllodes* Lour. would replace

Phrynium Willdenow (1797); *Columella* Lour. or *Lagenula* Lour. would replace *Cayratia* A. L. de Jussieu (1823); *Dasus* Lour. would replace *Lasianthus* Jack (1823); *Botria* Lour. would replace *Ampelocissus* Planchon (1887); *Polia* Lour. would replace *Polycarpaea* Lamarck (1792); *Campylus* Lour. (1790) might replace *Tinospora* Miers (1851); *Baryxylum* Lour. would replace *Peltophorum* Walpers (1842); *Aulacia* Lour. would replace *Micromelum* Blume (1825); *Bembix* Lour. would replace *Ancistrocladus* Wallich (1832); *Melodorum* Lour. would replace *Mitrephora* Hooker f. & Thomson (1855); *Barbula* Lour. would replace *Caryopteris* Bunge (1835); *Diphaca* Lour. would replace *Ormocarpum* Beauvois (1806); *Sarcodum* Lour. would replace *Clianthus* Banks & Solander (1832); *Placus* Lour. would replace *Blumea* de Candolle (1832); *Ceraia* Lour. or *Callista* Lour. would replace *Dendrobium* Swartz (1799); *Rhaphis* Lour. would replace *Chrysopogon* Trinius (1820); *Polychroa* Lour. would replace *Pellionia* Gaudichaud (1826); *Vanieria* Lour. would replace *Cudrania* Trécul (1847); *Phyllaurea* Lour. would replace *Codiaeum* A. L. de Jussieu (1824); *Pselium* Lour. would replace *Pericampylus* Miers (1851); *Pedicellia* Lour. would replace *Mischocarpus* Blume (1825), and *Triceros* Lour. would replace *Turpinia* Ventenant (1803); of these *Campylus* and *Turpinia* are not yet included in the lists of *nomina generica conservanda*.

Furthermore, the following Loureiroan genera are older than those in the conserved lists, but in each case there are still older generic names for each group; *Cylindria* Lour. would replace *Linociera* Swartz (1791), except for the older *Mayepea* Aublet (1775); *Eystathes* Lour. would replace *Xanthophyllum* Roxb. (1814, 1819), except for the earlier *Palae* Adanson (1763); *Tralliana* Lour. would replace *Colubrina* L. C. Richard (1827), except for *Marcorella* Necker which was also published in 1790; *Diatoma* Lour. would replace *Carallia* Roxburgh (1814), except for the older *Karekandel* Adanson (1763); *Diceros* Lour. would replace *Limnophila* R. Brown (1810), except for the older *Ambulia* Lamarck (1783); *Grona* Loureiro would replace *Desmodium* Desvaux (1813), except for the earlier *Meibomia* Adanson (1763);[32] and *Nephroia* Loureiro would replace *Cocculus* de Candolle (1818), except for the earlier *Cebatha* Forsk., *Leaeba* Forsk., and *Epibaterium* Forsk. (1776). On a very strict interpretation of types *Derris* Lour. is actually a synonym of *Dalbergia* Linnaeus f. (1781).

As this study has been consummated under the general provisions of the International Code of Botanical Nomenclature, I have accepted the conserved names approved by the Vienna, Brussels, and Cambridge Botanical Congresses, even where, in my personal opinion, some of these should not have been included.

Of Loureiro's 185 new genera there are only seven cases where exactly the same groups have not been designated and described under other generic names. One hundred fifteen of these had been characterized and named by other botanists before 1790, a reflection on Loureiro's bibliographic work. Sixty-three were given new names by later authors chiefly for the reason that Loureiro's contemporaries and successors failed to realize that the groups they named and described as new after 1790 had already been named and characterized by Loureiro, a reflection on the bibliographic work of the later botanists. These seven genera are *Lasia*, *Triphasia*, *Thylachium*, *Centipeda*, *Blastus*, *Knema*, and *Limacia*, and for two of these variant spellings have been proposed. Loureiro's successors, notably Rafinesque, proposed 54 additional new generic names, all based wholly or in part on Loureiro's

---

[32] It should be noted that Schindler now retains *Meibomia* Adanson as generically distinct from *Desmodium* Desvaux.

original descriptions. A list of these with their reductions as far as their status can be determined, follows:

NEW GENERIC NAMES BASED WHOLLY OR IN PART ON LOUREIRO'S DESCRIPTIONS

Acrodryon Spreng. Syst. 1: 386. 1825. Types *A. angustifolium* Spreng. *l.c.* based on *Cephalanthus angustifolius* Lour. = *C. angustifolius* Lour., and *A. orientale* Spreng. *l.c.* based on *C. orientalis* Lour. = *C. occidentalis* Linn.

Agonon Raf. Sylva Tellur. 161. 1838. Type *A. umbellata* Raf. *l.c.* based on *Callicarpa umbellata* Lour. = *Premna umbellata* Schauer = ?

Aponoa Raf. Sylva Tellur. 84. 1838. Type *A. repens* Raf. *l.c.* based on *Columnea stellata* Lour. = ? *Limnophila chinensis* (Osbeck) Merr.

Askofake Raf. Fl. Tellur. 4: 108. 1838. Type *A. recurva* Raf. *l.c.* based on *Utricularia recurva* Lour. = *U. bifida* Linn.

Axolus Raf. Sylva Tellur. 61. 1838. Type *A. angustifolius* Raf. *l.c.* based on *Cephalanthus angustifolius* Lour. = *C. angustifolius* Lour.

Camaion Raf. Sylva Tellur. 75. 1838. Types *C. hirsuta* Raf. *l.c.* based on *Helicteres hirsuta* Lour. = *H. hirsuta* Lour. and *C. undulata* Raf. *l.c.* based on *H. undulata* Lour. = *Sterculia lanceolata* Cav.

Cleisostoma Raf. Fl. Tellur. 4: 80. 1838. Type *C. villosa* Raf. *l.c.* based on *Convolvulus aggregatus* Lour. = *Ipomoea pes-tigridis* Linn.

Crozophylla Raf. Sylva Tellur. 64. 1838. A new generic name for several species congeneric with *Phyllaurea* Lour. "*Phyllaurea* of Lour. is one of these." = *Codiaeum*.

Dectis Raf. Fl. Tellur. 2: 43. 1837. Based on *Solidago arborescens* Forst. f. (= *Olearia nitida* Hook. f.), *S. leucadendra* Forst. f. (or Willd.) (= *Senecio*), *S. rugosa* Mill., *S. spuria* Forst. f. (= *Commidendron spurium* DC.) and *S. decurrens* Lour. = *Solidago virgaurea* Linn.

Drosodendron M. Roem. Syn. 1: 138, 140. 1846. Type *D. rosmarinus* M. Roem. *l.c.* based on *Cedrela rosmarinus* Lour. = *Baeckea frutescens* Linn.

Drupifera Raf. Sylva Tellur. 140. 1838. Type *D. oleosa* Raf. *l.c.* based on *Camellia drupifera* Lour. = *Thea oleosa* Lour.

Eresimus Raf. Sylva Tellur. 61. 1838. Type *E. stellatus* Raf. *l.c.* based on *Cephalanthus stellatus* Lour. = *C. angustifolius* Lour.

Faulia Raf. Fl. Tellur. 2: 84. 1837. Types *F. verrucosa* Raf. *l.c.* based on *Ligustrum nepalense* Wall. var. *glabrum* Wall. and *F. odorata* Raf. *l.c.* based on *L. spicatum* Don (*L. nepalense* Wall.) both = *L. nepalense* Wall. "The *Lig. japonicum, Sinense* [Lour.] *lucidum* of Asia may belong to it [*Faulia*]."

Gandola (Rumph.) Raf. Sylva Tellur. 60. 1838. Type *G. nigra* Raf. (*Basella nigra* Lour.) = *B. rubra* Linn. A generic name taken from the pre-Linnaean Herbarium Amboinense of Rumphius.

Gilipus Raf. Sylva Tellur. 61. 1838. Type *G. montanus* Raf. *l.c.* based on *Cephalanthus montanus* Lour. = *C. occidentalis* Linn.

Helicanthera Roem. & Schult. Syst. 5: X. 1819. Variant spelling of *Helixanthera* Lour., no binomial = *Loranthus pentapetalus* Roxb. Danser, Bull. Jard. Bot. Buitenzorg III 11: 370. 1931, retains *Helixanthera* Lour. as a valid genus distinct from *Loranthus* Linnaeus.

Helicia Pers. Syn. 1: 214. 1805.   Type *Helicia parasitica* Pers. *l.c.* based on *Helixanthera parasitica* Lour. = *Loranthus pentapetalus* Roxb.

Hexacadica Raf. Sylva Tellur. 158. 1838.   Type *Hexacadica corymbosa* Raf. *l.c.* based on *Hexacadica cochinchinensis* Lour. = *Ilex cochinchinensis* (Lour.) Loesen.

Hexepta Raf. Sylva Tellur. 164. 1838.   Types *H. axillaris* Raf. *l.c.* based on *Coffea zanguebariae* Lour. and *H. racemosa* Raf. *l.c.* based on *C. racemosa* Lour., both species of *Coffea* as at present understood; the latter probably represents some other genus than *Coffea*.

Hippoxylon Raf. Sylva Tellur. 78. 1838.   Type *H. indica* Raf. *l.c.* based on *Bignonia indica* auctt. and *B. pentandra* Lour. = *Oroxylum indicum* (Linn.) Vent.

Hisutsua DC. Prodr. 6: 44. 1837.   Type *H. cantoniensis* DC. *l.c.* based on *Matricaria cantoniensis* Lour. = *Boltonia indica* (Linn.) Benth. (*Aster indicus* Linn.).   The generic name is from the Chinese name *hi su tsu* as given by Loureiro.   Type in the herbarium of the Paris Museum.

Icosinia Raf. Sylva Tellur. 75. 1838.   Type *I. paniculata* Raf. *l.c.* based on *Helicteres paniculata* Lour. = *Sterculia* sp.

Isgarum Raf. Fl. Tellur. 3: 46. 1837.   Type *I. didymum* Raf. *l.c.* based on *Salsola didyma* Lour. = ?

Juchia M. Roem. Syn. 2: 48. 1846.   Type *J. hastata* M. Roem. *l.c.* based on *Bryonia hastata* Lour. = *Melothria heterophylla* (Lour.) Cogn.

Jürgensia Spreng. Anleit. 2(2): 806. 1818, Syst. 3: 50. 1826.   Type *J. anguifera* Spreng. Syst. 3: 50. 1826, based on *Medusa anguifera* Lour. = *Rinorea anguifera* (Lour.) O. Ktz. Given in Bentham & Hooker f.   Gen. Pl. 1: 970. 1867 as *Jurgensia*.

Lagansa (Rumph.) Raf. Sylva Tellur. 110. 1838.   Type *L. alba* Raf. *l.c.* based on *Cleome icosandra* "Linn. Lour." = *Polanisia icosandra* (Linn.) W. & A.   A generic name taken from the pre-Linnaean Herbarium Amboinense of Rumphius.

Lasipana Raf. Sylva Tellur. 21. 1838.   Type *L. tricuspis* Raf. *op. cit.* 22, based on *Echinus trisulcus* Lour. = *Mallotus paniculatus* (Lam.) Muell.-Arg.

Loureira Meisn. Gen. Comm. 53. 1837.   Type *L. cochinchinensis* Meisn. *l.c.* based on *Toluifera cochinchinensis* Lour. = *Glycosmis cochinchinensis* (Lour.) Pierre.

Munnickia Raf. Fl. Tellur. 4: 99, 125. 1838.   Type *Bragantia* [*racemosa* Loureiro]; no new binomial published.   The new name was proposed because of the earlier *Bragantia* Vandelli (1771) which is a synonym of *Gomphrena* Linn.

Nemanthera Raf. Fl. Tellur. 4: 80. 1838.   Type *N. bufalina* Raf. *l.c.* based on *Convolvulus bufalinus* Lour. = *Ipomoea* (vel *Merremia*) sp.

Nephroica Miers in Ann. Mag. Nat. Hist. II 7: 42. 1851, Contrib. Bot. 3: 260. 1871. Type *Nephroica sarmentosa* Miers *l.c.* based on *Nephroia sarmentosa* Lour. = *Cocculus sarmentosus* (Lour.) Diels.

Olofuton Raf. Sylva Tellur. 108. 1838.   Type *O. racemosum* Raf. *l.c.* based on *Capparis cantoniensis* Lour., of which *C. pumila* Champ. and *C. sciaphila* Hance are synonyms.

Opanea Raf. Sylva Tellur. 106. 1838.   Types *Myrtus trinervia* Sm. and *M. billiardieri* [*M. billiardiana* HBK.] "also the 2d sp. of *Opa* of Loureiro"; *M. trinervia* Sm. = *Rhodamnia trinervia* Blume and *M. billiardiana* HBK. = *Myrcia billiardiana* DC.   The second species of *Opa* of Loureiro is *Opa metrosideros* = *Raphiolepis indica* (Linn.) Lindl.

Petalotoma DC. Prodr. **3**: 294. 1838. Type *P. brachiata* DC. *l.c.* based on *Diatoma brachiata* Lour. = *Carallia brachiata* (Lour.) Merr. The new generic name was proposed because de Candolle himself had used the name *Diatoma* for a different genus.

Piarimula Raf. Fl. Tellur. **2**: 102. 1837. Type *P. chinensis* Raf. *l.c.* based on *Phyla chinensis* Lour. = *Lippia nodiflora* (Linn.) L. C. Rich.

Pogalis Raf. Fl. Tellur. **3**: 15. 1837. Type *P. barbata* Raf. *l.c.* based on *Polygonum barbatum* Linn., with *P. tinctoria* Raf. (*Polygonum tinctorium* Lour., or Aiton?) and *P. tomentosa* Raf. (*Polygonum tomentosum* Willd.).

Pseudiosma A. de Juss. in Mém. Mus. Hist. Nat. [Paris] **12**: 519. 1825. Type *P. asiatica* Juss. *l.c.* based on *Diosma asiatica* Lour. = ?

Selas Spreng. Syst. **2**: 216. 1825. Type *Selas lanceolatum* Spreng. *l.c.* based on *Gela* [*lanceolata* Lour.] = *Acronychia pedunculata* (Linn.) Miq.

Semetor Raf. Sylva Tellur. 69. 1838. Type *S. arborea* Raf. *l.c.* based on *Aspalathus arboreus* Lour. = *Derris heptaphylla* (Linn.) Merr.

Septilia Raf. Fl. Tellur. **4**: 68. 1838. Type *Septas repens* Lour. = *Bramia monnieri* (Linn.) Pennell. No binomial is published under *Septilia*.

Silamnus Raf. Sylva Tellur. 60. 1838. Type *S. procumbens* Raf. *op. cit.* 61 based on *Cephalanthus procumbens* Lour. = ?

Spirospatha Raf. Fl. Tellur. **4**: 8. 1838. Type *S. occulta* Raf. *l.c.* based on *Calla occulta* Lour. = *Homalomena* sp.

Stigmatanthus Roem. & Schult. Syst. **5**: 225. 1819. Type *Stigmatanthus cymosus* Roem. & Schult. *l.c.* based on *Stigmatanthus cymosus* Lour. = *Morinda umbellata* Linn.

Stylis Poir. in Lam. Encycl. Suppl. **5**: 260. 1817. Type *S. chinensis* Poir. *l.c.* based on *Stylidium chinense* Lour. = *Alangium chinense* (Lour.) Rehd.

Theaphylla Raf. Sylva Tellur. 138. 1838. A new generic name for *Thea* Linnaeus, based on several species of *Thea* and *Camellia*, including *Theaphylla cantoniensis* Raf. *op. cit.* 139, based on *Thea cantoniensis* Lour. = *Thea chinensis* Linn.; *Theaphylla oleifera* Raf. *l.c.* = *Thea oleosa* Lour.; and *Theaphylla annamensis* Raf. *l.c.* based on *Thea cochinchinensis* Lour. = *Thea chinensis* Linn. *Theaphylla* was proposed, without a binomial in Raf. Med. Feb. **2**: 267, 1830; in the Sylva Telluriana it appears as *Theaphyla*.

Trachytella DC. Syst. **1**: 410. 1818. Types *T. actaea* DC. *l.c.* based on *Actaea aspera* Lour. and *T. calligonum* DC. *l.c.* based on *Calligonum asperum* Lour., both = *Tetracera scandens* (Linn.) Merr. (*T. sarmentosa* Vahl).

Traxilisa Raf. Sylva Tellur. 161. 1838. Type *T. aspera* Raf. *op. cit.* 162 based on *Calligonum asperum* Lour. = *Tetracera scandens* (Linn.) Merr.

Tremotis Raf. Sylva Tellur. 59. 1838. Type *Tremotis cordata* Raf. *l.c.* = *Ficus auriculata* Lour. (*F. roxburghii* Wall.).

Triclanthera Raf. Sylva Tellur. 108. 1838. Types *T. corymbosa* Raf. *l.c.* based on *Capparis magna* Lour. = *Crataeva magna* (Lour.) DC. and *T. falcata* Raf. *l.c.* based on *Capparis falcata* Lour. = *Crataeva religiosa* Forst. f.

Triolobus Raf. Sylva Tellur. 110. 1838. Type *T. cordata* Raf. *op. cit.* 111 based on *Triphaca* [*africana* Lour.] = *Sterculia africana* (Lour.) Merr. (*S. triphaca* R. Br.).

Triplandra Raf. Sylva Tellur. 62. 1838. Type *T. lanata* Raf. *op. cit* 63 based on *Croton lanatum* Lour. = *C. lasianthus* Pers.

Triplosperma G. Don Gen. Syst. **4**: 134. 1838. Type *T. cochinchinensis* G. Don *l.c.* based on *Stapelia cochinchinensis* Lour. = *Hoya cochinchinensis* Roem. & Schult.

Tripodanthera M. Roem. Syn. 2: 48. 1846. Type *T. cochinchinensis* M. Roem. *l.c.* based on *Bryonia cochinchinensis* Lour. = *Gymnopetalum cochinchinense* (Lour.) Kurz.

Zelonops Raf. Fl. Tellur. 2: 102. 1837. Type *Z. pusilla* Raf. *l.c.* based on " *Phenix pusilla* Lour. Gaertn. t. 24, *Ph. farinifera* Roxb. Cor. t. 74, Smith et auctoris" = (as to *P. pusilla* Lour.) *Phoenix loureirii* Kunth.

Of these 54 generic names, based wholly or in part on Loureiro's published descriptions, except *Hisutsua* DC. which was based on an authentic specimen of *Matricaria cantoniensis* Lour., a few remain that are still of doubtful status, doubtful even as to the family to which they belong. I have been unable to place *Callicarpa umbellata* Lour. on which *Agonon* Raf. is wholly based; *Isgarum* Raf., based wholly on the description of *Salsola didyma* Lour., is doubtful as to its family but is possibly a representative of the Chenopodiaceae; *Pseudiosma* A. Juss., based on the description of *Diosma asiatica* Lour., remains unplaced as to its family, and *Silamnus* Raf., based on the description *Cephalanthus procumbens* Lour., is in the same category as *Agonon*, *Pseudiosma*, and *Isgarum*.

Thus out of 54 new generic names proposed by Loureiro's successors 4 remain as of entirely doubtful status, yet all of the 185 named and described by Loureiro as new have been placed with sufficient accuracy either as valid genera or as synonyms of those described by other authors.

LOUREIRO'S TREATMENT OF LINNAEAN GENERA

In his interpretation of many Linnaean genera, Loureiro made numerous grave errors, ascribing to certain generic names species totally unrelated to the groups as defined by Linnaeus. In other cases he was correct in the interpretation of certain Linnaean binomials, both as to the genus and the species, but was totally wrong as to generic positions of other species that he placed in the same group, as illustrated by *Juglans*. Of the three species described, *J. regia* is a form of the Linnaean species; *J. camirium* is the euphorbiaceous *Aleurites moluccana* Willd.; and *J. catappa* is the combretaceous *Terminalia catappa* Linn., but Loureiro's binomial in this case was not based on the earlier one of Linnaeus.

As illustrations of absolute errors in the interpretation of Linnaean genera there may be cited: *Euonymus* (Celastraceae), one species, *E. chinensis* Lour. = *Gymnopetalum cochinchinense* (Lour.) Kurz (Cucurbitaceae); *Primula* (Primulaceae), two species, *P. mutabilis* Lour. = *Hydrangea opuloides* (Lam.) K. Koch (Saxifragaceae), the other, *P. sinensis* Lour., a species of entirely doubtful status but no primulaceous plant; *Plectronia* (correctly Oliniaceae, not Rubiaceae), one species, *P. chinensis* Lour. = *Acanthopanax trifoliatus* (Linn.) Merr. (Araliaceae); *Crassula* (Crassulaceae), one species, *C. pinnata* (non Linn.) = *Eurycomia longifolia* Jack (Simarubaceae); *Santalum* (Santalaceae), one species, *S. album* (non Linn.) = *Dysoxylum loureiri* Pierre (Meliaceae); *Thuja* (Pinaceae), one species, *T. orientalis* (non Linn.) = *Glyptostrobus pensilis* (Abel) K. Koch (Pinaceae); *Reseda* (Resedaceae), two species, *R. chinensis* Lour. and *R. cochinchinensis* Lour., both = *Hypericum japonicum* Thunb. (Guttiferae); *Toluifera* (Leguminosae), one species, *T. cochinchinensis* Lour. = *Glycosmis cochinchinensis* Pierre (Rutaceae); *Melanthium* (Liliaceae), one species, *M. cochinchinense* Lour. = *Asparagus cochinchinensis* (Lour.) Merr. (Liliaceae); *Cedrela* (Meliaceae), one species, *C. rosmarinus* Lour. = *Baeckea frutescens* Linn. (Myrtaceae); *Juncus* (Juncaceae), one species, *J. bulbosus* Lour. = *Eleusine indica* Gaertn. (Gramineae), while a true *Juncus*, *J. effusus* Linn., was described as *Scirpus capsularis* Lour.; *Actaea*

(Ranunculaceae), one species, *A. aspera* Lour. = *Tetracera scandens* (Linn.) Merr. (Dilleniaceae); *Ervum* (Leguminosae), one species, *E. hirsutum* (non Linn.) = *Flemingia macrophylla* (Willd.) O. Ktz. (Leguminosae); *Lantana* (Verbenaceae), one species, *L. racemosa* Lour. = *Gmelina racemosa* (Lour.) Merr. (Verbenaceae); *Penaea* (Penaeaceae), two species, one *P. nitida* Lour. = *Gluta nitida* (Lour.) Merr. (Anacardiaceae), the other *P. scandens* Lour., status unknown; *Gardenia* (Rubiaceae), one species, *G. volubilis* Lour. = *Ichnocarpus volubilis* (Lour.) Merr. (Apocynaceae); *Hottonia* (Gentianaceae), one species, *H. litoralis* Lour. = *Catharanthus roseus* (Linn.) G. Don (Apocynaceae); *Varronia* (Boraginaceae), one species, *V. sinensis* Lour. = *Cordia dichotoma* Forst. f. (Boraginaceae), the same species also described as *Argyreia arborea* Lour. (Convolvulaceae); *Phyteuma* (Campanulaceae), two species, *P. bipinnata* Lour. and *P. cochinchinensis* Lour., both = *Sambucus javanica* Reinw. (Caprifoliaceae); *Scabiosa* (Dipsaceae), one species, *S. cochinchinensis* Lour. = *Elephantopus scaber* Linn. (Compositae); *Bobartia* (Cyperaceae, one species, *B. indica* (non Linn.) = *Rynchospora wightiana* Steud. (Cyperaceae); *Anthoxanthum* (Gramineae), one species, *A. pulcherrimum* Lour. = *Centotheca latifolia* (Osb.) Trin. (Gramineae); *Phleum* (Gramineae), one species, *P. cochinchinense* = *Ophiurus cochinchinensis* (Lour.) Merr. (Gramineae); *Ischaemum* (Gramineae), one species, *I. importunum* Lour. = *Panicum repens* Linn. (Gramineae); *Gaura* (Onagraceae), one species, *G. chinensis* Lour. = *Haloragis chinensis* (Lour.) Merr. (Haloragaceae); *Anagyris* (Leguminosae), one species, *A. foetida* Lour. = *Sophora japonica* Linn. (Leguminosae); *Salvadora* (Salvadoraceae), two species, *S. biflora* Lour. and *S. capitulata* Lour., both = *Streblus asper* Lour. (Moraceae); *Coccoloba* (Polygonaceae), two species, *C. cymosa* Lour. and *C. asiatica* Lour., both = *Polygonum chinense* Linn. (Polygonaceae); *Vateria* (Dipterocarpaceae), one species, *V. flexuosa* Lour. = *Mischocarpus flexuosus* (Lour.) Merr. (Sapindaceae); *Pistacia* (Anacardiaceae), one species, *P. oleosa* Lour. = *Schleichera oleosa* (Lour.) Merr. (Sapindaceae); *Digitalis* (Scrophulariaceae), two species, *D. sinensis* Lour. = *Adenosma glutinosum* (Lour.) Druce, and *D. cochinchinensis* Lour. = *Centranthera cochinchinensis* (Lour.) Merr. (Scrophulariaceae); *Ruellia* (Acanthaceae), two species, *R. antipoda* Linn. = *Ilysanthes antipoda* (Linn.) Merr. (Scrophulariaceae), *R. ciliaris* (non Linn.) = *Torenia peduncularis* Benth. (Scrophulariaceae), and *Elaeocarpus* (Elaeocarpaceae), one species, *Elaeocarpus integerrimus* Lour. = *Ochna integerrima* (Lour.) Merr. (Ochnaceae).

It is noted from an examination of the preceding data that a high percentage of Loureiro's new genera had already been described under other names previous to 1790. From the last discussion it will be correctly inferred that Loureiro's attempts to interpret the genera of other authors must be looked on with suspicion as to their correctness, and consequently all his descriptions, good and bad, must be critically scanned. As it is thus clear that in many cases his interpretations of Linnaean genera cannot be trusted; likewise it is found that very frequently little trust can be placed in his assignment of his new genera to the Linnaean classes and orders, as it is not infrequently found that they represent units remote from the major Linnaean groups in which he placed them.

LOUREIRO'S SPECIES

An examination of Loureiro's work shows that he frequently described the same species twice and sometimes three or even four times, often under different generic names, as did Blanco in his "Flora de Filipinas." This was due in part to his putting too much faith in the sexual system of classification of Linnaeus, not realizing that the number of stamens and carpels often vary in the same species; partly due to his failing to recognize that a fruiting specimen placed in one genus actually represented the same species as a flowering specimen placed in some other group; partly to his placing staminate specimens of dioecious species in one genus, and pistillate specimens of the same species in another genus; partly to his treatment of slightly different forms of the same species, particularly in the case of cultivated plants, as representing distinct species; and undoubtedly in large part because he did not consistently preserve herbarium material for purposes of comparison, therefore being obliged to rely too much on his own memory. Of the 1292 species described it has been possible to make about 135 reductions to those otherwise described in the same work, so that the total number of distinct species considered in the "Flora Cochinchinensis" would at most be but about 1157.

In his interpretation of Linnaean genera he made numerous grave errors as noted above, and in interpreting Linnaean species he was wrong in about 374 out of 663 cases, his error of interpretation in the case of species being slightly over 56 per cent, as compared with Blanco's misinterpretation of approximately 60 per cent of the species ascribed to Linnaeus in the "Flora de Filipinas." In all cases of misinterpretation on the part of Loureiro, as with Blanco, the specific names are considered to be invalid; any other treatment of them is not only illogical, but would entail a very large number of changes in accepted binomials for well-known species which it is otherwise unnecessary to make.

It seems to be evident, from a study of Loureiro's descriptions, that at some time he had actual specimens in practically every case, either when he prepared his notes or when he wrote the final description. Many of the descriptions are excellent, in fact distinctly superior to those of many of those prepared by Loureiro's contemporaries in Europe. In other cases they are short, incomplete, indefinite, and sometimes very inaccurate. He admits, in some cases, that data were added partly from memory and, occasionally, that certain data were taken from illustrations in Chinese medical books. Many of his most unsatisfactory descriptions are manifestly based on fragmentary or incomplete specimens, and most of such material was certainly secured by him from Chinese herbalists or dealers in medicinal plants. In some cases his descriptions are in part based on such material as he had for examination, in part on the Linnaean descriptions cited, and, in a few cases, in part on the pre-Linnaean illustrations that he thought represented the species he had in hand. That he actually had specimens of some kind in most cases is evidenced by the fact that he cites Anamese and Chinese names for nearly all of the Indo-Chinese and Chinese plants described. It is obvious, from a most cursory examination of his work, that in many cases where he cites illustrations in Rheede's "Hortus Malabaricus" and Rumphius' "Herbarium Amboinense" as representing his species, he gravely erred, the illustrations sometimes appertaining to genera and species totally different from those he actually described. In no case can a Loureiroan species be actually typified by a cited pre-Linnaean illustration, even, as in the case of some of the references to Rumphius, where Loureiro took his specific name from Rumphius for the plant he actually described, as his descriptions sometimes apply not only to plants in different genera from those represented by Rumphius, but even in different families, as illustrated by *Thysanus palala* Lour. = *Cnestis palala* (Lour.) Merr. (Connaraceae); but *Palala secunda* Rumph., whence Loureiro took his specific name, is the myristicaceous *Horsfieldia sylvestris* Warb.

The errors that Loureiro made in misinterpreting pre-Linnaean descriptions and illustrations were due to conditions under which he worked and are no worse than similar ones made by Linnaeus himself, as well as by Loureiro's contemporaries and immediate successors; in fact a considerable number of Loureiro's errors in placing pre-Linnaean descriptions and illustrations were merely copied from Linnaeus' works.

Hooker (p. 11) has already noted the faulty terminology used by Loureiro, a fact that renders the correct interpretation of some of his genera and species peculiarly difficult. That he did not use certain technical terms in the same sense as did his contemporaries must constantly be kept in mind. The difficulties are further enhanced by the fact that Loureiro's descriptions are frequently general rather than definite, and almost never are they comparative; measurements are rarely given; certain characters that are now deemed to be essential in diagnoses of new species were not considered to be of sufficient importance even to note; he is not definite in many cases in distinguishing between simple and compound leaves, alternate and opposite ones, inferior and superior ovaries, and free or united petals. It is a well-known fact that can be proven by an examination of Loureiro's extant types, that in many cases his descriptions of both genera and species are erroneous—not because Loureiro deliberately erred but because he seriously misinterpreted various morphological characters.

In some cases it is certain that Loureiro had only flowering specimens, and yet, particularly in those cases where he placed his species in Linnaean genera, he described the fruits in general terms. This usually does not mean that his descriptions were based on material from two unrelated species and that therefore the name should be ignored on the specious claim that the species was based on a mixture of material, *but rather that Loureiro deliberately added generalized fruit characters to make his description conform to the characters of the Linnaean genera in which the species were erroneously placed.* A manifest case of many of this type is represented by the supposedly rubiaceous *Gardenia volubilis* Lour. = *Ichnocarpus volubilis* (Lour.) Merr. of the Apocynaceae, where fortunately we are sure of what Loureiro intended, as his type, a flowering specimen, is still extant.

In other cases his descriptions are based on material originating from unrelated plants, a notable example being the genus *Baryxylum*, based on flowers of one species (a *Peltophorum*), the fruits of one, perhaps two, unrelated genera (*Gymnocladus, Intsia*), and possibly the leaves of a third genus, if we may trust the specimens in the herbarium of the British Museum received from Loureiro. *Aloexylum agallochum* Lour. was based in part on plants that are apparently identical with *Aquilaria agallocha* Roxb. of the Thymelaeaceae, and in part on some leguminous tree. *Convolvulus mammosus* Lour. is in part *Ipomoea batatas* Poir. and in part *Dioscorea esculenta* (Lour.) Burkill. The description of *Equisetum arvense* Linn. was based in part on material apparently representing that species, in part, and as to its Chinese name and uses, on *Ephedra sinica* Stapf.

There are a few remaining genera of doubtful status that were based by later authors on various species described by Loureiro. These are *Agonon, Isgarum*, and *Silamnus* of Rafinesque, and *Pseudiosma* A. Jussieu. They unquestionably represent known groups described by other botanists under other names. They remain unrecognizable either because of the misinterpretation of essential morphological characters, errors of observation in preparing the descriptions and thus errors in fact, incomplete data, or because the description of what was taken to represent a single species was actually based on material representing different and perhaps not even closely related ones. The same comments also apply to most or all of the remaining species that have as yet not been accurately placed as to their proper families and genera.

Taxonomists working in Europe and America, without a first-hand knowledge of the flora of such a region as Indo-China, do not always realize that the bulk of the material examined by the early pioneers in the oriental tropics, whether from India, Ceylon, Java, the Philippines, or Indo-China, came from low altitudes in the immediate vicinity of settlements and not from the more remote primary forests of the interior, as the latter regions were relatively inaccessible in the early years of European exploration and colonization. A high percentage of all botanical material collected in the settled areas at low altitudes in the Old World tropics invariably consists of common, widely distributed, well-known species rather than of local endemics. Thus the flora of Kwangtung Province, especially in the general vicinity of Canton whence Loureiro secured most of his Chinese material, is distinctly well known. It is certain that there has been no radical change in the nature and constituent species of this flora during the past 150 years, except in the naturalization of plants introduced within the period indicated; yet a number of binomials and some generic names based by Loureiro or his successors on Canton material remain of wholly doubtful status. These I have not been able to determine even by the tedious process of elimination, simple forms of which are illustrated by the following cases. By examining the description critically and then scanning the list of all species known from Kwangtung Province, the problem of the identity of *Drosera umbellata* Lour. was easily solved; the only known Kwangtung species in any family that conforms at all to the characters of *Drosera umbellata* Lour. is *Androsace saxifragifolia* Bunge, and as early as 1848 Planchon had indicated that Loureiro's species was an *Androsace*. A comparison of Loureiro's description with specimens of Bunge's species from Canton shows conclusively that the two species are identical. No botanist has ever suggested a reduction of *Gaura chinensis* Lour. which from the description is manifestly not a *Gaura*. By application of the method of elimination this proves to be the same as *Haloragis scabra* Benth.; Loureiro's description is an admirable one for this plant. *Antidesma scandens* Lour. similarly considered proves to be staminate *Humulus japonicus* Sieb. & Zucc., quite as *Salvadora biflora* Lour· and *S. capitulata* Lour. prove to be no other than the genus and species described by Loureiro himself as *Streblus asper*. In other cases the application of this method gives no productive results, and here I am forced to the conclusion that Loureiro's descriptions are erroneous in certain essential details. In such instances it is exceedingly difficult to differentiate between the accurate and inaccurate parts of a single description. Loureiro's published descriptions, therefore, like Blanco's, must be interpreted rather than always accepted as strictly accurate.

Many botanists have not realized that in a high percentage of the descriptions published by such authors as Loureiro and Blanco technical errors occur, due to misinterpretation of morphological characters, errors of observation, errors in transcription, and other causes. The bibliographic botanist has been prone to coin new binomials, and to make numerous transfers of specific names because of discrepancies noted in comparing published descriptions. Most of the new generic names and a high percentage of the new binomials typified by Loureiro's descriptions can thus be accounted for. The changes were for the most part proposed in good faith, but frequently without a clear understanding of the situation. Thus both Loureiro and Blanco described *Polyozus* as having *bipinnate leaves;*

in both cases they had specimens of the rubiaceous genus *Canthium*, and misinterpreted the distichously arranged branchlets and the distichously arranged leaves as representing *bipinnate leaves*, not an unnatural interpretation as those who are familiar with *Canthium* in the field will testify; both rather frequently described branchlets with distichously arranged simple leaves as *pinnate leaves*, again not at all an unnatural error.   In other cases in species having pinnate leaves the compound character is not mentioned and the leaves are described, at least by inference, as simple.   Because Blanco (Fl. Filip. 9. 1837) described *Nyctanthes sambac* as having *pinnate leaves*, Hasskarl 27 years later proposed the new binomial *Jasminum blancoi* for it; assuming that Blanco's description was technically correct, the species could not be *Nyctanthes sambac* Linn. which has simple leaves, hence the new name.   But Blanco merely misinterpreted the branchlets with their characteristic distichously arranged leaves as representing a *pinnate leaf.*   As late as 1932 we find Doctor Kobuski [33] taking exception to my reduction of *Jasminum blancoi* Hassk. (*Nyctanthes sambac* Blanco) to *Jasminum sambac* Ait. because he did not realize that Blanco's description of the leaves as pinnate was false.   He overlooked the fact, as Hasskarl did, that Blanco himself (Fl. Filip. ed. 2, 6. 1845) noticing his own error eliminated the reference to pinnate leaves in his description (1845) of *Nyctanthes sambac!*   But here again, one familiar with the Philippine flora would have been on his guard, for *sampagita* is an absolutely fixed and unvarying name in the Philippines, universally applied to the widely cultivated *Jasminum sambac* Ait. and never applied to any other plant; it is just as fixed a name for *Jasminum sambac* in the Philippines as is hawthorn for *Crataegus oxyacantha* Linn. in England, or mayflower for *Epigaea repens* Linn. in New England; and no matter how inaccurate, incomplete, or otherwise unsatisfactory Blanco's description may have been, the local name itself would definitely place the species.   Yet Doctor Kobuski suggests that *Jasminum blancoi* Hassk. (*Nyctanthes sambac* Blanco) must be a species of *Jasminum* in the group with *J. grandiflorum* Linn., a group unrepresented in the Philippines by either native or, in Blanco's time, introduced species!

Should a genus or species be eliminated merely because such errors are present in the -original description?   Personally I think not.   I would no more eliminate a Loureiroian or Blancoan species because its original description contained palpable errors due to false interpretations of morphological characters, than I would eliminate *Volkameria japonica* Thunb. which Thunberg described as a large tree, although it is actually only a small shrub. The logical action in such cases would seem to be the retention of the name, if the name be a valid one for a recognized species, with a proper correction of the error or errors contained in the original description.

THE BEARING OF LOCAL NAMES ON THE INTERPRETATION OF LOUREIRO'S SPECIES

Some of Loureiro's Chinese descriptions were based on material from northern China and one at least on material as far west as Yunnan.   Many of the descriptions based on specimens reported to have come from parts of China never visited by Loureiro are peculiarly imperfect, indefinite, or even inaccurate, and were manifestly based on more or less broken and fragmentary material secured from herbalists.   The various Indo-China-China species which can definitely be placed from Loureiro's descriptions, and which do not grow as far south as Indo-China, were likewise based on specimens secured from herbalists,

[33] Journ. Arnold Arb. 13: 172. 1932.

material imported into Indo-China from China, just as the Chinese today import their own drug plants into Singapore, Manila, San Francisco, New York, or any other city where large colonies of Chinese exist.

It seems probable that in some, perhaps many cases, he erred in citing Chinese names and in recording the reputed uses of certain species described. It is suspected that many of the Chinese names, even those compiled at Canton, are really Mandarin; some, however, are definitely Cantonese.

Loureiro's own field work in Kwangtung was strictly limited, as he was not permitted to proceed beyond the suburbs of the city of Canton. He definitely states (Fl. Cochinch. Introd. XI) that much of his botanical material was brought to him by the Chinese. His statements regarding Chinese species as to the size of the plants, if other than small herbs, their habits and habitats, must be interpreted with this fact in view; that is, that Loureiro in many cases never saw the plants growing in nature but had merely branches or branchlets and had to depend on hearsay evidence for certain data.

He realized, and so stated, that the Chinese names cited by him in many cases were untrustworthy. Here, as in other countries, local names for those species of distinct economic importance are much more constant and more to be trusted than are the vernacular names of numerous small herbs, grasses, sedges, weeds, and other plants of little use in the economy of the natives.

In Loureiro's text Anamese names are indicated by α, Chinese by β, African by γ, and Indian by δ. New genera and species are indicated by the conventional sign †. It is unfortunate that Loureiro did not give his Chinese vernacular names in Chinese characters, as this course would have rendered them much more dependable, much more easily located, and infinitely better as clues to what he intended to describe than do his transliterations, giving Portuguese values to the letters used.

It is a well-known fact that the proper transliteration of Chinese sounds through European values of Arabic letters is distinctly difficult. For these sounds Loureiro used certain diacritical marks, and for certain sound values that he could not indicate by normal letters and diacritical marks, he used mutilated types of the letters b, d, o, and u, as explained on page xv (ed. Willdenow, page xx) of the introduction to the Flora Cochinchinensis. In practice it has been found difficult to pronounce many of the local Chinese names so that they become intelligible to Chinese residents of Canton. Thus Loureiro's Chinese names have been of comparatively little value as clues to the identity of the species to which he ascribed them. It is judged that some of them are "false names" or, what is more unlikely, names that have become obsolete. On the other hand, many of the Anamese names cited by him are still in current use for the same species to which he applied them, and not infrequently supply confirmatory evidence in connection with interpretations herein made.

Since Loureiro lived at Hue, the capital of Cochinchina, for nearly thirty-six years, it may be safely assumed that most of the Indo-China species described by him were based on material secured in the immediate vicinity of that city. In a few cases Hue is definitely cited as the locality and occasionally other places are indicated. As most of Blanco's Philippine species should be interpreted largely from material representing species now growing in those provinces contiguous to Manila so should most of Loureiro's Indo-China species of doubtful status be interpreted largely on the basis of those plants now growing at Hue and in its general neighborhood. Hue, then, is the *locus classicus* for many of Loureiro's species; yet the importance of this locality as a region needing intensive botanical exploration has been curiously overlooked, and many other parts of Indo-China have been much more extensively and intensively explored from a botanical standpoint than has Hue.

One of the reasons why my original manuscript of 1919 was withheld from publication for so long was my full realization that what was most needed in reference to the solution of many problems raised by Loureiro was intensive and extensive botanical collections from Hue and vicinity with notes as to habitats, relative abundance of the various species, economic uses, and local names. It was not until after I left the Orient at the end of 1923 that an opportunity presented itself when, on my recommendation, this region was selected by Chaplain and Mrs. J. C. Clemens, who made extensive collections there from May to July, 1927, and Mr. Roy Squires, who made similar but smaller collections near Hue, January to May, 1927. These two collections of Anamese material have been of the very greatest value to me in connection with my present attempt to determine the status of the numerous enigmatic genera and species of Loureiro. In many cases actual plants from the vicinity of Hue, studied in association with Loureiro's descriptions, led definitely to the interpretation of some hitherto doubtful species. I am confident that had the local names been consistently recorded, the collections already available would have solved the problems of other species that are still of doubtful status. The local name of a plant, used with discretion and understanding, very frequently supplies the clue from which the identity of a hitherto doubtful species can be determined, particularly in such cases as Loureiro provided when he placed his new species in totally wrong groups or when he misinterpreted various characters or published descriptions that were erroneous in essential details. I am under special obligations to Doctor Aug. Chevalier of Paris, who kindly loaned to me for study a special collection of Anamese plants prepared by de Pirey in 1919, with special reference to local names cited by Loureiro. These in many cases supplied confirmatory evidence which materially assisted in placing some of Loureiro's species.

Fortunately Loureiro was careful to record native names. Even if he did make errors in this field, it is distinctly to his credit that he consistently compiled these data, even as it is regrettable that some modern botanists fail to realize their utility; many ignore even such local names as are recorded on herbarium labels. Even in some standard floras where common names are consistently given, one is frequently impressed by the fact that many of the common names given are coined ones, while many such names in current use are not even recorded.

Modern systematists, in interpreting the obscure species of early authors, in general do not attach sufficient importance to the local names of plants. Unfortunately the rich collections in the great herbaria are particularly poor in reference to both the local names and the economic uses of the plants they contain as herbarium specimens, so that in many cases no blame can be attached to the systematist for the fact that the necessary data are not available to him. It is perhaps more the fault of our herbarium systems than of the individual worker. We pride ourselves on the assumption that the Latin binomial is theoretically fixed, but, as all systematists know, theory and fact are not in agreement. Due to one cause or another, changes in binomials have been exceedingly frequent in modern taxonomic work, particularly in the past forty years.

While many local names are used for a particular species over a very wide culture area, others are over small geographic areas.   This has been found to hold for the Amboina species described by Rumphius previous to 1700, published between 1747 and 1755; also for the Philippine species described by Blanco from about 1815 to 1845, and for many of Loureiro's Indo-China and Chinese species.   Some local names are used in a generic rather than in a specific sense; others recorded in literature—and perhaps many of these—are names made up to suit the occasion by the individual consulted; and some names that are recorded in literature are doubtless obsolete.

The significance of the name is sometimes important, also knowledge as to whether it is used for a definite species or is loosely applied to a number of perhaps unrelated ones having certain obvious characters in common.   Thus, the Philippine names *malakafé* (literally false coffee) and *malabayábas* (literally false guava) are coined names based on fancied resemblances to coffee and to the guava, hence of little value.   The name *sampinít* is applied to various unrelated spiny plants such as *Rubus*, *Caesalpinia* and *Pterolobium*, and even to *Hibiscus surattensis*, representing the Rosaceae, the Leguminosae and the Malvaceae.   The Chinese name *ye tau* (literally wild bean) is used in Kwangtung for various leguminous plants, including herbs, vines, shrubs, and trees.

To one unfamiliar with the Anamese language it is, of course, difficult to evaluate local plant names, for frequently these names have definite and often descriptive meanings. The Anamese names are difficult to use because of their construction and because in alphabetized lists one never knows whether or not a "descriptive" term has been used or not as the first part of a name.   Thus in Loureiro's work about 40 of his Anamese names begin with the word *co* (= herb); over 400 begin with *cây* (= tree); about 30 with *hoa* (= flower); over 20 with *nam* (= south); about 60 with *rau* (= vegetable), etc., most of the names consisting of from two to three or four separate words.

Local names should always be used with caution, and specimens bearing these must be critically compared with the original description of the species it is suspected the specimen may represent.   Unfortunately, some authors fail to realize that *the name must be used in association with the published description.*   Thus A. Chevalier, in proposing various reductions of Loureiro's species (Cat. Pl. Jard. Bot. Saigon 1919), placed too much trust in the local names, not realizing the necessity of taking into consideration the descriptive data given by Loureiro as a check on the correctness of a reduction based on the modern use of a local name.   In several cases he transferred Loureiro's specific names to genera totally different from the ones represented by the original descriptions, and in some cases even to families remote from the ones in which they properly belong.

Loureiro, like all others who have recorded local names of plants, sometimes recorded the wrong name for the plant described and in some cases he misapplied economic uses as well as local names.   It is always a safe assumption never to accept an identification based on a local name unless the plant bearing the local name conforms reasonably well with the characters indicated in the original description.

The botanist with little or no field experience in regions like Indo-China, Malaysia, and the Philippines is apt to underrate the value of native names as indicating definite species.   Many of these are invariably applied to individual species; they have been so used for centuries and will continue to be so used for centuries to come regardless of our nomenclatural vagaries in the use and application of scientific binomials.   In very numerous cases in dealing with the botanical work of such authors as Blanco and Loureiro, the native names recorded by them are of the very greatest significance, supplying most important clues to the identity of this or that species perhaps erroneously placed as to their genera or inadequately or even erroneously described. Very numerous cases could be cited in Loureiro's work where the local name alone has proved the open sesame to the identity of this or that species, sometimes when all other methods of botanical detective work have failed; and in numerous other cases have supplied dependable corroborative evidence to support specific interpretations. Thus may be cited such cases as the following where, but for the native name, it is highly probable that a considerable number of the species could not definitely be placed.

Crassula pinnata Lour. = *Eurycoma longifolia* Jack; *Cicca racemosa* Lour. = *Cicca acida* (Linn.) Merr.; *Vitis labrusca* Lour. = *Ampelocissus martini* Planch.; *Myrtus androsaemoides* Lour. = *Eugenia bullockii* Hance; *Dartus perlarius* Lour. = *Maesa perlarius* (Lour.) Merr. (*M. sinensis* A. DC.); *Genista scandens* Lour. = *Caesalpinia nuga* Ait.; *Mercurialis indica* Lour. = *Claoxylon hainanense* Pax & Hoffm.; *Pistacia oleosa* Lour. = *Schleichera oleosa* (Lour.) Merr. (*S. trijuga* Willd.); *Elaeocarpus integerrimus* Lour. = *Ochna integerrima* (Lour.) Merr. (*O. harmandii* Lecomte); *Ischaemum importunum* Lour. = *Panicum repens* Linn.; *Phleum cochinchinense* Lour. = *Ophiurus cochinchinensis* (Lour.) Merr. (*O. monostachyus* Presl); *Aira arundinacea* Lour. = *Phragmitis maxima* (Forsk.) Merr.; *Arundo multiplex* Lour. = *Bambusa multiplex* (Lour.) Raeusch. (*B. nana* Roxb.); *Santalum album* Lour. = *Dysoxylum loureiri* Pierre; *Salvadora capitulata* Lour. and *S. biflora* Lour. = *Streblus asper* Lour.; *Lagenula pedata* Lour. = *Columella pedata* Lour. = *Cayratia pedata* (Lour.) Juss.; *Ilex aquifolium* Lour. = *Taxotrophis macrophylla* (Blume) Boerl.; *Pimela nigra* Lour. = *Canarium pimela* Koenig and *P. alba* Lour. = *Canarium album* (Lour.) Raeusch., where the Cantonese names *O lam* and *Pa lam* more definitely identify the species than do any technical descriptions yet published; *Marcanthus cochinchinensis* Lour. = *Mucuna cochinchinensis* (Lour.) A. Chevalier; *Mimosa stellata* Lour. = *Cephalanthus angustifolius* Lour.; *Baccaurea cauliflora* Lour. = *B. ramiflora* Lour. (*B. sapida* Muell.-Arg.); and *B. sylvestris* Lour. (*B. annamensis* Gagnep.).

These are but a few of the numerous cases where the significance of the local name has been discussed in this work under individual species, as supplying the clue to the identity of this or that one, or as corroborative evidence supporting the accepted reductions. The local names used in Loureiro's time in Indo-China are in perhaps the majority of cases there still applied to the same species for which he recorded them. Their significance cannot safely be ignored. Field work prosecuted in Anam, particularly in the vicinity of Hue, with special reference to the local names, will unquestionably yield material and data that will in turn elucidate additional Loureiroan species which I have not been able definitely to place in this work on the basis of material and information available to me.

Concluding Remarks

In interpreting species based on pre-Linnaean descriptions and illustrations, such as those of Rheede and Rumphius, as well as those described by such post-Linnaean authors as Blanco and Loureiro, the modern systematist is not always consistent. Sometimes individual species have been interpreted from botanical material originating in regions remote from those that supplied the original specimens; not infrequently, amplified descriptions of later authors based on erroneously identified specimens represent species very different from the one intended by the original author.   In interpreting a Chinese, an Indo-Chinese, a Mozambique, or an East African species described by Loureiro, it becomes essential to take into consideration the present day vegetation of these regions and to attempt an interpretation of Loureiro's species on the basis of those now growing in or near the original localities.   It is distinctly not safe to identify a Loureiroian Chinese or Indo-Chinese species with material representing a species known only from Java or India or some other region remote from the type locality.

In this introduction I have attempted to outline some of the difficulties the modern taxonomist encounters when he attempts to interpret descriptions based on material no longer extant, and particularly when a certain percentage of the descriptions are manifestly not technically correct.   In the detailed discussions of individual species in the following presentation, I have given further data, where it has been considered necessary, to explain the acceptance of this or that name, and to justify my present interpretation of the individual species concerned.   It is not to be expected that my interpretations are always correct; it is hoped that in the majority of cases they are reasonably so.

No attempt has been made to give the full synonymy, with all the more important literature references where later authors have considered Loureiro's species.   In general such synonyms as are necessary to explain the accepted name are given, and a serious attempt has been made to detect and to record all synonyms actually based on Loureiro's original descriptions.   In spite of an extensive examination of the widely scattered literature bearing on this subject, it is not to be expected that all of these have been detected; those actually recorded exceed 750.

The taxonomist who critically compares the individual references with those in standard and other works will note a number of discrepancies between accepted authorities for certain binomials and their places of publication as between this work and standard literature.   Botanical literature is replete with erroneous and incomplete references, but in this work, wherever it has been possible to do so, the original sources have been consulted, and the references as given have been carefully checked as to authority, volume, page, and date of issue.   In somewhat over 6,700 references given in this work very few have not thus actually been verified, these representing the few cases where the original works have not been accessible to me.

The original manuscript on which this work was based was completed in Manila in 1919, as noted on page 4.   Beginning in August, 1931, the manuscript was entirely rewritten, greatly amplified, and each species was critically reconsidered.   The total number of species described by Loureiro has been reduced to 1157 by reduction to synonymy, where Loureiro manifestly described the same species a second, third, or even a fourth time, and under different names.   While four new generic names proposed by later authors but based on certain species described by Loureiro, and twenty-three of Loureiro's species remain unplaced as to their proper families, most of his other genera and species have been placed with reasonable accuracy and the synonymy adjusted in accordance with my present understanding of each case.   As a better understanding of the individual species considered is gained, some further changes in nomenclature are to be expected, and there naturally will be some changes made as the concept of generic limits changes.   Of the Loureiro species that have been placed as far as their proper families are concerned, about 140, including cellular cryptogams, I consider to be indeterminable as to the species; of these about 80 can be safely referred to the genera they actually represent. This is a distinctly better showing than I had hoped it possible to make when work on this project was initiated. Doubtless from time to time in the future some of these unplaced species can be located as additional data and material become available, but the above number represents the residue that I cannot dispose of on the basis of the unsatisfactory original descriptions alone and such other data as are now available to me.

For convenience Loureiro's original arrangement by the Linnaean classes and orders has been broken down, and the genera and species, as far as it has been possible to place them in accordance with modern ideas, have been arranged in approximately the sequence of families and genera in Engler and Prantl's "Die natürlichen Pflanzenfamilien."

I am under obligation to numerous individuals in China, Indo-China, the United States, England, Sweden, the Netherlands, France, Germany, Switzerland, Portugal, and Mozambique, who have assisted me in various ways by supplying data, material and information. I am under particular obligations to the botanists at the British Museum, Natural History, and those at the Muséum d'Histoire Naturelle, Paris, where the extant Loureiro types are preserved, for the numerous courtesies extended to me in comparing Loureiro's types with authentic material representing species described by other authors, and for critical notes on Loureiro's specimens.

NEW YORK, September 1, 1934.

THALLOPHYTA

ALGAE

Loureiro described 11 species of algae, all of which were placed under Linnaean binomials. It is doubtful if in any case he had material representing the Linnaean species listed. His species are unrecognizable from the short indefinite descriptions alone and are here enumerated alphabetically under the binomials used by Loureiro, with comments on a few of the species. Material collected with special reference to the local names cited by Loureiro might lead to accurate identification of some of the forms described.

Conferva corallina (non Linn.) Lour. Fl. Cochinch. 690. 1790, ed. Willd. 848. 1793, Anamese *thach hoa*, Chinese *xĕ hõa*.

" Habitat in Cochinchina, inhaerens scopulis, qui aqua marina lavantur." This is apparently one of the Rhodophyceae, certainly not a *Conferva*. It typifies *Ceramium loureirii* Ag. (Sp. Alg. 2: 155. 1828) regarding which de Toni (Syll. Alg. 4: 1459. 1903) under *C. japonicum* Okam. states: " Quid sit *Ceramium Loureirii* Ag. Sp. II, p. 155, Kuetz. Sp. p. 688 (non Lightfoot), plane ignoro."

Conferva dichotoma (non Linn.) Lour. Fl. Cochinch. 690. 1790, ed. Willd. 847. 1793.

" Habitat in aquis stagnantibus in Cochinchina." The very short description applies to some species of *Chara* or *Nitella*. This typifies *Cladophora ? dichotoma* de Toni (Syll. Alg. 1: 353. 1889), de Toni crediting this binomial to Agardh (Syst. Alg. 121. 1824); but Agardh used only Loureiro's original binomial *Conferva dichotoma*, with the comment: " Videtur novi generis."

Conferva litoralis (non Linn.) Lour. Fl. Cochinch. 689. 1790, ed. Willd. 847. 1793 (*littoralis*), Anamese *raong bai bien*.

" Habitat ad litora Cochinchinae." Apparently one of the filamentous Chlorophyceae is represented. This typifies *Conferva ? loureiri* Ag. (Syst. Alg. 115. 1824) and *Cladophora ? loureiri* de Toni (Syll. Alg. 1: 353. 1889); de Toni erroneously credits the latter binomial to Agardh.

Fucus aculeatus (non Linn.) Lour. Fl. Cochinch. 688. 1790, ed. Willd. 846. 1793.

" Habitat ad litora Cochinchinensia." The imperfect description apparently applies to some species of *Sargassum*.

Fucus granulatus (non Linn.) Lour. Fl. Cochinch. 688. 1790, ed. Willd. 846. 1793, Anamese *raong bien*.

" Habitat ad litora Cochinchinae." Apparently some species of *Sargassum* was intended.

Fucus inflatus (non Linn.) Lour. Fl. Cochinch. 688. 1790, ed. Willd. 845. 1793, Anamese *raong thiá thiá*.

" Habitat in paludibus Cochinchinae." The description applies to some species of *Sargassum*.

Fucus natans (non Linn.) Lour. Fl. Cochinch. 688. 1790, ed. Willd. 845. 1793, Anamese *raong noi*.

" Habitat in pelago Cochinchinensi." The description applies to some species of *Sargassum*. *S. litoreum* Rumph. (Herb. Amb. **6**: 188. *pl. 76. f. 2*), cited by Loureiro as representing the species, is *S. bacciferum* Agardh.

Fucus saccharinus (non Linn.) Lour. Fl. Cochinch. 689. 1790, ed. Willd. 847. 1793.

" Habitat ad litora maris Sinici." Indeterminable.

Fucus tendo (non Linn.) Lour. Fl. Cochinch. 689. 1790, ed. Willd. 846. 1793.

" Habitat in Oceano Sinensi." Indeterminable.

Fucus uvarius (non Linn.) Lour. Fl. Cochinch. 688. 1790, ed. Willd. 845. 1793, Anamese *cây raong*.

" Habitat in pelago Cochinchinensi." Indeterminable.

Ulva pisum (non Linn.) Lour. Fl. Cochinch. 687. 1790, ed. Willd. 844. 1793, Anamese *rêu bot*.

" Habitat in fossis, & paludibus Cochinchinae, in aquis fluctuans." No *Ulva* is represented, but one of the filamentous Chlorophyceae or Cyanophyceae; it is most inadequately described.

FUNGI

Loureiro described 27 species of fungi of which 8 were proposed as new, all others being referred to Linnaean binomials. Except in a few cases, the species are unrecognizable from the short and imperfect descriptions. Probably in no case did he correctly interpret the Linnaean species, and in most cases the fungi actually described by Loureiro and ascribed to Linnaean binomials do not even belong in the genera in which Loureiro placed them. The species are here arranged alphabetically under Loureiro's binomials, with comments in the few cases where the species can, with safety, be referred to their proper places in our system of classification. Probably material from which some of the species could be definitely placed could be secured by collections made with special reference to nativè names and indicated habitats.

Agaricus androsaceus (non Linn.) Lour. Fl. Cochinch. 691. 1790, ed. Willd. 850. 1793, Anamese *nam rom*.

" Habitat in paleis Orizae coacervatis, ac putrescentibus, in Cochinchina." Probably a representative of some other genus than *Agaricus*.

Agaricus arecarius Lour. Fl. Cochinch. 692. 1790, ed. Willd. 850. 1793, Anamese *nam cau*.

" Habitat in caudice Palmae Arecae in Cochinchina." Probably some other genus than *Agaricus* is represented.

Agaricus campanulatus (non Linn.) Lour. Fl. Cochinch. 691. 1790, ed. Willd. 849. 1793, Anamese *nam chuóng*.

" Habitat in agris Cochinchinae." Probably some other genus than *Agaricus* is represented.

Agaricus deliciosus (non Linn.) Lour. Fl. Cochinch. 691. 1790, ed. Willd. 849. 1793, Anamese *nam dee*, Chinese *hiàm xuĕn*.

" Habitat loca agrestia Cochinchinae, & Chinae." An edible form of *Agaricus* or of some allied genus is represented by the inadequate description.

Agaricus equestris (non Linn.) Lour. Fl. Cochinch. 691. 1790, ed. Willd. 849. 1793, Anamese *nam cút ngua.*

" Habitat in acervis stercoris equini in Cochinchina." Apparently no *Agaricus* is described but rather a representative of some allied genus.

Agaricus fimetarius (non Linn.) Lour. Fl. Cochinch. 691. 1790, ed. Willd. 849. 1793, Anamese *nam cút tlâu.*

" Habitat in sterquiliniis bubalorum in Cochinchina." Probably no *Agaricus* but rather a species of some allied genus is represented by the inadequate description.

Agaricus integer (non Linn.) Lour. Fl. Cochinch. 690. 1790, ed. Willd. 848. 1793, Anamese *nam môi*, Chinese *kiún.*

" Habitat loca agrestia Cochinchinae, & Chinae: saepe etiam in hortis." This is probably *Amanita manginiana* Har. & Pat. as interpreted by Demange (Bull. Écon. Indochine **22**: 599. 1919).

Agaricus ovatus Lour. Fl. Cochinch. 692. 1790, ed. Willd. 850. 1793, Anamese *nam tlúng.*

" Habitat in sylvis Cochinchinae." Apparently no *Agaricus* is represented by the short description.

Agaricus ramosus Lour. Fl. Cochinch. 692. 1790, ed. Willd. 850. 1793, Anamese *nam cây muc.*

" Habitat in Cochinchina, in truncis arborum vetustate corruptis." This is apparently no *Agaricus* but a representative of some allied genus. *Fungus igneus* Rumph. (Herb. Amb. **6**: 130. *pl. 56. f. 5*) discussed by Loureiro under *Agaricus ramosus*, is undeterminable.

Boletus canalium Lour. Fl. Cochinch. 693. 1790, ed. Willd. 852. 1793, Anamese *nam xoi.*

" Habitat in Cochinchina, in tubis ligneis, pluviam tectorum evacuantibus." Demange (Bull. Écon. Indochine **22**: 603. 1919) gives the Anamese name *nam voi* for *Boletus castaneus* Bull. However Loureiro's description does not apply to *Boletus* but apparently appertains to some species of *Polystictus*. It is the whole basis of *Polystictus canalium* Fries (Epicr. 437. 1836–38).

Boletus igniarius (non Linn.) Lour. Fl. Cochinch. 693. 1790, ed. Willd. 851. 1793, Anamese *nam júa.*

" Habitat in caudice fruticis Pandani, in Cochinchina." This is clearly the same as the very common *Polystictus sanguineus* (Linn.) Fries, conspicuous because of its red color, and a species that commonly occurs on dead *Pandanus* stems. *Polyporus pandani* Fries (Syst. Mycol. **1**: 377. 1821) was based on Loureiro's description and thus safely becomes a synonym of *Polystictus sanguineus* Fries.

Boletus suberosus (non Linn.) Lour. Fl. Cochinch. 692. 1790, ed. Willd. 851. 1793, Anamese *nam moúc.*

" Habitat in arboribus sylvestribus." From the description some species of *Hexagonia* may be represented.

Boletus versicolor (non Linn.) Lour. Fl. Cochinch. 693. 1790, ed. Willd. 852. 1793, Anamese *nam cui.*

" Habitat in lignis putrescentibus, in Cochinchina." The description applies to one of the smaller species of *Polyporus* or *Polystictus*.

Clathrus campana Lour. Fl. Cochinch. 694. 1790, ed. Willd. 853. 1793, Anamese *nam răn.*

" Habitat circa hortos Cochinchinae cito putrescens, & foetidus." I have no suggestions to offer as to the generic identity of this species. Fries (Syst. Mycol. **2**: 285. 1822) states: " Albissimus, novum genus, nisi *Phallus.*"

Clavaria muscoides (non Linn.) Lour. Fl. Cochinch. 696. 1790, ed. Willd. 856. 1793, Anamese *louc giác thê*, Chinese *lū kiŏ tsái.*

" Habitat in scopulis, & aggeribus prope mare in China, & Cochinchina." I can make no suggestion as to the generic disposition of this species; it is apparently no *Clavaria.*

Clavaria ophioglossoides (non Linn.) Lour. Fl. Cochinch. 696. 1790, ed. Willd. 855. 1793.

" Habitat in ruderibus, in Cochinchina." This is probably no *Clavaria*, but I can suggest no generic disposition for it.

Clavaria pistillaris (non Linn.) Lour. Fl. Cochinch. 696. 1790, ed. Willd. 855. 1793, Anamese *nam cút boi*, Chinese *mŏ cū tsái.*

" Habitat plerumque in acervis stercoris elephantini in Cochinchina, & China." This is probably not a *Clavaria*, but I cannot suggest a generic disposition for it. Fries places it as a doubtful synonym of *Cauloglossum elatum* Fries (Syst. Mycol. **3**: 61. 1829).

Helvella amara Lour. Fl. Cochinch. 695. 1790, ed. Willd. 854. 1793, Anamese *nam tràm.*

" Habitat plerumque in arbore Melaleuca, in Cochinchina." *Leotia amara* Fries (Syst. Mycol. **2**: 27. 1822) was based entirely on Loureiro's description; this may or may not be the correct generic disposition of Loureiro's species.

Helvella mitra (non Linn.) Lour. Fl. Cochinch. 694. 1790, ed. Willd. 853. 1793, Anamese *nam rách.*

" Habitat in truncis arborum putrescentibus, in Cochinchina." This is clearly not the Linnaean species as noted by Willdenow: " Fungus ab Helvella mitra diversissimus." *Fungus elatus digitatus* Rumph. (Herb. Amb. **6**: 129. *pl. 57. f. E*), cited by Loureiro as representing his species, is *Ganoderma cochlear* (Nees) Merr. Fries (Syst. Mycol. **2**: 22. 1822) states: " Ne genere convenit."

Helvella pineti (non Linn.) Lour. Fl. Cochinch. 695. 1790, ed. Willd. 854. 1793, Anamese *nam goúc.*

" Habitat in Cochinchina, lateraliter adhaerens arborum truncis." The very short description is suggestive of *Polyporus.*

Hydnum auriscalpium (non Linn.) Lour. Fl. Cochinch. 693. 1790, ed. Willd. 852. 1793, Anamese *nam tlăng nhám.*

" Habitat in Cochinchina, ad arborum radices supra terram exertus." This may or may not be a *Hydnum*. Loureiro's description typifies *Hydnum orientale* Fries (Syst. Mycol. **1**: 407. 1821, Epicr. 510. 1836–38) this binomial having been based wholly on Loureiro's data.

Lycoperdon glomeratum Lour. Fl. Cochinch. 697. 1790, ed. Willd. 856. 1793.

" Habitat in sylvis Cochinchinae: non edulis." This is apparently not a *Lycoperdon.* *Tuber sampadarium* Rumph. (Herb. Amb. **6**: 123), cited by Loureiro as representing his species, typifies *Polygaster sampadarius* Fries, a genus and species of doubtful status known only from Rumphius' description.

Lycoperdon lamellatum Lour. Fl. Cochinch. 696. 1790, ed. Willd. 856. 1793, Anamese *nam cuc.*

" Habitat in sylvis, & sepibus Cochinchinae: non edulis." The description applies to *Daldinia*, and apparently to *Daldinia concentrica* (Bolton) Ces. & de Not.; certainly no *Lycoperdon* is represented.

Mucor glaucus (non Linn.) Lour. Fl. Cochinch. 697. 1790, ed. Willd. 857. 1793, Anamese *moúc xanh tlái hu.*

" Habitat in diversis pomis, quando putredine corrumpuntur." The very short description apparently applies to one of the blue moulds of the genus *Aspergillus.*

Mucor mucedo (non Linn.) Lour. Fl. Cochinch. 697. 1790, ed. Willd. 857. 1793, Anamese *moúc bánh hu*, Chinese *múi.*

" Habitat in placentis & oryza cocta diu relictis, in Cochinchina, & China." The very short description applies to one of the common moulds apparently of the genus *Mucor.*

Mucor sphaerocephalus (non Linn.) Lour. Fl. Cochinch. 697. 1790, ed. Willd. 857. 1793, Anamese *moúc dá.*

" Habitat in Cochinchina, insidens lapidibus, & lignis." The very short description apparently applies to one of the moulds of the genus *Mucor.*

Peziza auricula (non Linn.) Lour. Fl. Cochinch. 695. 1790, ed. Willd. 855. 1793, Anamese *nam meò*, Chinese *mŏ lh.*

" Habitat in Cochinchina, & China, adhaerens arborum truncis vetustis." The description applies to *Auricularia* and probably *A. porphyrea* Lév. or *A. polytricha* Lév., as interpreted by Demange (Bull. Écon. Indochine **22**: 594. 1919), is represented.

Phallus impudicus (non Linn.) Lour. Fl. Cochinch. 694. 1790, ed. Willd. 853. 1793, Anamese *nam chó.*

" Habitat ad sepes in Cochinchina." Loureiro placed *Phallus daemonum* Rumph. (Herb. Amb. **6**: 131. *pl. 56. f. 7*) as a synonym of his species; it is *Dictyophora phalloidea* Desv., and Loureiro's description does not apply to it. It may be that Loureiro's species is the same as *Ithyphallus aurantiacus* (Mont.) Fisch., as interpreted by Demange (Bull. Écon. Indochine **22**: 604. *f. 15.* 1919), for which the Anamese name *nam lo cho* is given.

LICHENES

Loureiro described 10 lichens, all but one of which he referred to Linnaean binomials; but in all probability he had no representatives of any of the Linnaean species listed. The species are unrecognizable from the short and imperfect descriptions.

Byssus candellaris (non Linn.) Lour. Fl. Cochinch. 687. 1790, ed. Willd. 844. 1793, Anamese *bôt vàng dính cây.*

There is no description other than the original Linnaean descriptive sentence with the observation that there were other species of *Byssus* in Indo-China.

Lichen arecarius Lour. Fl. Cochinch. 685. 1790, ed. Willd. 842. 1793, Anamese *rêu cây cau.*

" Habitat in hortis Cochinchinae," with a further note that it grows on the trunks of the betel-nut palm, *Areca catechu* Linn. It is apparently a species of *Pyxine*.

Lichen ericetorum (non Linn.) Lour. Fl. Cochinch. 685. 1790, ed. Willd. 841. 1793, Anamese *rêu rùng*.

" Habitat in sylvis Cochinchinae." I can make no suggestion as to the genus represented by the very short and indefinite description.

Lichen fagineus (non Linn.) Lour. Fl. Cochinch. 685. 1790, ed. Willd. 841. 1793, Anamese *rêu bót tláng*.

" Habitat in sylvis Cochinchinae." I can make no suggestion as to the genus represented by the very short and indefinite description.

Lichen imbricatus Lour. Fl. Cochinch. 686. 1790, ed. Willd. 842. 1793, Anamese *rêu báy cá*.

" Habitat frequens in dumetis Cochinchinae, fruticibus adrepens." From the description this is certainly not a lichen; apparently some jungermanniaceous plant is represented.

Lichen pulmonarius (non Linn.) Lour. Fl. Cochinch. 686. 1790, ed. Willd. 842. 1793, Anamese *thien hoa*, Chinese *tien hõa*.

" Habitat in arboribus sylvestribus, senescentibus Cochinchinae, & Chinae." There is no reason for considering that Loureiro had material representing the Linnaean species, but he apparently did have some species of *Sticta* to which genus *Lichen pulmonarius* Linn. belongs.

Lichen rangiferinus (non Linn.) Lour. Fl. Cochinch. 686. 1790, ed. Willd. 843. 1793, Anamese *rêu gác nai*.

" Habitat loca sterilia Cochinchinae." From the description I would suspect *Stereocaulon* to be represented rather than *Cladonia*, the latter genus being the one to which the Linnaean species belongs.

Lichen roccella (non Linn.) Lour. Fl. Cochinch. 686. 1790, ed. Willd. 843. 1793, Anamese *rêu tle muc*.

" Habitat in scopulis, & arundinibus Bambu putrescentibus in Cochinchina." It is suspected that some fungus, rather than a lichen, is represented by Loureiro's very imperfect description. *Alga coralloides* Rumph. (Herb. Amb. **6**: 181. *pl. 76. f. A–C.*) cited by Loureiro as representing the species, must be excluded as it represents a true alga, *Gracillaria lichenoides* Harv.

Lichen tartareus (non Linn.) Lour. Fl. Cochinch. 685. 1790, ed. Willd. 842. 1793, Anamese *rêu dá*, Chinese *tan*.

" Habitat loca saxosa in Cochinchina, & China." From the very short description I can suggest no reduction of this.

Lichen usnea (non Linn.) Lour. Fl. Cochinch. 687. 1790, ed. Willd. 843. 1793, Anamese *rêu chi rôi*.

" Habitat in arboribus sylvestribus Cochinchinae." The description applies to some species of *Usnea*. *Muscus capillaris* Rumph. (Herb. Amb. **6**: 89. *pl. 40. f. 2*) cited by Loureiro as representing his species, is also an *Usnea* but may or may not represent the form Loureiro had.

BRYOPHYTA

MUSCI

The five apparently true mosses described by Loureiro, two described as new, three referred to Linnaean binomials, are here enumerated in alphabetical sequence. I fail to recognize any of the species because of the short, inadequate, and indefinite descriptions.

Bryum truncatulum (non Linn.) Lour. Fl. Cochinch. 684. 1790, ed. Willd. 840. 1793, Anamese *rêu doùng.*
  " Habitat in agris, & fossis Cochinchinae."

Bryum undulatum (non Linn.) Lour. Fl. Cochinch. 683. 1790, ed. Willd. 840. 1793, Anamese *rêu doung*, Chinese *sièn.*
  " Habitat in agris Cochinchinae, & Chinae."

Bryum viridulum (non Linn.) Lour. Fl. Cochinch. 684. 1790, ed. Willd. 840. 1793, Anamese *rêu xanh.*
  " Habitat in agris Cochinchinae."

Fontinalis heterophylla Lour. Fl. Cochinch. 684. 1790, ed. Willd. 841. 1793.
  " Habitat in arborum truncis in Cochinchina." No *Fontinalis* is represented. The description is suggestive of *Selaginella* in reference to the " ramulis supremis quadragoins, fructificantibus," but the last part of the description may have been based on a mixture of material.

Porella imbricata Lour. Fl. Cochinch. 683. 1790, ed. Willd. 839. 1793, Anamese *râu bac.*
  " Habitat loca humida in Cochinchina." No *Porella* is represented by the imperfect description.

Sphagnum simplicissimum Lour. Fl. Cochinch. 683. 1790, ed. Willd. 839. 1793, Anamese *rêu thành gach.*
  " Habitat in Cochinchina, innascens muris, & truncis arborum tempore pluvio." No *Sphagnum* is represented by the imperfect description. It may be that Loureiro had specimens of the very common *Barbula orientalis* Broth., which, however, is scarcely two inches long.

HEPATICAE

Lycopodium complanatum (non Linn.) Lour. Fl. Cochinch. 682. 1790, ed. Willd. 838. 1793, Anamese *rêu them nhà.*
  " Habitat hyberno tempore lateritiis inhaerens in Cochinchina." The description, at least as to the habit and vegetative characters, applies to some one of the scale mosses, Jungermanniaceae.

Lichen imbricatus Lour. (see p. 55) is apparently some jungermanniaceous plant.

PTERIDOPHYTA

HYMENOPHYLLACEAE

Trichomanes Linnaeus

Trichomanes hirsutum (non Linn.) Lour. Fl. Cochinch. 681. 1790, ed. Willd. 837. 1793.

" Habitat loca agrestia Cochinchinae." It is possible that Loureiro had some hymenophyllaceous plant, although the habitat and the description, in part, as to the entire plant being hirsute, with pinnate fronds and oblong, obtuse alternate, pilose pinnae, scarcely indicates *Trichomanes* or *Hymenophyllum*. It is suspected that Loureiro had sterile material of small *Dryopteris*, adding the fruit characters to make his description conform to the characters of the genus in which he erroneously placed it.

CYATHEACEAE

Cibotium Kaulfuss

Cibotium barometz (Linn.) J. Sm. in Hook. Lond. Journ. Bot. 1: 437. 1842.

Polypodium barometz Linn. Sp. Pl. 1092. 1753; Lour. Fl. Cochinch. 675. 1790 (*baromez*), ed. Willd. 829. 1793, Anamese *câu tích*, Chinese *kèu tsiĕ*.

" Habitat in sylvis montanis Cochinchinae, & Chinae." The well-known and widely distributed Linnaean species was correctly interpreted by Loureiro, as indicated by his very clear description of the characteristic indumentum of the basal parts " pilis densissimis, tenuibus, rufis tota vestita."

Cyathea Smith

Cyathea sp.

Polypodium arboreum (non Linn.) Lour. Fl. Cochinch. 676. 1790, ed. Willd. 831. 1793, Anamese *cây nhum*.

" Habitat in sylvis Cochinchinae, in ultima provincia ad Austrum sita prope Cambodiam." The description is that of a *Cyathea*, or perhaps of *Alsophila* for those who retain the latter as a distinct genus. It is, however, manifestly not the same as *Palmifilix postium* Rumph. (Herb. Amb. 6: 63) to which Loureiro ascribed *plate 27* of Rumphius' work. The Rumphian illustration represents *Cyathea rumphii* Desv. (Prodr. 323. 1827) which name should probably replace *Cyathea amboinensis* (v.A.v.R.) Merr. (*Alsophila amboinensis* v.A.v.R.) for the Moluccan species.

POLYPODIACEAE

Dryopteris Adanson

Dryopteris sp.

Polypodium fragrans (non Linn.) Lour. Fl. Cochinch. 675. 1790, ed. Willd. 829. 1793, Anamese *rau deón loung*.

" Habitat loca inculta Cochinchinae." The short and imperfect description apparently applies to some species of *Dryopteris* but no further identification seems possible.

Dryopteris sp.

? Polypodium scolopendrioides (non Linn.) Lour. Fl. Cochinch. 674. 1790, ed. Willd. 827. 1793.

" Habitat loca inculta Cochinchinae." The description is very short and incomplete. From the first illustration cited, Plukenet (Almag. 152. *pl. 290. f. 1*), after Linnaeus, it seems probable that Loureiro may have had specimens of some species of *Dryopteris*. The reference to Plumier (Fil. *pl. 91*) was also copied from Linnaeus and does not represent the species Loureiro described. Linnaeus (Sp. Pl. ed. 2, 1544. 1763) cites *Filix jamaicensis simpliciter pinnatis* Plukenet (Almag. 152. *p. 290. f. 1*) as representing *Polypodium scolopendrioides* Linn.: the Linnaean species is the West Indian *Dryopteris scolopendrioides* (Linn.) O. Ktz.

Dryopteris sp.

? Polypodium venosum Lour. Fl. Cochinch. 674. 1790, ed. Willd. 828. 1793.

No habitat or locality is indicated, but from Loureiro's indication under α (native name not given) it seems probable that he had Indo-China material as the Greek letter alpha is used by him to indicate Anamese names. *Dryopteris* is suggested as the genus represented.

Polystichum Roth

Polystichum varium (Linn.) Presl Epim. 57. 1851.

Polypodium varium Linn. Sp. Pl. 1090. 1753; Lour. Fl. Cochinch. 675. 1790, ed. Willd. 829. 1793, Anamese *rau deón tlon.*

" Habitat loca saxosa in Cochinchina, & China." The description is very short but it seems probable that Loureiro may have had material representing the Linnaean species, the type of which was a specimen collected by Osbeck, near Canton, China.

Egenolfia Schott

Egenolfia appendiculata (Willd.) J. Sm. Ferns Brit. For. 111. 1866.

Acrostichum appendiculatum Willd. Sp. Pl. **5**: 114. 1810.

Polybotrya appendiculata J. Sm. in Hook. Journ. Bot. **4**: 150. 1842.

Asplenium trichomanoides (non Linn.) Lour. Fl. Cochinch. 678. 1790, ed. Willd. 833. 1793.

" Habitat loca saxosa in Cochinchina." If *Asplenium trichomanes* Linn. were a low altitude fern in Indo-China, I should be willing to accept Loureiro's interpretation of the Linnaean species as correct, as the species is of very wide geographic distribution. I believe, however, that he had a small sterile form of *Egenolfia appendiculata* J. Sm., similar to *Clemens 4364* from Tourane, and that he added the expression " fructificationes squamosae, lineares in disco frondis sparsae " to make his description conform to the generic characters of *Asplenium;* or that he took these data from Tournefort (Inst. 539. *pl. 315*), this illustration representing a species of *Polypodium*, which Loureiro erroneously cites as illustrating his species. It may be noted that *Asplenium trichomanoides* Linn. is apparently an error for *A. trichomanes* Linn. The former binomial does not appear in Christensen's Index Filicum and the only places I have found it are in Linnaeus (Syst. Nat. ed. 12, **2**: 690. 1767, ed. 13 (Gmelin) **2**: 1303. 1796) where it is a manifest error for *A. trichomanes* Linn.

Nephrolepis Schott

Nephrolepis cordifolia (Linn.) Presl Tent. Pterid. 79. 1836.

Polypodium cordifolium Linn. Sp. Pl. 1089. 1753.

Asplenium bulbosum Lour. Fl. Cochinch. 678. 1790, ed. Willd. 833. 1793, Anamese *cây maóng tlâu.*

" Habitat in montibus Cochinchinae." This doubtful reduction follows the suggestion of Christensen (Ind. Fil. 104. 1906). I know of no other Asiatic fern that remotely agrees with Loureiro's description therefore this suggested reduction is accepted as possibly correct; this in spite of the statement " fructificationes in disco lineolis obliquis, parallelis " and other discrepancies in the short description.

Microlepia Presl

Microlepia speluncae (Linn.) Moore. Ind. Fil. XCIII. 1857.

Polypodium speluncae Linn. Sp. Pl. 1093. 1753; Lour. Fl. Cochinch. 677. 1790, ed. Willd. 831. 1793.

" Habitat loca umbrosa in Cochinchina." There is little in Loureiro's short description on which to base a judgment. As far as it goes the description conforms to the characters of the common and widely distributed Linnaean species

Asplenium Linnaeus

Asplenium sp. ?

Asplenium nodosum (non Linn.) Lour. Fl. Cochinch. 678. 1790, ed. Willd. 832. 1793.

" Habitat agros incultos Cochinchinae." As far as the description goes, an *Asplenium* seems to be represented by it. In my original manuscript of 1919 *A. macrophyllum* Sw. was suggested as a possibility, but Doctor E. B. Copeland thinks this not possible. *Lingua cervina nodosa* Plum. (Fil. Amer. *pl. 108*) cited by Loureiro as representing his species, is a sterile fern remote from the one Loureiro so inadequately described.

Adiantum Linnaeus

Adiantum caudatum Linn. Mant. 2: 308. 1771; Lour. Fl. Cochinch. 680. 1790, ed. Willd. 835. 1793.

" Habitat loca humida Cochinchinae." Loureiro's description applies unmistakably to the very common and widely distributed Linnaean species.

Adiantum capillus veneris Linn. Sp. Pl. 1096. 1753; Lour. Fl. Cochinch. 681. 1790, ed. Willd. 836. 1793, Anamese *cây duôi chôn.*

" Habitat in puteis, fontibus, & aliis locis humidis Cochinchinae." Loureiro's short description apparently applies to the Linnaean species.

Adiantum flabellulatum Linn. Sp. Pl. 1095. 1753; Lour. Fl. Cochinch. 680. 1790, ed. Willd. 836. 1793, Chinese *tiet qūat tsào.*

" Habitat suburbia Cantoniensia apud Sinas." The description applies unmistakably to the Linnaean species which is common in the vicinity of Canton. The Linnaean type is a specimen collected by Osbeck near Canton.

Pteris Linnaeus

Pteris ensiformis Burm. f. Fl. Ind. 230. 1768.

Pteris caudata (non Linn.) Lour. Fl. Cochinch. 680. 1790, ed. Willd. 835. 1793.

" Habitat ad rudera, & muros veteres in Cochinchina." The description and the indicated habitat clearly indicate that the form described is the common and widely distributed *Pteris ensiformis* Burm. f.

Pteris quadriaurita Retz. Obs. **6**: 38. 1791.

Pteris biaurita (non Linn.) Lour. Fl. Cochinch. 679. 1790, ed. Willd. 835. 1793.

" Habitat loca inculta Cochinchinae." The description apparently applies to the very common and widely distributed *Pteris quadriaurita* Retz.

Pteris vittata Linn. Sp. Pl. 1074. 1753; Lour. Fl. Cochinch. 679. 1790, ed. Willd. 834. 1793.

" Habitat fossas, & loca humida in Cochinchina, & China." The description does not apply in all respects to *Pteris vittata*, notably in the size of the fronds, 5 feet long; yet clearly a *Pteris* is described. *Lonchitis amboinica recta major* Rumph. (Herb. Amb. **6**: 70. *pl. 30. f. 1*), cited by Loureiro as a synonym, is *Blechnum orientale* Linn., and Loureiro's statement as to the size may have been derived from Rumphius; or he may have had sterile specimens of *Blechnum orientale*, adding the fructification characters to make his description conform to *Pteris*. The type of *Pteris vittata* Linn. was from the vicinity of Canton, China. This oriental form is usually known as *Pteris longifolia* Linn. but the latter name appertains to an American plant; see Hieronymus (Hedwigia **54**: 284. 290. 1914).

Pteris sp.

? Asplenium hemionitis (non Linn.) Lour. Fl. Cochinch. 677. 1790, ed. Willd. 832. 1793.

" Habitat circa muros in hortis Cochinchinae." C. Christensen, Index Filicum, suggests *Polypodium* of the section *Selliguea* and in a list supplied to me has indicated *P. ampelideum* Christ. as a possibility; but the indicated habitat eliminates this. The frond form seems to eliminate *Asplenium* and *Athyrium* and the habitat also is not that of representatives of these genera, at least of those species that occur in the vicinity of Hue. It is suspected that Loureiro had sterile juvenile material of *Pteris ensiformis* Burm. f., adding the infructescence characters to make the description conform to the characters of *Asplenium;* or the description may in part have been based on previously published ones of *Asplenium hemionitis* Linn.

Drymoglossum Presl

Drymoglossum piloselloides (Linn.) Presl Tent. Pterid. 227. *pl. 10. f. 5, 6.* 1836; C. Chr. in Dansk Bot. Arkiv **6**: 86. *pl. 12. f. 4, 5. pl. 13. f. 4.* 1929.

Pteris piloselloides Linn. Sp. Pl. ed. 2, 1530. 1763; Lour. Fl. Cochinch. 678. 1790, ed. Willd. 833. 1793.

Acrostichum heterophyllum (non Linn.) Lour. Fl. Cochinch. 673. 1790, ed. Willd. 826. 1793.

Loureiro's specimens of both were from Cochinchina, the first: " Habitat loca saxosa Cochinchinae," the second: " Habitat in sylvis Cochinchinae." Both descriptions, short as they are, apply closely to *Drymoglossum* and certainly represent some species of this small genus, and, probably *D. piloselloides* (Linn.) Presl, which is common and widely distributed in the Old World tropics and which occurs near Hue where Loureiro lived (*Clemens 3333*, det. C. Christensen).   Both might with equal propriety, as far as the descriptions go, be referred to *D. heterophyllum* (Linn.) C. Chr., which, however, as pointed out by C. Christensen, is confined to India and Ceylon.

Polypodium Linnaeus

Polypodium longissimum Blume Enum. Pl. Jav. **2**: 127. 1828.

> *Polypodium simile* (non Linn.) Lour. Fl. Cochinch. 674. 1790, ed. Willd. 828. 1793, Chinese *kŭ tsúi pù.*

"Habitat apud Sinas, agreste."   The very short and imperfect description applies to *Polypodium longissimum* Blume, as far as it goes, a species that occurs in southern China, but it applies equally well to allied species such as *P. schneideri* Christ.   This reduction to *P. longissimum* Blume is based more on the similarity of *Lonchitis altissima* Sloane (Hist. Jam. **1**: 77. *pl. 32*) to Blume's species, than on Loureiro's actual description, Sloane's figure being cited by Loureiro as representing the species.

Polypodium loriceum (non Linn.) Lour. Fl. Cochinch. 674. 1790, ed. Willd. 828. 1793.

"Habitat agreste in Cochinchina."   Probably some species of *Polypodium* is represented but the description is too short and indefinite to determine which; *P. lehmannii* Mett. is suggested.

Polypodium scolopendria Burm. f. Fl. Ind. 232. 1768.

> *Polypodium phymatodes* Linn. Mant. **2**: 306. 1771; Lour. Fl. Cochinch. 673. 1790, ed. Willd. 827. 1793.
>
> *Polypodium biforme* Lour. Fl. Cochinch. 673. 1890, ed. Willd. 827. 1793.

For *Polypodium phymatodes* Loureiro states: "Habitat in Cochinchina."   His interpretation was correct.   *Polypodium biforme* described as new, "Habitat in sylvis Cochinchinae," is manifestly a form of the same species.   *Polypodium scolopendria* Burm. f. is the oldest binomial for this common, widely distributed, protean species.

Cyclophorus Desvaux

Cyclophorus lanceolatus (Linn.) Alston in Journ. Bot. **69**: 102. 1931.

> *Acrostichum lanceolatum* Linn. Sp. Pl. 1067. 1753; Lour. Fl. Cochinch. 672. 1790, ed. Willd. 826. 1793, Anamese *cây kim luon.*
>
> *Polypodium adnascens* Sw. Syn. 25, 222. *pl. 2, f. 2.* 1806.
>
> *Cyclophorus adnascens* Desv. in Berl. Mag. **5**: 300. 1811.

"Habitat in hortis, & sylvis Cochinchinae, arboribus inhaerens."   The description applies unmistakably to the very common and widely distributed species commonly known as *Cyclophorus adnascens* Desv. but which Alston has recently shown should be known as *Cyclophorus lanceolatus* (Linn.) Alston.   Bonaparte (Notes Pterid. **7**: 136. 1918) thought that Loureiro's description might refer to *Cyclophorus acrostichoides* (Forst.) Presl or *Leptochilus linnaeanus* Fée; the cited habitat "arboribus inhaerens" eliminates the latter as a possibility and the description of the fronds as 8 inches long would seem to eliminate the former.   *Cyclophorus lanceolatus* (Linn.) Alston is common in the vicinity of Hue, *Clemens 3177*, and *C. acrostichoides* (Forst.) Presl also occurs there.

Acrostichum Linnaeus

Acrostichum aureum Linn. Sp. Pl. 1069. 1753.

> *Asplenium scolopendrium* (non Linn.) Lour. Fl. Cochinch. 677. 1790, ed. Willd. 832. 1793, Anamese *cây ráng lá.*

"Habitat loca humida in Cochinchina." Christensen (Ind. Fil. 131. 1906) reduced this to *Asplenium nidus* Linn. with the characters of which Loureiro's description does not at all agree. I judge from Loureiro's statement that the leaves are relatively non-inflammable and durable when used as thatch, which, taken with Bonaparte's statement (Notes Pterid. 7: 22. 1918) that the fronds of *Acrostichum aureum* Linn. are so used, that this Linnaean species was the one Loureiro had in mind; and that he saw only sterile specimens and added the statement "fructificationes squamosae, in lineas obliquas dispositae" to make his description conform to the characters of the genus in which he erroneously placed it. The statement regarding the fronds being "crispis, inaequalibus, apice multifidis" was manifestly taken from the illustrations of Plukenet (Phyt. *pl. 248, f. 2*) and Tournefort (Inst. *pl. 320*) cited by Loureiro; these represent abnormal forms of the European *Phyllitis scolopendrium* (Linn.) Newm.

Polypodiaceae of Uncertain Generic Status

Pteris lanceolata (non Linn.) Lour. Fl. Cochinch. 679. 1790, ed. Willd. 834. 1793.

"Habitat loca inculta Cochinchinae." The description is short and imperfect: fronds simple, lanceolate to sub-triangular, cordate, subentire, glabrous, the fructifications apical, marginal. I know no fern having this particular combination of characters. This description was manifestly taken largely from *Lingua cervina foliis acutis* Plumier (Descr. Pl. Amer. 28. *pl. 40.* 1693) erroneously cited by Loureiro as representing his species. It is suspected that Loureiro had sterile specimens of some entirely different fern and added the fruit characters from Plumier's figure. *Pteris lanceolata* Linn. is the tropical American *Paltonium lanceolatum* Presl.

Polypodium repandum Lour. Fl. Cochinch. 673. 1790, ed. Willd. 826. 1793, Anamese *côt toái bô.* Chinese *kū tsúi pù.*

> *Polypodium loureiri* Kostel. Allgem. Med.-Pharm. Fl. 1: 57. 1831 (based on *Polypodium repandum* Lour.).

"Habitat agreste apud Sinas." I do not recognize this from Loureiro's description, which I judge may have been based on some herbaceous flowering plant with certain fern characters added; the "radix ovata" and the distinction between radical and cauline leaves do not appear to be fern characters. It may be that Loureiro's description was taken from the same Chinese book from which he took the supposed medicinal qualities of the plant.

SCHIZAEACEAE

Lygodium Swartz

Lygodium polystachyum Wall. List. no. 177. 1828, *nomen nudum;* Moore in Gard. Chron. 671. 1859.

> *Adiantum scandens* Lour. Fl. Cochinch. 681. 1790, ed. Willd. 837. 1793, Anamese *cây baong baong,* non *Lygodium scandens* Sw.

" Habitat loca plana & inculta Cochinchinae." In my original manuscript of 1919, Loureiro's species was placed as a synonym of *Lygodium japonicum* (Thunb.) Sw., which was Bonaparte's disposition of it (Notes Pterid. 7: 138. 1918) but there are too many discrepancies between the description of *Adiantum scandens* Lour. and the characters of Thunberg's species to warrant this reduction. The description is short and somewhat indefinite. The pinnules are described as cuneiform, which does not apply to Wallich's species, or for that matter to any species of *Lygodium*. *Clemens 3723* from the general vicinity of Hue represents *Lygodium polystachyum* Wall. Some points in the description suggest *Lindsaya* but the expression " denticulis marginalibus fructificationes amplectentibus " does not apply to *Lindsaya*.

Lygodium flexuosum (Linn.) Sw. in Schrad. Journ. 1800(2): 106. 1801.

Ophioglossum flexuosum Linn. Sp. Pl. 1063. 1753.

Ophioglossum scandens (non Linn.) Lour. Fl. Cochinch. 672. 1790, ed. Willd. 825. 1793, Anamese *thach vi deei*, Chinese *xĭ úi tân*.

" Habitat loca plana & fluminum ripas in China, & Cochinchina." The description agrees closely with the characters of the Linnaean species, although it is possible that it includes, in part, *Lygodium japonicum* (Thunb.) Sw. *Adianthum volubile minus* Rumph. (Herb. Amb. **6**: 76. *pl. 32. f. 2, 3*) cited as a synonym, is correctly placed.

OPHIOGLOSSACEAE

Ophioglossum Linnaeus

Ophioglossum nudicaule Linn. f. Suppl. 443. 1781.

Ophioglossum lusitanicum (non Linn.) Lour. Fl. Cochinch. 672. 1790, ed. Willd. 825. 1793, Anamese *thach vi*, Chinese *xĕ úi*.

Ophioglossum loureirianum Presl Suppl. 55. 1845 (based on *O. lusitanicum* Lour.).

" Habitat loca plana, & humida Cochinchinae, & Chinae tempore autumnali, ac cito periens." I believe this to be the small Indo-Malaysian form currently referred to *Ophioglossum nudicaule* Linn. f. Bonaparte (Notes Pterid. 7: 136. 1918) considered that Loureiro correctly interpreted *Ophioglossum lusitanicum* Linn. which seems to me to be improbable because this species does not occur in Asia, being essentially a Mediterranean type.

EQUISETACEAE

Equisetum Linnaeus

Equisetum arvense Linn. Sp. Pl. 1061. 1753; Lour. Fl. Cochinch. 671. 1790, ed. Willd. 823. 1793, pro parte.

" Habitat incultum in agris Sinensibus." The description, in part, the local names *ma hoàng* and *mâ hoâm*, and the indicated medicinal uses appertain to *Ephedra*, and probably to *E. sinica* Stapf (see p. 67). *Equisetum arvense* Linn. is common in northern China but it is probable that that part of Loureiro's description which applies to *Equisetum* was taken from the Linnaean description.

Equisetum hyemale Linn. Sp. Pl. 1062. 1753; Lour. Fl. Cochinch. 671. 1790, ed. Willd. 824. 1793, Anamese *mouc tăc*, Chinese *mŏ cĕ*.

" Habitat loca paludosa Sinarum." It is probable that Loureiro's material was se-cured from herbalists. The Linnaean species occurs in northern China and Loureiro's data conform to its characters as far as the description goes. The only species known from southern China is *Equisetum debile* Roxb. which is frequently confused with *E. ramosissi-mum* Desf., but Loureiro's description does not apply at all to either of these.

LYCOPODIACEAE

Lycopodium Linnaeus

Lycopodium cernuum Linn. Sp. Pl. 1103. 1753; Lour. Fl. Cochinch. 682. 1790, ed. Willd. 838. 1793, Anamese *thoung dat.*

" Habitat in collibus sylvestribus Cochinchinae." Loureiro apparently had speci-mens of the Linnaean species, which is very common and widely distributed in the Indo-Malaysian region.

Lycopodium phlegmaria Linn. Sp. Pl. 1101. 1753; Lour. Fl. Cochinch. 682. 1790, ed. Willd. 837. 1793, Anamese *rêu cây.*

" Habitat in truncis arborum Cochinchinensium." The description apparently applies to the Linnaean species.

GYMNOSPERMAE

CYCADACEAE

Cycas Linnaeus

Cycas inermis Lour. Fl. Cochinch. 632. 1790, ed. Willd. 776. 1793, Anamese *cây san tué;* Chev. & Poilane in Journ. Bot. Appl. **4**: 472. 1924; Tandy in Journ. Bot. **65**: 281. 1927; Leandri in Lecomte Fl. Gén. Indo-Chine **5**: 1091. 1931.

Cycas siamensis Miq. subsp. *inermis* Schuster in Pflanzenreich 99(IV-1): 80. 1932 (based on *Cycas inermis* Lour.).

" Habitat agrestis, cultaque ob pulchritudinem in Cochinchina, & China." Loureiro's type, a leaf specimen, is preserved in the herbarium of the British Museum. Tandy states that it is not *Cycas revoluta* Linn., where it was placed by Dyer (Journ. Linn. Soc. Bot. **26**: 559. 1902) but that it resembles *C. rumphii* Miq. and *C. circinalis* Linn. Tandy ex-presses the opinion that *C. micholitzii* Dyer may well be a form of *C. inermis* Lour. Le-andri, in his treatment of the Cycadaceae of Indo-China (Lecomte Fl. Gén. Indo-Chine **5**: 1085–1092. 1931) admits 10 species of *Cycas*, including *C. circinalis* L., *C. rumphii* Miq., *C. micholitzii* Dyer, and *C. inermis* Lour., but several of these are imperfectly known; he apparently overlooked Tandy's note on Loureiro's type. The illustrations cited by Lou-reiro represent both *Cycas revoluta* Linn. and *C. rumphii* Miq. Miquel (Anal. Bot. Ind. **2**: 28. *pl. 4.* 1851) misinterpreted Loureiro's species, as the form he described and illustrated is *Cycas revoluta* Linn. Schuster's recent monographic treatment of the genus (Pflanzen-reich 99(IV-1): 64–84. 1932) is distinctly unsatisfactory. On page 81 he cites *Cycas in-ermis* Lour. excl. syn., as a synonym of *Cycas revoluta* Linn., yet on the preceding page he makes *Cycas inermis* Lour. the name-bringing synonym of *C. siamensis* Miq. subsp. *iner-mis* (Lour.) Schuster, citing as representing it " Cochinchina: nahe Saigon (Loureiro III.

1887 in Herb. Barbey-Boissier)," and a specimen cultivated at Buitenzorg. The late Dr. Chodat informed me in February, 1934, that not only was there no Loureiro specimen of this species in the Barbey-Boissier herbarium, but that there was no material from him in the entire collection. Schuster was apparently unaware of Loureiro's extant type in the British Museum herbarium and of Tandy's published note on it. The critical note by Chevalier and Poilane should be consulted; they record the modern Anamese name as *cay xuong tè*.

PINACEAE

Pinus (Tournefort) Linnaeus

Pinus merkusii Jungh. & de Vriese Pl. Nov. Ind. Bat. 5. *pl. 2*. 1845, Bot. Zeit. **4**: 13. 1846; Hickel in Lecomte Fl. Gén. Indo-Chine **5**: 1077. 1931.

Pinus sylvestris (non Linn.) Lour. Fl. Cochinch. 579. 1790, ed. Willd. 709. 1793, Anamese *cây thoung*, Chinese *sūm*.

" Habitat agrestis in montibus Cochinchinae, & Chinae: colitur etiam." The Chinese form Loureiro had unquestionably represents *Pinus massoniana* Lamb., as this is the only species of the genus growing near Canton. I assume that Loureiro's short description and longer discussion were based primarily on the Indo-China form, which can scarcely be other than *Pinus merkusii* Jung. & de Vriese, where Shaw (Gen. Pinus 58. 1914) and Hickel place it.

Cunninghamia [34] R. Brown

Cunninghamia lanceolata (Lamb.) Hook. in Curtis's Bot. Mag. **54**: *pl. 2743*. 1827.

Pinus lanceolata Lamb. Gen. Pinus 52. *pl. 34*. 1803.

Abies lanceolata Poir. in Lam. Encycl. **6**: 523. 1804.

Belis jaculifolia Salisb. in Trans. Linn. Soc. **8**: 315. 1807.

Cunninghamia sinensis R. Br. ex Richard Conif. 80. *pl. 18. f. 3*. 1826.

Cunninghamia jaculifolia Druce in Rept. Bot. Exch. Club. Brit. Isles **4**: 618. 1917.

Pinus abies (non Linn.) Lour. Fl. Cochinch. 579. 1790, ed. Willd. 710. 1793, Anamese *cây thoung taù*, Chinese *xān mŏ*.

" Habitat frequentissima in provinciis Australibus imperii Sinensis." The description manifestly applies to the well-known *Cunninghamia lanceolata* Hook. *Dammara alba foemina* Rumph. (Herb. Amb. **2**: 175. *pl. 57*) cited by Loureiro as a synonym represents the very different *Agathis alba* (Lam.) Foxw.

Glyptostrobus Endlicher

Glyptostrobus pensilis (Abel) K. Koch Dendrol. 2(2): 191. 1873.

Thuja pensilis Abel in Staunton Embassy China 436. 1797.

Thuja lineata Poir. in Lam. Encycl. Suppl. **5**: 303. 1817.

Taxodium heterophyllum Brongn. in Ann. Sci. Nat. **30**: 184. 1833.

Glyptostrobus heterophyllus Endl. Syn. Conif. 70. 1847.

Glyptostrobus lineatus Druce in Rept. Bot. Exch. Club Brit. Isles **4**: 624. 1917.

Thuja orientalis (non Linn.) Lour. Fl. Cochinch. 580. 1790, ed. Willd. 712. 1793.

---

[34] *Cunninghamia* R. Brown (1826), conserved name, Vienna Code; an older one is *Belis* Salisbury (1807).

" Habitat frequens in China; colitur raro in Cochinchina." The short description clearly applies to the *shui tsung* or water pine which is common in the vicinity of Canton. It is commonly known as *Glyptostrobus heterophyllus* Endl.

Thuja [35] Linnaeus

Thuja orientalis Linn. Sp. Pl. ed. 2, 1422. 1763.

> *Biota orientalis* Endl. Syn. Conif. 47. 1847; Hickel in Lecomte Fl. Gén. Indo-Chine 5: 1082. 1931.
>
> *Cupressus thyoides* (non Linn.) Lour. Fl. Cochinch. 580. 1790, ed. Willd. 711. 1793, Anamese *trác bá diep.*

" Habitat in China, & Cochinchina." The correctness of this reduction of the form so briefly and inadequately described is probable, in view of the wide distribution and common occurrence of the Linnaean species in southern China; Hickel records it as often cultivated in Tonkin.

Cupressus (Tournefort) Linnaeus

Cupressus torulosa D. Don. Prodr. Fl. Nepal. 55. 1825; Hickel in Lecomte Fl. Gén. Indo-Chine 5: 1081. 1931.

> *Cupressus sempervirens* (non Linn.) Lour. Fl. Cochinch. 580. 1790, ed. Willd. 711. 1793, Anamese *cây duong*, Chinese *pĕ xú.*

" Habitat in China: inde in Cochinchinam delata, cultaque." In my original manuscript of 1919 I placed this as a probable synonym of *Cupressus funebris* Endl., partly because of Farges' record of *peechou* as the Chinese name for Endlicher's species. Loureiro, however, does not mention the characteristic elongated pendulous branchlets; his description " ramis patentibus, brevibus, multis, frondes lineares quadragonae, sparsae " applying better to *Cupressus torulosa* D. Don than to *C. funebris* Endl. Hickel records Don's species from Anam but does not indicate whether the tree is native or planted.

Juniperus (Tournefort) Linnaeus

Juniperus chinensis Linn. Mant. 1: 127. 1767; Lour. Fl. Cochinch. 636. 1790, ed. Willd. 781. 1793, Anamese *bien bá tung;* Hickel in Lecomte Fl. Gén. Indo-Chine 5: 1085. 1931.

> *Juniperus barbadensis* (non Linn.) Lour. Fl. Cochinch. 636. 1790, ed. Willd. 781. 1793, Anamese *nhit bien tung.*

For both species Loureiro states: " Habitat culta in Cochinchina, a Sinis oriunda." *Juniperus chinensis* Linn. was apparently interpreted correctly by Loureiro and the plant he described under *J. barbadensis* Linn. as having bluish-green leaves can scarcely be other than a form of the same species. *Juniperus chinensis* Linn. is the only representative of the genus credited to Indo-China by Hickel who, however, cites no local name but speaks of it as frequently cultivated.

---

[35] Frequently spelled *Thuya*. The original form is here retained; see Sprague, Kew Bull. 363. 1928.

GNETACEAE

Ephedra (Tournefort) Linnaeus

Ephedra sinica Stapf in Kew Bull. 133. 1927.

> *Equisetum arvense* (non Linn.) Lour. Fl. Cochinch. 671. 1790, ed. Willd. 823. 1793, pro parte, Anamese *mâ hoâm*, Chinese *ma hoàng*.

"Habitat incultum in agris Sinensibus." The description, in part, the local names, and the indicated medicinal uses appertain to *Ephedra* and probably to the true *ma huang*, *E. sinica* Stapf. In part it is apparently *Equisetum*, and probably *E. arvense* Linn. (see p. 63). Loureiro certainly secured his material from herbalists.

Gnetum Linnaeus

Gnetum indicum (Lour.) Merr. Interpret. Herb. Amb. 77. 1917 (based on *Abutua indica* Lour.).

> *Abutua indica* Lour. Fl. Cochinch. 630. 1790, ed. Willd. 775. 1793, Anamese *cây sót*, *cây gám*.
>
> *Gnetum montanum* Marcgraf in Bull. Jard. Bot. Buitenzorg III **10**: 466. *pl. 8.* 1930; Leandri in Lecomte Fl. Gén. Indo-Chine **5**: 1057. 1931.

"Habitat in sylvis Cochinchinae & in aliis Indiae locis." In accepting Loureiro's specific name I apparently erred in reducing to this species *Gnetum latifolium* Blume and *Gnetum funiculare* Brongn. Marcgraf in his recent monographic treatment of the genus (Bull. Jard. Bot. Buitenzorg III **10**: 407–511. 1930) considered *Abutua indica* Lour. = *Gnetum indicum* Merr. to be undeterminable species. Loureiro's material consists of three sheets, all sterile, preserved in the herbarium of the British Museum. I consider it certain that *Gnetum montanum* Marcgraf is the same as *Abutua indica* Lour., and in my opinion Loureiro's specific name should be retained. It is to be noted that Marcgraf cites specimens as representing *Gnetum montanum* from Cochinchina (Gaudichaud, Tourane), Hainan, and Kwangtung Province, China (forma *parvifolium* Warb.). He is followed by Leandri (Lecomte Fl. Gén. Indo-Chine **5**: 1058. 1931). The latter author fails to account for *Abutua indica* Lour., admitting 5 species in the general group with it, namely *G. montanum* Marcgraf, *G. latifolium* Blume, *G. macrostachyum* Hook. f., *G. leptostachyum* Blume, and *G. formosum* Marcgraf; among these Loureiro's species definitely belongs to the first one. I unhesitatingly refer *Clemens 3345*, from the vicinity of Tourane near Hue, the classical locality, to *Gnetum indicum* (Lour.) Merr.; it is also *G. montanum* Marcgr.

ANGIOSPERMAE

MONOCOTYLEDONEAE

TYPHACEAE

Typha Linnaeus

Typha angustifolia Linn. subsp. **javanica** (Schnizl.) Graebn. in Pflanzenreich 2 (IV–8): 13. 1900.

> *Typha javanica* Schnitzl. ex Zoll. Syst. Verzeich. Ind. Archipel. Pflanz. 77. 1854, *nomen nudum;* Rohrb. in Verh. Bot. Ver. Brandenb. **11**: 98. *f. 8.* 1869, *descr.*

Typha latifolia (non Linn.) Lour. Fl. Cochinch. 552. 1790, ed. Willd. 675. 1793, Anamese *bô hoàng*, Chinese *pu hoâm*.

"Habitat in paludibus Chinae, & Cochinchinae." Loureiro's description is imperfect and indefinite. The form he had was without doubt the widely distributed Indo-Malaysian one currently referred to the above variety of the Linnaean species.

PANDANACEAE

Pandanus (Rumphius) Linnaeus f.

Pandanus humilis Lour. Fl. Cochinch. 603. 1790, ed. Willd. 740. 1793, Anamese *júa rùng*.

   Pandanus pierrei Martelli in Bull. Soc. Bot. Ital. 303. 1904, Webbia **4**(1): 27. *pl. 40. f. 4–7.* 1913.

   "Habitat agrestis in Cochinchina: amat loca montana, & petrosa." *Pandanus humilis* Rumph. (Herb. Amb. **4**: 143. *pl. 76*), cited by Loureiro as synonym, must be excluded as it represents an entirely different species, *P. polycephalus* Lam. Similarly *Kaida taddi* Rheed. (Hort. Malabar. **2**: *pl. 6*) must be excluded as it apparently represents *Pandanus tectorius* Parkinson. Loureiro's description was based on an actual specimen from Indo-China, not on *Pandanus humilis* Rumph.; his name is valid and should be accepted.

Pandanus laevis Lour. Fl. Cochinch. 604. 1790, ed. Willd. 741. 1793, Anamese *lá buon, lá khai*.

   Corypha laevis A. Chev. Cat. Pl. Jard. Bot. Saigon 66. 1919, quoad syn. Lour. (based on *Pandanus laevis* Lour.).

   "Habitat in sylvis Cochinchinae: amat loca deserta." *Pandanus moschatus seu laevis* Rumph. (Herb. Amb. **4**: 147), cited by Loureiro as a synonym, must be excluded as it represents *Pandanus tectorius* Parkinson var. *laevis* (Kunth) Warb. From Loureiro's description I judge his species to be a representative of the section *Keura*, but from his description of the fruits as small, scarcely *P. tectorius* Parkinson. A. Chevalier (Cat. Pl. Jard. Bot. Saigon 66. 1919) states that *lá buon* is *Corypha laevis* (Lour.) A. Chev. (*C. lecomtei* Becc.), and while Loureiro may have erroneously ascribed the local name *lá buon* and the indicated uses of *Corypha* to *Pandanus*, yet his entire description applies to *Pandanus* and not at all to *Corypha*. I cannot therefore accept Chevalier's interpretation and agree with Lecomte (Not. Syst. **4**: 61. 1923) that *Pandanus laevis* Lour. is a true *Pandanus*. Loureiro's specific name cannot replace *Corypha lecomtei* Becc. as nothing in the description indicates a palm.

Pandanus tectorius Parkinson Voy. South Seas H. M. S. Endeavor 46. 1773; Warb. in Pflanzenreich **3**(IV–9): 46. 1900.

   Pandanus odoratissimus Linn. f. Suppl. 424. 1781; Lour. Fl. Cochinch. 603. 1790, ed. Willd. 739. 1793, Anamese *cây júa*.

   Pandanus loureirii Gaudich. Bot. Voy. Bonite *pl. 22. f. 13.* 1839–52.

   Pandanus tectorius Parkins. var. *loureirii* Martelli in Webbia **4**(1): 34. 1913.

   Pandanus odoratissimus Linn. f. var. *loureirii* Martelli in Univ. Calif. Publ. Bot. **12**: 373. 1930.

   "Habitat agrestis in Cochinchina, & China: colitur etiam ad ducendas sepes agrorum et viarum, ad pascendosque elephantes domesticos: amat loca arenosa." Loureiro's description for the most part applies to the very common littoral species that occurs every-where near the seashore in the Indo-Malaysian region, and one that must have been very familiar to him; yet his description of the fruits as 1-seeded is either an error of observation on his part, or due to the inclusion of some other species, possibly *P. ceratostigma* Martelli. *Pandanus verus* Rumph. (Herb. Amb. **4**: 139. *pl. 74*) cited by Loureiro as a synonym, represents the species as here interpreted; *P. spurius* Rumph. (op. cit. 142. *pl. 75*) also cited as a synonym, represents *Pandanus robinsonii* Merr. For a discussion of the synonymy see Martelli, U., "*Pandanus odoratissimus*" o "*Pandanus tectorius*" (Nuov. Giorn. Bot. Ital. **36**: 328–337. 1929) who adopts the binomial *Pandanus odoratissimus* Linn. f. making *P. tectorius* a variety of it, *P. odoratissimus* Linn. f. var. *tectorius* Martelli (*op. cit.* 336). As Solander's description was not published and Parkinson is given as the authority for the binomial as published in 1773, I believe *Pandanus tectorius* Parkinson should stand and that if a variety be represented by the form described by the younger Linnaeus, it should be given varietal status. *Pandanus loureirii* Gaudich., no description published, is an illustration of a single drupe based on material collected by Gaudichaud probably at Tourane, near Hue. Martelli described it in 1930 from *Clemens 4343*, collected at Tourane, near Hue, " omnipresent, roadsides, dune thickets, hedges."

POTAMOGETONACEAE

Potamogeton (Tournefort) Linnaeus

Potamogeton octandrum Poir. in Lam. Encycl. Suppl. **4**: 534. 1816 (based on *Hydrogeton heterophyllum* Lour.).

Hydrogeton heterophyllum Lour. Fl. Cochinch. 244. 1790, ed. Willd. 301. 1793, Anamese *raong hai thú lá;* Moore in Journ. Bot. **63**: 255. 1925.

Potamogeton javanicus Hassk. in Verh. Natuurk. Ver. Nederl. Ind. **1**: 26. 1856; Graebn. in Pflanzenreich **31**(IV–11): 46. *f. 14. A-C.* 1907.

" Habitat fluviis, & paludibus Cochinchinae." *Hydrogeton* has been generally accepted as a synonym of *Potamogeton*, but Graebner (Pflanzenreich **31**(IV–11): 142. 1907) failed to account for Loureiro's species, placing it among the *species excludendae vel incertae.* Arthur Bennett reported to me April 4, 1919, that he had located Loureiro's type in the herbarium of the British Museum, his statement being quoted by Moore (Journ. Bot. **63**: 255. 1925) and Moore confirms Bennett's conclusion that *Hydrogeton heterophyllum* Lour. is identical with *Potamogeton javanicus* Hassk. Loureiro's specific name being invalidated in *Potamogeton* by the previous use of the same name by six different authors, Poiret's binomial stands. Loureiro was misled by the broad connective and described the 8 anther-cells as 8 anthers; there are but 4 stamens.

APONOGETONACEAE

Aponogeton Linnaeus f.

Aponogeton natans (Linn.) Engl. & Krause in Pflanzenreich **24**(IV–13): 11. 1906.

Saururus ? natans Linn. Mant. **2**: 227. 1771.

Aponogeton monostachyon Linn. f. Suppl. 214. 1781.

? *Zannichellia tuberosa* Lour. Fl. Cochinch. 543. 1790, ed. Willd. 662. 1793, Anamese
mach môn nam.

"Habitat loca humida Cochinchinae, non frequens."   Ascherson & Graebner, in excluding this from the Potamogetonaceae (Pflanzenreich 31(IV–11): 158. 1907) thought that
some genus and species of the Araceae might be represented, although there is little in the
description that would indicate an araceous plant.   The few data given regarding the
floral structure do not agree with *Aponogeton*, but the habit and other characters agree
closely with this genus.   Erroneous observations on the part of Loureiro are suspected.

ALISMACEAE

Sagittaria (Ruppius) Linnaeus

Sagittaria sagittifolia Linn. Sp. Pl. 993. 1753; Lour. Fl. Cochinch. 570. 1790, ed. Willd.
698. 1793, Anamese *cây mác tláng.*

"Habitat paludes, & loca caenosa Cochinchinae."   Loureiro's description applies to
the Linnaean species as that is currently interpreted.   A form of this is very extensively
cultivated for food in southern China, apparently the one illustrated in Curtis's Bot. Mag.
39: *pl. 1631.* 1814 as *Sagittaria sinensis* Sims.   Sims there cites *Sagittaria sagittifolia* Lour.
as a synonym of *S. sinensis.*

HYDROCHARITACEAE

Vallisneria (Micheli) Linnaeus

Vallisneria spiralis Linn. Sp. Pl. 1015. 1753; Gagnep. in Lecomte Fl. Gén. Indo-Chine **6**:
7. *f. 4.* 1908.

Physkium natans Lour. Fl. Cochinch. 663. 1790, ed. Willd. 814. 1893, Anamese *raong
mái cheò;* Moore in Journ. Bot. **63**: 290. 1925.

Vallisneria physcium Juss. ex Spreng. Syst. **3**: 900. 1826 (based on *Physkium natans*
Lour.).

"Habitat in fluviis lenti cursus in Cochinchina."   Moore (Journ. Bot. **63**: 290. 1925)
has given an extensive note on Loureiro's type which is preserved in the herbarium of the
British Museum, quoting Jussieu's opinion, which definitely places *Physkium* as a synonym of *Vallisneria*, and Britten's statement to the effect that Jussieu, "Note sur le genre
Physkium de Loureiro" (Ann. Mus. Hist. Nat. (Paris) **9**: 402–404. 1807) does not actually
form the binomial ascribed to him by Sprengel.   Britten's comment "where Sprengel
derived the name is not apparent," refers to Jussieu as authority for the binomial, as
Sprengel cites "*Physcium natans* Lour." as a synonym; what he intended was that it is
not apparent why Sprengel quotes Jussieu as the authority for the binomial which Jussieu
himself did not publish.

GRAMINEAE

Zea Linnaeus

Zea mays Linn. Sp. Pl. 971. 1753; Lour. Fl. Cochinch. 550. 1790, ed. Willd. 672. 1793,
Anamese *cây báp,* Chinese *páo túc leâm.*

"Habitat late culta in Cochinchina, & China."   This is the common maize or Indian
corn, the Linnaean species being correctly interpreted by Loureiro.

Coix Linnaeus

Coix lachryma-jobi Linn. Sp. Pl. 972. 1753.

Coix lachryma Linn. Syst. ed. 10, 1261. 1759; Lour. Fl. Cochinch. 551. 1790, ed. Willd. 673. 1793, Anamese *cây bo bo, i'di nhon,* Chinese *ý ỳ gîn.*

Coix agrestis Lour. Fl. Cochinch. 551. 1790, ed. Willd. 674. 1793, Anamese *bo bo hoang.*

For the first Loureiro states: " Frequenter culta in Cochinchina & China "; the second: " Habitat agrestis in locis humidis: nec esculenta." Both are but forms of the Linnaean species, the one Loureiro described as new having smaller involucres than the cultivated one; it is probably the same as *Coix lachryma-jobi* Linn. var. *puellarum* (Balansa) E. G. & A. Camus (Lecomte Fl. Gèn. Indo-Chine 7: 220. 1922).

Imperata Cyrillo

Imperata cylindrica (Linn.) Beauv. Agrost. Expl. Pl. *pl. 5. f. 1.* 1812; E. G. & A. Camus in Lecomte Fl. Gén. Indo-Chine 7: 231. 1922.

Lagurus cylindricus Linn. ed. 10, 878. 1759.

Saccharum koenigii Retz. Obs. **5**: 16. 1789.

Saccharum spicatum (non Linn.) Lour. Fl. Cochinch. 53. 1790, ed. Willd. 67. 1793, Anamese *tranh co,* Chinese *mâo kĕn.*

" Habitat loca montana in Cochinchina." The brief description, and particularly the notes on the economic uses of the species, clearly indicate the very common and widely distributed lalang grass. *Gramen caricosum* Rumph. (Herb. Amb. **6**: 17. *pl. 7. f. 2A*) cited as a synonym, is correctly placed. This form is currently referred to the var. *koenigii* (Retz.) Durand & Schinz.

Rottboellia [36] Linnaeus f.

Rottboellia exaltata Linn. f. Suppl. 114. 1781; E. G. & A. Camus in Lecomte Fl. Gén. Indo-Chine 7: 380. 1922.

Stegosia cochinchinensis Lour. Fl. Cochinch. 51. 1790, ed. Willd. 65. 1793, Anamese *co tranh.*

" Nullibi eam vidi praeterquam in Cochinchina." The description applies unmistakably to the common and widely distributed *Rottboellia exaltata* Linn. f. Loureiro's type is preserved in the herbarium of the British Museum, Natural History.

Saccharum Linnaeus

Saccharum officinarum Linn. Sp. Pl. 54. 1753; Lour. Fl. Cochinch. 52. 1790, ed. Willd. 66. 1793, Anamese *mià, cam giá,* Chinese *càn ché.*

" Habitat, et colitur abundantissime in omnibus provinciis regni Cochinchinensis: simul in aliquibus imperii Sinensis, sed minori copia." They are all forms of the common sugar cane. Under the species Loureiro briefly characterizes three varieties, *Saccharum album, miá lau; S. rubrum, miá mung;* and *S. elephantium, miá boi;* all are forms of the variable *Saccharum officinarum* Linn.

---

[36] *Rottboellia* Linnaeus f. (1779), conserved name, Vienna Code; an older one is *Manisuris* Linnaeus (1771). The original spelling is *Rottbŏllia* in the definition of the genus, p. 114, and *Rottbŏlla,* p. 114, where the species are considered.

Saccharum arundinaceum Retz. Obs. **4**: 14. 1786; E. G. & A. Camus in Lecomte Fl. Gén. Indo-Chine **7**: 241. 1922.

Saccharum jaculatorium Lour. Fl. Cochinch. 53. 1790, ed. Willd. 67. 1793, Anamese *cây lau.*

Imperata jaculatoria Poir. ex Roem. & Schult. Syst. **2**: 289. 1817 (based on *Saccharum jaculatorium* Lour.).

" Habitat agreste in Cochinchina." Loureiro's species was placed by Hackel (DC. Monog. Phan. **6**: 99. 1889) as a doubtful synonym of *Imperata exaltata* Brong., where it cannot possibly belong, not only on account of Loureiro's technical description, but also because of his statement that the culms were used for making javelins. It is, without doubt, referable to *Saccharum arundinaceum* Retz. E. G. & A. Camus do not account for Loureiro's species, or the synonym based upon it, in their treatment of the grasses of Indo-China. The binomial *Imperata jaculatoria* is usually credited to Poiret (Lam. Encycl. Suppl. **2**: 70. 1811) but does not appear there; Poiret uses only the combination *Saccharum jaculatorium* Lour.

Saccharum spontaneum Linn. Mant. **2**: 183. 1771; Lour. Fl. Cochinch. 52. 1790, ed. Willd. 65. 1793, Anamese *cây lách;* E. G. & A. Camus in Lecomte Fl. Gén. Indo-Chine **7**: 239. 1922.

" Habitat circa paludes, & loca humida Cochinchinae frequenter." Loureiro probably had specimens of the very common and widely distributed Linnaean species, yet his description of the leaves as smooth and the " pilis calycinis duriusculis " is distinctly not good for *Saccharum spontaneum* Linn. The Anamese names *lach* and *cay lach* are today among those used to designate this characteristic species.

Ophiurus Gaertner f.

Ophiurus cochinchinensis (Lour.) comb. nov.

Phleum cochinchinense Lour. Fl. Cochinch. 48. 1790, ed. Willd. 61. 1793, Anamese *co choung.*

Paspalum cochinchinense Poir. in Lam. Encycl. Suppl. **4**: 316. 1816; Steud Syn. **1**: 33. 1854 (based on *Phleum cochinchinense* Lour.).

Ophiurus monostachyus Presl Rel. Haenk. **1**: 330. 1830; E. G. & A. Camus in Lecomte Fl. Gén. Indo-Chine **7**: 373. 1922.

" Habitat spontaneum in Cochinchina." Loureiro's description unmistakably applies to the grass commonly known as *Ophiurus monostachyus* Presl. *Paspalum* was suggested by Willdenow in a footnote in his edition of Loureiro's Flora Cochinchinensis. Steudel, who erroneously credited the binomial *Paspalum cochinchinense* to Willdenow, queried: " an forsan *Rottboelliae* species? " The species is represented by *Squires 165*, Anamese *ka luung*, from the general vicinity of Hue. It extends from Indo-China to southeastern China, the Philippines, and the Marianne Islands. It is not uncommon, and is frequently rather abundant, but as it grows among other grasses, it is very readily overlooked because of its inconspicuous narrow spikes.

Andropogon Linnaeus

Andropogon aciculatus Retz. Obs. **5**: 22. 1789.

> *Rhaphis* [37] *trivialis* Lour. Fl. Cochinch. 553. 1790, ed. Willd. 676. 1793, Anamese *co may*.
>
> *Chrysopogon aciculatus* Trin. Fund. Agrost. 188. 1820; E. G. & A. Camus in Lecomte, Fl. Gén. Indo-Chine **7**: 333. 1922.
>
> *Andropogon acicularis* Willd. Sp. Pl. 4(2): 906. 1805.
>
> *Rhaphis acicularis* Desv. Opusc. 69. 1831.

" Habitat ubique prope vias, hominibus valde incommoda in Cochinchina, & China: quia vestibus adhaerens taediose avellitur, cum excuti nequeat." One familiar with the vegetation of the Old World tropics would scarcely need more than the last part of the statement above quoted to settle the status of the species intended. This grass is a very common one in open places, and is a distinct pest in many regions. *Gramen aciculatum* Rumph. (Herb. Amb. **6**: 13. *pl. 5. f. 1*) cited by Loureiro as a synonym, is correctly placed.

Andropogon citratus DC. Cat. Hort. Monspel. 78. 1813.

> *Cymbopogon citratus* Stapf in Kew Bull. 357. *pl.* 1906; A. G. & A. Camus in Lecomte Fl. Gén. Indo-Chine **7**: 338. 1922.
>
> *Andropogon schoenanthus* (non Linn.) Lour. Fl. Cochinch. 646. 1790, ed. Willd. 793. 1793, Anamese *lá sa*, Chinese *máo hiàm*.

" Habitat culta in hortis Cochinchinae, & Chinae." Loureiro's description unmistakably applies to the lemon grass, which is commonly planted throughout the Old World tropics. *Schoenanthum amboinicum* Rumph. (Herb. Amb. **5**: 181. *pl. 72. f. 2*) is correctly placed as a synonym.

Andropogon sorghum (Linn.) Brot. Fl. Lusit. **1**: 88. 1804, var.

> *Holcus saccharatus* Linn. Sp. Pl. 1047. 1753; Lour. Fl. Cochinch. 645. 1790, ed. Willd. 792. 1793, Anamese *cây mach maoc*.

" Habitat cultus in Cochinchina, & China. Hic etiam occurrit Holcus sorghum panicula coarctata, erecta." The form that Loureiro described was one with diffuse panicles. From the data available it is impossible to carry the reduction to any of the very numerous described varieties or forms of this polymorphous species. *Sorghum Battari* Rumph. (Herb. Amb. **5**: 194. *pl. 75. f. 1*), cited by Loureiro as a synonym, is a form or variety of the Linnaean species.

Thysanolaena Nees

Thysanolaena maxima (Roxb.) O. Ktz. Rev. Gen. Pl. 794. 1891; E. G. & A. Camus in Lecomte Fl. Gén. Indo-Chine **7**: 468. 1922.

> *Agrostis maxima* Roxb. Fl. Ind. **1**: 319. 1820.
>
> *Thysanolaena agrostis* Nees in Edinb. New Philos. Mag. **18**: 180. 1835.
>
> *Arundo epigejos* (non Linn.) Lour. Fl. Cochinch. 55. 1790 (*epigegos*), ed. Willd. 70. 1793, Anamese *cây trai*.

" Habitat in sylvis Cochinchinae." This is in all probability the correct disposition of the plant Loureiro so inadequately described, the only part of the description that does not apply being the seeds described as " papposo," perhaps taken from Linnaeus. This

---

[37] *Chrysopogon* Trinius (1820), conserved name, Cambridge Code, for those who recognize this group as a generic segregate from *Andropogon; Rhaphis* Loureiro (1790) is the oldest name for this group.

common and conspicuous grass is one Loureiro would scarcely have overlooked; the habitat agrees, the inflorescences before full maturity are " subcoarctata," and the local name is significantly like *cây xay*, cited by E. G. & A. Camus as one of the local names for *Thysanolaena maxima* O. Ktz. Mez has adopted the binomial *Thysanolaena procera* (Retz.) Mez (ex Jan. in Bot. Arch. Mez **1**: 27. 1922) based on *Agrostis procera* Retz. (Obs. **4**: 19. 1786) for this species, but Retzius' description does not apply at all to *Thysanolaena*, notably in the indicated size, the narrow much shorter leaves, villous nodes, hirsute spikelets, and very different glumes. A recent examination of the type in Retzius' herbarium, kindly loaned to me for study, shows that Hubbard, who also examined it, was correct in referring it to *Eriochloa* as *E. procera* (Retz.) Hubbard (Kew Bull. 256. 1930) (*E. ramosa* O. Ktz.). Rendle (Journ. Linn. Soc. Bot. **36**: 391. 1904) considered that Loureiro correctly interpreted *Arundo epigejos* Linn. = *Calamagrostis epigejos* Roth, but no *Calamagrostis* is known from Indo-China, and the Asiatic species are not sylvan ones.

Digitaria Scopoli

Digitaria sp.

> *Agrostis radiata* Lour. Fl. Cochinch. 50. 1790, ed. Willd. 63. 1793, Anamese *co pháo luói*.

" Habitat spontanea in Cochinchina." Willdenow in a footnote states that this is a *Chloris*, but no species of this or any allied genera in Cochinchina conforms to Loureiro's description. I suspect that what he had in mind was the species admitted by E. G. & A. Camus (Lecomte Fl. Gén. Indo-Chine **7**: 401. 1922) as *Digitaria horizontalis* Willd.; but they do not account for *Agrostis radiata* Lour. The height of the plant, 6 feet, as indicated by Loureiro, is against this disposition of his species.

Panicum Linnaeus

Panicum auritum Presl Rel. Haenk. **1**: 305. 1830.

> *Sacciolepis aurita* E. G. & A. Camus in Lecomte Fl. Gén. Indo-Chine **7**: 459. 1922.
>
> *Agrostis indica* (non Linn.) Lour. Fl. Cochinch. 49. 1790, ed. Willd. 63. 1793, Anamese *co lung*.

Loureiro's specimens were from Cochinchina. The size and habit of the plant eliminate the possibility of this being the same as *Agrostis indica* Linn. = *Sporobolus indicus* R. Br., and Loureiro indicates his own doubt as to the correctness of his identification. E. G. & A. Camus make no attempt to account for Loureiro's species in their treatment of the grasses of Indo-China (Lecomte Fl. Gén. Indo-Chine **7**: 202–650. 1922–23). *Panicum auritum* Presl may be the species Loureiro attempted to describe.

Panicum crus-galli Linn. Sp. Pl. 56. 1753.

> *Echinochloa crus-galli* Beauv. Agrost. 53. 1812; E. G. & A. Camus in Lecomte Fl. Gén. Indo-Chine **7**: 425. 1922.
>
> *Panicum crus-corvi* Linn. Syst. ed. 10, 870. 1759; Lour. Fl. Cochinch. 46. 1790, ed. Willd. 59. 1793, Anamese *co dô dôt*.

No locality is given, but the Anamese name indicates an Indo-China specimen. The species so imperfectly described is doubtless a form of the polymorphous *Panicum crusgalli* Linn., *sensu latiore*. Loureiro himself notes: " Haec herba quasi media inter Crus

Corvi, & Crus Galli, ab ambabus adhuc differens est." It may be noted that E. G. & A. Camus cite *có do dôt* as one of the local names of this species without, however, citing Loureiro. *Panicum crus-corvi* Linn. is by some authors maintained as a species distinct from *P. crus-galli* Linn.

Panicum indicum Linn. Mant. **1**: 184. 1767.

> *Sacciolepis indica* Chase in Proc. Biol. Soc. Washington **21**: 8. 1908; E. G. & A. Camus in Lecomte Fl. Gén. Indo-Chine **7**: 461. 1922.
>
> *Agrostis capillaris* (non Linn.) Lour. Fl. Cochinch. 50. 1790, ed. Willd. 63. 1793, Anamese *co deé*.

"Habitat in Cochinchina." The very short description is unsatisfactory, but the "spicate" panicle seems to indicate a species of *Panicum* of the section *Sacciolepis*. The height of the plant, 4 feet, is not a serious objection to this identification, as *Panicum indicum* Linn. in luxuriant forms may reach this size; the "spiculis longis, tenuissimis" is a more serious objection. E. G. & A. Camus make no attempt to identify Loureiro's species in their treatment of the grasses of Indo-China (Lecomte Fl. Gén. Indo-Chine **7**: 202-650. 1922-23).

Panicum miliaceum Linn. Sp. Pl. 58. 1753; Lour. Fl. Cochinch. 47. 1790, ed. Willd. 59. 1793.

"Habitat Pekini, & aliis locis Sinarum." Loureiro's description applies to the Linnaean species, the common millet, a widely cultivated cereal in China.

Panicum miliare Lam. Ill. **1**: 173. 1791; E. G. & A. Camus in Lecomte Fl. Gén. Indo-Chine **7**: 448. 1922.

> *?Milium effusum* (non Linn.) Lour. Fl. Cochinch. 49. 1790, ed. Willd. 62. 1793, Anamese *cây co gao*.

"Nascitur incultum inter segetes oryzae montanae in Cochinchina." The description apparently applies to some species of *Panicum* with diffuse panicles. *P. miliare* Lam. as interpreted by E. G. & A. Camus (including *P. psilopodium* Trin.) probably is the species that Loureiro so inadequately described.

Panicum pilipes Nées in Hook. Journ. Bot. Kew Gard. Miscel. **2**: 97. 1850 *nomen nudum;* Nees & Arn. ex Büse in Miq. Pl. Jungh. 376. 1854 (descr.).

> *Cyrtococcum pilipes* A. Camus in Bull. Mus. Hist. Nat. (Paris) **27**: 118. 1921; E. G. & A. Camus in Lecomte Fl. Gén. Indo-Chine **7**: 463. 1922.
>
> *Agrostis minima* (non Linn.) Lour. Fl. Cochinch. 50. 1790, ed. Willd. 64. 1793.

"Habitat in Cochinchina." The description is so short and imperfect as to be almost valueless for purposes of determining what grass Loureiro had in mind; he expresses doubt as to the correctness of his interpretation of the Linnaean species. It is very probable that *Panicum pilipes* Nees is the species intended. E. G. & A. Camus make no attempt to identify Loureiro's species in their treatment of the Gramineae of Indo-China.

Panicum repens Linn. Sp. Pl. ed. 2, 87. 1762; E. G. & A. Camus in Lecomte Fl. Gén. Indo-Chine **7**: 450. 1922.

> *Ischaemum importunum* Lour. Fl. Cochinch. 646. 1790, ed. Willd. 794. 1793, Anamese *co oúng*.

" Habitat agreste, & frequens in Cochinchina: herba odiosa agricolis, quia difficulter eradicatur, & facile iterum pullulat." Hackel, in his monograph of the Andropogoneae (DC. Monog. Phan. **6**: 253. 1889) left this as " omnino dubitum " under the excluded species. The description conforms closely with the characters of the very common *Panicum repens* Linn., which, as Loureiro notes, is difficult to eradicate because of its well developed rhizomes. E. G. & A. Camus fail to account for Loureiro's species in their treatment of the Gramineae of Indo-China, yet they cite the Anamese names *có ông* and *có gung* under *Panicum repens* Linn., corroborative evidence that this reduction of *Ischaemum importunum* Lour. is correct.

Oplismenus Beauvois

Oplismenus compositus (Linn.) Beauv. Agrost. 54. 1812; E. G. & A. Camus in Lecomte Fl. Gén. Indo-Chine **7**: 429. 1922.

Panicum compositum Linn. Sp. Pl. 57. 1753.

Panicum hirtellum (non Linn.) Lour. Fl. Cochinch. 47. 1790, ed. Willd. 59. 1793, Anamese *co khé*.

No locality is given but the Anamese name indicates an Indo-China specimen. The description conforms in all essentials to the very common *Oplismenus compositus* Beauv. *Panicum hirtellum* Burm. f. (Fl. Ind. 24. *pl. 12. f. 1.* 1768), cited by Loureiro as representing his species, is the very different *Oplismenus burmanni* Beauv.

Setaria [38] Beauvois

Setaria italica (Linn.) Beauv. Agrost. 51. 1812; E. G. & A. Camus in Lecomte Fl. Gén. Indo-Chine **7**: 476. 1922.

Panicum italicum Linn. Sp. Pl. 56. 1753; Lour. Fl. Cochinch. 46. 1790, ed. Willd. 58. 1793, Anamese *cây khê*, Chinese *siào mě*, *sŏ*.

No locality is given but Loureiro doubtless observed the Italian millet in both China and Indo-China, as it is commonly planted in both countries. *Panicum indicum s. botton* (Rumph. Herb. Amb. **5**: 202. *pl. 75. f. 2*), cited as a synonym by Loureiro, is correctly placed.

Setaria geniculata (Lam.) Beauv. Agrost. 51. 1812.

Panicum geniculatum Lam. Encycl. **4**: 727 (err. typ. 737). 1797.

Panicum flavum Nees in Mart. Fl. Bras. **2**: 238. 1829.

Setaria flava Kunth Rev. Gram. **1**: 46. 1829.

Panicum polystachion Linn. Syst. ed. 10, 870. 1759; Lour. Fl. Cochinch. 46. 1790, ed. Willd. 58. 1793, Anamese *co sâu rom*, non *Setaria polystachya* Schrad.

Setaria aurea Hochst. ex A. Braun in Flora **24**: 276. 1841; E. G. & A. Camus in Lecomte Fl. Gén. Indo-Chine **7**: 473. 1922.

No locality is given but Loureiro's specimens were manifestly from Indo-China as he cites an Anamese name. The Rumphian figure cited as *Gramen caricosum vulpinum* (Herb. Amb. **6**: 18. *pl. 7. f. 2 B*) is a good representation of *Setaria geniculata* (Lam.) Beauv. and typifies *Panicum polystachion* Linn. The type of *Panicum geniculatum* Lam. was from Guadalupe in the West Indies, and Hitchcock (North Am. Fl. **17**: 320. 1931) who, including

---

[38] *Setaria* Beauvois (1812), non Acharius (1798), conserved name, Cambridge Code in place of *Chaeochloa* Scribner (1897).

varieties, cites about seventy synonyms, limits it to North and South America, or at least gives no Old World distribution for it.   As he interprets the species I cannot see why the very common and widely distributed Old World form should not be included.   E. G. & A. Camus retain it as a distinct species, *S. aurea* Hochst.

Setaria palmifolia (Koenig) Stapf in Journ. Linn. Soc. Bot. **42**: 186. 1914.

Panicum palmaefolium Koenig in Naturforsch. **23**: 208. 1788.

Panicum latifolium (non Linn.) Lour. Fl. Cochinch. 47. 1790, ed. Willd. 60. 1793.

Agrostis plicata Lour. Fl. Cochinch. 51. 1790, ed. Willd. 64. 1793, Chinese *sam soŭc tsao*.

Loureiro's *Panicum latifolium* was from Cochinchina, and his type of *Agrostis plicata* was from China: " Habitat in suburbiis urbis Cantoniensis Sinarum."   Both descriptions apply to the very common, widely distributed and rather variable *Setaria palmifolia* (Koenig) Stapf, a species already overburdened with numerous synonyms.   It is common both in Indo-China and in the vicinity of Canton, China.

Pennisetum L. C. Richard

Pennisetum glaucum (Linn.) R. Br. Prodr. 195. 1810; Hitch. in Am. Journ. Bot. **2**: 300. 1915; Chase *op. cit.* **8**: 48. 1921.

Panicum glaucum Linn. Sp. Pl. 56. 1753.

Pennisetum typhoideum Rich. in Pers. Syn. **1**: 72. 1805.

Phleum africanum Lour. Fl. Cochinch. 48. 1790, ed. Willd. 61. 1793.

" Habitat in ora Orientali Africae, contra Mozambicum."   Loureiro gives the local name as *muxoreira;* his description applies unmistakably to the pearl millet which is extensively cultivated in tropical Africa.   I follow Stuntz (U. S. Dept. Agr. Bur. Pl. Ind. Invent. Seeds & Plants Import. **31**: 21, 84. 1914), Hitchcock (Am. Journ. Bot. **2**: 300. 1915), and Chase, " The Linnaean concept of the pearl millet " (Am. Journ. Bot. **8**: 41–49. 1921) in their interpretations of the type of *Panicum glaucum* Linn., rather than Stapf (Kew Bull. 147–149. 1928).

Pennisetum alopecuroides (Linn.) Spreng. Syst. **1**: 303. 1825.

Panicum alopecuroides Linn. Sp. Pl. 55. 1753.

Pennisetum compressum R. Br. Prodr. 195. 1810.

Alopecurus hordeiformis (non Linn.) Lour. Fl. Cochinch. 48. 1790, ed. Willd. 60. 1793.

" Habitat spontaneus in Cochinchina."   Loureiro's description applies closely to the species commonly known as *Pennisetum compressum* R. Br. which is abundant in southeastern Asia.   The type of *Panicum alopecuroides* Linn. is an actual specimen from China, probably collected by Osbeck near Canton; *Pennisetum compressum* R. Br. represents the same species.   Rendle (Journ. Linn. Soc. Bot. **36**: 343. 1904) referred *Alopecurus hordeiformis* Lour. to *Perotis latifolia* Ait. = *P. indica* (Linn.) O. Ktz.

Spinifex Linnaeus

Spinifex littoreus (Burm. f.) Merr. in Philip. Journ. Sci. Bot. **7**: 229. 1912; E. G. & A. Camus in Lecomte Fl. Gén. Indo-Chine **7**: 490. 1923.

Stipa littorea Burm. f. Fl. Ind. 29. 1768.

Stipa spinifex Linn. Mant. **1**: 34. 1767.

Spinifex squarrosus Linn. Mant. **2**: 300. 1771; Lour. Fl. Cochinch. 647. 1790, ed. Willd. 794. 1793, Anamese *co chay bai bien.*

" Habitat ad maris litora in Cochinchina, & China." This characteristic species was correctly interpreted by Loureiro. *Cyperus littoreus* Rumph. (Herb. Amb. **6**: 6. *pl. 2. f. 2*), cited as a synonym, is correctly placed.

Oryza Linnaeus

Oryza sativa Linn. Sp. Pl. 333. 1753; Lour. Fl. Cochinch. 214. 1790, ed. Willd. 266. 1793, Anamese *luá,* Chinese *mêu, hô.*

Oryza communissima Lour. *l.cc.* 215, 267, Anamese *luá chính muà.*

Oryza glutinosa Lour. *l.cc* , Anamese *luá nêp,* Chinese *nó.*

Oryza montana Lour. *l.cc.,* Anamese *luá rêy.*

Oryza praecox Lour. *l.cc.,* Anamese *luá tháng tlám.*

Loureiro's descriptions were all based on cultivated forms of the common rice and all are manifestly to be reduced to *Oryza sativa.* *Oryza montana* is the so-called upland rice, grown without irrigation; *O. glutinosa* is a form the grains of which, when cooked, are very soft and glutinous; *O. praecox* is a short-awned form; *O. communissima* is a long-awned form more or less adapted to growth in brackish water. Some authors, following Loureiro, have assumed that the various species proposed by him as distinct should be maintained as valid ones. There is, however, no justification for this; they are merely a few of the myriad varieties of *O. sativa* cultivated in the Old World tropics.

Eleusine Gaertner

Eleusine indica (Linn.) Gaertn. Fruct. **1**: 8. 1788; E. G. & A. Camus in Lecomte Fl. Gén. Indo-Chine **7**: 542. 1923.

Cynosurus indicus Linn. Sp. Pl. 72. 1753; Lour. Fl. Cochinch. 59. 1790, ed. Willd. 75. 1793, Anamese *co chi tiá.*

Juncus bulbosus (non Linn.) Lour. Fl. Cochinch. 213. 1790, ed. Willd. 265. 1793, Anamese *co mang châu.*

Juncus loureiroanus Schult. & Schult. f. Syst. **7**: 238. 1829 (based on *Juncus bulbosus* Lour.).

For *Cynosurus indicus* no locality is given but the Anamese name cited indicates an Indo-China specimen. The description applies to the very common Linnaean species. *Gramen vaccinum mas* (erroneously referred to as *Gramen vaccinum foemina*) Rumph. (Herb. Amb. **6**: 9. *pl. 4. f. 1*) is correctly placed as a synonym, as is *Gramen dactyloides* Burm. (Thes. Zeyl. 106. *pl. 47. f. 1*). For *Juncus bulbosus* Loureiro states: " Habitat passim ad vias, & hortos Cochinchinae minus cultos." His imperfect description manifestly applies to *Eleusine indica* Gaertn., the " calyx 6-phyllus, foliolis distiche imbricatis " being the six distichous glumes of the individual spikelets. Buchenau (Pflanzenreich **25**(IV-36): 263. 1906) merely enumerates *Juncus loureiroanus* Schult. & Schult. f. under the *species excludendae,* without indicating what it may be, other than that no *Juncus* was represented by the description.

Dactyloctenium Willdenow

Dactyloctenium aegyptium (Linn.) Richter Pl. Europ. **1**: 68. 1889; Merr. Interpret. Herb. Amb. 94. 1917, Enum. Philip. Fl. Pl. **1**: 86. 1923.

Cynosurus aegyptius Linn. Sp. Pl. 72. 1753; Lour. Fl. Cochinch. 59. 1790, ed. Willd. 75. 1793, Anamese *co chi tláng.*

Dactyloctenium aegyptiacum Willd. Enum. Hort. Berol. 1029. 1809; E. G. & A. Camus in Lecomte Fl. Gén. Indo-Chine 7: 544. 1923.

No locality is given but the Anamese name cited indicates an Indo-China specimen. Loureiro's description applies to the Linnaean species. *Gramen vaccinum* Rumph. (Herb. Amb. 6: 9. *pl. 4. f. 2*), cited by Loureiro as a synonym, is correctly placed.

Leptochloa Beauvois

Leptochloa chinensis (Linn.) Nees in Syll. Ratisb. 1: 4. 1824.

Poa chinensis Linn. Sp. Pl. 69. 1753; Lour. Fl. Cochinch. 54. 1790, ed. Willd. 69. 1793, Anamese *co duói phung;* E. G. & A. Camus in Lecomte Fl. Gén. Indo-Chine 7: 545. 1923.

Poa decipiens R. Br. Prodr. 181. 1810.

Leptochloa decipiens Druce Rept. Bot. Exch. Club Brit. Isles 4: 632. 1917.

" Habitat in agris Cochinchinae." Loureiro's description conforms entirely with the characters of the Linnaean species and to no other grass known from Indo-China. The Linnaean type was a specimen collected by Osbeck near Canton, China.

Phragmites Trinius

Phragmites maximus (Forsk.) Chiov. in Nuov. Giorn. Bot. Ital. 26: 80. 1919.

Arundo maxima Forsk. Fl. Aeg.-Arab. 24. 1775.

Arundo phragmites Linn. Sp. Pl. 81. 1753.

Arundo vulgaris Bauh. ex Lam. Fl. Fr. 3: 615. 1778.

Phragmites communis Trin. Fund. Agrost. 134. 1820.

Phragmites vulgaris Crép. Man. Fl. Belg. ed. 2, 345. 1866; Pilger in Perk. Frag. Fl. Philip. 147. 1904; Merr. Enum. Philip. Fl. Pl. 1: 87. 1923; E. G. & A. Camus in Lecomte Fl. Gén. Indo-Chine 7: 550. 1923.

Trichoon phragmites Schinz & Thell. in Viert. Nat. Ges. Zürich 53: 587. 1908; Nieuwl. in Am. Midl. Nat. 3: 332. 1914.

Arundo donax (non Linn.) Lour. Fl. Cochinch. 55. 1790, ed. Willd. 69. 1793, Anamese *cây cuoi.*

Aira arundinacea (non Linn.) Lour. Fl. Cochinch. 54. 1790, ed. Willd. 68. 1793, Anamese *cây sây.*

For *Aira arundinacea* Loureiro states: " Habitat coenosa, Cochinchinae." The description in general, although very inadequate, applies sufficiently well to *Phragmites* to warrant this reduction. The local names given by E. G. & A. Camus for *Phragmites karka* Trin., who make no attempt to identify Loureiro's species, are *sây* and *cay ke sây*, which corroborate this reduction, Loureiro's local name being *cây sây. Canna palustris* Rumph. (Herb. Amb. 4: 20. *pl. 5*), cited by Loureiro as representing his species, is a form of *Phragmites maximus* (Forsk.) Chiov. For *Arundo donx* Loureiro states: " Habitat loca humida in Cochinchina." This he described as 10 feet high, which is rather more a character of *Phragmites karka* Trin. than of *P. maximus* Chiov. *Phragmites karka* Trin. may be but a form of the very widely distributed and variable *P. maximus* (Forsk.) Chiov. No agrostologist has as yet clearly indicated constant distinctive characters between the two supposed species; Hooker's long note (Fl. Brit. Ind. **7**: 304. 1897) should be consulted. In explanation of the acceptance of Forskål's specific name for this collective species, it is to be noted that it was published in 1775, and Muschler (Man. Fl. Egypt **1**: 116. 1912) places it as a synonym of *Phragmites communis* Trin. var. *isiaca* (Del.) Coss. The binomial *Arundo vulgaris* appears in Lamarck (Fl. Fr. **3**: 615. 1778) there accredited to Bauhlin (Theatr. 69); it is three years later than Forskål's name. As to *Phragmites vulgaris* (Lam.) which I had previously accepted, after Pilger, taking Trinius as the authority from Index Kewensis, Trinius does not publish this (Fund. Agrost. 134. 1820) as indicated in Index Kewensis; the only binomial there is *Phragmites communis* Trin. Chiovenda's publication of the accepted binomial is merely: "*Phragmites maxima* (Forsk.) (= *Ph. isiacus* Coss.)." Crépin was apparently the first botanist to publish the combination *Phragmites vulgaris*.

Centotheca [39] Desvaux

Centotheca latifolia (Osbeck) Trin. Fund. Agrost. 141. 1820.

> *Holcus latifolius* Osbeck Dagbok Ostind. Resa 247. 1757; Linn. Syst. ed. 10, 1305. 1759, Sp. Pl. ed. 2, 1486. 1763.
>
> *Cenchrus lappaceus* Linn. Sp. Pl. ed. 2, 1488. 1763.
>
> *Centosteca lappacea* Desv. in Nuov. Bull. Soc. Philom. **2**: 189. 1810.
>
> *Centotheca lappacea* Desv. in Journ. Bot. Desvaux **1**: 70. 1813; E. G. & A. Camus in Lecomte Fl. Gén. Indo-Chine **7**: 577. 1923.
>
> *Poa latifolia* Forst. f. Prodr. 8. 1786.
>
> *Anthoxanthum pulcherrimum* Lour. Fl. Cochinch. 29. 1790, ed. Willd. 36. 1793, Ana-mese *co thía thía*.

"Habitat in agris et hortis Cochinchinae, incultum." Moore's examination of Loureiro's type in the herbarium of the British Museum (Journ. Bot. **63**: 246. 1925) adds another synonym to the already over-long list for this characteristic and widely distributed species, E. G. & A. Camus enumerating no less than 21 synonyms. No attempt was made by them to account for *Anthoxanthum pulcherrimum* Lour. in their treatment of the grasses of Indo-China (Lecomte Fl. Gén. Indo-Chine **7**: 202–650. 1922–23). It should be noted in passing that Moore is in error in following Hooker f. in citing *Poa malabarica* Linn. as a synonym of this *Centotheca*. *Poa malabarica* Linn. if interpreted by the Hortus Malabricus illustration cited, which I consider to be the type, is the same as the grass currently referred to *Diplachne fusca* (Linn.) Beauv. = *Diplachne malabarica* (Linn.) Merr. in Bull. Torr. Bot. Club **60**: 635. 1933; it is certainly not the same as *Panicum arnottianum* as I suspected when I proposed for the latter the new binomial *Panicum malabaricum* (Linn.) Merr. (Philip. Journ. Sci. **4**: Bot. 248. 1909). If interpreted by the specimen collected by Osbeck near Canton, which was in the Linnaean herbarium in 1753, and was named *Poa malabarica* Linn. by Linnaeus himself, then *Poa malabarica* Linn. = *Hemigymnia malabarica* Henr. (Meded. Rijks Herb. Leiden **61**: 12. 1930) = *Ottochloa malabarica* Dandy (Journ. Bot. **69**: 55. 1931). It is to be noted in reference to the binomial *Centotheca latifolia* (Linn.) Trin., that in originally publishing this name Trinius cites only the synonym

---

[39] Desvaux (Nuov. Bull. Soc. Philom. **2**: 187–190. 1810) proposed three new genera of grasses, *Heterosteca, Calosteca,* and *Centosteca;* in the somewhat amplified republication of the paper three years later (Journ. Bot. Desvaux **1**: 63–77. 1813) he left *Heterosteca* unchanged but altered *Calosteca* and *Centosteca* to *Calotheca* and *Centotheca* respectively.

Cenchrus lappaceus Linn. but in a later publication (Mém. Acad. St. Pétersb. Math. Phys. Nat. **1**: 358. 1830), as indicated to me by Doctor Hitchcock, he cites under *Uniola lappacea* Trin. (= *Centotheca lappacea* = *C. latifolia*), *Holcus latifolius* Linn. In my preliminary manuscript of 1919, working from Loureiro's description alone and influenced by his discussion of *Gramen fumi* Rumph. (Herb. Amb. **6**: 11. *pl. 4. f. 3*) I erroneously referred Loureiro's species to *Eragrostis amabilis* (Linn.) W. & A. which is the species represented by Rumphius' illustration.

Triticum Linnaeus

For wheat, *Triticum*, Loureiro, Fl. Cochinch. 59. 1790, ed. Willd. 75. 1793, gives the Anamese name *luá mì* and the Chinese name *mĕ*.

Hordeum Linnaeus

For barley, *Hordeum*, Loureiro, Fl. Cochinch. 59. 1790, ed. Willd. 75. 1793, gives the Anamese name *mach nha*, and the Chinese name *mêu*. He gives no generic descriptions and no binomials under either *Triticum* or *Hordeum*, stating: " Tam Tritici, quam Hordei variae species, & vulgares quidem nascuntur in China, quarum flores non examinavi."

Bambusa Schreber [40]

Bambusa bambos (Linn.) Druce in Rept. Bot. Exch. Club Brit. Isles **4**: 608. 1917.
  Arundo bambos Linn. Sp. Pl. 81. 1753.
  Bambos arundinacea Retz. Obs. **5**: 24. 1789.
  Bambusa arundinacea Willd. Sp. Pl. **2**: 245. 1799; E. G. & A. Camus in Lecomte Fl. Gén. Indo-Chine **7**: 606. 1923.
  Bambos maxima Poir. ex Steud. Nomencl. 100. 1821, ed. 2, **1**: 183. 1840 (based on *Arundo maxima* Lour.).
  Bambusa maxima Poir. ex Steud. Syn. **1**: 331. 1854 (based on *Arundo maxima* Lour.).
  Arundarbor maxima O. Ktz. Rev. Gen. Pl. 761. 1891 (based on *Arundo maxima* Lour.).
  Arundo maxima Lour. Fl. Cochinch. 58. 1790, ed. Willd. 74. 1793, Anamese *tle lang nga*, non *Arundo maxima* Forsk. (1775).

" Habitat praecipua inculta ingentes tractus fluminis Lavii a magno portu Bassac usque ad metropolim Cambodiae." The reference to *Arundarbor maxima* Rumph. (Herb. Amb. **4**: 12) must be excluded as this certainly does not represent the very spiny bamboo Loureiro so inadequately described; the Rumphian reference typifies *Bambusa excelsa* Miq., a species of doubtful status. Material collected by M. Parraut in central Anam, as *cay la nga* and *la nga*, an excessively spiny bamboo growing in large tufts, undoubtedly represents Loureiro's species, it being one in the group with *Bambusa spinosa* Roxb. (*B. blumeana* Schult. f.) and *B. flexuosa* Munro. *Melocanna excelsa* Roep. ex Trin. (Clav. Agrost. 105. 1822) was based on *Arundarbor maxima* Rumph. (Herb. Amb. **4**: 12) Loureiro's binomial being excluded: " *Arundo maxima* Lour. non convenit." Munro (Trans. Linn. Soc. **26**: 121. 1868) interprets *Bambusa maxima* Poir. as based solely on the Rumphian reference, stating: " Arundo maxima Lour. *l.c.* 74. ' spinis densissimis horrida differe videtur,' " but Poiret's description was based on Loureiro's. He does not publish the binomial *Bambusa maxima* (Lam. Encycl. **8**: 704. 1808) merely stating under the doubtful species: "*Bam-*

[40] *Bambusa* Schreber and *Bambos* Retzius were both published in 1789.

bos (arundo maxima) Lour. Flor. Coch., vol. 1, page 74." I place Loureiro's species as a synonym of *Bambusa bambos* largely on the basis of his statement: " Ista est ominum arundinum maxima longitudine, et crassitudine; spinisque densissimis horrida." It may be that he had only a large form of *Bambusa spinosa* Roxb. (*B. blumeana* Schult. f.), particularly in his reference to his species of Chinese forms, as *Bambusa bambos* (Linn.) Druce is not known from southern China while *B. spinosa* Roxb. is very common there.

Bambusa agrestis (Lour.) Poir. in Schult. & Schult. f. Syst. 7: 1344. 1830; Steud. Syn. 1: 330. 1854; Munro in Trans. Linn. Soc. 26: 117. 1868; E. G. Camus Bamb. 131. 1913; E. G. & A. Camus in Lecomte Fl. Gén. Indo-Chine 7: 609. 1923 (based on *Arundo agrestis* Lour.).
> *Arundo agrestis* Lour. Fl. Cochinch. 57. 1790, ed. Willd. 72. 1793, Anamese *tle rí.*
> *Bambusa agrestis* Raeusch. Nomencl. ed. 3, 103. 1797, *nomen nudum.*
> *Bambos agrestis* Poir. in Lam. Encycl. 8: 704. 1808; Steud. Nomencl. 100. 1821 (based on *Arundo agrestis* Lour.).
> *Bambusa flexuosa* Munro in Trans. Linn. Soc. 26: 101. 1868; E. G. Camus Bamb. 130. pl. *74B.* 1913; E. G. & A. Camus in Lecomte Fl. Gén. Indo-Chine 7: 608. 1923.

" Habitat agrestis in montibus, locisque desertis, & aridis per totam Cochinchinam: puto, quod etiam in China." *Arundarbor spinosa* Rumph. (Herb. Amb. 4: 14. pl. *3*), cited by Loureiro as a synonym, represents *Bambusa spinosa* Roxb. (*B. blumeana* Schult. f.). E. G. & A. Camus admit *Bambusa agrestis* Poir. as a valid species with a short description compiled from Loureiro; they think it may be near *B. flexuosa* Munro. In view of Loureiro's statement that the species occurs in all parts of Cochinchina, it is manifest that among the three spiny bamboos definitely known from Indo-China, Loureiro's species must represent either *B. arundinacea* Retz. = *B. bambos* (Linn.) Druce, *B. blumeana* Schult. f. = *B. spinosa* Roxb., or *B. flexuosa* Munro. There is much confusion in the distinctive characters of this small group of Asiatic spiny bamboos, including *Bambusa bambos* (Linn.) Druce (*B. arundinacea* Willd.), *B. spinosa* Roxb. (*B. blumeana* Schult. & Schult. f.), *B. flexuosa* Munro (type from Kwangtung), and *B. stenostachya* Hack. (type from Formosa). It is suspected that *Bambusa flexuosa* Munro may prove to be but a form of *B. spinosa* Roxb.; at least the distinctive characters given by E. G. & A. Camus are not convincing. Raeuschel (Nomencl. ed. 3, 103. 1797) lists the binomial *Bambusa agrestis* but failed to indicate its name-bringing synonym; doubtless *Arundo agrestis* Lour. was its basis.

Bambusa fax (Lour.) Poir. ex Steud. Nomencl. ed. 2, 1: 183. 1840 (based on *Arundo fax* Lour.).
> *Arundo fax* Lour. Fl. Cochinch. 58. 1790, ed. Willd. 74. 1793, Anamese *tle núa.*
> *Bambos fax* Poir. ex Steud. Nomencl. 100. 1821 (based on *Arundo fax* Lour.).
> *Beesha fax* Schult. & Schult. f. Syst. 7: 1336. 1830 (based on *Arundo fax* Lour.).
> *Melocanna humilis* Roep.[41] ex Trin. Clav. Agrost. 105. 1822 (based in part on *Arundo fax* Lour.).
> *Beesha humilis* Kunth Enum. 1: 434. 1833 (based in part on *Arundo fax* Lour.).

---

[41] *Melocanna humilis* Roep. and *Beesha humilis* Kunth are essentially based on *Arundarbor cratium* Rumph. (Herb. Amb. 4: 5), *Arundo fax* Lour., after Loureiro himself, being erroneously cited as a synonym under both.

Loureiro's very brief description was based on an Indo-China specimen. *Arundarbor cratium* Rumph. (Herb. Amb. **4**: 5), cited by Loureiro as a synonym, must be excluded as it represents *Schizostachyum brachycladum* Kurz = *Schizostachyum lima* (Blanco) Merr. *Melocanna humilis* Kurz, as interpreted and described by Gamble (Ann. Bot. Gard. Calcutta **7**: 120. *pl. 106*. 1896) and E. G. Camus (Bamb. 180. 1913) has nothing to do with Loureiro's species. *Bambusa fax* is not published by Poiret (Lam. Encycl. **8**: 704. 1808); he merely states under the doubtful species: "*Bambos* (arundo fax) Lour. Flor. Cochinch., vol. 1, p. 74."

Bambusa multiplex (Lour.) Raeusch.[42] Nomencl. ed. 3, 103. 1797, *nomen nudum;* Schult. & Schult. f. Syst. **7**: 1350. 1830 (based on *Arundo multiplex* Lour.); Merr. Enum. Philip. Fl. Pl. **1**: 94. 1923.
   Arundo multiplex Lour. Fl. Cochinch. 58. 1790, ed. Willd. 73. 1793, Anamese *cây hóp.*
   Ludolphia glaucescens Willd. in Ges. Naturf. Fr. Berl. Mag. **2**: 320. 1808.
   Bambusa nana Roxb. Hort. Beng. 25. 1814, *nomen nudum,* Fl. Ind. ed. 2, **2**: 199. 1832; E. G. & A. Camus in Lecomte Fl. Gén. Indo-Chine **7**: 598. 1923.
   Arundinaria glaucescens Beauv. Agrost. 144, 152. 1812.
   Bambusa glaucescens Sieb. ex Munro in Trans. Linn. Soc. **26**: 89. 1868, in syn.
   Arundarbor multiplex O. Ktz. Rev. Gen. Pl. 761. 1891 (based on *Arundo multiplex* Lour.).

" Habitat loca culta in provinciis borealibus Cochinchinae, ex eaque plantantur sepes ad divisionem hortorum." Loureiro's species has remained one of doubtful status from the time his description was published. Messrs. Dodo and Parraut kindly secured material for me under the local name *cây hóp,* and this, which agrees with Loureiro's description, proves to be the species commonly known as *Bambusa nana* Roxb. It is commonly planted for hedges in China and in the Indo-Malaysian region, the culms sometimes reaching a height of 4 to 5 m., but more commonly only 2 to 3 m. high. *Arundarbor tenuis* Rumph. (Herb. Amb. **4**: 1. *pl. 1*), cited by Loureiro as representing his species, must be excluded as it is the entirely different *Bambusa atra* Lindl. (*B. rumphiana* Kurz). E. G. & A. Camus (Lecomte Fl. Gén. Indo-Chine **7**: 611. 1923) admit *Bambusa multiplex* Raeusch. with a short description compiled from Loureiro, stating: " Proche du *B. nutans* Wall."

Bambusa spinosa Roxb. Hort. Beng. 25. 1814; Merr. Interpret. Herb. Amb. 97. 1917, Enum. Philip. Fl. Pl. **1**: 94. 1923.
   Bambusa spinosa Blume ex Nees in Flora **8**: 580. 1825.
   Bambusa blumeana Schult. & Schult. f. Syst. **7**: 1343. 1830; E. G. & A. Camus in Lecomte Fl. Gén. Indo-Chine **7**: 607. 1923.
   Arundo bambu (non *Arundo bambos* Linn.) Lour. Fl. Cochinch. 56. 1790, Anamese *tle vuòn,* Chinese *yĕ chŏ.*
   Arundo bambos (non Linn.) Lour. Fl. Cochinch. ed. Willd. 70. 1793.

" Habitat ubique culta in Cochinchina, & China, Tunkino, & Cambodia." *Tabaxir mambu arbor* Rheed. (Hort. Malabar. 1: 25. *pl. 16*), cited by Loureiro, represents *Bambusa arundinacea* Willd. = *B. bambos* (Linn.) Druce. *Arundarbor vasaria,* Bulu Java Rumph. (Herb. Amb. **4**: 8) may represent *Bambusa vulgaris* Schrad. *Arundo bambos* Linn. is *Bam-*

---

[42] Raeuschel (Nomencl. ed. 3, 103. 1797) merely gives the name *Bambusa multiplex;* he does not indicate on what binomial it was based.

busa bambos (Linn.) Druce.   The indicated characters and uses of *Arundo bambu* Lour. seem to apply to the very common, spiny species which is widely distributed in southeastern Asia and Malaysia and which is generally known as *Bambusa blumeana* Schult. & Schult. f.   Specimens collected in Anam by M. Parraut as *tle viron* or *tre viron* (i.e., garden bamboo), now more commonly known as *tre nha* (i.e., house bamboo), agree with Loureiro's description, and, as far as can be determined from sterile material, with the Malaysian *Bambusa spinosa* Roxb. (*B. blumeana* Schult. f.); two collections made by M. Dodo in northern Anam as *tre hoa* also represent the same species.   *Bambusa spinosa* Roxb. (Hort. Beng. 25. 1814) was based wholly on *Arundarbor spinosa* Rumph. (Herb. Amb. **4**: 14. *pl. 3*); see Merrill, E. D. (Interpret. Herb. Amb. 97. 1917).   Probably Roxburgh intended to name the common Indian spiny bamboo, *Bambusa arundinacea* Willd. = *B. bambos* (Linn.) Druce, but what he actually and validly named in 1814 is *B. blumeana* Schult. & Schult. f.

Bambusa tabacaria (Lour.) Schult. & Schult. f. Syst. **7**: 1351. 1830; Steud. Syn. **1**: 331. 1854; Munro in Trans. Linn. Soc. **26**: 121. 1868; E. G. Camus Bamb. 134. 1913; E. G. & A. Camus in Lecomte Fl. Gén. Indo-Chine **7**: 611. 1923 (based on *Arundo tabacaria* Lour.).

   Bambos tabacaria Poir. ex Steud. Nomencl. 100. 1821 (based on *Arundo tabacaria* Lour.).

   Arundo tabacaria Lour. Fl. Cochinch. 58. 1790, ed. Willd. 74. 1793, Anamese *oúng thaóng.*

   Arundarbor tabacaria O. Ktz. Rev. Gen. Pl. 761. 1891 based on *Arundo tabacaria* Lour.).

   Loureiro's very short description was based on a Cochinchina specimen.   He discusses *Arundarbor spiculorum* Rumph. (Herb. Amb. **4**: 7) under his species, but this must be excluded as it typifies *Schizostachyum longinodis* Miq. and is perhaps a form of *S. brachycladum* Kurz = *Schizostachyum lima* (Blanco) Merr.   Poiret does not publish the binomial *Bambusa tabacaria* (Lam. Encycl. **8**: 705. 1808) merely stating under the doubtful species: "*Bambos* (arundo tabacaria) Lour. Fl. Cochinch., vol. 1, page 74."   Further field work in Indo-China will be necessary to throw any further light on this very imperfectly known species, which is doubtless described under some other name in the Camus' treatment of the bamboos of Indo-China.

Dendrocalamus Nees

Dendrocalamus flagellifer Munro in Trans. Linn. Soc. **26**: 150. 1868; E. G. & A. Camus in Lecomte Fl. Gén. Indo-Chine **7**: 637. 1923.

   ? Arundo piscatoria Lour. Fl. Cochinch. 55. 1790, ed. Willd. 70. 1793, Anamese *cây trúc.*

   ? Calamagrostis piscatoria Steud. Syn. **1**: 193. 1854 (based on *Arundo piscatoria* Lour.).

   " Habitat agrestis, cultaque in Cochinchina."   Loureiro's species is not mentioned by Munro or by Camus in their monographic treatments of the Bambusae, although Loureiro's description is of a bamboo.   His statement regarding the leaves " omnium arundinum minima " clearly indicates a bamboo, as he was comparing them with those of *Phragmites* and *Thysanolaena.*   He notes that the culms were especially valued for making fishing rods.   Three sterile specimens collected in Anam by Messrs. Dodo and Parraut under the local names *truc* and *cay truc* all represent a single species and this apparently a *Dendrocalamus* similar to *D. flagellifer* Munro.

Dendrocalamus sp.

Arundo mitis Lour. Fl. Cochinch. 57. 1790, ed. Willd. 73. 1793, Anamese *cây mò ho.*

Bambusa mitis Raeusch.[43] Nomencl. ed. 3, 103. 1797, *nomen nudum.*

Bambos mitis Poir. in Lam. Encycl. **8**: 704. 1808 (based on *Arundo mitis* Lour.).

Bambusa mitis Schult. & Schult. f. Syst. **7**: 1351. 1830; Steud. Syn. **1**: 330. 1854; Munro in Trans. Linn. Soc. **26**: 119. 1868; E. G. Camus Bamb. 132. 1913; E. G. & A. Camus in Lecomte Fl. Gén. Indo-Chine **7**: 611. 1923 (based on *Arundo mitis* Lour.).

Phyllostachys mitis A. & C. Rivière in Bull. Soc. Acclim. III **5**: 697. 1878 (based on *Arundo mitis* Lour.).

" Habitat culta, non frequens in agris, & saepibus Cochinchinae." *Arundarbor fera* Rumph. (Herb. Amb. **4**: 16. *pl. 4*), cited by Loureiro as a synonym, must be excluded as I take this to represent *Bambusa vulgaris* Schrad. Specimens from Anam collected by Messrs. Dodo and Parraut under the local name *cay lo-o*, which agree with Loureiro's description, apparently represent a species of *Dendrocalamus*, but I am unable to place it more definitely from the data at present available.

GRAMINEAE OF UNCERTAIN GENERIC STATUS

Agrostis odorata Lour. Fl. Cochinch. 50. 1790, ed. Willd. 64. 1793, Anamese *cây hoa co.*

" Habitat prope litora in Cochinchina." The description is short and indefinite and certainly does not appertain to the genus *Agrostis.* Loureiro states that the whole plant was esteemed by the Anamese on account of its fragrance, and that it was used for perfuming clothes. E. G. & A. Camus make no attempt to identify it in their treatment of the Gramineae of Indo-China (Lecomte Fl. Gén. Indo-Chine **7**: 202–650. 1912–1913). M. Crevost kindly had material collected for me under Loureiro's Anamese name from the beach at Cana and at Long Hai. Both specimens represent *Andropogon pertusus* (L.) Willd. M. Crevost states that the grass is no longer used for the purposes indicated by Loureiro. It is suspected that Loureiro actually did have basal leaf tufts of some of the aromatic species of *Andropogon*, adding erroneous *Agrostis* characters to his description.

Arundo dioica Lour. Fl. Cochinch. 55. 1790, ed. Willd. 70. 1793, Anamese *cây bac mai.*

Calamagrostis dioica Steud. Syn. **1**: 193. 1854 (based on *Arundo dioica* Lour.).

" Habitat in sylvis Cochinchinae." Three sterile collections by M. Dodo and one by M. Parraut in central and northern Anam as *cay bac mai* and *tre bac mai* represent three different species of Bambusae. One is the same as *cay truc* and is *Arundo piscatoria* Lour., the culms being noted by M. Parraut as being especially useful as fishing rods because of their exceptional strength and flexibility. The other two probably represent two different species of *Dendrocalamus*, as far as can be determined from sterile material. Loureiro's species may be some coarse grass in a genus unrelated to the Bambusae.

Nardus indica (non Linn.) Lour. Fl. Cochinch. 44. 1790, ed. Willd. 56. 1793, Anamese *cam tung huong*, Chinese *cām sām hiàm.*

" Habitat in montibus Occidentalibus imperii Sinensis." Loureiro had only fragmentary material, unquestionably secured from an herbalist; he saw no flowers. From the very brief description it is impossible to determine the status of what he tried to describe,

---

[43] Raeuschel here lists *Bambusa mitis* but failed to cite the binomial on which it was based.

although it is safely not *Nardus indica* Linn. ≡ *Microchloa indica* (Linn.) O. Ktz. (*M. setacea* R. Br.). From the local names cited and the long discussion of the medicinal uses of the plant, it may at some time be possible to determine its status; it may prove not to belong in the Gramineae.

Phalaris zizanoides (non Linn.) Lour. Fl. Cochinch. 49. 1790, ed. Willd. 62. 1793, Anamese *co chuóc mât.*

"Habitat spontanea in Cochinchina." The description is altogether too indefinite to warrant a reduction. In some respects *Andropogon* is suggested.

CYPERACEAE

Cyperus (Micheli) Linnaeus

Cyperus articulatus Linn. Sp. Pl. 44. 1753; ? Lour. Fl. Cochinch. 41. 1790, ed. Willd. 53. 1793; E. G. Camus in Lecomte Fl. Gén. Indo-Chine **7**: 66. 1912.

"Habitat in locis humidis Cochinchinae." The description is very short and imperfect and may or may not appertain to the Linnaean species. Camus admits the species in his treatment of the Cyperaceae of Indo-China solely on Clarke's Siam record and that of Loureiro; he apparently saw no specimens from Indo-China from which one would infer that Loureiro may not have had specimens representing the Linnaean species.

Cyperus compressus Linn. Sp. Pl. 46. 1753; Lour. Fl. Cochinch. 43. 1790, ed. Willd. 54. 1793, Anamese *co cú lép;* E. G. Camus in Lecomte Fl. Gén. Indo-Chine **7**: 57. 1912.

"Habitat spontanea in Cochinchinae " (under no. 6, p. 43). Loureiro describes the styles as 2-fid, which, if correct, would indicate a species *Pycreus.* On the whole, judging from the short description, it is probable that Loureiro had specimens of the very common and widely distributed Linnaean species. Camus, without citing Loureiro as its source, gives the Anamese name *ca cu lep* for *Cyperus compressus* Linn.

Cyperus malaccensis Lam. Ill. **1**: 146. 1791; E. G. Camus in Lecomte Fl. Gén. Indo-Chine **7**: 63. 1912.

Cyperus elatus (non Linn.) Lour. Fl. Cochinch. 42. 1790, ed. Willd. 54. 1793, Anamese *lác làm chiéo.*

"Habitat spontanea in Cochinchina " (under no. 6, p. 43), and in the description "planta palustris." The description is short and imperfect, but probably *Cyperus malaccensis* Lam. is the species he had in hand. Loureiro notes that the split stems were used in making mats.

Cyperus pilosus Vahl Enum. **2**: 354. 1806; E. G. Camus in Lecomte Fl. Gén. Indo-Chine **7**: 62. 1912.

Cyperus monti (non Linn. f.) Lour. Fl. Cochinch. 42. 1790, ed. Willd. 53. 1793, Anamese *cây lác tlòn.*

"Habitat in Cochinchina." The description is short and imperfect. *Cyperus pilosus* Vahl is undoubtedly the form represented by it, certainly not *C. monti* Linn. f., as the description, as far as it goes, applies unmistakably to Vahl's species, which is a very common and widely distributed one.

Cyperus rotundus Linn. Sp. Pl. 45. 1753; Lour. Fl. Cochinch. 42. 1790, ed. Willd. 53. 1793, Anamese *co cú, huog phu*, Chinese *hiàm phú cù*; E. G. Camus in Lecomte Fl. Gén. Indo-Chine **7**: 69. 1912.

" Habitat ubique in Cochinchina, & China." Loureiro's description applies unmistakably to the very common and widely distributed Linnaean species, this being also verified by his description of the characteristic tubers and their medicinal uses.

Cyperus uncinatus Poir. in Lam. Encycl. **7**: 247. 1806; E. G. Camus in Lecomte Fl. Gén. Indo-Chine **7**: 47. 1912.

Cyperus pumilus (non Linn.) Lour. Fl. Cochinch. 43. 1790, ed. Willd. 54. 1793, Anamese *co cú chít.*

" Habitat spontanea in Cochinchina." The short description, especially in the phrase " glumis mucronatis, apice reflexis " applies to *Cyperus uncinatus* Poir. rather than to *Cyperus pumilus* Linn. = *Pycreus nitens* Nees.

Kyllinga Rottboell

Kyllinga monocephala Rottb. Descr. Ic. Pl. 13. *pl. 4.* 1773; E. G. Camus in Lecomte Fl. Gén. Indo-Chine **7**: 25. 1912.

Scirpus cephalotes Jacq. Hort. Vind. **1**: 42. *pl. 97.* 1770, non Linn. 1762.

Kyllinga cephalotes Druce Rept. Bot. Exch. Club Brit. Isles **4**: 630. 1917.

Schoenus coloratus (non Linn.) Lour. Fl. Cochinch. 41. 1790, ed. Willd. 52. 1793, Anamese *co dĕù dĕù.*

" Habitat ubique in hortis, viis, & agris Cochinchinae." Willdenow notes that Loureiro's description appertained to a *Kyllinga*. I believe the species to be the very common and widely distributed *K. monocephala* Rottb., rather than the equally common and closely allied *K. brevifolia* Rottb. because Loureiro describes the inflorescences as white; in *K. brevifolia* Rottb. they are usually green. *Gramen capitatum* Rumph. (Herb. Amb. **6**: 8. *pl. 3. f. 2*), cited by Loureiro as representing his species, is *Kyllinga monocephala* Rottb.

Scirpus (Tournefort) Linnaeus

Scirpus erectus Poir. in Lam. Encycl. **6**: 761. 1804; E. G. Camus in Lecomte Fl. Gén. Indo-Chine **7**: 136. 1912.

Scirpus supinus (non Linn.) Lour. Fl. Cochinch. 43. 1790, ed. Willd. 55. 1793, Anamese *cây năn thâp.*

No locality is given but Indo-China is indicated by the local name cited. The description is not particularly good but seems to agree better with the characters of *Scirpus erectus* Poir. than with those of the Linnaean species. Both *Scirpus supinus* Linn. and *S. erectus* Poir. are apparently common in Indo-China and both occur near Hue.

Heleocharis [44] R. Brown

Heleocharis dulcis (Burm. f.) Trin. ex Henschel Vita Rumph. 186. 1833; Merr. Interpret. Herb. Amb. 104. 1917, Enum. Philip. Fl. Pl. **1**: 119. 1923.

[44] Robert Brown's original spelling was *Eleocharis*. The corrected form *Heleocharis* is accepted in accordance with the additions to the list of conserved names adopted by the Cambridge Botanical Congress.

Andropogon dulcis Burm. f. Fl. Ind. 219. 1768.

Hippuris indica Lour. Fl. Cochinch. 16. 1790, ed. Willd. 21. 1793, Anamese *cây năn.*

Eleocharis plantaginea R. Br. Prodr. 224. 1810; E. G. Camus in Lecomte Fl. Gén. Indo-Chine **7**: 82. *f. 12. 3–6.* 1912 (*Heleocharis*).

Scirpus plantaginoides Rottb. Descr. Nov. Pl. 45. 1773.

Scirpus plantagineus Retz. Obs. **5**: 14. 1789.

Eleocharis plantaginoidea W. F. Wight in Contr. U. S. Nat. Herb. **9**: 268. 1905.

Eleocharis indica Druce Rept. Bot. Exch. Brit. Isles **4**: 621. 1917 (based on *Hippuris indica* Lour.).

" Habitat loca aquosa Cochinchinae." Loureiro's description clearly applies to the wild, tuber-bearing, widely distributed Indo-Malaysian form, described and figured by Rumphius as *Cyperus dulcis* (Herb. Amb. **6**: 7. *pl. 3. f. 1*) and which Loureiro cites as a synonym of his species. The Rumphian description is the whole basis of the binomial here accepted. *Heleocharis tuberosa* (Roxb.) Schult. (Roem. & Schult. Mant. **2**: 86. 1824) which is the extensively cultivated *maa tai* (water chestnut) of Kwangtung Province, is probably a form of the same species, improved by selection and cultivation for its edible tubers. This cultivated form apparently produces flowers only rarely.

Fimbristylis [45] Vahl

Fimbristylis miliacea (Linn.) Vahl Enum. **2**: 287. 1806; E. G. Camus in Lecomte Fl. Gén. Indo-Chine **7**: 115. 1912.

Scirpus miliaceus Linn. Syst. ed. 10, 868. 1759; Lour. Fl. Cochinch. 43. 1790, ed. Willd. 55. 1793, Anamese *co rát.*

No locality is given, but the native name clearly indicates Indo-China material. Apparently Loureiro had a somewhat dwarfed form of the very common Linnaean species. Burman's illustration (Fl. Ind. 22. *pl. 9. f. 2.* 1768), cited by Loureiro, is a good representation of *Fimbristylis miliacea* Vahl.

Fimbristylis setacea Benth. in Hook. Lond. Journ. Bot. **2**: 239. 1843; E. G. Camus in Lecomte Fl. Gén. Indo-Chine **7**: 96. 1912.

Eriocaulon setaceum (non Linn.) Lour. Fl. Cochinch. 60. 1790, ed. Willd. 77. 1793, Anamese *co chát.*

" Habitat loca humida Cochinchinae." The description is sufficiently definite to indicate a *Fimbristylis* of the section *Neodichelostachys* but the species might with almost equal propriety be referred to *F. acuminata* Vahl, *F. setacea* Benth., *F. nutans* Vahl, or *F. polytrichoides* Vahl, all of which occur in Indo-China. *Gramen polytrichum* Rumph. (Herb. Amb. **6**: 17. *pl. 7. f. 1*), cited by Loureiro as representing his species, is rather clearly *Fimbristylis setacea* Benth., and I have so placed the species Loureiro described under *Eriocaulon setaceum.*

[45] *Fimbristylis* Vahl (1806) conserved name, Vienna Code; an older one is *Iria* L. C. Richard (1805), changed to *Iriha* O. Kuntze (1891).

Rynchospora [46] Vahl

Rynchospora rubra (Lour.) Makino in Bot. Mag. Tokyo **17**: 180. 1903; Merr. Enum.
Philip. Fl. Pl. **1**: 130. 1923 (based on *Schoenus ruber* Lour.).
Schoenus ruber Lour. Fl. Cochinch. 41. 1790, ed. Willd. 52. 1793, Anamese *co duoi luon.*
Rynchospora wallichiana Kunth Enum. **2**: 289. 1837; C. B. Clarke in Hook. f. Fl. Brit.
Ind. **6**: 668. 1893; E. G. Camus in Lecomte Fl. Gén. Indo-Chine **7**: 145. 1912.
Morisia wallichii Nees in Edinb. New Philos. Journ. **17**: 265. 1834.
Rynchospora wallichii K. Schum. Fl. Kais. Wilh. Land 25. 1889.
" Habitat spontaneus in Cochinchina." Loureiro's short description applies unmistakably to the very common and widely distributed species currently known as *Rynchospora wallichiana* Kunth, and his specific name, being older than that of Kunth, should be retained.

Rynchospora wightiana (Nees) Steud. Syn. Pl. Glum. **2**: 148. 1855; C. B. Clarke in Hook. f.
Fl. Brit. Ind. **6**: 669. 1893; E. G. Camus in Lecomte Fl. Gén. Indo-Chine **7**: 146.
pl. 1. f. B 3–5. 1912.
Haplostylis wightiana Nees in Nov. Act. Acad. Nat. Cur. **19**: Suppl. **1**: 101. 1843.
Bobartia indica (non Linn.) Lour. Fl. Cochinch. 45. 1790, ed. Willd. 58. 1793, Anamese
co gà.
No locality is given but the Anamese name cited clearly indicates that Loureiro based his description on Indo-China material. The Linnaean species is the iridaceous *Moraea spathacea* Willd. of South Africa (Trimen Fl. Ceyl. **5**: 24. 1900). Loureiro's description applies unmistakably to *Rynchospora wightiana* Steud. No previous reduction of *Bobartia indica* Lour. has been suggested.

Scleria Bergius

Scleria terrestris (Linn.) Fassett in Rhodora **26**: 159. 1924.
Zizania terrestris Linn. Sp. Pl. 991. 1753.
Diaphora cochinchinensis Lour. Fl. Cochinch. 578. 1790, ed. Willd. 709. 1793, Anamese
lách khên.
Scleria cochinchinensis Druce in Rept. Bot. Exch. Club Brit. Isles **4**: 646. 1917; Pfeiffer
in Fedde Repert. **26**: 263. 1929 (based on *Diaphora cochinchinensis* Lour.).
Scleria elata Thw. Enum. Pl. Zeyl. 353. 1864; E. G. Camus in Lecomte Fl. Gén. Indo-
Chine **7**: 167. 1912.
Olyra orientalis Lour. Fl. Cochinch. 552. 1790, ed. Willd. 674. 1793, Anamese *cây lách
khên.*
For *Scleria cochinchinensis* Loureiro states: " Habitat inculta in agris Cochinchinae."
His description unmistakably applies to the species generally known as *Scleria elata* Thw.,
for which much older names are available. The transfer to *Scleria* as *S. cochinchinensis*
was made independently by Druce and by Pfeiffer twelve years apart. Loureiro's type is
preserved in the herbarium of the British Museum. For *Olyra orientalis* Loureiro states:

[46] *Rhynchospora* (*Rynchospora* Vahl 1806, corr. Willdenow 1809), conserved name, Vienna Code; an
older one is *Triodon* L. C. Richard (1805). The original spelling is here retained; see Sprague, Kew Bull.
359. 1928.

" Habitat agrestis in Cochinchina." C. B. Clarke reduced this to *Scleria lithosperma* Sw. in which he was followed by E. G. Camus (Lecomte Fl. Gén. Indo-Chine 7: 161. 1912). Loureiro's description, however, applies to a very different species in its size, 4 feet high, in its " aristate " lower glumes, and especially in its fruit characters, " trigono-rotundum, asperum . . . magnum " which are not characters of Swartz's species. Its Anamese name *cây lách khen* is essentially the same as that of *Scleria terrestris* Fassett; possibly some other coarse species is represented such as *Scleria scrobiculata* Nees or *S. multifoliata* Boeckl.

PALMAE

Phoenix Linnaeus

Phoenix loureiri Kunth Enum. **3**: 257. 1841 (based on *P. pusilla* Lour.).

> *Phoenix pusilla* (non Gaertn.) Lour. Fl. Cochinch. 614. 1790, ed. Willd. 753. 1793, Anamese *cây cha la.*
>
> *Zelonops pusilla* Raf. Fl. Tellur. **2**: 102. 1837 (based on *Phoenix pusilla* Lour.).
>
> *Phoenix humilis* Royle var. *loureirii* Becc. Malesia **3**: 348, 379, 382. *pl. 44. f. 16–17.* 1890 (based on *P. pusilla* Lour.).

" Habitat agrestis in montibus *Côn mít*, 6 leucis distantibus a metropoli Cochinchinae Huaeâ. Amat loca petrosa prope rivos." Beccari (Webbia **3**: 238. 1910) retains *Phoenix roebelinii* O'Brien as distinct from *P. humilis* Royle, but is uncertain whether Loureiro's species belongs with one or with the other. Blatter (Palms of British India and Ceylon 20. 1926) places *Phoenix roebelini* O'Brien as a synonym of *P. humilis* Royle var. *loureirii* Becc. Chevalier (Rev. Bot. Appl. **3**: 837–839. 1923) considers *Phoenix roebelinii* O'Brien, which was introduced into cultivation from Indo-China, to be specifically distinct. His note and Miévilles' appended description, based on material collected in 1922, should be consulted. As a species the proper name is *Phoenix loureirii* Kunth; as a variety *Phoenix humilis* Royle var. *loureirii* Becc.

Rhapis Linnaeus f.

Rhapis cochinchinensis (Lour.) Mart. Hist. Nat. Palm. **3**: 254. 1849 (based on *Chamaerops cochinchinensis* Lour.); Becc. in Webbia **3**: 245. 1910; A. Chevalier in Bull. Écon. Indochine 22: 499. 1919.

> *Chamaerops cochinchinensis* Lour. Fl. Cochinch. 657. 1790, ed. Willd. 808. 1793, Anamese *cây lui.*
>
> *Rhapis laosensis* Becc. in Webbia **3**: 225. 1910.

" Habitat in sylvis Cochinchinae." Beccari thought that Loureiro's description might have been based on a mixture of material, leaves of a *Livistona* and flowers of *Rhapis*. I do not accept this, but agree with Chevalier who has given a full description of Loureiro's species, reducing to it *Rhapis laosensis* Becc.

Licuala (Rumphius) Wurmb

Licuala spinosa Wurmb in Verh. Bat. Genoots. **2**: 469. 1780; Thunb. in Vet. Akad. Handl. Stockholm 287. 1782, Nov. Gen. Pl. **3**: 70. 1782; Becc. in Webbia **3**: 240. 1910.

> *Corypha pilearia* Lour. Fl. Cochinch. 213. 1790, ed. Willd. 265. 1793, Anamese *cây lá lip.*

Licuala pilearia Blume Rumphia **2**: 42. 1836 (based on *Corypha pilearia* Lour.).
Licuala spinosa Wurmb var. *cochinchinensis* Becc. Malesia **3**: 74. 1886.
" Habitat in sylvis Cochinchinae." I am not certain that *Licuala pilearia* (Lour.)
Blume is properly referable to *L. spinosa* Wurmb, as *Clemens 3356, 4444*, from Mount
Bana, near Tourane, and *Squires 203* from Hue, may represent Loureiro's species. The
material is rather too imperfect for me to determine whether or not *L. spinosa* Wurmb is
represented by these specimens. *Robinson 1400*, from Nhatrang, some distance south of
Hue, apparently represents Wurmb's species. In proposing the variety *cochinchinensis*
Beccari stated that it scarcely differed from the species, and later (Webbia **3**: 240. 1910)
reduced it to *Licuala spinosa* Wurmb. Loureiro cites *Licuala* Rumph. (Herb. Amb. **1**:
44. *pl. 9*) = *Licuala rumphii* Blume and probably = *Licuala spinosa* Wurmb, as a synonym of his *Corypha pilearia*.

Livistona R. Brown

Livistona saribus (Lour.) Merr. ex A. Chevalier in Bull. Écon. Indochine **21**: 501. 1919
(based on *Corypha saribus* Lour.).
Corypha saribus Lour. Fl. Cochinch. 212. 1790, ed. Willd. 263. 1793, Anamese *cây tlo.*
Livistona cochinchinensis Mart. Hist. Nat. Palm. **3**: 319. 1849 (based on *Corypha saribus* Lour.); Becc. in Webbia **3**: 241. 1910, Philip. Journ. Sci. **14**: 340. 1919.
Saribus cochinchinensis Blume Rumphia **2**: 49. 1836 (based on *Corypha saribus* Lour.).
" Habitat in sylvis Cochinchinae." A species extending from Indo-China to the Malay Peninsula and Luzon, allied to *Livistona rotundifolia* Mart. but distinct. Loureiro
took his specific name from *Saribus* Rumph. (Herb. Amb. **1**: 42. *pl. 8*) (= *Livistona rotundifolia* Mart.), which he erroneously cites as a synonym of his *Corypha saribus*. Loureiro's
species, however, is typified by the Indo-China specimen described, not by the Rumphian
synonym cited.

Hyphaene Gaertner

Hyphaene coriacea Gaertn. Fruct. **1**: 28. *pl. 10. f. 2.* 1788; C. H. Wright in Thiselton-Dyer
Fl. Trop. Afr. **8**: 119. 1902.
Corypha africana Lour. Fl. Cochinch. 213. 1790, ed. Willd. 264. 1793.
" Habitat in sylvis Africae Orientalis, nec alibi a me visa." This is the current reduc-
tion of Loureiro's species and is apparently correct. The species is recorded only from
the Mozambique District and Madagascar. Loureiro cites the local name *mulale*.

Borassus Linnaeus

Borassus aethiopum Mart. in Münch. Gel. Anzeig. 639. 1838, 46. 1839; Hist. Nat. Palm.
3: 221. 1849; Becc. in Webbia **4**: 325. 1913.
Borassus flabellifer Linn. var. *aethiopum* Warb. in Engler Pfl. Ost.-Afr. B. 20, C. 130.
1895; Wright in Thiselton-Dyer Fl. Trop. Afr. **8**: 117. 1902.
Borassus flabelliformis (non Murr.) Lour. Fl. Cochinch. 618. 1790, ed. Willd. 758.
1793.
" Habitat tam agrestis, quam culta in ora orientali Africae, puto quod nascatur etiam
in sylvis Cochinchinae, sed ibi mihi non obvia." Loureiro cites the East African local
name *murume* for this species. In all probability the form he observed there is the one
retained by Beccari as a distinct species, *Borassus aethiopium* Mart. It should be noted

however that Martius' type was from tropical West Africa; the East African form may prove to be *Borassus aethiopium* Mart. var. *bagamojensis* Becc., type from German East Africa. The Cochinchina plant casually mentioned by Loureiro is probably *B. flabellifer* Linn.

Borassus flabellifer Linn. Sp. Pl. 1187. 1753; Becc. in Webbia 4: 304. 1914.

> *Borassus tunicata* Lour. Fl. Cochinch. 619. 1790, ed. Willd. 760. 1793.
>
> *Pholidocarpus tunicatus* H. Wendl. ex Jackson Ind. Kew. 3: 502. 1894 (based on *Borassus tunicatus* Lour.).

"Habitat in India ad oras regnorum Decan et Guzerate." Loureiro cites the local name *tarfulim*. His description applies to the Palmyra palm which is common and widely distributed in India. Beccari in his study of the genus *Borassus* (Webbia 4: 293–385. 1914) fails to account for Loureiro's binomial. It is to be noted that Wendland did not form the binomial *Pholidocarpus tunicatus* credited to him in Index Kewensis. He merely states under *Borassus tunicata* Lour. (Kerch. Palm., p. 235), "vide *Pholidocarpus*." The name does not appear under *Pholidocarpus*, p. 253, nor in the index. *Borassus tunicata* Lour. has nothing to do with *Pholidocarpus*.

Plectocomia Martius

Plectocomia sp.

> *? Calamus petraeus* Lour. Fl. Cochinch. 209. 1790, ed. Willd. 260. 1793, Anamese *mây dá*.
>
> *Flagellaria petraea* Lour.[47] ex Gomes in Mem. Acad. Sci. Lisb. Cl. Sci. Mor. Pol. Bel.-Let. n. ser. 4(1): 27. 1868.
>
> *Palmijuncus petraeus* O. Ktz. Rev. Gen. Pl. 733. 1891 (based on *Calamus petraeus* Lour.).

"Habitat in sylvis Cochinchinae." Loureiro's description is short and very imperfect. He states, however, that the stems are the thickest of all of the rattan species known to him. Beccari (Ann. Bot. Gard. Calcutta 11(1): 501. 1908) thinks that it may be a species of *Korthalsia* or *Plectocomia* on account of Loureiro's description of the spadix as long and terminal, which, if true, would exclude the species from *Calamus*. Manifestly the indicated large size excludes *Korthalsia* from consideration. Later Beccari (Webbia 3: 245. 1910) states: "Forse identificabile con la *Plectocomia Pierreana* Becc." *Palmijuncus calapparius* Rumph. (Herb. Amb. 5: 98. *pl. 51*), cited by Loureiro as representing his species, must be excluded as it represents *Daemonorops calapparius* Blume.

Calamus Linnaeus

Calamus amarus Lour. Fl. Cochinch. 210. 1790, ed. Willd. 261. 1793, Anamese *mây dáng*.

> *Calamus tenuis* Roxb. Hort. Beng. 73. 1814, *nomen nudum*, Fl. Ind. ed. 2, 3: 780. 1832; Becc. in Ann. Bot. Gard. Calcutta 11(1): 262. 1908.
>
> *Palmijuncus amarus* O. Ktz. Rev. Gen. Pl. 733. 1891 (based on *Calamus amarus* Lour.).

"Habitat in sylvis Cochinchinae." The status of this species is somewhat doubtful, although in all probability *Calamus tenuis* Roxb. is the same, in which case Loureiro's spe-

---

[47] A Loureiro herbarium name here first published by Gomes.

cific name should be adopted.   Beccari (Ann. Bot. Gard. Calcutta 11(1): 262. 1908) places *C. amarus* Lour. as a doubtful synonym of *C. tenuis* Roxb. making the following comment: " I also regard *C. tenuis* as identical with *Calamus amarus* of Loureiro, judging from sterile specimens collected by Pierre in Cochinchina and labelled with the same indigenous name as is assigned by Loureiro to *C. amarus* "; and later (Webbia **3**: 245. 1910): " Forse corrisponde al *Calamus tenuis* Roxb." Roxburgh's species extends from India to Burma and Indo-China. A small part of an infructescence is among the Loureiro specimens in the herbarium of the British Museum.

Calamus dioicus Lour. Fl. Cochinch. 211. 1790, ed. Willd. 262. 1793, Anamese *mây tăt*.
    Calamus salicifolius Becc. in Rec. Bot. Surv. India **2**: 206. 1902, Ann. Bot. Gard. Calcutta **11**(1): 91, 279. *pl. 103*. 1908.
    " Habitat in sylvis Cochinchinae, prope flumina, & litora." Beccari's detailed description and illustration of *Calamus dioicus* (Ann. Bot. Gard. Calcutta **11**(1): 195. *pl. 51*. 1908) was based on Indo-China specimens collected by Pierre, but I believe his interpretation is erroneous and that some other name will have to be adopted for it. On the other hand *Calamus salicifolius* Becc., represented by *Clemens 3100, 4485*, in thickets, from Loureiro's classical region, conforms closely to his description, and I believe Loureiro's binomial should be applied to the plant he described. *Palmijuncus equestris* Rumph. (Herb. Amb. **5**: 110. *pl. 56*) and *P. viminalis* Rumph. (*op. cit.* 108. *pl. 55. f. 2*), cited by Loureiro as doubtfully representing his species, must be excluded. The first is *Calamus equestris* Willd. and the second is *C. viminalis* Willd., both Moluccan species that do not occur in Asia.

Calamus rudentum Lour. Fl. Cochinch. 209. 1790, ed. Willd. 260. 1793, Anamese *mây saong;* Becc. in Ann. Bot. Gard. Calcutta **11**(1): 139. *pl. 13*. 1908.
    " Habitat in sylvis Cochinchinae." Beccari's ample description and his illustration are based on specimens collected in Indo-China by Pierre, and it may be assumed that he is correct in his interpretation of the species. *Palmijuncus albus* Rumph. (Herb. Amb. **5**: 102. *pl. 53*), cited by Loureiro as a synonym, must be excluded as it represents the Moluccan *Calamus albus* Pers.

Calamus scipionum Lour. Fl. Cochinch. 210. 1790, ed. Willd. 260. 1793, Anamese *heò tàu;* Becc. in Ann. Bot. Gard. Calcutta **11**(1): 371. *pl. 156*. 1908.
    Palmijuncus scipionum O. Ktz. Rev. Gen. Pl. 733. 1891 (based on *Calamus scipionum* Lour.).
    " Habitat imprimis in sylvis Malaiorum, ex utraque parte freti Malacensis: unde abundanter in Sinas, & in Europam exportatur." While Loureiro cites the Anamese name *heò tàu*, this was probably derived from commercial sources for there is no evidence that he had Indo-China specimens. His description was probably based on specimens from the Malay Peninsula and doubtless Beccari's interpretation of it is correct; it is known from the Malay Peninsula, Billiton and Borneo.

Daemonorops Blume

Daemonorops sp.
    Calamus verus Lour. Fl. Cochinch. 210. 1790, ed. Willd. 261. 1793, Anamese *mây nuóc, mây ra.*

" Habitat in sylvis tam montanis, quam planis Cochinchinae." This species is known only from Loureiro's description. Beccari (Ann. Bot. Gard. Calcutta 11(1): 39. 1908) states that by its short spadix and oblong spathe, taking into consideration also the Rumphian illustration cited, it might be that a *Daemonorops* of the section *Piptospatha* is represented, and by the leaflets a species near *D. didymophyllus* Becc. He thinks that a species may be represented not secured by modern botanists and collectors. Later (Webbia 3: 245. 1910) he states: " É probabilmente una specie di *Daemonorops.*" *Palmijuncus verus latifolius* Rumph. (Herb. Amb. 5: 106. *pl. 54*), cited by Loureiro as representing his species, is to be excluded as it represents either *Calamus rumphii* Blume or *C. pisicarpus* Blume.

Caryota Linnaeus

Caryota mitis Lour. Fl. Cochinch. 569. 1790, ed. Willd. 697. 1793, Anamese *cây dung dinh.*
  Caryota sobolifera Wall. in Mart. Hist. Nat. Palm. 3: 194. *pl. 107. f. 2.* 1849.
  " Habitat in sylvis Cochinchinae." A characteristic soboliferous species, well described by Loureiro. It is probably the smallest species in the genus and is often planted for ornamental purposes. It extends from Indo-China to Burma, the Sunda Islands and Palawan. A specimen from Loureiro, named however by Dryander, is preserved in the herbarium of the British Museum.

Arenga [48] La Billardière

Arenga pinnata (Wurmb) Merr. Interpret. Herb. Amb. 119. 1917.
  Saguerus pinnatus Wurmb in Verh. Bat. Genootsch. 1: 351. 1779.
  Borassus gomutus Lour. Fl. Cochinch. 618. 1790, ed. Willd. 759. 1793, Anamese *cây duác.*
  Sagus gomutus Perr. in Mém. Soc. Linn. Paris 3: 142. 1824.
  Arenga saccharifera Labill. in Mém. Inst. Paris 4: 209. 1801.
  Arenga gamuto Merr.[49] in Philip. Journ. Sci. 9: Bot. 63. 1914.
  " Habitat in sylvis Cochinchinae." *Borassus gomutus* Lour. is clearly the same as the common sugar palm currently known as *Arenga saccharifera* Labill. *Saguerus sive Gomutus* Rumph. (Herb. Amb. 1: 57. *pl. 13*), cited by Loureiro as representing his species, is correctly placed.

Didymosperma H. Wendland & Drude

Didymosperma caudata (Lour.) Wendl. & Drude in Kerch. Palm. 243. 1878 (based on *Borassus caudatus* Lour.); Becc. Malesia 3: 97. 1889; Webbia 3: 200. 1910.
  Borassus caudatus Lour. Fl. Cochinch. 619. 1790, ed. Willd. 760. 1793, Anamese *cây duói chuot.*
  Wallichia caudata Mart. Hist. Nat. Palm. 3: 315. 1849 (based on *Borassus caudata* Lour.).
  Blancoa caudata O. Ktz. Rev. Gen. Pl. 727. 1891 (based on *Borassus caudatus* Lour.).

[48] *Arenga* Labillardière (1803), conserved name, Vienna Code; an older one is *Saguerus* (Rumphius) Adanson (1763). Originally published as *Areng.*
  [49] This binomial was based on *Saguerus gamuto* Houtt. "Handl. I. 410. t. 4. f. 2" as given in Index Kewensis. The binomial does not appear in Houttuyn's work. He merely described the Sagueerboom giving the Malay names as *gamutu* and *gamoto* and the Portuguese name as *saguiero. Sagus gomutus* Perr. was published independently and with no reference to *Borassus gomutus* Lour.

" Habitat in sylvis Cochinchinae." Beccari (Webbia **3**: 201. 1910) has published a detailed description of this species based on a series of specimens from various parts of Indo-China.

Pinanga Blume

Pinanga sylvestris (Lour.) Becc. Malesia **3**: 143. 1886 (based on *Areca sylvestris* Lour.).

Areca sylvestris Lour. Fl. Cochinch. 568. 1790, ed. Willd. 696. 1793, Anamese *cau rùng*.

Pinanga cochinchinensis Blume Rumphia **2**: 77. 1836 (based on *Areca sylvestris* Lour.), Bull. Néerl. 65. 1838; Becc. in Webbia **3**: 237. 1910.

Seaforthia sylvestris Blume in Mart. Hist. Nat. Palm. **3**: 185. 1838 (based on *Areca sylvestris* Lour.).

Seaforthia cochinchinensis Mart. Hist. Nat. Palm. **3**: 313. 1845 (based on *Areca sylvestris* Lour.).

Ptychosperma cochinchinensis Miq. Fl. Ind. Bat. **3**: 23. 1855 (based on *Areca sylvestris* Lour.).

" Habitat in sylvis Cochinchinae." Willdenow, in a footnote, referred this to *Areca globulifera* Lam. = *Pinanga globulifera* (Lam.) Merr. (Interpret. Herb. Amb. 122. 1917), which is correct as to *Pinanga oryzaeformis* Rumph. (Herb. Amb. **1**: 40. *pl. 5. f. 2 B*) which Loureiro cites under his species, but which is not the species he actually described. *Pinanga sylvestris* (Lour.) Becc. is known only from Indo-China.

Areca Linnaeus

Areca catechu Linn. Sp. Pl. 1189. 1753 (err. typ. *cathecu*); Lour. Fl. Cochinch. 567. 1790, ed. Willd. 695. 1793, Anamese *cây cau, bình lang*, Chinese *pín lām*.

"'Habitat vastissime culta in omnibus praediis Cochinchinae, tam maritimis, quam montanis; raro in China Australi." The Linnaean species was correctly interpreted by Loureiro, and the pre-Linnaean synonyms cited by him are correctly placed. It is the common betel-nut palm.

Cocos Linnaeus

Cocos nucifera Linn. Sp. Pl. 1188. 1753; Lour. Fl. Cochinch. 566. 1790, ed. Willd. 692. 1793; Anamese *cây dùà*, Chinese *yâi xú*.

" Habitat culta in terris planis, non paludosis, Cochinchinae: minus frequenter apud Sinas in Insula Hai-nan." It is the common coconut palm, which was correctly interpreted by Loureiro.

Nypa Wurmb

(*Nipa* Thunberg)

Nypa fruticans Wurmb in Verh. Bot. Genoots. **1**: 349. 1779.

Nipa fruticans Thunb. in Vet. Akad. Nya Handl. Stockh. **3**: 231. 1782.

Cocos nypa Lour. Fl. Cochinch. 567. 1790, ed. Willd. 694. 1793, Anamese *cây dùà núoc*.

" Habitat frequentissima in aquis salsis ad maris litora, & ostia fluminum caenosa Cochinchinae, Cambodiae, Philippinarum, freti Malaccensis, &c." The common nipa palm was correctly interpreted by Loureiro. It is generally known as *Nipa fruticans* Thunb.

ARACEAE

Pothos Linnaeus

Pothos scandens Linn. Sp. Pl. 968. 1753; Lour. Fl. Cochinch. 532. 1790, ed. Willd. 650. 1793, Anamese *cây ráy leo*.

" Habitat in sylvis Cochinchinae." Loureiro's description apparently applies to the Linnaean species which extends from India to Indo-China, Sumatra, Java, and Borneo. Mrs. Clemens collected it near Hue.

Pothos repens (Lour.) Druce Rept. Bot. Exch. Club Brit. Isles **4**: 641. 1917; Merr. in Philip. Journ. Sci. **15**: 228. 1919 (based on *Flagellaria repens* Lour.).

Flagellaria repens Lour. Fl. Cochinch. 212. 1790, ed. Willd. 263. 1793, Anamese *mây baóc bò cây*.

Pothos loureirii Hook. & Arn. Bot. Beechey's Voy. 220. 1836 (based on *Flagellaria repens* Lour.); Engl. in Pflanzenreich 21(IV–23B): 35. *f. 15.* 1905.

Pothos terminalis Hance in Ann. Sci. Nat. V Bot. **5**: 247. 1866.

" Habitat agrestis in locis planis Cochinchinae." Loureiro's description applies unmistakably to the species commonly known as *Pothos loureirii* Hook. & Arn. *Adpendix duplo folio* Rumph. (Herb. Amb. **5**: 490. *pl. 184. f. 1*), cited by Loureiro as representing his species, I take to represent the allied *Pothos hermaphroditus* (Blanco) Merr. (*P. longifolius* Presl).

Acorus Linnaeus

Acorus gramineus Soland. in Ait. Hort. Kew. **1**: 474. 1789.

Acorus terrestris Spreng. Syst. **2**: 118. 1825 (based on *Acorus calamus* Lour.).

Acorus calamus (non Linn.) Lour. Fl. Cochinch. 208. 1790, ed. Willd. 259. 1793, Anamese *thach xuog bô*, Chinese *xě chăm pú*.

" Habitat in montibus, & locis petrosis Cochinchinae, & Chinae." The habitat is the typical one for *Acorus gramineus* Soland., a species which occurs gregariously and in great abundance on thin soil covering ledges and boulders in the beds of small mountain streams in Kwangtung Province. *Acorum terrestre* Rumph. (Herb. Amb. **5**: 178), cited by Loureiro as a synonym, is the form cultivated in the Malaysian region and is *Acorus calamus* Linn.

Acorus calamus Linn. Sp. Pl. 324. 1753.

Orontium cochinchinense Lour. Fl. Cochinch. 208. 1790, ed. Willd. 258. 1793, Anamese *thuy xuong bô*, Chinese *xuì chăm pú*.

" Habitat in paludibus, & locis aquosis Cochinchinae, & Chinae, non fluctuans." The description clearly applies to *Acorus calamus* Linn. *Acorum palustre* Rumph. (Herb. Amb. **5**: 178. *pl. 72. f. 1*), cited by Loureiro, represents the Linnaean species.

Lasia Linnaeus

Lasia spinosa (Linn.) Thwaites Enum. Pl. Zeyl. 336. 1864; Engl. in Pflanzenreich 48(IV–23C): 24. *f. 9.* 1911.

Dracontium spinosum Linn. Sp. Pl. 967. 1753.

Pothos lasia Roxb. Fl. Ind. **1**: 458. 1820 (quoad syn. Lour.).

Lasia loureiri Schott in Bonplandia **5**: 125. 1857 (based on *Lasia aculeata* Lour.).

Lasia aculeata Lour. Fl. Cochinch. 81. 1790, ed. Willd. 103. 1793, Anamese *cu chaóc gai*.

" Habitat loca plana, & humida Cochinchinae."   A well-known species extending from India and Ceylon to Indo-China, Sumatra, Borneo and Java.   Loureiro's description conforms to the characters of the Linnaean species; his type is preserved in the herbarium of the British Museum.

Homalomena Schott

Homalomena sp.

Calla occulta Lour. Fl. Cochinch. 532. 1790, ed. Willd. 651. 1793, Anamese *cây ôí*.

Zantedeschia occulta Spreng. Syst. **3**: 765. 1826 (based on *Calla occulta* Lour.).

Spirospatha occulta Raf. Fl. Tellur. **4**: 8. 1838 (based on *Calla occulta* Lour.).

" Habitat loca humida Cochinchinae."   Loureiro's description seems clearly to represent a species of *Homalomena* and might refer to one of the several species credited to Indo-China.   Engler (Pflanzenreich **55**(IV–23D): 59. 1912) reduced *Zantedeschia occulta* Spreng. to *Homalomena aromatica* Schott, but failed to account for *Calla occulta* Lour. on which Sprengel's binomial was based.   *Dracunculus amboinicus* Rumph. (Herb. Amb. **5**: 322. *pl. 111. f. 2*), cited by Loureiro, with doubt, as representing his species, is *Homalomena cordata* (Houtt.) Schott.

Alocasia Necker

Alocasia cucullata (Lour.) Schott Melet. 18. 1832 (based on *Arum cucullatum* Lour.); Oesterr. Bot. Wochenbl. **4**: 410. 1854, non Engler in DC. Monog. Phan. **2**: 498. 1879, non Engler & Krause in Pflanzenreich **71**(IV–23E): 77. 1920.

Arum cucullatum Lour. Fl. Cochinch. 536. 1790, ed. Willd. 656. 1793, Chinese *chim mī vú*.

Caladium cucullatum Pers. Syn. **2**: 575. 1807 (based on *Arum cucullatum* Lour.).

Colocasia cucullata Schott ex Kunth Enum. **3**: 38. 1841 (based on *Arum cucullatum* Lour.).

" Habitat in suburbiis Cantoniensibus."   The synonymy given above includes only those binomials based on Loureiro's original one.   N. E. Brown (Journ. Linn. Soc. Bot. **36**: 183. 1903) records the species from Szechuan and Hainan.   It is clear that Engler's description, which was based on specimens from India, Ceylon, and Burma, does not apply to Loureiro's species, and the figure given by Engler and Krause represents a species remote from the one Loureiro described.   From Loureiro's description of the leaves as peltate it is suspected that some other genus than *Alocasia* is represented, perhaps *Colocasia*.

Alocasia macrorrhiza (Linn.) Schott Melet. 18. 1832; Engler & Krause in Pflanzenreich **71**(IV–23E): 84. 1920.

Arum macrorrhizum Linn. Sp. Pl. 965. 1753.

Colocasia indica Kunth Enum. **3**: 39. 1841 (based on *Arum indicum* Lour.).

Alocasia indica Schott in Oesterr. Bot. Wochenschr. **4**: 410. 1854 (based on *Arum indicum* Lour.).

Arum indicum Lour. Fl. Cochinch. 536. 1790, ed. Willd. 655. 1793, Anamese *ráy cây*.

" Habitat cultum in Cochinchina."   *Arum indicum sativum* Rumph. (Herb. Amb. **5**: 308. *pl. 106*), cited by Loureiro as a synonym, and the very definite description, rather clearly indicate that *Arum indicum* Lour. is the very common widely distributed *Alocasia macrorrhiza* Schott, which is found both cultivated and wild throughout the Indo-Malaysian region.   Engler (DC. Monog. Phan. **2**: 494. 1879) and Engler & Krause (Pflanzenreich

71(IV–23E): 69. 1920) adopted *Colocasia indica* (Lour.) Hassk. [Kunth] as the proper name, but *Colocasia* has peltate leaves and Loureiro's description otherwise does not apply to this genus; they saw only Javan specimens.   I believe that this Javan form should receive a new name, perhaps to be derived from *Caladium giganteum* Blume (Cat. Gew. Buitenzorg 103. 1823) which is there described as having peltate leaves.

Alocasia sp.
> *? Arum arisarum* (non Linn.) Lour. Fl. Cochinch. 535. 1790, ed. Willd. 655. 1793, Anamese *ráy hoang.*
> *? Calyptrocoryne cochinchinensis* Schott Prodr. 105. 1860 (based on *Arum arisarum* Lour.).
> *? Typhonium cochinchinense* Blume Rumphia 1: 135. 1835 (based on *Arum arisarum* Lour.).

" Habitat in sylvis, & sepibus Cochinchinae."   I do not recognize this species.   It is difficult to determine just how much of Loureiro's description was based on actual specimens and how much on the illustration of *Arisarum latifolium* Hill (Herb. Brit. *pl. 48. f. 11*) which he erroneously cites as illustrating his species.   Engler (Pflanzenreich 73(IV–23F): 93. 1920) in excluding the species from *Arum* refers it to *Theriophonum wightii* Blume, yet in his treatment of *Theriophonum* (*op. cit.* 104–108) he does not cite Loureiro's binomial under any of the species admitted, and all known representatives of the genus are from India.

Alocasia sp.
> *? Arum macrorrhizum* (non Linn.) Lour. Fl. Cochinch. 535. 1790, ed. Willd. 654. 1793, Anamese *ráy tláng,* Chinese *dea vú.*

" Habitat loca plana, praesertim humida in Cochinchina, & China."   From the description this may be an *Alocasia* but it can scarcely be the Linnaean species.   The description of the leaves as peltate suggests *Colocasia* rather than *Alocasia.*

Colocasia Schott

Colocasia esculentum (Linn.) Schott Melet. 1: 18. 1832.
> *Arum esculentum* Linn. Sp. Pl. 965. 1753; Lour. Fl. Cochinch. 535. 1790, ed. Willd. 654. 1793, Anamese *cây môn,* Chinese *hài yú.*
> *Arum colocasia* Linn. *l.c.*; Lour. Fl. Cochinch. 534. 1790, ed. Willd. 653. 1793, Anamese *ráy bac hà.*
> *Arum sagittifolium* (non Linn.) Lour. Fl. Cochinch. 534. 1790, ed. Willd. 653. 1793, Anamese *ráy tiá,* Chinese *tái leí thâu.*

Arum esculentum as interpreted by Loureiro was from Cochinchina and China: " Habitat frequentissime, maxime ad ripas fluminum, & paludum margines in aqua non profunda."   *A. colocasia:* " Habitat cultum in Cochinchina "; and *A. sagittifolium:* " Habitat cultum, incultumque in Cochinchina, & China."   All three descriptions apparently appertain to forms of the very common and variable taro, *Colocasia esculentum* Schott.

Typhonium Schott

Typhonium divaricatum (Linn.) Decne. in Nouv. Ann. Mus. Hist. Nat. Paris 3: 367. 1834.
> *Arum divaricatum* Linn. Sp. Pl. ed. 2, 1369. 1763.

Arum trilobatum (non Linn.) Lour. Fl. Cochinch. 534. 1790, ed. Willd. 652. 1793, Anamese *nam tinh.*

"Habitat in Cochinchina." Both *Typhonium divaricatum* Decaisne and *T. trilobatum* Schott are recorded from Indo-China. I judge that Loureiro's description best applies to *T. divaricatum.* His idea of what he took to be *Arum trilobatum* Linn. was probably based for the most part on *Arisarum amboinicum* Rumph. (Herb. Amb. **5**: 319. *pl. 110. f. 2*) which he cites, after Linnaeus, as representing that species; the Rumphian reference, however, is a synonym of *Typhonium divaricatum* (Linn.) Decne., a variable species extending from Ceylon to southern China and the Moluccas.

Arisaema Martius

Arisaema sp.

Arum pentaphyllum (non Linn.) Lour. Fl. Cochinch. 533. 1790, ed. Willd. 652. 1793, Anamese *nam tinh taù,* Chinese *tiĕn nӑn sīn.*

"Habitat incultum in China." The Linnaean species is one of wholly doubtful status, having been based on illustrations in pre-Linnaean literature, that were probably based on material from some part of Asia. There is no reason for assuming that Loureiro had specimens representing the Linnaean species. It is suspected that his very poor description was based on fragmentary material secured from an herbalist.

Pinellia Tenore

Pinellia ternata (Thunb.) Ten. ex Breitenb. in Bot. Zeit. **37**: 687. *f. 1–4.* 1879; Makino in Bot. Mag. Tokyo **15**: 135. 1901; Druce in Rept. Bot. Exch. Club Brit. Isles **4**: 640. 1917.

Arum ternatum Thunb. Fl. Jap. 233. 1784.

Arum triphyllum (non Linn.) Lour. Fl. Cochinch. 533. 1790, ed. Willd. 652. 1793, Anamese *bán ha taù,* Chinese *puón hía.*

Arum dracontium (non Linn.) Lour. Fl. Cochinch. 533. 1790, ed. Willd. 651. 1793, Anamese *cu chӑóc, bán ha,* Chinese *puón hía.*

Arisaema cochinchinense Blume Rumphia **1**: 107. 1835 (based on *Arum dracontium* Lour.).

Pinellia cochinchinensis W. F. Wight in U. S. Dept. Agr. Bur. Pl. Ind. Bull. **142**: 35. 1909 (based on *Arum dracontium* Lour.).

Arisaema loureiri Blume Rumphia **1**: 108. 1835 (based on *Arum triphyllum* Lour.).

Pinellia tuberifera Ten. in Att. Accad. Sci. Napol. **4**: 57. 1839.

Arum triphyllum as described by Loureiro was observed by him "incultum in China, & Cochinchina," and *A. dracontium,* "incultum in hortis & agris Cochinchinae & Chinae." Engler (DC. Monog. Phan. **2**: 566, 567. 1879) referred the former with doubt to *Pinellia tuberifera* Tenore, and the latter to *P. wawrae* Engler; in the Pflanzenreich (**73**(IV–23F): 224. 1920) *P. wawrae* is reduced to *P. pedatisecta* Schott. It is to be noted that the latter species is known only from northern China, while Loureiro's specimens were from Indo-China and presumably Kwangtung Province in southern China; they may have been secured from an herbalist. The descriptions apply to a single species and the correctness of this view is in a measure verified by the native names cited by Loureiro. The species is generally known under the comparatively recent name *Pinellia tuberifera* Tenore.

Pistia Linnaeus

Pistia stratiotes Linn. Sp. Pl. 963. ·1753; Engl. in Pflanzenreich **73**(IV–23F): 259. *f. 63.* 1920.

Zala asiatica Lour. Fl. Cochinch. 405. 1790, ed. Willd. 492. 1793, Anamese *beò phù binh,* Chinese *fêu pêng.*

" Habitat fluctuans in fluminibus lenti cursus in Cochinchina, & China." This reduction of Loureiro's genus and species is manifestly correct. *Plantago aquatica* Rumph. (Herb. Amb. **6**: 177. *pl. 74. f. 2*) and the other pre-Linnaean synonyms cited by Loureiro, are correctly placed.

LEMNACEAE

Lemna Linnaeus

Lemna paucicostata Hegelm. Lemn. 139. *pl. 8.* 1868.

Lemna minor (non Linn.) Lour. Fl. Cochinch. 550. 1790, ed. Willd. 671. 1793, Anamese *bèo cám.*

" Habitat ubique in aquis stagnantibus in Cochinchina." As *Lemna paucicostata* Hegelm. is the common Indo-Malaysian representative of this genus I have assumed that this is the form Loureiro had.

FLAGELLARIACEAE

Flagellaria Linnaeus

Flagellaria indica Linn. Sp. Pl. 333. 1753; Lour. Fl. Cochinch. 211. 1790, ed. Willd. 262. 1793, Anamese *mây baóc.*

Flagellaria catenata Lour.[50] ex Gomes in Mem. Acad. Sci. Lisb. Cl. Sci. Mor. Pol. Bel.-Let. n.s. **4**(1): 27. 1868.

" Habitat agrestis in locis planis Cochinchinae." The common Linnaean species was correctly interpreted by Loureiro. *Palmijuncus laevis* Rumph. (Herb. Amb. **5**: 120. *pl. 59. f. 1*) is correctly placed as a synonym.

ERIOCAULACEAE

Eriocaulon Linnaeus

Eriocaulon sexangulare Linn. Sp. Pl. 87. 1753; Lecomte Fl. Gén. Indo-Chine **7**: 15. 1912.

Eriocaulon quadrangulare (non Linn.) Lour. Fl. Cochinch. 60. 1790, ed. Willd. 76. 1793, Anamese *co dùi coùng,* Chinese *koŭc san tsào.*

" Habitat in hortis, & agris Cochinchinae ubique obvia. In China similem habitu vidi, sed florum non examinavi." This disposition of the form Loureiro described is not entirely satisfactory, but in general the characters given by Loureiro conform fairly well with those of the Linnaean species as interpreted by Lecomte.

---

[50] A Loureiro herbarium name here first published by Gomes.

BROMELIACEAE

Ananas (Tournefort) Adanson

Ananas comosus (Linn.) Merr. Interpret. Herb. Amb. 133. 1917; Mez in Pflanzenreich **100** (IV–32): 102. *f. 29*. 1934.

Bromelia comosa Linn. in Stickman Herb. Amb. 21. 1754; Amoen Acad. **4**: 130. 1759.

Bromelia ananas Linn. Sp. Pl. 285. 1753; Lour. Fl. Cochinch. 192. 1790, ed. Willd. 237. 1793, Anamese *tlái thom*.

Ananas sativus Schult. & Schult. f. Syst. **7**: 1283. 1830.

Ananassa sativa Lindl. Bot. Reg. **13**: sub *pl. 1068*. 1827.

" Habitat in magna copia in agris & hortis Cochinchinae." The Linnaean species was correctly interpreted by Loureiro. The pre-Linnaean synonyms cited by him are correctly placed. This is the common pineapple.

COMMELINACEAE

Commelina (Plumier) Linnaeus

Commelina nudiflora Linn. Sp. Pl. 41. 1753.

Commelina communis (non Linn.) Lour. Fl. Cochinch. 38. 1790, ed. Willd. 48. 1793, Anamese *rau tlai ăn*.

Lechea chinensis Lour. Fl. Cochinch. 60. 1790, ed. Willd. 76. 1793, Chinese *chăt yú tsào*.

Commelina loureirii Kunth Enum. **4**: 60. 1843 (based on *Lechea chinensis* Lour.).

For *Commelina communis* Loureiro states " in Cochinchina edulis tam cocta, quam cruda," but otherwise cites no locality; *Commelina nudiflora* Linn. is somewhat used as a vegetable. It seems evident therefore that Loureiro had specimens of this species and not of *C. communis* Linn., the latter not being a tropical plant. For *Lechea chinensis* Loureiro states: " Habitat prope Cantonem Sinarum." The description applies to *Commelina nudiflora* Linn.

Commelina africana Linn. Sp. Pl. 41. 1753; ? Lour. Fl. Cochinch. 39. 1790, ed. Willd. 49. 1793.

" Habitat in ora orientali Africae." It is very doubtful if Loureiro had specimens of the Linnaean species which, although recorded by Clarke from Mozambique Territory (Thiselton-Dyer Fl. Trop. Afr. **8**: 45. 1902), yet from data given by him, it is not there a low altitude species. Loureiro's description is not sufficiently definite to warrant reduction of the form he attempted to describe to any of the 34 species of the genus credited to the Mozambique District.

Commelina benghalensis Linn. Sp. Pl. 41. 1753; Lour. Fl. Cochinch. 39. 1790, ed. Willd. 49. 1793, Anamese *rau tlai loung*.

Commelina cucullata Linn. Mant. **2**: 176. 1771; Lour. Fl. Cochinch. 39. 1790, ed. Willd. 49. 1793, Anamese *rau tlai tlâu*.

There is but a single species represented by the two that Linnaeus proposed, and Loureiro was apparently correct in his interpretations. Both were from Cochinchina, the first " spontanea," the second " inculta in hortis." *Commelina benghalensis* Linn. is very common and widely distributed, growing in waste places throughout the Indo-Malaysian region.

Aneilema R. Brown

Aneilema medicum (Lour.) Kostel. Allgem. Med.-Pharm. Fl. 1: 127. 1831 (based on *Commelina medica* Lour.); R. Br. ex C. B. Clarke in DC. Monog. Phan. 3: 202. 1881.

Aneilema loureirii Hance in Journ. Bot. 6: 250. 1868 (based on *Commelina tuberosa* Lour.); C. B. Clarke in DC. Monog. Phan. 3: 201. 1881.

Commelina medica Lour. Fl. Cochinch. 40. 1790, ed. Willd. 50. 1793, Anamese *cu eó chum, mach môn doung*, Chinese *mĕ mûen tūm*.

Commelina tuberosa (non Linn.) Lour. Fl. Cochinch. 40. 1790, ed. Willd. 50. 1793, Anamese *cu eó rai*.

Commelina edulis Stokes Bot. Mat. Med. 2: 184. 1812 (based on *C. tuberosa* Lour.).

For *Commelina medica* Loureiro states: " Habitat in China & Cochinchina," and for *C. tuberosa:* " Habitat ubique in Cochinchina, in locis humidis." I consider that both descriptions apply to a single species. It is to be noted that R. Brown did not actually publish the binomial *A. medicum* in the Prodromus as is indicated by Clarke, but merely indicates that *Commelina medica* belongs in the genus *Aneilema;* Kosteletzky actually made the transfer.

Cyanotis [51] D. Don

Cyanotis loureiriana (Schult. & Schult. f.) Merr. in Lingnaam Agr. Rev. 1(2): 61. 1923, Lingnan Sci. Journ. 5: 45. 1927.

Tradescantia loureiriana Schult. & Schult. f. Syst. 7: 1178. 1830 (based on *Tradescantia geniculata* Lour.).

Tradescantia geniculata (non Jacq.) Lour. Fl. Cochinch. 193. 1790, ed. Willd. 239. 1793, Anamese *rau éo tia.*

Cyanotis geniculata C. B. Clarke in DC. Monog. Phan. 3: 260. 1881 (based on *Tradescantia geniculata* Lour.).

" Habitat inculta in agris, & sepibus Cochinchinae." C. B. Clarke considered this to be an imperfectly known species of *Cyanotis*, the genus to which it certainly belongs. Recently collected Hainan material apparently represents Loureiro's species, which is allied to but apparently distinct from *Cyanotis papilionacea* (Linn. f.) Schult. & Schult. f.

Cyanotis cristata (Linn.) Schult. & Schult. f. Syst. 7: 1150. 1830.

Commelina cristata Linn. Sp. Pl. 42. 1753.

Commelina zanonia (non Linn.) Lour. Fl. Cochinch. 40. 1790, ed. Willd. 51. 1793, Anamese *rau rio.*

From the local name cited, Loureiro's specimens were from Cochinchina. His description conforms sufficiently well with the characters of the common and widely distributed *Cyanotis cristata* Schult. & Schult. f. to warrant this reduction.

Cyanotis vaga (Lour.) Schult. & Schult. f. Syst. 7: 1153. 1830 (based on *Tradescantia vaga* Lour.).

Tradescantia vaga Lour. Fl. Cochinch. 193. 1790, ed. Willd. 239. 1793, Chinese *xít koăt houng.*

[51] *Cyanotis* D. Don (1825), conserved name, Vienna Code; older ones are *Tonningia* Necker (1790) and *Zygomenes* Salisbury (1812).

Cyanotis barbata D. Don Prodr. Fl. Nepal. 46. 1825.

" Habitat inculta Cantone Sinarum." Loureiro's species, generally considered one of doubtful status, has apparently been interpreted by some authors from Blume's description and Javan specimens of *T. vaga* Blume (Enum. Pl. Jav. **1**: 5. 1827) which Blume does not there describe as a new species but ascribes to Loureiro. It seems clear, however, that Blume misinterpreted Loureiro's species. The only *Cyanotis* from the vicinity of Canton that agrees with Loureiro's description is the form currently referred to *C. barbata* D. Don, and the only possible objection to interpreting *Tradescantia vaga* Lour. as identical with *Cyanotis barbata* D. Don is Loureiro's description " spathis diphyllis " without indicating the numerous imbricated bracts; yet in his note he states " habitat vero designat Tradescantiam cristatum Linnaei " which would indicate that *Cyanotis barbata* D. Don is the form he intended; *C. cristata* (Linn.) Schult. & Schult. f. is not known from Kwangtung Province.

Floscopa Loureiro

Floscopa scandens Lour. Fl. Cochinch. 193. 1790, ed. Willd. 238. 1793, Anamese *deei hoa chôi;* C. B. Clarke in DC. Monog. Phan. **3**: 265. 1881.

" Habitat agrestis in montibus Cochinchinae." This is a well-known, widely distributed species, the type of the genus. Loureiro's type is preserved in the herbarium of the British Museum, where it was examined by C. B. Clarke.

PONTEDERIACEAE

Monochoria Presl

Monochoria vaginalis (Burm. f.) Presl Rel. Haenk. **1**: 128. 1827; Cherfils in Lecomte Fl. Gén. Indo-Chine **6**: 818. 1934.

Pontederia vaginalis Burm. f. Fl. Ind. 80. 1768.

Pontederia cordata (non Linn.) Lour. Fl. Cochinch. 198. 1790, ed. Willd. 245. 1793, Anamese *boung mác cây.*

Pontederia loureiriana Schult. & Schult. f. Syst. **7**: 1145. 1830 (based on *Pontederia cordata* Lour.).

Monochoria loureirii Kunth Enum. **4**: 135. 1843 (based on *Pontederia cordata* Lour.).

" Habitat in paludibus Cochinchinae." Loureiro's description applies unmistakably to the common and widely distributed *Monochoria vaginalis* (Burm. f.) Presl; *Pontederia cordata* Linn. is an American species.

Monochoria hastata (Linn.) Solms in DC. Monog. Phan. **4**: 523. 1883.

Pontederia hastata Linn. Sp. Pl. 288. 1753; Lour. Fl. Cochinch. 199. 1790, ed. Willd. 246, 1793, Anamese *roung mác lá.*

Monochoria hastaefolia Presl Rel. Haenk. **1**: 128. 1827; Cherfils in Lecomte Fl. Gén. Indo-Chine **6**: 822. 1934.

" Habitat prope ripas fluminis interfluentis metropolim Cochinchinae." Loureiro's description conforms sufficiently well with the characters of the Linnaean species, which is one of wide distribution in the Old World tropics.

PHILYDRACEAE

Philydrum Banks

Philydrum lanuginosum Banks in Gaertn. Fruct. 1: 62. 1788; Cherfils in Lecomte Fl. Gén. Indo-Chine 6: 831. 1934.

Garciana cochinchinensis Lour. Fl. Cochinch. 15. 1790, ed. Willd. 20. 1793, Anamese *cây bôn bôn*, Chinese *ti'ēn lúm*.

" Habitat loca humida Cochinchinae. Unde anno 1774 per me in Europam missa. Postea eandem examinavi Cantone Sinarum." Specimens from Loureiro are preserved in the herbaria of the British and the Paris Museums. A monotypic genus, extending from Burma to southern China and Australia.

JUNCACEAE

Juncus (Tournefort) Linnaeus

Juncus effusus Linn. Sp. Pl. 326. 1753.

Scirpus capsularis Lour. Fl. Cochinch. 44. 1790, ed. Willd. 55. 1793, Anamese *tím bóc, dang tâm*, Chinese *tem sin tsao*.

Juncus ? bracteatus Stokes Bot. Nat. Med. 2: 289. 1812 (based on *Scirpus capsularis* Lour.).

Juncus capsularis Steud. Syn. Pl. Glum. 2: 309. 1855 (based on *Scirpus capsularis* Lour.).

" Habitat frequenter in China: invenitur etiam in Cochinchina." Willdenow first noted that Loureiro's description appertained to *Juncus* rather than to a cyperaceous plant. Although Buchenau (Pflanzenreich 25(IV–36): 263. 1906) enumerates *Juncus capsularis* Steud. among the " species excludendae," it can scarcely be other than a form of the very common and widely distributed *Juncus effusus* Linn. There is a specimen of *Scirpus capsularis* Lour. in the herbarium of the British Museum endorsed by Dryander " Cochinchina J. de Loureiro " but it has no label and it is not checked in the British Museum copy of the Flora Cochinchinensis.

STEMONACEAE

Stemona Loureiro

Stemona tuberosa Lour. Fl. Cochinch. 404. 1790, ed. Willd. 490. 1793, Anamese *cây bach bó*, Chinese *pĕ pú tsào;* Gagnep. in Lecomte Fl. Gén. Indo-Chine 6: 746. 1934.

Roxburghia gloriosoides Roxb. Pl. Coromandel 1: 29. *pl. 32.* 1795.

Roxburghia stemona Steud. Nomencl. ed. 2, 2: 475. 1841 (based on *Stemona tuberosa* Lour.).

" Habitat inculta in Cochinchina, & China." *Stemona tuberosa* Lour. has by some authors been given a very wide geographic range in the Indo-Malaysian region but in this " collective species " several distinct ones are included. *Ubium polypoides* Rumph. (Herb. Amb. 5: 364. *pl. 129*), cited as a synonym by Loureiro, probably represents the allied *Stemona moluccana* C. H. Wright. Schlechter (Notizbl. Bot. Gart. Berlin 9: 194. 1924) restricts Loureiro's species to India, Assam, Burma, and southern China. A Loureiro specimen listed as being among the plants received from him has not been located in the herbarium of the British Museum.

LILIACEAE

Dianella Lamarck

Dianella ensifolia (Linn.) DC. in Red. Lil. **1**: *pl. 1*. 1802; Gagnep. in Lecomte Fl. Gén. Indo-Chine **6**: 784. 1934.

Dracaena ensifolia Linn. Mant. **1**: 63. 1767; Lour. Fl. Cochinch. 197. 1790, ed. Willd. 243. 1793, Anamese *cây huong lâu*.

" Habitat in locis agrestibus Cochinchinae." Loureiro's description clearly applies to the Linnaean species, which is widely distributed in southern Asia. *Gladiolus odoratus indicus* Rumph. (Herb. Amb. **5**: 185. *pl. 73*), cited by Loureiro, after Linnaeus, as representing this species, represents the allied *Dianella odorata* Blume.

Hemerocallis Linnaeus

Hemerocallis fulva Linn. Sp. Pl. ed. 2. 462. 1762; Lour. Fl. Cochinch. 205. 1790, ed. Willd. 254. 1793, Anamese *kĭm châm hõa*, Chinese *rau hién*.

Gloriosa luxurians Lour.[52] ex Gomes in Mem. Acad. Sci. Lisb. Cl. Sci. Pol. Mor. Bel.-Let. n.s. **4**(1): 29. 1868.

" Habitat culta in Cochinchina, & China." The description applies to the Linnaean species, as currently interpreted. A specimen from Loureiro is preserved in the herbarium of the Paris Museum.

Aloe (Tournefort) Linnaeus

Aloe vera Linn. var. **chinensis** (Haw.) Berger in Pflanzenreich **33**(IV–38–III–II): 230. 1908.

Aloe barbadensis Mill. var. *chinensis* Haw. Suppl. Succ. 45. 1819.

Aloe chinensis Baker in Curtis's Bot. Mag. **103**: *pl. 6301*. 1877.

Aloe perfoliata (non Linn.) Lour. Fl. Cochinch. 203. 1790, ed. Willd. 252. 1793, Anamese *cây nha dam, lu hôi*, Chinese *lû hôei*.

" Habitat agrestis in vastis arenariis regni Champavae." There are no indigenous species of this genus in tropical Asia. Berger reports the above variety of *Aloe vera* from India, southern China and Formosa; I think it highly probable that this is the correct disposition of the form that Loureiro described.

Allium (Tournefort) Linnaeus

Allium sativum Linn. Sp. Pl. 296. 1753; Lour. Fl. Cochinch. 201. 1790, ed. Willd. 249. 1793, Anamese *cây toi*, Chinese *suón*.

" Colitur ubique in Cochinchina, & China." Loureiro correctly interpreted the Linnaean species, the common garlic.

Allium cepa Linn. Sp. Pl. 300. 1753; Lour. Fl. Cochinch. 201. 1790, ed. Willd. 249. 1793, Anamese *cây hành*, Chinese *tsûm xĭ*.

" Colitur abundanter in Cochinchina, & China." Loureiro's description applies unmistakably to the Linnaean species, the common onion.

Allium ascalonicum Linn. Amoen. Acad. **4**: 454. 1759; ? Lour. Fl. Cochinch. 202. 1790, ed. Willd. 250. 1793, Anamese *cây nén*.

---

[52] A Loureiro herbarium name here first published by Gomes.

" Colitur in Cochinchina, nec ibi rarum." Loureiro may have been correct in his interpretation of the Linnaean species, but his description is too incomplete to confirm the identification.

Allium chinense G. Don in Mem. Wern. Soc. **6**: 23. 1827 (based on *Allium odorum* Lour.).

> *Allium odorum* (non Linn.) Lour. Fl. Cochinch. 203. 1790, ed. Willd. 251. 1793, Anamese *kieù, khío,* Chinese *he taù, phi thê.*
> ? *Allium triquetrum* (non Linn.) Lour. Fl. Cochinch. 202. 1790, ed. Willd. 250. 1793, Anamese *kieu,* Chinese *kiái, kiao thêu.*

For *A. odorum* Loureiro states: " Habitat in China, & Cochinchina," the inference from the statement under uses being that the plant was cultivated. He states that the plant was intermediate between *Allium cepa* and *A. sativum* in odor and flavor and was used for flavoring foods. Loureiro's description is the basis of G. Don's binomial, but the exact status of the species is not clear. Wright (Journ. Linn. Soc. Bot. **36**: 120. 1903) reduced *A. triquetrum* Lour. to *A. chinense* G. Don, but it is probable that this is an error. Loureiro states: " Colitur in Cochinchina, & China," and explains that the plant was used for culinary purposes. These cultivated forms of *Allium* need critical study on the basis of fresh material secured under the local names cited by Loureiro. Gagnepain (Lecomte Fl. Gén. Indo-Chine **6**: 813. 1934) reduces both of Loureiro's species to *Allium thunbergii* G. Don but does not account for *A. chinense* G. Don.

Allium uliginosum G. Don in Mem. Wern. Soc. **6**: 60. 1832.

> *Allium tuberosum* Roxb. Hort. Beng. 24. 1814, *nomen nudum,* Fl. Ind. ed. 2, **2**: 141. 1832, non Rottl. 1825.
> *Allium angulosum* (non Linn.) Lour. Fl. Cochinch. 203. 1790, ed. Willd. 251. 1793, Anamese *cây he,* Chinese *kieù tsai.*

" Colitur in Cochinchina, & China." This is very extensively cultivated in the vicinity of Canton, the local name on recently collected material appearing as *kau tsoi;* in the Philippines, where it is also cultivated by the Chinese, it is known as *kuchay.* This white-flowered species is represented by a very excellent colored drawing in the Calcutta herbarium, prepared under Roxburgh's direction, a copy of which I have examined; it is the plant known in Canton as *kau tsoi.* Roxburgh's specific name is apparently invalidated by *Allium tuberosum* Rottl. ex Spreng. Syst. **2**: 38. 1825. Gagnepain (Lecomte Fl. Gén. Indo-Chine **6**: 813. 1934) reduces *Allium uliginosum* G. Don and *A. tuberosum* Roxb. to *A. odorum* Linn.

Lilium (Tournefort) Linnaeus

Lilium longiflorum Thunb. in Trans. Linn. Soc. **2**: 333. 1794.

> *Lilium candidum* (non Linn.) Lour. Fl. Cochinch. 207. 1790, ed. Willd. 256. 1793, Anamese *bach hap hõa,* Chinese *pĕ hó.*

" Colitur in China, & Cochinchina." Loureiro probably saw specimens of *Lilium longiflorum* Thunb. It is improbable that he had specimens of the allied *Lilium brownii* F. E. Br. as he describes the flowers as " albissimi." Gagnepain (Lecomte Fl. Gén. Indo-Chine **6**: 808. 1934) places Loureiro's species as a synonym of *L. brownii* F. E. Brown.

Lilium concolor Salisb. Parad. Lond. *pl. 47.* 1806.

> *Lilium camschatcense* (non Linn.) Lour. Fl. Cochinch. 207. 1790, ed. Willd. 257. 1793, Anamese *lon dièo tàu,* Chinese *chū tán hõa.*

" Habitat in China: in Cochinchina rarius." Wright (Journ. Linn. Soc. Bot. **36**: 136. 1903) considered that Loureiro correctly interpreted the Linnaean species and admitted *Fritillaria camschatcensis* (Linn.) Ker as a Chinese species solely on the basis of Loureiro's record. Loureiro's description however is fairly good for *Lilium concolor* Salisb. in spite of his description of the leaves as verticillate. It is probable that he saw only fragmentary material secured from an herbalist, as the species does not occur naturally in southern China nor in Indo-China. It is interesting to note that Elwes (A monograph of the genus Lilium, sub *pl. 18.* 1880) states that *Lilium concolor* Salisb. was introduced into England in 1804, probably from some Canton garden.

Lilium tigrinum Ker in Curtis's Bot. Mag. **31**: *pl. 1237.* 1809.

> *Lilium pomponium* (non Linn.) Lour. Fl. Cochinch. 207. 1790, ed. Willd. 257. 1793, Chinese *cuôn tán hõa.*
> " Colitur Cantone Sinarum." Loureiro's description clearly applies to the tiger lily.

Scilla Linnaeus

Scilla sinensis (Lour.) Merr. in Philip. Journ. Sci. **15**: 229. 1919.

> *Ornithogalum sinense* Lour. Fl. Cochinch. 206. 1790, ed. Willd. 255. 1793, Chinese *tiēn suón.*
> *Convallaria chinensis* Osbeck Dagbok Ostind. Resa 220. 1757.
> *Barnardia scilloides* Lindl. Bot. Reg. **12**: *pl. 1029.* 1826.
> *Scilla chinensis* Benth. Fl. Hongk. 373. 1861.
> *Dracaena alliaria* Lour.[53] ex Gomes in Mem. Acad. Sci. Lisb. Cl. Sci. Pol. Bel.-Let. n.s. **4**(1): 29. 1868.
> *Scilla scilloides* Druce Rept. Bot. Exch. Club Brit. Isles **4**: 646. 1917.
> " Habitat Cantone Sinarum." Loureiro's description definitely applies to the species

currently known as *Scilla chinensis* Benth., a fairly common one in open grassy places near Canton. Bentham's species was not, however, based on *Convallaria chinensis* Osbeck, the original description of the latter being very inadequate, consisting only of the phrase " foliis linearibus, corollis sexpartis." Loureiro's type is preserved in the herbarium of the Paris Museum.

Cordyline [54] Commerson

Cordyline fruticosa (Linn.) A. Cheval. Cat. Pl. Jard. Bot. Saigon 66. 1919.

> *Convallaria fruticosa* Linn. in Stickman Herb. Amb. 16. 1754, Amoen. Acad. **4**: 126. 1759, Syst. ed. 10, 984. 1759.
> *Asparagus terminalis* Linn. Sp. Pl. ed. 2, 450. 1762.
> *Cordyline terminalis* Kunth in Abh. Acad. Berl. 30. 1820; Gagnep. in Lecomte Fl. Gén. Indo-Chine **6**: 801. 1934.
> *Dracaena ferrea* Linn. Syst. ed. 12, 246. 1766; Lour. Fl. Cochinch. 196. 1790, ed. Willd. 242. 1793, Anamese *cây phăt duu,* Chinese *tsiét tsào.*
> *Taetsia fruticosa* Merr. Interpret. Herb. Amb. 137. 1917; Enum. Philip. Fl. Pl. **1**: 205. 1923.

[53] A Loureiro herbarium name here first published by Gomes.

[54] *Cordyline* Commerson (1789), non Adanson (1768), conserved name, Vienna Code; an older one is *Terminalis* Rumphius (1744, 1755; O. Kuntze 1891). In my Interpret. Herb. Amb. 137. 1917, I accepted *Taetsia* Medikuss the generic name for this group, which is not permissible under the International Code.

" Habitat in Cochinchina, & China tam culta, quam agrestis." A specimen from Loureiro is preserved in the herbarium of the Paris Museum and one also in the herbarium of the British Museum. The description applies to this very common Linnaean species. *Dracaena ferrea* Linn. is apparently only a color form of *Cordyline fruticosa* (Linn.) A. Cheval. although retained by Baker as a variety, *Cordyline terminalis* Kunth var. *ferrea* (Linn.) Baker (Journ. Linn. Soc. Bot. **14**: 540. 1875). *Terminalis rubra* Rumph. (Herb. Amb. **4**: 80. *pl. 34. f. 2*), cited by Loureiro as a synonym, is correctly placed.

Pleomele Salisbury

Pleomele cochinchinensis (Lour.) Merr. ex Gagnep. in Bull. Soc. Bot. France **81**: 287. 1934, in syn. (based on *Aletris cochinchinensis* Lour.).

Aletris cochinchinensis Lour. Fl. Cochinch. 204. 1790, ed. Willd. 253. 1793, Anamese *cây boùng boùng*.

Dracaena loureiroi Gagnep. in Bull. Soc. Bot. France **81**: 287. 1934 (based on *Aletris cochinchinensis* Lour.); Lecomte Fl. Gén. Indo-Chine **6**: 796. *f. 78, 1–5*. 1934.

" Habitat in Cochinchina in hortis culta: puto, quod etiam inculta." Without specimens from the region where Loureiro lived, I originally determined this to be a *Pleomele* near *P. angustifolia* (Roxb.) N. E. Br. With a specimen before me from Tourane, *Clemens 4048*, which unquestionably represents Loureiro's species, I do not hesitate to adopt his specific name for this Anamese form. It is not closely allied to *P. angustifolia* N. E. Br. differing notably in the much wider, differently shaped leaves.

Asparagus (Tournefort) Linnaeus

Asparagus cochinchinensis (Lour.) Merr. in Philip. Journ. Sci. **15**: 230. 1919; Gagnep. in Lecomte Fl. Gén. Indo-Chine **6**: 780. 1934.

Melanthium cochinchinense Lour. Fl. Cochinch. 216. 1790, ed. Willd. 268. 1793, Anamese *thien môn doung*, Chinese *tiēn mûen tūm*.

Anguillaria cochinchinensis Spreng. Syst. **2**: 147. 1825 (based on *Melanthium cochinchinense* Lour.).

Asparagopsis sinica Miq. in Journ. Bot. Néerl. **1**: 90. 1861.

Asparagus sinicus C. H. Wright ex Forbes & Hemsl. in Journ. Linn. Soc. Bot. **36**: 103. 1903; Merr. in Sunyatsenia **1**: 8. 1930.

" Habitat frequens in sepibus aridis in Cochinchina, & China." For a general discussion of the problems involved in reference to *Asparagus sinicus* C. H. Wright and *A. lucidus* Lindl. see Sunyatsenia (**1**: 8–9. 1930). *Asparagus lucidus* Lindl. is the common form with cladodes 2.5 to 3.5 cm. long, as confirmed by an examination of Lindley's type at Cambridge. This, and the form known as *A. sinicus* C. H. Wright, occurs in the vicinity of Hue and in Kwangtung Province, *Clemens 3930* representing *A. sinicus* C. H. Wright, and *Clemens 4486, Squires 332* representing *A. lucidus* Lindl. I now interpret Loureiro's species as the form having short cladodes, about 1 cm. in length (*Asparagus sinicus* C. H. Wright) from Loureiro's statement " folia linearia, triquetra, stellato-terna, *minuscula*, inaequalis "; he does not indicate the length of the cladodes. Loureiro's description of the fruit as a capsule may be ignored, as this term was manifestly used to make his description conform to the characters of the genus in which he erroneously placed the species.

Disporum Salisbury

Disporum cantoniense (Lour.) Merr. in Philip. Journ. Sci. **15**: 229. 1919 (based on *Fritillaria cantoniensis* Lour.).

Fritillaria cantoniensis Lour. Fl. Cochinch. 206. 1790, ed. Willd. 255. 1793, Chinese *lin nì hōa*.

Uvularia chinensis Ker in Curtis's Bot. Mag. **23**: *pl. 916*. 1806.

Disporum pullum Salisb. in Trans. Hort. Soc. London **1**: 331. 1812; Gagnep. in Lecomte Fl. Gén. Indo-Chine **6**: 783. 1934.

Disporum chinense O. Ktz. Rev. Gen. Pl. 708. 1891.

"Colitur Cantone Sinarum." C. H. Wright (Journ. Linn. Soc. Bot. **36**: 136. 1903) admits *Fritillaria cantoniensis* Lour. with the following comment: "A doubtful plant, supposed by Gawler [Ker] to be the same as *Uvularia chinensis*, which is now reduced to *Disporum pullum* Salisb." Loureiro's description is ample and applies unmistakably to Salisbury's species. I am not certain that all of the synonyms placed by Hooker f. (Fl. Brit. Ind. **6**: 360. 1892) under *Disporum pullum* Salisb. are correctly placed.

Liriope Loureiro

Liriope spicata Lour. Fl. Cochinch. 201. 1790, ed. Willd. 248. 1793, Anamese *taóc tien*, Chinese *mac lân;* L. H. Bailey Gent. Herb. **2**: 33. *f. 6*. 1929; Rodr. in Lecomte Fl. Gén. Indo-Chine **6**: 664. 1934.

Ophiopogon spicatus Lodd. Bot. Cab. **7**: *pl. 694*. 1822.

Liriope spicata Lour. var. *minor* C. H. Wright in Journ. Linn. Soc. Bot. **36**: 79. 1903.

"Habitat frequenter culta, incultaque in Cochinchina, & China." Loureiro's type is preserved in the herbarium of the Paris Museum, and Bailey's illustration, cited above, is a photographic reproduction of it. Bailey retains Loureiro's species as a valid one distinct from *L. graminifolia* (L.) Baker to which most authors have reduced it.

Smilax (Tournefort) Linnaeus

Smilax bauhinioides Kunth Enum. **5**: 243. 1850; C. DC. Monog. Phan. **1**: 180. 1878; Gagnep. in Lecomte Fl. Gén. Indo-Chine **6**: 768. 1934.

Smilax caduca (non Linn.) Lour. Fl. Cochinch. 622. 1790, ed. Willd. 764. 1793, Anamese *cây sam com.*

Smilax incerta Kunth Enum. **5**: 263. 1850 (based on *Smilax caduca* Lour.).

Smilax anamitica O. Ktz. Rev. Gen. Pl. 715. 1891; Gagnep. in Lecomte Fl. Gén. Indo-Chine **6**: 771. 1934.

"Habitat in sylvis Cochinchinae." Loureiro's description conforms with the characters of *Smilax bauhinioides* Gaudich., the type of which was from Tourane, a duplicate of which I have examined. Probably Kunth's failure to recognize that *Smilax caduca* Lour., for which he proposed a new name *S. incerta*, was really the form he described as *S. bauhinioides*, was due to his lack of appreciation of the fact that Tourane, type locality of the latter, is but a short distance from Loureiro's classical locality Hue. The species is represented by *Kuntze 3810* from Tourong = Tourane, type of *Smilax anamitica* O. Ktz. in the herbarium of the New York Botanical Garden, and *Clemens 3887* from the same locality.

Smilax china Linn. Sp. Pl. 1029. 1753; Lour. Fl. Cochinch. 622. 1790, ed. Willd. 763. 1793, Anamese *cây khúc khác, thô phuc linh,* Chinese *thù fū lin.*

" Habitat in collibus silvaticis Cochinchinae, & Chinae." The description conforms fairly well with the characters of the Linnaean species as the latter is currently interpreted. It is common in the immediate vicinity of Canton.

Smilax corbularia Kunth Enum. **5**: 262. 1850 (based on *Smilax pseudochina* Lour.); Gagnep. in Lecomte Fl. Gén. Indo-Chine **6**: 759. 1934.

Smilax pseudochina (non Linn.) Lour. Fl. Cochinch. 623. 1790, ed. Willd. 765. 1793, Anamese *kim kang rê.*

Smilax hypoglauca Benth. Fl. Hongk. 369. 1861; C. DC. Monog. Phan. **1**: 61. 1878.

" Habitat in montibus Cochinchinae." C. de Candolle (Monog. Phan. **1**: 211. 1878) enumerates *Smilax corbularia* Kunth among the " inextricables." The species is represented by *Clemens 3335, 3430, 3857,* from Tourane and vicinity and these seem to be in all respects *Smilax hypoglauca* Benth. Loureiro does not mention the characteristic glaucous character of the lower surface of the leaves and his description of the lower leaves as cordate is probably due to his attempt to make his description conform to the Linnaean description, which calls for cordate leaves. The leaves are 3- to 5-nerved; Loureiro describes them as 3-nerved.

Smilax lanceaefolia Roxb. Hort. Beng. 72. 1814, *nomen nudum,* Fl. Ind. ed. 2, **3**: 792. 1832; C. DC. Monog. Phan. **1**: 57. 1878; Gagnep. in Lecomte Fl. Gén. Indo-Chine **6**: 767. 1934.

Smilax lanceolata (non Linn.) Lour. Fl. Cochinch. 623. 1790, ed. Willd. 764. 1793, Anamese *cây chaóng chaóng.*

" Habitat agrestis in Cochinchina." The description is short but rather definite and agrees with the characters of Roxburgh's species.

Smilax perfoliata Lour. Fl. Cochinch. 622. 1790, ed. Willd. 763. 1793, Anamese *kim kang mo, tì giai;* A. Chev. in Bull. Écon. Indochine **21**: 327. 1918; Gagnep. in Lecomte Fl. Gén. Indo-Chine **6**: 761. 1934.

Smilax ocreata C. DC. Monog. Phan. **1**: 191. 1878.

" Habitat in collibus silvaticis Cochinchinae." Doctor Chevalier (Bull. Écon. Indochine **21**: 327. 1918) has given an amplified description of Loureiro's species, which up to that time was considered to be one of doubtful status, although he did not recognize that Loureiro's species was the same as the one later described as *Smilax ocreata* C. DC. The specific name was derived not from the perfoliate leaves but from the very conspicuous and characteristic perfoliate stipules which Loureiro describes. It is represented by *Clemens 3623* from Tourane, near the classical locality.

AMARYLLIDACEAE

Crinum Linnaeus

Crinum asiaticum Linn. Sp. Pl. 292. 1753; Lour. Fl. Cochinch. 197. 1790, ed. Willd. 244. 1793, Anamese *cây chuói nuóc,* Chinese *màn sȳ làn;* Gagnep. in Lecomte Fl. Gén. Indo-Chine **6**: 688. 1934.

Crinum cochinchinense M. Roem. Syn. **4**: 71. 1847 (based on *Crinum asiaticum* Lour.).

" Habitat in loca humida in Cochinchina." Loureiro's description conforms to the characters of the widely distributed Linnaean species. It is common along or near the seashore throughout the Indo-Malaysian region. *Radix toxicaria* Rumph. (Herb. Amb. **6**: 155. *pl. 69*), cited by Loureiro as illustrating the species, is correctly placed.

Crinum latifolium Linn. Sp. Pl. 291. 1753; Gagnep. in Lecomte Fl. Gén. Indo-Chine **6**: 686, *fig. 69, 1–4*. 1934.

Amaryllis zeylanica Linn. Sp. Pl. 293. 1753.

Crinum zeylanicum Linn. Syst. ed. 12, 236. 1768; Lour. Fl. Cochinch. 198. 1790, ed. Willd. 245. 1793, Anamese *toi loi*, Chinese *sān toát*.

Crinum loureirii M. Roem. Syn. **4**: 85. 1847 (based on *Crinum zeylanicum* Lour.).

" Habitat loca arenosa in Cochinchina, & China." *Tulipa javana* Rumph. (Herb. Amb. **5**: 306. *pl. 105*), cited by Loureiro as a synonym, represents the Linnaean species. From this cited illustration and Loureiro's description it seems probable that Loureiro may have had specimens of *Crinum latifolium* Linn. On the other hand the Linnaean species is not recorded from China and in all probability occurs in Indo-China only as an introduced and cultivated plant. It may be that he had a small form of *C. asiaticum* Linn., which varies greatly in size as to its vegetative characters.

Lycoris Herbert

Lycoris radiata (L'Hérit.) Herb. App. 20. 1821.

Amaryllis radiata L'Herit. Sert. Angl. 10. 1788.

? *Amaryllis sarniensis* (non Linn.) Lour. Fl. Cochinch. 200. 1790, ed. Willd. 247. 1793, Chinese *hiūien tsào*, Anamese *tuyen thao*.

" Habitat, ob pulchritudinem culta in Sinis." I believe Loureiro misinterpreted *Amaryllis sarniensis* Linn. and that the plant he described is a form of *Lycoris radiata* Herb., which is widely cultivated for ornamental purposes. The statements " floribus paniculatis," " petalis . . . luteis intus rubro punctatis," and " stigma 3-fidum " are against this suggested reduction, as *Lycoris radiata* Herb. normally has red flowers; *L. aurea* (L'Hérit) Herb. has yellow flowers and this is the only species of *Lycoris* recorded by Dunn and Tutcher from Kwangtung. Nor does the description conform any better with the characters of *Hemerocallis*, the flowers of which are more or less paniculate.

Polianthes Linnaeus

Polianthes tuberosa Linn. Sp. Pl. 316. 1753; Lour. Fl. Cochinch. 204. 1790, ed. Willd. 253. 1793 (*Polyanthes*), Anamese *hoa huê*.

" Habitat ubique in Cochinchinae hortis." This is the common tuberose, the Linnaean species being correctly interpreted by Loureiro. *Amica nocturna* Rumph. (Herb. Amb. **5**: 285. *pl. 98*) is correctly placed by Loureiro as a synonym.

Hypoxis Linnaeus

Hypoxis aurea Lour. Fl. Cochinch. 200. 1790, ed. Willd. 248. 1793; Gagnep. in Lecomte Fl. Gén. Indo-Chine **6**: 678. 1934.

" Habitat in colle arenoso *Son Koung* in Cochinchina." It is a well-known species extending from India to Japan and southward to Luzon and Java.

Curculigo Gaertner

Curculigo capitulata (Lour.) O. Ktz. Rev. Gen. Pl. 703. 1891 (based on *Leucoium capitulatum* Lour.).

Leucoium capitulatum Lour. Fl. Cochinch. 199. 1790, ed. Willd. 246. 1793, Anamese *hùynh lon.*

Curculigo recurvata Dryand. in Ait. Hort. Kew. ed. 2, **2**: 253. 1811; Gagnep. in Lecomte Fl. Gén. Indo-Chine **6**: 681. 1934.

Molineria recurvata Herb. Amaryl. 84. 1837; Brackett in Rhodora **25**: 161. *f. 17.* 1923 (Contr. Gray Herb. **69**: 161).

" Habitat inculta, non frequens, in Cochinchina." This species is widely distributed in the Indo-Malaysian region, being commonly known as *Cucurligo recurvata* Dryand.

TACCACEAE

Tacca [55] J. R. & G. Forster

Tacca pinnatifida J. R. & G. Forst. Char. Gen. 70. *pl. 35.* 1776; Lour. Fl. Cochinch. 300. 1790, ed. Willd. 368. 1793, Anamese *cây nua.*

" Habitat frequens in Cochinchina culta in hortis, & agris: etiam in China." Loureiro's description applies to Forster's species, which is widely distributed in the Old World tropics.

DIOSCOREACEAE

Dioscorea (Plumier) Linnaeus

Dioscorea alata Linn. Sp. Pl. 1033. 1753; Lour. Fl. Cochinch. 623. 1790, ed. Willd. 765. 1793, Anamese *khoai tía,* Chinese *yú thâu;* Prain & Burkill in Lecomte Fl. Gén. Indo-Chine **6**: 735. 1934.

Dioscorea eburina Lour. Fl. Cochinch. 625. 1790, Anamese *khoai ngà.*

Dioscorea eburnea Willd. in Lour. Fl. Cochinch. ed. Willd. 767. 1793, Anamese *khoai ngà.*

For *Dioscorea alata* Linn., which Loureiro correctly interpreted, he states: " Habitat culta in Cochinchina, & China, in multisque Indiae locis "; and for *D. eburina*: " Habitat agrestis, cultaque in Cochinchina." Mr. I. H. Burkill thinks that *D. eburina* Lour. is a race of *D. alata* Linn., the specific name being merely a Latin translation of the local one referring to the shape and color of the tubers, which resemble the ivory tusks of the elephant. *Khoai nga* is considered by Burkill (Gard. Bull. Straits Settlem. **3**: 207. 1924) and a photographic reproduction of the tubers of this race is given by him (*op. cit.* **1**: *pl. 5.* 1917).

Dioscorea cirrhosa Lour. Fl. Cochinch. 625. 1790, ed. Willd. 767. 1793, Anamese *khoai leng;* Prain & Burkill in Journ. As. Soc. Bengal n.s. **10**: 31. 1914; Prain & Burkill in Lecomte Fl. Gén. Indo-Chine **6**: 738. 1934.

Dioscorea rhipogonoides Oliv. in Hook. Ic. **19**: *pl. 1868.* 1889, excl. ♀.

Dioscorea bonnetii A. Chev. in Bull. Écon. Indochine **21**: 328. 1918.

---

[55] *Tacca* J. R. & G. Forster (1776), conserved name, Brussels Code; an older one is *Leontopetaloides* Boehmer (1760).

" Habitat in sylvis Cochinchinae." Loureiro's type is preserved in the herbarium of the British Museum. As Prain and Burkill note, this is a very characteristic species. It is known from Formosa, Hongkong, Kwangtung and Indo-China. Mr. Burkill calls my attention to the fact that Oliver's illustration of *D. rhipogonoides* is based on a flowering specimen of *D. cirrhosa* Lour. with the fruits of a different species added. Knuth (Pflanzenreich **87**(IV–43): 288. 1924) has certainly misinterpreted the species in his reference to it of Philippine and Ceram material.

Dioscorea esculenta (Lour.) Burkill in Gard. Bull. Straits Settlem. **1**: 396. *pl. 7*. 1917 (based on *Oncus esculentus* Lour.); Merr. Interpret. Herb. Amb. 147. 1917; Knuth in Pflanzenreich **87**(IV–43): 189. 1924; Prain & Burkill in Lecomte Fl. Gén. Indo-Chine **6**: 713. 1934.

   Oncus esculentus Lour. Fl. Cochinch. 194. 1790, ed. Willd. 240. 1793, Anamese *khoai buu*.

   Convolvulus mammosus Lour. Fl. Cochinch. 108. 1790, ed. Willd. 132. 1793, pro parte, Anamese *khoai tù*.

   Dioscorea fasciculata Roxb. Fl. Ind. ed. 2, **3**: 801. 1832.

   Dioscorea tiliaefolia Kunth Enum. **5**: 401. 1850.

   Oncorhiza esculenta Pers. ex Jackson Ind. Kew. **3**: 346. 1894 (based on *Oncus esculentus* Lour.).

For *Oncus esculentus* Loureiro states: " Habitat in sylvis Cochinchinae," and his type is preserved in the herbarium of the British Museum. It may be noted that Persoon does not actually make the combination *Oncorhiza esculenta* (Syn. **1**: 374. 1805) credited to him in Index Kewensis. He merely enumerates the species as " *Oncus* (*Onchorhiza*) *esculentus*." For *Convolvulus mammosus* Loureiro states: " Habitat frequenter cultus in agris Cochinchinae." It was apparently based on a mixture of material, i.e., as to the vegetative characters and flowers, a form of *Ipomoea batatas* Poir., and as to its Anamese name and its tubers, *Dioscorea esculenta* (Lour.) Burkill. *Battata mammosa* Rumph. (Herb. Amb. **5**: 370. *pl. 131*), which Loureiro cites as a synonym and from which he took his specific name, was apparently based in part on *Operculina turpethum* (Linn.) S. Manso and in part on the tubers of some species of *Dioscorea*. Doctor A. Chevalier, writing from Saigon in November, 1918, informs me that after studying Loureiro's descriptions and investigating the native names, he finds that *khoai buu*, *khoai lo* and *khoai tu* are Anamese names still in use for cultivated forms or varieties of *Dioscorea esculenta* (Lour.) Burkill. Gagnepain and Courchet's reduction of *Convolvulus mammosus* Lour., and the binomials based upon it, *Merremia mammosa* Hall. f., and *Ipomoea mammosa* Choisy (Lecomte Fl. Gén. Indo-Chine **4**: 254. 1915) as doubtful synonyms of *Ipomoea gomezii* C. B. Clarke, does not appear to me to be well taken, as it is very unlikely that a convolvulaceous plant, other than *Ipomoea batatas* Poir., common in cultivation and yielding edible tubers, would have been overlooked by all the numerous field workers in Indo-China during the past sixty years. They record *Ipomoea gomezii* C. B. Clarke from Indo-China on the basis of a single collection from Pulu Condor, which is out of range as far as Loureiro's species is concerned. Mr. Burkill calls my attention to the fact that the British Museum specimen of *Oncus esculentus* Lour. is a specimen with pistillate flowers, but Loureiro mistook the bracts and bracteoles for the calyx, the sepals and petals for the corolla, the staminodes for stamens, and added erroneous fruit characters. Diels (Pflanzenreich 46(IV–94): 61. 1910) cites "' *Oncus esculentus* Lour.' in Herb. Sprengel " as a synonym of *Tiliacoria acuminata* (Lam.) Hook. f. & Th., but the specimen, *fide* Diels in lit., was one collected by Rottler and merely represents a misidentification of Loureiro's species. Mr. Burkill agrees with me that Loureiro certainly had tubers of *D. esculenta* (Lour.) Burkill which he described in *Convolvulus mammosus*, the rest of this description being probably based on *Ipomoea batatas* Poir.

Dioscorea persimilis Prain & Burkill in Journ. As. Soc. Bengal n.s. **4**: 454. 1908, **10**: 39.
1914; Knuth in Pflanzenreich **87**(IV–43): 267. 1924; Prain & Burkill in Lecomte Fl. Gén. Indo-Chine **6**: 732. 1934.

> *Dioscorea oppositifolia* (non Linn.) Lour. Fl. 624. 1790, ed. Willd. 766. 1793, Anamese *khoai mài, son duoc,* Chinese *xān yŏ.*

" Habitat in sylvis Cochinchinae, & Chinae, amatque loca argillacea & petrosa." This is Prain and Burkill's identification of Loureiro's concept of *Dioscorea oppositifolia* Linn. (Kew Bull. 66. 1925) which is doubtless the correct disposition of it. The species is apparently common in southern China and occurs also in Indo-China. Mr. Burkill calls my attention to the fact that *khoai mai* is described as a wild plant with delicate edible tubers, and feels confident that its identification as *Dioscorea persimilis* Prain & Burkill is correct. The Chinese name *xan yo,* i.e., *shan yu* (mountain yam) is usually applied to *Dioscorea opposita* Thunb. (*D. batatas* Decne.).

Dioscorea sp.

> *Dioscorea aculeata* (non Linn.) Lour. Fl. Cochinch. 625. 1790, ed. Willd. 768. 1793, Anamese *khoai lô.*

" Habitat agrestis in Cochinchina." For this Loureiro cites as a synonym, after Linnaeus, *Combilium* Rumph. (Herb. Amb. **5**: 357. pl. 126) which Linnaeus himself erroneously referred to *Dioscorea aculeata;* the Linnaean species is the form that appears in current literature as *Dioscorea wallichii* Hook. f. *Combilium* Rumph. is *Dioscorea esculenta* (Lour.) Burkill. There is a staminate specimen of *khoai la* from Loureiro in the herbarium of the British Museum, with detached elliptic-ovate leaves. Mr. Burkill informs me that it represents a species near *Dioscorea cirrhosa* Lour., not determinable at present, in the absence of complete material. As noted under *Dioscorea esculenta* (Lour.) Burkill, Doctor Chevalier reports *khoai lo* as one of the local names for *Dioscorea esculenta* (Lour.) Burkill. On the basis of Loureiro's reference of *Combilium* Rumph. to this species, I had formerly thought that *Dioscorea aculeata* (non Linn.) Lour. was the same as *D. esculenta* Burkill, but this seems not to be the case.

IRIDACEAE

Belamcanda [56] Adanson

Belamcanda chinensis (Linn.) DC. in Red. Lil. **3**: *pl. 121.* 1807; Gagnep. in Lecomte Fl. Gén. Indo-Chine **6**: 675. 1934.

> *Ixia chinensis* Linn. Sp. Pl. 36. 1753; Lour. Fl. Cochinch. 36. 1890, ed. Willd. 46. 1793, Anamese *ré quat, xa căn,* Chinese *xĕ càn.*

[56] *Belamcanda* Adanson (1763), conserved name, Vienna Code; an older one is *Gemmingia* Heister (1759).

Belamcanda punctata Moench Meth. 529. 1794.

Pardanthus chinensis Ker in Koenig & Sims Ann. Bot. 1: 247. 1805.

" Habitat culta, incultaque in Cochinchina, & China." The description unmistakably applies to the widely cultivated Linnaean species.

Gladiolus (Tournefort) Linnaeus

Gladiolus sp.

Gladiolus undulatus (non Linn.) Lour. Fl. Cochinch. 36. 1790, ed. Willd. 45. 1793.

" Habitat Cantone Sinarum, in hortis cultis." The description is clearly that of a *Gladiolus*, but it does not seem to me to apply to the Linnaean species = *Gladiolus cuspidatus* Jacq. That *Gladiolus* was cultivated in Canton in the last quarter of the 18th century is not surprising in view of the considerable amount of traffic between Europe and Canton at that time via the Cape of Good Hope.

MUSACEAE

Musa Linnaeus

Musa nana Lour. Fl. Cochinch. 644. 1790, ed. Willd. 791. 1793, Anamese *chuói duŭ;* K. Schum. in Pflanzenreich 1(IV–45): 19. 1900; Gagnep. in Lecomte Fl. Gén. Indo-Chine 6: 142. 1932.

Musa cavendishii Lamb. in Paxt. Mag. Bot. 3: 51. *1 pl.* 1837; K. Schum. in Pflanzenreich 1(IV–45): 17. 1900.

Judging from the native name cited, Loureiro had specimens from Indo-China. His description applies to the dwarfed form extensively cultivated in some parts of the world and known as the Chinese or dwarf banana and technically as *Musa cavendishii* Lamb.; this is currently accepted as a valid species, yet it does not appear to be other than a form or variety of the common banana, *Musa paradisiaca* Linn. *Musa humilis* Perr. (Mém. Soc. Linn. Paris 3: 131. 1824), cited by K. Schumann as a doubtful synonym of *Musa cavendishii* Lamb., must be excluded as it represents *Musa paradisiaca* Linn. var. *humilis* (Perr.) Merr. (Enum. Philip. Fl. Pl. 1: 224. 1923) illustrated by Teodoro (Philip. Journ. Sci. Bot. 10: 392. *pl. 18. f. 1–5.* 1915) as *Musa humilis* Perr.

Musa paradisiaca Linn. subsp. **seminifera** (Lour.) K. Schum. in Pflanzenreich 1(IV–45): 21. 1900 (based on *Musa seminifera* Lour.).

Musa sapientum Linn. subsp. *seminifera* Baker in Ann. Bot. 7: 213. 1893 (based on *Musa seminifera* Lour.).

Musa seminifera Lour. Fl. Cochinch. 644. 1790, ed. Willd. 791. 1793, Anamese *chúoi dá* (var. 1), *chúoi sú* (var. 2), *chúoi màt* (var. 3).

The three varieties are so briefly characterized as to be recognizable only by the native names listed. All have fruits with numerous seeds. Conventionally the seeded forms and those whose pulp is edible only after cooking, the plantains, are frequently referred to *Musa sapientum* Linn., and the seedless ones, fruits edible without cooking, the bananas, to *Musa paradisiaca* Linn., but there seems to be no justification for recognizing two species here. The older Linnaean binomial is here adopted as the group name *Musa paradisiaca* Linn. dating from 1753; *M. sapientum* Linn. was published in 1759.

Musa paradisiaca Linn. Sp. Pl. 1043. 1753, var.

> *Musa corniculata* Lour. Fl. Cochinch. 644. 1790, ed. Willd. 791. 1793, Anamese *chúoi boi;* K. Schum. Pflanzenreich 1(IV–45): 21. 1900.

One of the numerous forms of the common banana. There is little or no reason for considering that *Musa corniculata* Rumph. (Herb. Amb. **5**: 130) represents the same form which Loureiro described, or for considering that *Musa corniculata* Lour. represents a distinct species as K. Schumann has done. Gagnepain (Lecomte Fl. Gén. Indo-Chine **6**: 141. 1932) retains Loureiro's species, after K. Schumann, as a valid one, taking his brief description from Loureiro.

Musa paradisiaca Linn. subsp. [var.] **sapientum** O. Ktz. var. **odorata** (Lour.) K. Schum. in Pflanzenreich 1(IV–45): 20. 1900 (based on *Musa odorata* Lour.).

> *Musa sapientum* Linn. var. *odorata* Baker in Ann. Bot. **7**: 212. 1893 (based on *Musa odorata* Lour.).
>
> *Musa odorata* Lour. Fl. Cochinch. 644. 1790, ed. Willd. 791. 1793, Anamese *chúoi bà huong* (var. 1), *chúoi tieo* (var. 2), *chúoi moi* (var. 3).
>
> *Musa sapientum* Linn. var. *cochinchinensis* Quis. in Philip. Agr. Rev. **12**(3): 56. 1919.

The three varieties so briefly characterized by Loureiro as to be recognizable only by the native names cited are manifestly but cultivated forms of the common banana.

Musa uranoscopos Lour. Fl. Cochinch. 645. 1790, ed. Willd. 792. 1793, Anamese *chúoi tàu.*

> *Musa coccinea* Andr. Bot. Repos. **1**: *pl. 47.* 1799; K. Schum. in Pflanzenreich 1(IV–45): 23. 1900; Gagnep. in Lecomte Fl. Gén. Indo-Chine **6**: 140. 1932.

From the local name cited by Loureiro it appears that his specimens were from Indo-China. His description is manifestly based on actual specimens, not on the pre-Linnaean *Musa uranoscopos* Rumph. (Herb. Amb. **5**: 137), cited as a synonym, and the source of his specific name. In my work on Rumphius (p. 149) I referred this to *Musa paradisiaca* Linn. but I am now convinced that what Rumphius actually had was a form of *Musa fehi* Bert. as he describes it as having erect inflorescences; the term *uranoscopos* used by Rumphius is merely a translation of the Malay *toncat langit*, literally meaning to look at or watch the sky. Loureiro's description applies unmistakably to the species with red bracts cultivated for ornamental purposes and currently known as *Musa coccinea* Andr., but not at all to the form Rumphius described. The binomial *Musa uranoscopos* Lour. is valid under all rules for this particular species (*M. coccinea* Andr.).

ZINGIBERACEAE

Zingiber Adanson

Zingiber officinale Rosc. in Trans. Linn. Soc. **8**: 348. 1807; Gagnep. in Lecomte Fl. Gén. Indo-Chine **6**: 82. 1908.

> *Amomum zingiber* Linn. Sp. Pl. 1. 1753; Lour. Fl. Cochinch. 2. 1790, ed. Willd. 2. 1793, Anamese *cây gùng, sinh kùong*, Chinese *sēm kiām*.

"Habitat in Cochinchina, & China, ubique cultum." Loureiro's specimen, in the herbarium of the Paris Museum, represents the Linnaean species, and his description conforms to its characters. *Zingiber majus* Rumph. (Herb. Amb. **5**: 156. *pl. 66. f. 1*), cited by Loureiro, after Linnaeus, as a synonym, is correctly placed. It may be noted, however, that Gagnepain (Lecomte Fl. Gén. Indo-Chine 6: 84. 1908) places *Amomum zingiber* Lour. as a synonym of *Zingiber zerumbet* Sm., yet under *Z. officinale* Rosc. he cites the local name "*Cay gûng* (Lour. Fl. Cochinch. p. 2)." It is suspected that *Amomum zingiber* in this case was a *lapsus calami* for *A. zerumbet*.

Zingiber zerumbet (Linn.) Smith Exot. Bot. 2: 105. *pl. 112.* 1805; Gagnep. in Lecomte Fl. Gén. Indo-Chine 6: 84. 1908.

> *Amomum zerumbet* Linn. Sp. Pl. 1. 1753; Lour. Fl. Cochinch. 2. 1790, ed. Willd. 3. 1793, Anamese *ngai xanh, ngai mat tlòi.*

"Habitat incultum, cultumque in Cochinchina." Loureiro's specimen in the herbarium of the Paris Museum represents the common and widely distributed Linnaean species, and the description applies to it. *Lampujum* Rumph. (Herb. Amb. 5: *pl. 64. f. 1*) is correctly placed as a synonym.

Languas Retzius

(*Alpinia* auct. plur., non Linn.)

Languas galanga (Linn.) Merr. in Lingnan Sci. Journ. 5: 51. 1927.

> *Maranta galanga* Linn. Sp. Pl. ed. 2, 3. 1762; Sw. Obs. 8. 1791.
>
> *Amomum galanga* Lour. Fl. Cochinch. 5. 1790, ed. Willd. 7. 1793, Anamese *cây rièng, cao luong kuong,* Chinese *cào leâm kiām.*
>
> *Alpinia galanga* Willd. Sp. Pl. 1: 12. 1797; Rosc. in Trans. Linn. Soc. 8: 345. 1807; Gagnep. in Lecomte Fl. Gén. Indo-Chine 6: 87. 1908.

"Habitat tam culta, quam agrestis in Cochinchina, & China." Loureiro's description apparently applies to the Linnaean species. *Galanga major* Rumph. (Herb. Amb. 5: 143. *pl. 63*), cited by Loureiro as representing the species, is correctly placed. Swartz, credited in most modern literature as the author of *Alpinia galanga* (Obs. 8. 1791) does not make this combination. He merely states "MARANTA *Galanga.* Obs. *Alpiniae* forte species: cfr. fig. Rumph. amb. 5. t. 63. f. 6 "; Willdenow made the transfer in 1797.

Languas globosa (Lour.) Burkill in Kew Bull. 26. 1930 (based on *Amomum globosum* Lour.).

> *Amomum globosum* Lour. Fl. Cochinch. 4. 1790, ed. Willd. 6. 1793, Anamese *mé tlé,* Chinese *tsāo keu.*
>
> *Alpinia globosa* Horan. Prodr. Monog. Scit. 34. 1862; Gagnep. in Lecomte Fl. Gén. Indo-Chine 6: 90. 1908 (based on *Amomum globosum* Lour.).

"Habitat in montibus Cochinchinae, & Chinae." Gagnepain has given an ample description of the species based on specimens from Tonkin, his interpretation undoubtedly being correct. K. Schumann (Pflanzenreich 20(IV-46): 368. 1904) left it among the "species haud satis notae" in the section *Bintalua.*

Languas sp.

> *Amomum hirsutum* Lour. Fl. Cochinch. 5. 1790, ed. Willd. 6. 1793, Anamese *mé tlé bà.*
>
> *Alpinia hirsuta* Horan. Prodr. Monog. Scit. 34. 1862; K. Schum. in Pflanzenreich 20 (IV-46): 368. 1904 (based on *Amomum hirsutum* Lour.).

"Habitat in sylvis Cochinchinae." The species is clearly a *Languas* but from Loureiro's description alone I am unable to place it among the 18 species of this genus (as *Alpinia*) admitted by Gagnepain for Indo-China, who does not mention it in his treatment of the Zingiberaceae (Lecomte Fl. Gén. Indo-Chine 6: 25-121. 1908). K. Schumann

(Pflanzenreich 20(IV–46): 368. 1904), placed it in the section *Bintalua* as one of the " species haud satis notae " with a brief description compiled from Loureiro's original one.

Languas sp.

> *Amomum medium* Lour. Fl. Cochinch. 4. 1790, ed. Willd. 5. 1793, Anamese *thao qua*, Chinese *tsăo quo.*

" Habitat in provincia Yū nän imperii Sinensis, ad occasum provinciae Cantoniensis." Loureiro described the stem, leaves and fruits, but saw no flowers. Willdenow in a footnote states: " Fortasse Languas vulgare *Koenigii l.c.* p. 64. eadem est planta." C. H. Wright (Journ. Linn. Soc. Bot. 36: 70. 1903) admits *Amomum medium* Lour. with the comment that it has been doubtfully referred to *Alpinia alba* Rosc., which K. Schumann reduces to *A. galanga* Willd. The description manifestly appertains to a species of *Languas* (*Alpinia*), but beyond that its identity is uncertain. It may well represent *Languas galanga* (Linn.) Merr.

Amomum Linnaeus

Amomum repens Sonner. Voy. Ind. 2: 240. *pl. 136.* 1782; Gagnep. in Lecomte Fl. Gén. Indo-Chine 6: 107. 1908.

> *Amomum cardamomum* (non Linn.) Lour. Fl. Cochinch. 3. 1790, ed. Willd. 4. 1793, Anamese *bach dâu khâu*, Chinese *pě téu keu.*

" Habitat agreste in regno Cambodia, Cochinchinae tributario." Loureiro's description apparently applies to Sonnerat's species as interpreted by Gagnepain, not to the Linnaean species which, as currently interpreted, is an *Elettaria.* *Cardamomum minus* Rumph. (Herb. Amb. 5: 152. *pl. 65. f. 1*), cited by Loureiro as a synonym, is perhaps *Amomum cardamomum* Willd.

Amomum villosum Lour. Fl. Cochinch. 4. 1790, ed. Willd. 4. 1793, Anamese *sa nhon*, Chinese *sŏ xā mí.*

> *Amomum echinosphaera* K. Schum. in Bot. Jahrb. 27: 322. 1899, Pflanzenreich 20(IV–46): 248. 1904; Gagnep. in Bull. Soc. Bot. France 49: 257. 1902, Lecomte Fl. Gén. Indo-Chine 6: 105. 1908.

" Habitat agreste in montibus Cochinchinae." There seems to be no doubt as to the identity of *Amomum echinosphaera* K. Schum. with *A. villosum* Lour. Loureiro's rather inappropriate name invalidates *Amomum villosum* Blume, the currently accepted name for a Malaysian species; it was derived from the fruit characters which he describes as: " Pericarpium . . . exterius obsessum villis multis, crassis," it being really densely echinulate. *Globba crispa* Rumph. (Herb. Amb. 6: 137. *pl. 61*), cited by Loureiro as representing his species, must be excluded, as it is, in part, *Amomum roseum* Benth. & Hook. f., and in part *Amomum* sp., both different from the Indo-China form described by Loureiro.

Kaempferia Linnaeus

Kaempferia galanga Linn. Sp. Pl. 2. 1753; Lour. Fl. Cochinch. 12. 1790, ed. Willd. 15. 1793, Anamese *thien lien, tam nai*, Chinese *săn lây;* Gagnep. in Lecomte Fl. Gén. Indo-Chine 6: 49. 1908.

" Habitat in hortis Cochinchinae, & Chinae." The description applies to the well-known Linnaean species, one of wide distribution in tropical Asia.

Curcuma Linnaeus

Curcuma longa Linn. Sp. Pl. 2. 1753; Lour. Fl. Cochinch. 8. 1790, ed. Willd. 11. 1793, Anamese *ngê, kuong hùynh*, Chinese *kiām hoâm;* Gagnep. in Lecomte Fl. Gén. Indo-Chine **6**: 63. 1908.

" Habitat ubique culta, incultaque in Cochinchina, & China." Loureiro's description of this plant and his reference to its use as a condiment apply to the Linnaean species as currently interpreted. *Curcuma domestica* Rumph. (Herb. Amb. **5**: 164. *pl. 67*) (*err. cit. pl. 64*), which he cites, after Linnaeus, as representing the species, is correctly placed. Valeton (Merrill, Interpret. Herb. Amb. 163. 1917) interprets the Linnaean species from the reference to Hermann (Lugdb. 208. *pl. 209*) as probably being the same as *Curcuma aromatica* Salisb., but the second reference (Fl. Zeyl. 7) is *Curcuma longa* Linn. as most authors understand it; see Trimen (Journ. Linn. Soc. Bot. **24**: 133. 1887). The specimen in the Linnaean herbarium cannot possibly be the type, as it was not in the herbarium in 1753 or in 1754; it is listed in the 1767 enumeration, according to Jackson (Proc. Linn. Soc. **124**: Suppl. 66. 1912). My inclination is to interpret the Flora Zeylanica reference as the type as it was based on a specimen examined by Linnaeus and one that is still extant.

Curcuma rotunda Lour. Fl. Cochinch. 9. 1790, ed. Willd. 11. 1793, Anamese *ngai mio*, Chinese *pum ngô mêu.*

" Habitat frequens montes Cochinchinae, & Chinae." The description is, for the most part, of the leaves and the rhizomes only, as Loureiro states that he saw no flowers. While probably a *Curcuma* is represented, it cannot with any degree of certainty be referred to any of the seventeen species now known from Indo-China. The references to Rheede, Burman, and Rumphius may be ignored. *Manjella Kua* Rheede (Hort. Malabar. **11**: 17. *pl. 11*) represents *Kaemfera pandurata* Roxb. *Zerumbet majus* Rumph. (Herb. Amb. **5**: 168. *pl. 68*) is *Curcuma zedoaria* Rosc. or *C. viridiflora* Roxb. Willdenow in a footnote states: " Secundum *Clariss. Koenigium* est Amomi species." K. Schumann (Pflanzenreich **20**(IV–46): 114. 1904) gives a very short description compiled from Loureiro.

Curcuma zedoaria (Berg.) Rosc. Mondr. Pl. *pl. 109.* 1828; Gagnep. in Lecomte Fl. Gén. Indo-Chine **6**: 67. 1908.

Amomum zedoaria Berg. Mat. Med. 4. 1778.

Curcuma pallida Lour. Fl. Cochinch. 9. 1790, ed. Willd. 12. 1793, Anamese *ngê hoang*, Chinese *san kiām hoâm.*

" Habitat agrestis in Cochinchina, & Cantone Sinarum." The description seems to apply to the very common and widely distributed *Curcuma zedoaria* Rosc. K. Schumann (Pflanzenreich **20**(IV–46): 115. 1904) gives merely a short description compiled from Loureiro, stating that the data given by Loureiro are insufficient from which to make an accurate identification.

Costus Linnaeus

Costus speciosus (Koenig) Sm. in Trans. Linn. Soc. **1**: 249. 1791.

Banksia speciosa Koenig in Retz. Obs. **3**: 75. 1783.

Amomum arboreum Lour. Fl. Cochinch. 7. 1790, ed. Willd. 9. 1793.

Costus loureiri Horan. Prodr. Monog. Scit. 38. 1862 (based on *Amomum arboreum* Lour.).

" Habitat in insula Samatra [Sumatra], versus plagam orientalem in sylva parum a litore distante." Manifestly a species of *Costus* is represented by the description. K. Schumann (Pflanzenreich 20(IV–46): 398. 1904) placed Loureiro's species as a synonym of *Costus speciosus* Sm. which may be the correct disposition of it, although in some details Loureiro's description does not agree very well with the characters of Koenig's species; the latter is, however, a common and widely distributed one in Malaysia and is the one Loureiro most likely would have observed.

CANNACEAE

Canna Linnaeus

Canna indica Linn. Sp. Pl. 1. 1753; Lour. Fl. Cochinch. 10. 1790, ed. Willd. 13. 1793, Ana-
    mese *ngai hoang*, Chinese *san kiām*.

" Habitat in collibus et agris incultis Cochinchinae, ac Chinae." Loureiro's description applies unmistakably to the common, widely distributed form with small red flowers that was introduced into the Old World tropics at an early date from America. *Cannacorus* Rumph. (Herb. Amb. **5**: 177. *pl. 71. f. 2*), cited by Loureiro as a synonym, is correctly placed.

MARANTACEAE

Donax Loureiro

Donax arundastrum Lour. Fl. Cochinch. 11. 1790, ed. Willd. 15. 1793, Anamese *cây lung;*
    Rolfe in Journ. Bot. **45**: 243. 1907.

" Habitat in sylvis Cochinchinae." K. Schumann in his monograph of the Maranta-ceae (Pflanzenreich 11(IV–48): 32. 1902) curiously misinterpreted the genus *Donax*, placing in it two species, *D. arundastrum* (non Lour.) and *D. virgata* K. Schum.; the first is *Schumannianthus dichotomus* Gagnep., and the second is *S. virgatus* Rolfe. *Actoplanes* K. Schum., described as a new genus, is *Donax* Lour. The complicated synonymy has been adjusted by Rolfe (Journ. Bot. **45**: 242–244. 1907) on the basis of an examination of Loureiro's type specimen in the herbarium of the British Museum. *Arundastrum* Rumph. (Herb. Amb. **4**: 22. *pl. 7*), cited by Loureiro as representing *Donax arundastrum* Lour., must be excluded as it represents *Donax cannaeformis* (Forst.) K. Schum.

Phrynium [57] Willdenow

Phrynium placentarium (Lour.) Merr. in Philip. Journ. Sci. **15**: 230. 1919 (based on *Phyllo-
    des placentaria* Lour.).
    Phyllodes placentaria Lour. Fl. Cochinch. 13. 1790, ed. Willd. 17. 1793, Anamese *lâ
        deaong*, Chinese *toung iep;* Moore in Journ. Bot. **63**: 246. 1925.
    Maranta placentaria A. Dietr. Sp. Pl. **1**: 30. 1831 (based on *Phyllodes placentaria* Lour.).
    Phrynium parviflorum Roxb. Fl. Ind. **1**: 7. 1820; K. Schum. in Pflanzenreich 11(IV–
        48): 54. 1902; Gagnep. in Lecomte Fl. Gén. Indo-Chine **6**: 134. 1932.

" Habitat loca umbrosa Cochinchinae, & Chinae." K. Schumann (Pflanzenreich 11 (IV–48): 53. 1902) cites *Phyllodes placentaria* Lour. as a doubtful synonym of *Phrynium*

---

[57] *Phrynium* Willdenow (1797), conserved name, Vienna Code; an older one is *Phyllodes* Loureiro (1790).

capitatum Willd.   Moore's examination of Loureiro's type in the herbarium of the British Museum (Journ. Bot. **63**: 246. 1925) definitely settles the matter, in that he confirms the conclusion I reached in 1919 on the basis of the descriptions and geographic distribution of the species involved.   Gagnepain cites *Phyllodes placentaria* Lour. as a synonym of *Phrynium parviflorum* Roxb., but curiously (p. 132) cites it also as a doubtful synonym of *P. capitatum* Willd.   Loureiro's generic name is older than Willdenow's but the latter is the conserved one.

ORCHIDACEAE

Habenaria Willdenow

Habenaria susannae (Linn.) R. Br. ex Spreng. Syst. **3**: 692. 1826.

> *Orchis susannae* Linn. Sp. Pl. 939. 1753; Lour. Fl. Cochinch. 522. 1790, ed. Willd. 638. 1793, Chinese *má chăc lân.*
>
> *Platanthera susannae* Lindl. Gen. Sp. Orch. Pl. 295. 1835.
>
> *Pecteilis susannae* Raf. Fl. Tellur. **2**: 38. 1836; Schltr. in Fedde Repert. Beih. **4**: 121. 1919.
>
> *Hemihabenaria susannae* Finet in Rev. Gén. Bot. **13**: 532. 1901.

" Habitat inculta prope Cantonem Sinarum."   Loureiro's description applies to the Kwangtung form currently referred to the Linnaean species.   *Flos susannae* Rumph. (Herb. Amb. **5**: 286. *pl. 99*), cited by Loureiro, after Linnaeus, as a synonym, actually typifies the Linnaean species, and because of the discontinuous distribution it is not unreasonable to expect that the southern China form may prove to be distinct from the Moluccan one; Schlechter confines *Pecteilis susannae* Raf. to Hongkong, Kwangtung, and Yunnan, and Gagnepain (Lecomte Fl. Gén. Indo-Chine **6**: 616. 1934) credits it to India, Indo-China, Siam, and China, yet its type was from the Moluccas and it still grows in Amboina.   R. Brown (Prodr. 312. 1810) does not actually make the transfer of the specific name to *Habenaria* as indicated in current literature.   The generic name *Hemihabenaria* Finet (Rev. Gén. Bot. **13**: 532. 1901) typified by *Orchis susannae* Linn. = *Habenaria susannae* R. Br., is an unnecessary one as the same group had already been characterized by Rafinesque as *Pecteilis* (Fl. Tellur. **2**: 38. 1836) based on *Orchis gigantea* Sm., *Orchis susannae* Linn., and *Orchis radiata* Pers. (*O. susannae* Thunb.).

Habenaria sp.

> *Orchis morio* (non Linn.) Lour. Fl. Cochinch. 523. 1790, ed. Willd. 639. 1793, Anamese *cu deái chôn.*

" Habitat in sylvis montanis Cochinchinae."   Apparently a species of *Habenaria* is represented by Loureiro's description, probably that represented by *Squires 1551*, from near Hue, in the herbarium of the University of California.

Habenaria sp. ?

> *Orchis latifolia* (non Linn.) Lour. Fl. Cochinch. 523. 1790, ed. Willd. 639. 1793, Anamese *hoùng món.*

" Habitat in sylvis planis Cochinchinae ad loca Borealia."   The description is inadequate.   *Habenaria* may or may not be the genus represented.   In any case the species Loureiro described is definitely not the Linnaean one.

Galeola Loureiro

Galeola nudifolia Lour. Fl. Cochinch. 521. 1790, ed. Willd. 636. 1793, Anamese *cây nu deei*.
  Epidendrum galeola Raeusch. Nomencl. ed. 3, 265. 1797 (based on *Galeola nudifolia*
    Lour.).
  Craniches nudifolia Pers. Syn. **2**: 511. 1807 (based on *Galeola nudifolia* Lour.).
  Vanilla pterosperma Lindl. in Wall. List no. 7402. 1832, *nomen nudum*.
  Erythrorchis kuhlii Reichb. f. Xen. Orch. **2**: *pl. 119*. 1862.
  Galeola kuhlii Reichb. f. Xen. Orch. **2**: 78. 1865.
  Galeola hydra Reichb. f. Xen. Orch. **2**: 77. 1865.
  Galeola pterosperma Schltr. in Bot. Jahrb. **45**: 386. 1911.
" Habitat in sylvis Cochinchinae." This is the type of the genus *Galeola*. I believe
Loureiro's species to be identical with the species generally known as *Galeola hydra* Reichb. f.
Ames (Merrill Enum. Philip. Fl. Pl. **1**: 263. 1924) has clearly shown that of the binomials
cited above, other than Loureiro's and those based on it, *G. kuhlii* Reichb. f. has priority.
The species extends from India to Hainan, the Philippines, Malay Peninsula, Java and
Sumatra. Two sheets without flowers are among the Loureiro specimens in the herbarium
of the British Museum. Gagnepain (Lecomte Fl. Gén. Indo-Chine **6**: 635. 1934) places
Loureiro's species as a doubtful synonym of *Galeola altissima* Rchb. f.

Spiranthes [58] L. C. Richard

Spiranthes aristotelia (Raeusch.) Merr. in Philip. Journ. Sci. **15**: 230. 1919 (based on *Aris-
    totelea spiralis* Lour.).
  Aristotelea spiralis Lour. Fl. Cochinch. 522. 1790, ed. Willd. 638, 1793, Chinese *hoân
    lûm*.
  Epidendrum aristotelia Raeusch. Nomencl. ed. 3, 265. 1797 (based on *Aristotelea spi-
    ralis* Lour.).
  Neottia sinensis Pers. Syn. **2**: 511. 1807 (based on *Aristotelea spiralis* Lour.).
  Spiranthes australis Lindl. in Bot. Reg. **10**: *sub pl. 823*. 1824.
  Spiranthes sinensis Ames Orch. **2**: 53. 1908, et in Merr. Enum. Philip. Fl. Pl. **1**: 268.
    1924 (based on *Aristotelea spiralis* Lour.).
  Spiranthes australis Lindl. var. *sinensis* Gagnep. in Lecomte Fl. Gén. Indo-Chine **6**:
    546. 1933.
" Habitat inculta prope Cantonem Sinarum." Loureiro's type is preserved in the
herbarium of the Paris Museum. The species is not uncommon in open grassy places near
Canton and is widely distributed in the Indo-Malaysian region. Loureiro's generic name
is invalidated by *Aristotelia* L'Hérit. (1784), while his specific name is invalidated in *Spiranthes* by *S. spiralis* Koch. Ames accepts Persoon's specific name *sinensis* and cites about
forty synonyms for this much-named species. He states that *Epidendrum aristotelia*
Raeusch. was a *nomen*, hence a binomial without standing; Raeuschel's footnote clearly
indicates that it was based on *Aristotelea spiralis* Lour. and in my opinion constitutes valid
publication.

[58] *Spiranthes* L. C. Richard (1818), conserved name, Vienna Code; older ones are *Gyrostachis* Persoon
(1807) and *Ibidium* Salisbury (1812).

Bletilla Reichenbach f.

Bletilla striata (Thunb.) Rchb. f. in Bot. Zeit. **36**: 75. 1878; Schltr. in Fedde Repert. **10**: 255. 1911.

Limodorum striatum Thunb. Fl. Jap. 28. 1784.

Epidendrum tuberosum (non Linn.) Lour. Fl. Cochinch. 523. 1790, ed. Willd. 639. 1793, Anamese *hoa lon tiá*.

Bletia hyacinthina R. Br. in Ait. Hort. Kew. ed. 2, **5**: 206. 1813; Gagnep. in Lecomte Fl. Gén. Indo-Chine **6**: 374. 1933.

Bletia striata Druce in Rept. Bot. Exch. Club Brit. Isles **4**: 609. 1917.

Polytoma inodora [59] Lour. ex Gomes in Mem. Acad. Sci. Lisb. Cl. Sci. Mor. Pol. Bel.-Let. n.s. 4(1): 30. 1868.

" Habitat cultum in hortis Cochinchinae, & Chinae." *Epidendrum tuberosum* Linn. is a species of doubtful status, being based in part on *Angraecum terrestre primum* Rumph. (Herb. Amb. **6**: 112. *pl. 52. f. 1*) which is also cited by Loureiro. The Rumphian illustration represents *Phaius amboinensis* Blume. A specimen from Loureiro, identified as representing *Epidendrum tuberosum*, is in the herbarium of the Paris Museum as *Polytoma inodora* Lour.

Phaius Loureiro

Phaius tankervilliae (Banks) Blume Mus. Bot. Lugd.-Bat. **2**: 177. 1856 (*tankervillii*).

Limodorum tankervilliae Banks ex L'Hérit. Sert. Angl. 28. 1788, ed. alt. 17. 1788 (*tancarvilleae*).

Limodorum tankervilliae Dryand. in Ait. Hort. Kew. **3**: 302. *pl. 12*. 1789.

Phaius grandifolius Lour. Fl. Cochinch. 529. 1790, ed. Willd. 647. 1793, Anamese *hác lon*.

" Habitat cultus in hortis Cochinchinae, & Chinae, pulchritudine floris aestimabilis." This is the type of the genus *Phaius*. It was first described in 1788 and again in 1789, from specimens cultivated in England introduced from China about 1778. The British Museum specimen listed as being among those received from Loureiro has not been located. Gagnepain (Lecomte Fl. Gén. Indo-Chine **6**: 384. 1933) admits only *P. Wallichii* Lindl. for Indo-China and cites *P. grandifolius* Lour. as a doubtful synonym of Lindley's species.

Dendrobium [60] Swartz

Dendrobium amabile (Lour.) O'Brien in Gard. Chron. III **46**: 393. 1909 (based on *Callista amabilis* Lour.).

Callista amabilis Lour. Fl. Cochinch. 519. 1790, ed. Willd. 634. 1793, Anamese *nhánh goi lon;* Kränzl. in Pflanzenreich **45**(IV–50–II B–21): 314. *f. 27*. 1910; Rolfe in Orch. Rev. **18**: 99, 242. 1910; Moore in Journ. Bot. **63**: 288. 1925.

Epidendrum callista Raeusch. Nomencl. ed. 3, 265. 1797 (based on *Callista amabilis* Lour.)

" Habitat in sylvis Cochinchinae, truncis arborum adhaerens." Loureiro's type is preserved in the herbarium of the British Museum and Kränzlin's illustration was drawn

[59] A Loureiro herbarium name here first published by Gomes.

[60] *Dendrobium* Swartz (1799, 1800), conserved name, Vienna Code; older ones are *Callista* Loureiro and *Ceraia* Loureiro (1790).

from this specimen.   Moore (Journ. Bot. **63**: 288. 1925) calls attention to the correspondence published appertaining to this species (Gard. Chron. III. **46**: 354, 393, 431. 1909, **47**: 19,
34. 1910).   Dammer there states definitely that Loureiro's type is a true *Dendrobium* and,
as he critically examined it, I can see no reason for attempting to retain *Callista* as a distinct genus.   Rolfe confirms Dammer's conclusions.   Gagnepain in his treatment of *Dendrobium* (Lecomte Fl. Gén. Indo-Chine **6**: 194–260. 1932) does not mention Loureiro's
species or any of the synonyms based upon it.

Dendrobium simplicissimum (Lour.) Kränzl. in Pflanzenreich 45(IV–50–II B–21): 235.
    1910, in nota (based on *Ceraia simplicissima* Lour.).
    Ceraia simplicissima Lour. Fl. Cochinch. 518. 1790, ed. Willd. 633. 1793, Anamese
    tach haoc, Chinese *xĕ hŏ.*
    Epidendrum ceraia Raeusch. Nomencl. ed. 3, 265. 1797 (based on *Ceraia simplicissima*
    Lour.).
    Dendrobium ceraia Lindl. Gen. Sp. Orch. Pl. 89. 1830; Schltr. in Fedde Repert. Beih.
    4: 207. 1919 (based on *Ceraia simplicissima* Lour.).
" Habitat in sylvis Cochinchinae, & Chinae, rupibus, ac arboribus inhaerens." Loureiro's description clearly indicates that he had specimens of a *Dendrobium* but the exact
status of his species is uncertain.   *Herba supplex quinta* Rumph. (Herb. Amb. **6**: 111. *pl. 51.
f. 2*), cited as representing a plant resembling *Ceraia simplicissima,* is *Dendrobium calceo-
lum* Roxb.   Lindley thought that Loureiro's species was one allied to *Dendrobium crumena-
tum* Sw.   Kränzlin in transferring Loureiro's specific name to *Dendrobium* at the same
time (op. cit. 234) cites *Ceraia simplicissima* Lour. as a doubtful synonym of *D. blumei*
Lindl. quoting: " Provinz der Philippinen: Manila (Cuming, Loureiro!) "; the exclamation
mark should mean that he had seen a specimen collected by Loureiro, but such a specimen
is not known to be extant and, if extant, it certainly did not come from the Philippines;
Kränzlin may have intended Llanos as the collector instead of Loureiro.   Gagnepain in
his treatment of *Dendrobium* (Lecomte Fl. Gén. Indo-Chine **6**: 194–260. 1932) overlooked
Loureiro's binomial and all synonyms based upon it.

Cymbidium Swartz

Cymbidium ensifolium (Linn.) Sw. in Nov. Act. Soc. Sci. Upsal. II **6**: 77. 1799; Gagnep.
    in Lecomte Fl. Gén. Indo-Chine **6**: 423. 1933.
    Epidendrum ensifolium Linn. Sp. Pl. 954. 1753; Lour. Fl. Cochinch. 524. 1790, ed.
    Willd. 640. 1793, Anamese *hoa lon taù,* Chinese *lân hōa.*
" Habitat curiose cultum in hortis Chinae, & Cochinchinae." Loùreiro apparently
was correct in his interpretation of the Linnaean species, the type of the latter being a
specimen collected by Osbeck at Canton.   A specimen from Loureiro is preserved in the
herbarium of the British Museum.

Renanthera Loureiro

Renanthera coccinea Lour. Fl. Cochinch. 521. 1790, ed. Willd. 637. 1793, Anamese *quách
    lon dieo;* Guill. in Lecomte Fl. Gén. Indo-Chine **6**: 530. 1933.
    Epidendrum renanthera Raeusch. Nomencl. ed. 3, 265. 1797 (based on *Renanthera
    coccinea* Lour.).

" Habitat in sylvis Cochinchinae per arbores repens."   A well-known species, now not uncommon in cultivation.   Loureiro's type, a specimen without flowers, is preserved in the herbarium of the British Museum where it was examined by Reichenbach f. (Flora **51**: 52. 1868).

Thrixspermum Loureiro

Thrixspermum centipeda Lour. Fl. Cochinch. 520. 1790, ed. Willd. 635. 1793, Anamese *nhánh gòi rit;* Guill. in Lecomte Fl. Gén. Indo-Chine **6**: 515. 1933.

Epidendrum thrixspermum Raeusch. Nomencl. ed. 3, 265. 1797 (based on *Thrixspermum centipeda* Lour.).

Sarcochilus centipeda Naves Novis. App. Blanco Fl. Filip. ed. 3, 238. 1882 (based on *Thrixspermum centipeda* Lour.).

" Habitat in Cochinchina, ad arbores sylvestres adrepens."   Loureiro's type is preserved in the herbarium of the British Museum.   Reichenbach f. (Flora **51**: 52. 1868) examined it, stating: " *Thrixspermum* ist eine Form aus der Verwandtschaft der *Dendrobium auriferum* Lindl., *Liparis serraeformis* Lindl., *Dendrocolla arachnites* Bl.   Hierher gehören auch *Sarcochilus*, der grössere Theil von *Dendrocolla, Gunnia, Chiloschista.*"   In reinstating the genus *Thrixspermum* he then transferred 34 species to it, many of which had been previously transferred by him (Xen. Orch. **2**: 120–123. 1867).

Aerides Loureiro

Aerides odoratum Lour. Fl. Cochinch. 525. 1790, ed. Willd. 642. 1793, Anamese *phaong lon,* Chinese *fūm lâu;* Guill. in Lecomte Fl. Gén. Indo-Chine **6**: 465. 1933.

Epidendrum aerides Raeusch. Nomencl. ed. 3, 265. 1797 (based on *Aerides odorata* Lour.).

Polytoma odorifera[61] Lour. ex Gomes in Mem. Acad. Sci. Lisb. Cl. Sci. Mor. Pol. Bel.-Let. n.s. **4**(1): 30. 1868.

" Habitat in sylvis Cochinchinae, & Chinae, arboribus inhaerens, vel ex illis pendula." This is the type of the genus and is a well-known species of wide distribution in tropical Asia.   Ames (Orchidaceae **2**: 250. 1908) gives a very complete list of synonyms including references to the numerous illustrations which have been published by various authors. Specimens from Loureiro are preserved in the herbaria of the British and Paris Museums.

DICOTYLEDONEAE

CASUARINACEAE

Casuarina (Rumphius) Adanson

Casuarina equisetifolia Forst. Char. Gen. 103. *pl. 52.* 1776.

Casuarina equisifolia Linn. Amoen. Acad. **4**: 143. 1759, *nomen.*

Casuarina africana Lour. Fl. Cochinch. 549. 1790, ed. Willd. 670. 1793.

" Habitat litora arenosa continentis Africae Orientalis."   The form Loureiro described is manifestly the widely distributed littoral *Casuarina equisetifolia* Forst. for which he cites as synonyms "*Casuarina equisetifolia* (Lin. Jun. Suppl. pag. 412. ex Forst. gen. n.

[61] A Loureiro herbarium name here first published by Gomes.

52) " and " *Casuarina litorea* (Rumph. Amb. 1. 4. cap. 50. tab. 57) " [Herb. Amb. 3: 86. *pl. 57*]. He states that these do not differ from the African form. The binomial *Casuarina equisifolia* was published by Linnaeus in 1759, but without a description other than the reference to Rumphius. The genus *Casuarina* was not formally characterized until 1776.

SAURURACEAE

Saururus (Plumier) Linnaeus

Saururus chinensis (Lour.) Baill. in Adansonia 10: 71. 1871 (based on *Spathium chinense* Lour.).

> *Spathium chinense* Lour. Fl. Cochinch. 217. 1790, ed. Willd. 270. 1793, Chinese *thong pin ngau.*
> *Saururus loureirii* Decne. in Ann. Sci. Nat. III Bot. 3: 102. 1845 (based on *Spathium chinense* Lour.).

" Habitat in locis paludosis, prope Cantonem Sinarum." *Saururus chinensis* Loud. (Hort. Brit. 144. 1830) is a *nomen nudum*, and the binomial appears again (Loudon Encycl. Pl. 298. 1866) with a six-word description without reference to any other diagnosis of the species. It is common in Kwangtung Province; the Cantonese name appears on recently collected material as *tong pin ngau.*

Houttuynia Thunberg

Houttuynia cordata Thunb. Fl. Jap. 234. *pl. 26*. 1784; C. DC. in Lecomte Fl. Gén. Indo-Chine 5: 60. *f. 7*. 1910.

> *Polypara cochinchinensis* Lour. Fl. Cochinch. 61. 1790, ed. Willd. 78. 1793, Anamese *rau giáp cá.*
> *Polypara cordata* O. Ktz. Rev. Gen. Pl. 565. 1891.

" Habitat in hortos Cochinchinenses." Loureiro's description conforms in all respects to the characters of Thunberg's species, which extends from India to Japan, Siam, Indo-China, and Formosa. Thunberg's original spelling of the generic name was *Houtuynia;* the correct form is *Houttuynia*. A fragmentary Loureiro specimen is preserved in the herbarium of the British Museum. O. Kuntze adopts *Polypara* Lour. as the valid name for this genus because of the earlier *Houttuynia* Houtt. (1780) = *Acidanthera* Hochst.

PIPERACEAE

Piper Linnaeus

Piper betle Linn. Sp. Pl. 28. 1753; Lour. Fl. Cochinch. 31. 1790, ed. Willd. 39. 1793, Anamese *cây tlâù*, Chinese *lâu yep;* C. DC. in Lecomte Fl. Gén. Indo-Chine 5: 74. 1910.

No locality is cited, but unquestionably Loureiro knew this commonly cultivated species as an Indo-China plant. His description and indicated uses clearly apply to the Linnaean species. The modern Anamese name is given by C. de Candolle as *jaou.*

Piper longum Linn. Sp. Pl. 29. 1753; Lour. Fl. Cochinch. 32. 1790, ed. Willd. 40. 1793, Anamese *cây lôt, tăt phăt*, Chinese *pipŏ;* C. DC. in Lecomte Fl. Gén. Indo-Chine 5: 71. 1910.

No locality is indicated by Loureiro but he undoubtedly had Indo-China specimens. His description apparently applies to the Linnaean species as interpreted and amply described by C. de Candolle, although it is possible that he may have had specimens of *Piper retrofractum* Vahl. *Piper longum* Rumph. (Herb. Amb. **5**: 333. *pl. 116. f. 1*), regarding which Loureiro states: " ubi fructus, non folia, cum nostris conveniunt " represents *Piper retrofractum* Vahl.

Piper nigrum Linn. Sp. Pl. 28. 1753; Lour. Fl. Cochinch. 30. 1790, ed. Willd. 37. 1793, Anamese *tieo bo, hô tieo*, Chinese *hū tsiāo;* C. DC. in Lecomte Fl. Gén. Indo-Chine **5**: 88. 1910.

No locality is cited but Loureiro undoubtedly knew this as an Indo-China plant, as it is common in cultivation. The Linnaean species, common pepper, was correctly interpreted by Loureiro.

Piper sylvestre Lour. Fl. Cochinch. 30. 1790, ed. Willd. 38. 1793, Anamese *tieo rùng.*

" Habitat in sylvis Cochinchinae, ad nullum usum, quem sciam, aptum." Because Loureiro discusses *Piper caninum* Rumph. (Herb. Amb. **5**: 49. *pl. 28. f. 2*) under *P. sylvestre:* " Suspicior idem esse Piper caninum Rumph. Amb. 1. 7. cap. 26. tab. 28. fig. 2," his species has been placed as a doubtful synonym of *Piper caninum* Blume. Whatever else it may be, it is certainly not Blume's species. It is probably one of the 37 species admitted by C. de Candolle in his treatment of the Piperaceae of Indo-China (Lecomte Fl. Gén. Indo-Chine **5**: 66–92. 1910) but I am unable to refer it to any one of these from the data at present available. C. de Candolle fails to account for it.

CHLORANTHACEAE

Chloranthus Swartz

Chloranthus spicatus (Thunb.) Makino in Bot. Mag. Tokyo **16**: 180. 1902.

Nigrina spicata Thunb. Nov. Gen. Pl. 59. 1783, Fl. Jap. 65. 1784.

Chloranthus inconspicuus Swartz in Philos. Trans. **78**: 359. *pl. 15.* 1787; Britten in Journ. Bot. **55**: 344. 1917; Lecomte Fl. Gén. Indo-Chine **5**: 95. 1910.

Creodus odorifer Lour. Fl. Cochinch. 89. 1790, ed. Willd. 112. 1793, Anamese *hoa sói.*

" Habitat in hortis Cochinchinae, culta ob gratis floris odorem." Loureiro's type is preserved in the herbarium of the British Museum. Britten (Journ. Bot. **55**: 344. 1917) gives a critical note on it and on a duplicate type of *Chloranthus inconspicuus* Swartz, indicating that they represent the same species. Thunberg's binomial supplies the oldest specific name. *Bunius sativa* Rumph. (Herb. Amb. **3**: 204. *pl. 131*), discussed by Loureiro as doubtfully representing his species, is *Antidesma bunius* (Linn.) Spreng. of the Euphorbiaceae. Lecomte does not account for Loureiro's binomial, giving no synonyms under any of the four species of *Chloranthus* admitted by him.

SALICACEAE

Salix (Tournefort) Linnaeus

Salix babylonica Linn. Sp. Pl. 1017. 1753; Lour. Fl. Cochinch. 609. 1790, ed. Willd. 747. 1793, Anamese *cây lieo lá tle*, Chinese *liêu xú.*

" Habitat frequens in China: colitur rarius in Cochinchina." Loureiro's description conforms to the characters of the Linnaean species which is commonly planted in China. The Cantonese name appears on recent collections as *lau shue*.

MYRICACEAE

Myrica Linnaeus

Myrica rubra Sieb. & Zucc. in Abh. Akad. Münch. 4(3): 230. 1846.

Morella rubra Lour. Fl. Cochinch. 548. 1790, ed. Willd. 670. 1793, pro parte (as to the Chinese plant), Chinese *yâm mûei, deang mai*.

" Culta in China." The description in part applies to the tree commonly cultivated in China for its edible fruit, generally known as *Myrica nagi* Thunb.; but *Myrica nagi* Thunb. is *Podocarpus nagi* (Thunb.) Makino. The correct name for this species is apparently *Myrica rubra* Sieb. & Zucc. Dode (Lecomte Fl. Gén. Indo-Chine 5: 933. 1929) considers the Indo-China and central China plant to be the same as *Myrica sapida* Wall. (Tent. Fl. Nepal. 59. *pl. 45*. 1826) yet curiously he fails to mention Loureiro's genus and species in his synonymy, although Loureiro's type is preserved in the herbarium of the Paris Museum. A. Chevalier (Bull. Écon. Indochine 21: 867. 1918) reduced *Morella rubra* Lour. to *Myrica integrifolia* Roxb.; and Dode placed *M. integrifolia* A. Chev. (non Roxb.) as *M. sapida* Wall. var. *chevalieri* Dode (Lecomte Fl. Gén. Indo-Chine 5: 934. 1929). *Myrica rubra* Sieb. & Zucc. was not based on *Morella rubra* Lour.

Myrica integrifolia Roxb. Hort. Beng. 71. 1814, *nomen nudum;* Fl. Ind. ed. 2, 3: 765. 1832; A. Cheval. in Rev. Bot. Appl. 2: 635. 1922.

Morella rubra Lour. Fl. Cochinch. 548. 1790, ed. Willd. 670. 1793, pro parte (as to the Indo-China plant), Anamese *deâu ruu*.

Myrica rubra A. Cheval. Cat. Pl. Jard. Bot. Saigon 66. 1919 (based on *Morella rubra* Lour.), non *Myrica rubra* Sieb. & Zucc.

Myrica sapida Wall. var. *chevalieri* Dode in Lecomte Fl. Gén. Indo-Chine 5: 934. 1929.

" Agrestis fruticosa, & multo minor [than the Chinese form, *M. rubra* Sieb. & Zucc.] in Cochinchina." This Indo-China form is apparently the one represented by Loureiro's specimen in the herbarium of the Paris Museum. From Loureiro's description this has been identified by Doctor A. Chevalier as representing *Myrica integrifolia* Roxb. but for which he proposed a new but invalid binomial *M. rubra* (Lour.) A. Cheval.; *Myrica rubra* S. & Z. is the correct name for the species commonly cultivated in China and Japan.

JUGLANDACEAE

Juglans Linnaeus

Juglans regia Linn. Sp. Pl. 997. 1753; Lour. Fl. Cochinch. 573. 1790, ed. Willd. 702. 1793, Anamese *cây hach dào*, Chinese *hŏ taô*.

" Habitat in provinciis Borealibus imperii Sinensis." The description applies to a form of the common walnut. Skan (Journ. Linn. Soc. Bot. 26: 493. 1899) places the form Loureiro described under *Juglans regia* Linn. var. *sinensis* A. DC.

FAGACEAE

Castanopsis Spach

Castanopsis indica (Roxb.) A. DC. in Journ. Bot. **1**: 182. 1863, Prodr. **16**(2): 109. 1864; Hickel & A. Camus in Lecomte Fl. Gén. Indo-Chine **5**: 1027. 1929.

Castanea indica Roxb. Fl. Ind. ed. 2, **3**: 643. 1832.

Castanea chinensis Spreng. Syst. **3**: 856. 1826 (based on *Fagus castanea* Lour.).

Quercus loureirii Hance in Journ. Linn. Soc. Bot. **10**: 201. 1868, in nota (based on *Fagus castanea* Lour.).

Castanopsis chinensis A. Chev. in Bull. Écon. Indochine **21**: 874. 1918 (based on *Fagus castanea* Lour.).

Castanopsis sinensis A. Chev. op. cit. 875 (based on *Fagus castanea* Lour.).

Fagus castanea (non Linn.) Lour. Fl. Cochinch. 571. 1790, ed. Willd. 699. 1793, Anamese *cây dee gai*, Chinese *lie tsù*.

" Habitat in sylvis montanis Cochinchinae: in China colitur Europeae aequalis." Loureiro's description was manifestly based on a Cochinchina plant; the Chinese form mentioned, but not described by him, may have been the Chinese chestnut *Castanea mollissima* Blume, although Hance thought that a nut occasionally sold in the Canton markets represented Loureiro's Chinese form and *Castanopsis chinensis* Hance. His careful description of *Castanopsis chinensis* (Journ. Bot. **12**: 243. 1874) was based wholly on Kwangtung specimens. Hickel & A. Camus (Lecomte Fl. Gén. Indo-Chine **5**: 1019. 1929) admit *Castanopsis chinensis* Hance as an Indo-China species solely on the assumption that Hance's reduction of *Fagus castanea* Lour. to *Castanopsis chinensis* Hance was correct. I believe that what Loureiro actually had was Indo-China specimens of the common *Castanopsis indica* (Roxb.) A. DC. in spite of his description of the leaves as glabrous; they are glabrous on the upper surface but distinctly pubescent beneath. The local names *dé (gié) gai, cây dè gai* and *gie gai* cited by Hickel & A. Camus for *Castanopsis indica* (Roxb.) A. DC. in a measure confirm this suggested reduction of *Fagus castanea* Lour. A. Chevalier, however, accepts the binomial *Castanopsis chinensis* (Spreng.) A. Cheval. for the Indo-China form described by Loureiro, it, however, being invalidated by *C. chinensis* Schottky (1912), based on *Quercus chinensis* Abel = *Castanopsis sclerophylla* (Lindl.) Schottky, and *C. chinensis* Hance (1868). If a valid species is represented here, distinct from *C. indica* A. DC., then the binomial *Quercus loureirii* Hance provides a specific name for it.

Castanopsis sp.

Fagus cochinchinensis Lour. Fl. Cochinch. 571. 1790, ed. Willd. 699. 1793, Anamese *xuong cá lón lâ*.

" Habitat in sylvis Cochinchinae." A. de Candolle (Prodromus **16**(2): 123. 1864) repeats Loureiro's description under *Fagus* and queries: " Forsan *Castanea?* aut *Castanopsis?* aut arbor diversissima ? " I am unable to suggest any further identification, except that probably a species of *Castanopsis* is represented. Hickel & A. Camus (Lecomte Fl. Gén. Indo-Chine **5**: 1007–1033. 1929–31) admit 47 species of *Castanopsis* for Indo-China, but they do not mention Loureiro's species.

Quercus (Tournefort) Linnaeus

Quercus cornea Lour. Fl. Cochinch. 572. 1790, ed. Willd. 700. 1793, Anamese *dee sùng,* Chinese *chū;* A. DC. Prodr. **16**(2): 90. 1864; Seem. Bot. Voy. Herald 413. *pl. 87.* 1857; Hemsl. in Hook. Ic. **27**: *pl. 2665.* 1900.

Pasania cornea Oersted in Vid. Med. Nat. Foren. Kjöb. II **8**: 83. 1867; Hickel & A. Camus in Lecomte Fl. Gén. Indo-Chine **5**: 1001. 1929 (based on *Quercus cornea* Lour.).

Synaedrys ossea Lindl. Nat. Syst. Bot. ed. 2, 441. 1836.

Quercus hainanensis Merr. in Philip. Journ. Sci. **23**: 239. 1923.

Lithocarpus hainanensis Chun in Journ. Arnold Arb. **8**: 21. 1927.

" Habitat in altis sylvis Cochinchinae, & Chinae." The status of this species is well known. Skan (Journ. Linn. Soc. Bot. **26**: 510. 1899) gives its range as southern China, Hainan, Indo-China, Borneo and Java. Hickel and Camus cite only Indo-China specimens, mentioning no extra-Indo-China range of the species.

Quercus concentrica Lour. Fl. Cochinch. 572. 1790, ed. Willd. 701. 1793, Anamese *dee bôp;* A. DC. Prodr. **16**(2): 94. 1864.

Pasania sabulicola Hickel & A. Camus in Ann. Sci. Nat. X Bot. **3**: 389. *f. 3, 11–13.* 1921; Lecomte Fl. Gén. Indo-Chine **5**: 970. *f. 112, 11–13.* 1930.

" Habitat in altis sylvis Cochinchinae." The indicated habitat is not good for the recently described species I here reduce to *Quercus concentrica* Lour., as *Pasania sabulicola* grows in thickets in sand dunes, not in the high forest, yet I believe *Pasania sabulicola* Hickel & Camus to be the same as *Quercus concentrica* Lour. To be noted, in comparing the descriptions, is Loureiro's term " incurva " applied to the leaves, expressed by Hickel & A. Camus " plicata, arcuata " and as " se pliant sur le sec, parallèlement à la nervure médiane et incurvées "; these authors do not mention Loureiro's species in their treatment of Fagaceae of Indo-China (Lecomte Fl. Gén. Indo-Chine **5**: 937–1033. 1929–31). *Clemens 4141,* from sand dunes, Tourane, near Hue, represents the species.

Quercus helferiana A. DC. Prodr. **16**(2): 101. 1864; King in Ann. Bot. Gard. Calcutta **2**: 35. *pl. 25 B.* 1889; Hickel & A. Camus in Lecomte Fl. Gén. Indo-Chine **5**: 958. 1929.

Quercus ilex (non Linn.) Lour. Fl. Cochinch. 571. 1790, ed. Willd. 700. 1793, Anamese *dee gao.*

" Habitat in sylvis Cochinchinae." A. de Candolle tentatively made this reduction of Loureiro's misinterpretation of the Linnaean species. Loureiro's description conforms closely to the characters of *Q. helferiana* A. DC. and the species is recorded from various localities in Anam and from other parts of Indo-China by Hickel & A. Camus; the latter authors do not mention *Quercus ilex* Lour., but it is unquestionably referable to *Q. helferiana* A. DC.

ULMACEAE

Celtis (Tournefort) Linnaeus

Celtis sinensis Pers. Syn. **1**: 292. 1805.

Streblus cordatus Lour. Fl. Cochinch. 615. 1790, ed. Willd. 755. 1793, Chinese *tsong xú,* non *Celtis cordata* Pers.

Trophis cordata Poir. in Lam. Encycl. **8**: 124. 1808 (based on *Streblus cordatus* Lour.). " Habitat circa Cantonem Sinarum." The description is short but it agrees with the characters of Persoon's species, except that the leaves are scarcely cordate. *Celtis sinensis* is very common in the vicinity of Canton and the Cantonese names *sheung see, sheung shue chi* and *seung sz shue*, appear on recently collected specimens, cognate forms of *tsong xu* as recorded by Loureiro. *Streblus cordatus* has been reduced to *Broussonetia papyrifera* Vent. but Loureiro's description does not conform to the characters of that species, particularly in the staminate inflorescences, while the Cantonese name of Ventenat's species is the very different *kuk muk*.

Celtis sp.

? Bosea cannabina Lour. Fl. Cochinch. 176. 1790, ed. Willd. 220. 1793, Anamese *cây rach.*

" Habitat in sylvis Cochinchinae." I believe that a species of *Celtis* is represented by Loureiro's description, perhaps *C. sinensis* Pers. or *C. cinnamomea* Lindl.; the statement " antherae subrotundae, curvaturis calycis defensa " is significantly like *Celtis* characters and the description otherwise agrees; yet I know of no *Celtis* that yields a stout bast fiber. It may be noted that Loureiro describes the leaves as " serrata " and as " integerrima " in the same line. Possibly a species of *Trema* is represented.

Trema Loureiro

Trema cannabina Lour. Fl. Cochinch. 563. 1790, ed. Willd. 689. 1793, Anamese *cây rach chiéo;* Moore in Journ. Bot. **63**: 288. 1925.

Celtis amboinensis Willd. Sp. Pl. 4(2): 997. 1805.

Sponia amboinensis Decne. in Nouv. Ann. Mus. Paris **3**: 498. 1834.

Trema amboinensis Blume [62] Mus. Bot. Lugd.-Bat. **2**: 61. 1856; Merr. Interpret. Herb. Amb. 187. 1917, Enum. Philip. Fl. Pl. **2**: 33. 1923.

Trema virgata Blume Mus. Bot. Lugd.-Bat. **2**: 59. 1856; Gagnep. in Lecomte Fl. Gén. Indo-Chine **5**: 686. 1928.

Sponia virgata Planch. in Ann. Sci. Nat. III Bot. **10**: 316. 1848.

" Habitat in sylvis Cochinchinae." Loureiro's type is preserved in the herbarium of the British Museum and it seems well to quote Moore's comment on it, in view of Gagnepain's recent misinterpretation of Loureiro's species, who places it as a doubtful synonym of *Trema velutina* Blume (Lecomte Fl. Gén. Indo-Chine **5**: 689. 1928). Moore (Journ. Bot. **63**: 288. 1925) states: " The late C. B. Robinson saw Loureiro's Museum specimens, and decided they were not referable to *T. amboinensis* Auct. as has been suspected. Merrill thinks *T. cannabina* should prove to be *T. amboinensis* Bl. (= *T. virgata* Bl.), and in this he is certainly correct." I personally examined Loureiro's type in 1930 and confirm Moore's decision. Gagnepain [63] (Bull. Soc. Bot. France **72**: 807. 1926) in referring *Trema cannabina* Lour. to *T. velutina* Blume, states: " Les feuilles tomenteuses au dire de Loureiro, excluent le *T. virgata*, les fruits jaunâtres excluent le *T. angustifolia* à fruits rouges; le *T.*

---

[62] This is not at all *Trema amboinensis* as that species is erroneously interpreted in modern literature. I refer *T. amboinensis* auct. plur., non (Willd.) Blume to *T. occidentalis* (Linn.) Blume.

[63] Gagnepain, F. Quelques espèces litigieuses du genre Trema. Bull. Soc. Bot. France **72**: 806–808. 1926.

politoria, à fruits noirs, et (la description de Loureiro) ne convient qu'au *T. velutina*." It is true that Loureiro describes the leaves of *T. cannabina* as " tomentosa," while those of *T. virgata* Blume are glabrous except for appressed hairs on the nerves beneath, yet at the same time he describes the leaf shape as " lanceolatis " in the diagnostic sentence, and as " ovato-lanceolata " in the description; the lanceolate leaf character applies to *T. virgata* Blume, not to *T. velutina* Blume. It is possible that the description as finally prepared was based on specimens representing more than one species, but in any case I prefer to interpret it by Loureiro's undubitable type, a specimen he unquestionably had in hand when he wrote the original description of the genus and species. With this interpretation the currently misapplied binomial *Trema amboinensis* Bl. falls into synonymy.

MORACEAE

Morus (Tournefort) Linnaeus

Morus alba Linn. Sp. Pl. 986. 1753; Lour. Fl. Cochinch. 555. 1790, ed. Willd. 678. 1793, Anamese *deâu taù*, Chinese *ɤín pĕ ɤú;* Gagnep. in Lecomte Fl. Gén. Indo-Chine **5**: 707. 1928.

Morus rubra (non Linn. ?) Lour. Fl. Cochinch. 555. 1790, ed. Willd. 679. 1793, Anamese *dèau moi.*

For *Morus alba* Loureiro states: " Habitat culta, incultaque in China, raro in Cochinchina; " his description seems to apply to the common mulberry, *Morus alba* Linn. For *M. rubra* he states: " Habitat agrestis apud Molos populos, Cochinchinae tributarios, ad occasum sitos: colitur etiam ad nutriendos bombyces, minus frequenter." This description is apparently referable to the red-fruited form of *Morus alba* Linn.

Morus australis Poir. in Lam. Encycl. **4**: 380. 1783; Hand.-Maz. Symb. Sin. **7**: 90. 1929.

Morus acidosa Griff. Notul. **4**: 388. 1854; Schneider in Sargent Pl. Wils. **3**: 297. 1916; Gagnep. in Lecomte Fl. Gén. Indo-Chine **5**: 709. *f. 83. 5–11.* 1928.

Morus longistyla Diels in Notes Bot. Gard. Edinb. **5**: 293. 1912.

Morus indica (non Linn.) Lour. Fl. Cochinch. 555. 1790, ed. Willd. 679. 1793, Anamese *deâu se dê tàm.*

" Habitat latissime culta in Cochinchinae agris, praecipue ad ripas fluminum." Loureiro's description agrees rather closely with the characters of *Morus acidosa* Griff. = *M. australis* Poir. which is common and widely distributed in Indo-China.

Malaisia Blanco

Malaisia scandens (Lour.) Planch. in Ann. Sci. Nat. IV Bot. **3**: 293. 1855; Blume Mus. Bot. Lugd.-Bat. **2**: 76. 1856 (based on *Caturus scandens* Lour.).

Caturus scandens Lour. Fl. Cochinch. 612. 1790, ed. Willd. 751. 1793, Anamese *cây di giéi.*

Trophis scandens Hook. & Arn. Bot. Beechey's Voy. 214. 1836 (based on *Caturus scandens* Lour.).

Malaisia tortuosa Blanco var. *scandens* Bureau in DC. Prodr. **17**: 222. 1853 (based on *Caturus scandens* Lour.); Gagnep. in Lecomte Fl. Gén. Indo-Chine **5**: 696. 1928.

Alchornea scandens Muell.-Arg. in DC. Prodr. **15**(2): 906. 1866 (based on *Caturus scandens* Lour.); Pax in Pflanzenreich **63**(IV–147–VII): 244. 1914.

" Habitat in sylvis Cochinchinae." Loureiro had only staminate specimens which led him to place his species in *Caturus*, a euphorbiaceous genus. His species is unquestionably identical with the Philippine *Malaisia tortuosa* Blanco. Loureiro's type is preserved in the herbarium of the British Museum, and Mr. J. E. Dandy who kindly re-examined it for me in September, 1931, reports it to be labelled by Mr. S. Le M. Moore as *Malaisia tortuosa* Blanco var. *scandens* Bureau. DePirey's specimen from Anam, *Chevalier 40211* bearing the local name *day giay* is the species as here interpreted. Pax apparently accepted Mueller's erroneous reference of the species to *Alchornea* as correct and without investigation.

Taxotrophis Blume

Taxotrophis macrophylla (Blume) Boerl. Handl. Fl. Nederl. Ind. **3**: 359. 1900; Merr. Enum. Philip. Fl. Pl. **2**: 38. 1923.

Streblus macrophyllus Blume Mus. Bot. Lugd.-Bat. **2**: 80. 1856.

Diplocos macrophylla Bureau in DC. Prodr. **17**: 216. 1873.

Ilex loureirii Steud. Nomencl. ed. 2, **1**: 802. 1840 (based on *Ilex aquifolium* Lour.).

Taxotrophis ilicifolia Vidal Rev. Pl. Vasc. Filip. 249. 1886; Hutch. in Kew Bull. 150. 1918; Gagnep. in Lecomte Fl. Gén. Indo-Chine **5**: 699. 1928.

Ilex aquifolium (non Linn.) Lour. Fl. Cochinch. 91. 1790, ed. Willd. 114. 1793, Anamese *ô rô cây.*

" Habitat in sylvis Cochinchinae." Loesener (Nov. Act. Acad. Leop.-Carol. Nat. Cur. **78**: 263. 1901) refers Loureiro's binomial to *Ilex aquifolium* Linn. var. *chinensis* Loesen., together with *I. loureirii* Steud., both with expressed doubt; but on page 495 he also lists *I. loureirii* Steud. as a species of doubtful status, noting that *I. aquifolium* Linn. does not occur in Indo-China. In my original manuscript of 1919 I suggested that what Loureiro actually had might be *Taxotrophis ilicifolia* Vidal and that the local name should eventually solve the problem; this species is not uncommon in Indo-China. It may be noted that one of the local names cited by Gagnepain for *Taxotrophis ilicifolia* Vidal is *o rô*, which is unquestionably Loureiro's *ô rô cây*. De Pirey collected three sheets for Dr. Chevalier, nos. *40212, bis* and *ter*, under the Anamese name *o ro* and these represent a juvenile form of *Prunus phaeosticta* (Hance) Max., the younger leaves sinuate-lobed and spiny-toothed as in *Ilex aquifolium*, the more mature leaves entire. Loureiro apparently described leaf specimens of the *Taxotrophis* as an *Ilex* and added *Ilex* flower and fruit characters from some published description of *Ilex aquifolium* to make his description conform to the characters of that species. In the herbarium of the British Museum is a Loureiro specimen named *Ilex aquifolium* but with no Loureiro label; this was identified by Dryander as *Acanthus ilicifolius* Linn., but *Ilex aquifolium* is not indicated in the British Museum copy of Loureiro's Flora Cochinchinensis as being among the plants received from him, and Loureiro's description does not apply to *Acanthus*. Steudel's specific name is older than Blume's, but for obvious reasons I do not here transfer it to *Taxotrophis*.

Streblus Loureiro

Streblus asper Lour. Fl. Cochinch. 615. 1790, ed. Willd. 754. 1793, Anamese *cây deó duói;*
Gagnep. in Lecomte Fl. Gén. Indo-Chine **5**: 712. 1928.
Trophis cochinchinensis Poir. in Lam. Encycl. **8**: 123. 1808 (based on *Streblus asper*
Lour.).
Salvadora biflora Lour. Fl. Cochinch. 88. 1790, ed. Willd. 110. 1793, Anamese *cây dúoi.*
Salvadora capitulata Lour. Fl. Cochinch. 87. 1790, ed. Willd. 110. 1793, Anamese *cây*
dúoi.

For *Streblus asper* Loureiro states: " Habitat in sylvis montanis Cochinchinae." It is
a widely distributed Indo-Malaysian monotypic genus, with its limits and relationships well
understood. A specimen listed as being among Loureiro's plants in the herbarium of the
British Museum has not been located. For *Salvadora biflora* and *S. capitulata*, species not
previously placed, Loureiro states: " Habitat utraque species ad sepes, & in sylvis Cochinchinae: nec satis una ad alia dignoscitur, nisi floreant." The descriptions of both manifestly
apply to the species otherwise described by Loureiro as a new genus and species, *Streblus*
asper, the two species of *Salvadora* having page priority. The correctness of this reduction
of these two species is verified by the Anamese names cited for the three species here combined and by Gagnepain's citation of *cây ruôi, duói, dui, guoi* and *ruôi* for *Streblus asper*
Lour.

Cudrania [63] Trécul

Cudrania cochinchinensis (Lour.) Kudo & Masamume in Ann. Rept. Taihoku Bot. Gard.
2: 27. 1932 (based on *Vanieria cochinchinensis* Lour.).
Vanieria cochinchinensis Lour. Fl. Cochinch. 564. 1790, ed. Willd. 691. 1793, Anamese
cây vang lô.
? Vanieria chinensis Lour. Fl. Cochinch. 565. 1790, ed. Willd. 691. 1793, Chinese *húng*
hõâng xtong.
Vanieria alternifolia Stokes Mat. Med. **4**: 381. 1812 (based on *Vanieria cochinchinensis*
Lour.).
Procris cochinchinensis Spreng. Syst. **3**: 846. 1826 (based on *Vanieria cochinchinensis*
Lour.).
? Procris cantoniensis Spreng. Syst. **3**: 846. 1826 (based on *Vanieria chinensis* Lour.).
Cudrania javanensis Trécul in Ann. Sci. Nat. III Bot. **8**: 123. 1847; Gagnep. in Lecomte
Fl. Gén. Indo-Chine **5**: 726. 1928.

For *Vanieria cochinchinensis* Loureiro states: " Habitat in dumetis Cochinchinae."
The genus *Vanieria* Lour., long considered to be a genus of doubtful status, is safely referable to *Cudrania* Trécul and is a much older name, but the latter is the conserved one.
Gagnepain cites *Vanieria cochinchinensis* Lour. as a doubtful synonym of *Cudrania javanensis* Tréc. but does not account for the synonyms based on Loureiro's binomial. *Kuntze*
3758 and *Clemens 4248* from Tourane represent Loureiro's species, the latter having been
identified by Gagnepain as *C. javanensis* Tréc. For *Vanieria chinensis* Loureiro states:
" Habitat dumetis provinciae Cantoniensis apud Sinas." This he describes as an unarmed
shrub 15 inches high with fascicled leaves and long peduncles. While these data do not

---

[63] *Cudrania* Trécul (1847), conserved name, Cambridge Code. *Vanieria* Loureiro (1790) is the oldest
name for the genus.

well apply to the more normal forms of *Cudrania cochinchinensis* Kudo & Masamume (*C. javanensis* Tréc.), yet it is possible that he had a form of this variable species which is common in thickets in Kwangtung Province.

Artocarpus Forster

Artocarpus integra (Thunb.) Merr. Interpret. Herb. Amb. 190. 1917, Enum. Philip. Fl. Pl. **2**: 41. 1923.

Radermachia integra Thunb. in Vet. Akad. Handl. Stockholm 254. 1776.

Artocarpus integrifolia Linn. f. Suppl. 412. 1781; Gagnep. in Lecomte Fl. Gén. Indo-Chine **5**: 732. 1928.

Artocarpus jaca Lam. Encycl. **3**: 209. 1789.

Polyphema jaca Lour. Fl. Cochinch. 546. 1790, ed. Willd. 667. 1793, Anamese *cây mít*, Chinese *yă xú, po lô mat.*

" Habitat frequentissime culta in tota Cochinchina: raro in China." Loureiro's ample description applies to the common jak fruit, generally known as *Artocarpus integrifolia* Linn. f. *Nanka*, i.e., *Soccus arboreus major* Rumph. (Herb. Amb. **1**: 104. *pl. 30*), cited by Loureiro as a synonym, is correctly placed. The specific name *jaca* is taken from Acosta, cited by Loureiro, after Clusius, and Loureiro's binomial was published independently of that of Lamarck, who used the same specific name.

Artocarpus champeden (Lour.) Spreng. Syst. **3**: 804. 1826 (based on *Polyphema champeden* Lour.).

Polyphema champeden Lour. Fl. Cochinch. 547. 1790, ed. Willd. 668. 1793, Anamese *cây mít nai.*

Artocarpus polyphema Pers. Syn. **2**: 531. 1807 (based on *Polyphema champeden* Lour.).

Saccus champeden O. Ktz. Rev. Gen. Pl. 633. 1891 (based on *Polyphema champeden* Lour.).

" Habitat in altis sylvis Cochinchinae. Colitur etiam, & Champeden vocatur a populis Malaiis circa fretum Malaccense habitantibus." The species is a well-known one and of wide distribution in Malaysia. Loureiro's type is preserved in the herbarium of the British Museum. *Soccus arboreus minor: Tsjampadaha* Rumph. (Herb. Amb. **1**: 107. *pl. 31*), cited by Loureiro as a synonym, is correctly placed. Gagnepain (Lecomte Fl. Gén. Indo-Chine **5**: 734. 1928) sub *Artocarpus styracifolia* Pierre, states: " L'*A. Polyphema* Pers. (Tréc. Bl. King et nombreux auteurs) serait le *Polyphema Champeden* Lour. Fl. Cochinch., p. 547. La plante de Loureiro ne se rapporte certainement à aucune espèce décrite ici. L'*A. Polyphema* Pers. n'existe pas dans notre domaine." Nevertheless, Loureiro's description applies to this species as currently interpreted. It was based on plants observed by him near Malacca and his reference to the Anamese *cây mít nai* and the occurrence of the species in the forests of Cochinchina was apparently based on Indo-China specimens that he had seen which he thought represented the same species as the Malayan form.

Ficus (Tournefort) Linnaeus

Ficus auriculata Lour. Fl. Cochinch. 666. 1790, ed. Willd. 819. 1793, Anamese *cây ba.*

Ficus roxburghii Wall. List no. 4508. 1831, *nomen nudum;* Brandis For. Fl. Brit. Ind. 422. 1874; Kurz For. Fl. Brit. Burma **2**: 460. 1877; King in Ann. Bot. Gard. Calcutta **1**: 168. *pl. 211.* 1888; Gagnep. in Lecomte Fl. Gén. Indo-Chine **5**: 806. 1928.

Ficus macrophylla Roxb. Hort. Beng. 66. 1814, Fl. Ind. ed. 2, **3**: 556. 1832, non Desf. *Tremotis cordata* Raf. Sylva Tellur. 59. 1838 (based on *Ficus auriculata* Lour.).

" Habitat culta in Cochinchina: puto, quod etiam agrestis." Loureiro's species has been reduced to *Ficus cunia* Ham., and King (Ann. Bot. Gard. Calcutta 1: 179. 1888) thought it to be the same as Hamilton's species. The description indicates a very different plant and unquestionably the species currently known as *Ficus roxburghii* Wall. It should be noted that the specific name *auriculata* was not taken from the leaf characters but from the nature of the characteristic scales surrounding the ostiole on the receptacle. Gagnepain ignores Loureiro's species entirely in his treatment of the genus (Lecomte Fl. Gén. Indo-Chine **5**: 740–828. 1928–29) yet *cây va*, which he cites as one of the local names for *Ficus roxburghii* Wall., in a measure confirms this interpretation of Loureiro's species. *Caprificus amboinensis esculenta* Rumph. (Herb. Amb. **3**: 145. *pl. 93*), discussed by Loureiro under his species, represents *Ficus racemifera* Roxb. The statement " racemis . . . erectis, terminalibus " is of course an error, as no *Ficus* presents such characters; Loureiro doubtless had detached cauline infructescences and assumed them to be terminal and erect. This species typifies Rafinesque's genus *Tremotis*.

Ficus benjamina Linn. Mant. **1**: 129. 1767; Gagnep. in Lecomte Fl. Gén. Indo-Chine **5**: 766. 1928.

> *Ficus indica* (non Linn.) Lour. Fl. Cochinch. 665. 1790, ed. Willd. 818. 1793, Anamese *cây sanh.*

" Habitat loca plana Cochinchina." Loureiro's description agrees with the characters of *Ficus benjamina* Linn. as far as it goes; the local name *cây sanh* is one of those cited by Gagnepain for it, and the species is common and widely distributed in Indo-China. It is represented by *Clemens 3837, 4393,* from Tourane and Hue.

Ficus carica Linn. Sp. Pl. 1059. 1753; Lour. Fl. Cochinch. 664. 1790, ed. Willd. 816. 1793, Anamese *sung taù*, Chinese *máo hōa qua.*

" Habitat culta in China: raro in Cochinchina, a Sinis delata." The description applies to the Linnaean species, the common cultivated fig.

Ficus heterophylla Linn. f. Suppl. 442. 1781; Gagnep. in Lecomte Fl. Gén. Indo-Chine **5**: 775. 1928.

> *Ficus maculata* (non Linn.) Lour. Fl. Cochinch. 666. 1790, ed. Willd. 819. 1793, Anamese *cây ngáy.*
>
> *Ficus septica* (non Burm. f.) Lour. Fl. Cochinch. 667. 1790, ed. Willd. 819. 1793, Anamese *cây laúc chó.*
>
> *? Ficus politoria* (non Lam.) Lour. Fl. Cochinch. 667. 1790, ed. Willd. 820. 1793, Anamese *cây bú chó, cây ngaong.*
>
> *? Ficus cannabina* Lour. Fl. Cochinch. 668. 1790, ed. Willd. 821. 1793, Anamese *cây giéi.*

In sequence as the Loureiroan binomials are given above, the indicated habitats are as follows: " frequenter ad sepes, & ripas fluminum in Cochinchina," " agrestis in Cochinchina," " in sylvis Cochinchinae," and " loca plana, & agrestia Cochinchinae." In spite of certain discrepancies noted below, I believe all are referable to the rather protean *Ficus heterophylla* Linn. f., at least as that species is interpreted by Gagnepain. *Ficus maculata*

Lour. and *F. septica* Lour. seem safely to be referable here; the latter was described by Loureiro as a new species although he added a reference to " Burm. Ind. pag. 226 " but without giving Burman's binomial, and one to Rumph. (Herb. Amb. 3: 153. *pl. 96*) which is *Ficus septica* Burm. f.; Burman's species has smooth leaves and is the one currently known as *Ficus leucantatoma* Poir. *Ficus politoria* Lour. was placed by Gagnepain as a doubtful synonym of *Ficus leekensis* Drake (Fl. Gén. Indo-Chine 5: 797). It was described by Loureiro as a new species, but the binomial is invalid, having been antedated by *Ficus politoria* Lam. (1788), its type a specimen from Madagascar. Nor is it the same as *Ficus ampelos* Burm. f. (Fl. Ind. 226. 1768) which Loureiro cites as a synonym. *Folium politorium* Rumph. (Herb. Amb. 4: 128. *pl. 63*), also cited as a synonym, is Burman's species. The description of the fruit as spicate must be an error, agreeing with neither *Ficus heterophylla* Linn. f. nor *F. leekensis* Drake. *Ficus cannabina* Lour., placed by Gagnepain as a definite synonym of *Ficus heterophylla* Linn. f., differs in its long-peduncled fruits; but I know of no rough-leaved *Ficus* in Indo-China having this character.

Ficus hispida Linn. f. Suppl. 442. 1781; Gagnep. in Lecomte Fl. Gén. Indo-Chine 5: 810. 1928.

> *Ficus sycomorus* (non Linn.) Lour. Fl. Cochinch. 664. 1790, ed. Willd. 816. 1793, Anamese *cây sung*.

" Habitat agrestis in Cochinchina." The species as described by Loureiro is clearly a representative of the section *Covellia*, but his data do not apply in all particulars to *Ficus hispida* Linn. f., notably in the description of the leaves as " integerrima . . . parva," characters that may have been taken by Loureiro from the pre-Linnaean illustrations of Matthiolus and Plukenet cited by him. The local name *cây sung* corresponds to *cây sum* cited by Gagnepain as one of the names of *Ficus hispida* Linn. f.

Ficus indica Linn. Sp. Pl. ed. 2. 1514. 1763; Gagnep. in Lecomte Fl. Gén. Indo-Chine 5: 778. 1928 (var. *gelderi* King).

> *Ficus benjamina* (non Linn.) Lour. Fl. Cochinch. 665. 1790, ed. Willd. 818. 1793, Anamese *cây kùa*.

" Habitat ad ripas fluminum in Cochinchina." Loureiro's incomplete description applies to *Ficus benjamina* Linn. about as well as it does to the not closely related *Ficus indica* Linn., yet I believe the latter to be the species he attempted to describe. *Varinga parvifolia* Rumph. (Herb. Amb. 3: 139. *pl. 90*), cited by Loureiro, after Linnaeus, represents *Ficus benjamina* Linn., not *F. indica* Linn. Gagnepain cites the local names *cây dua* and *cây gia* for *Ficus indica* Linn. var. *gelderi* King.

Ficus obtusifolia Roxb. Fl. Ind. ed. 2, 3: 546. 1832; Gagnep. in Lecomte Fl. Gén. Indo-Chine 5: 779. 1928.

> *Ficus benghalensis* (non Linn.) Lour. Fl. Cochinch. 665. 1790, ed. Willd. 817. 1793, Anamese *cây dĕa tlon lá*.

" Habitat agrestis loca plana Cochinchinae." I believe this to be the proper reduction of the species Loureiro inadequately described, in spite of his description of the fruit as " minutus." The data given by Loureiro otherwise apply to Roxburgh's species.

Ficus pumila Linn. Sp. Pl. 1060. 1753; Lour. Fl. Cochinch. 667. 1790, ed. Willd. 820. 1793, Anamese *deei xôp xôp;* Gagnep. in Lecomte Fl. Gén. Indo-Chine 5: 793. 1928.

" Habitat in sylvis Cochinchinae." This is Gagnepain's doubtful reference of Loureiro's description and the local name *sôp* cited by him is suggestive of the one given by Loureiro. It is suspected that Loureiro actually had specimens of the Linnaean species, adding erroneous data as to the " fructus subrotundus, parvus, croceus, spicis terminalibus," which are not characters of *Ficus pumila* Linn. *Varinga repens* Rumph. (Herb. Amb. 3: 134. *pl. 85*), discussed by Loureiro, apparently represents a species near *Ficus calophylla* Blume.

Ficus rumphii Blume Bijdr. 437. 1825; Gagnep. in Lecomte Fl. Gén. Indo-Chine 5: 768. 1928.

> *Ficus religiosa* (non Linn.) Lour. Fl. Cochinch. 664. 1790, ed. Willd. 817. 1793, Anamese *cây bô dê*.

" Habitat agrestis loca plana Cochinchinae." Loureiro's description conforms with the characters of Blume's species better than with the Linnaean one; the latter is chiefly a planted tree in Indo-China and is not recorded from Anam; the former is apparently common in Anam and its local names, given by Gagnepain, are *cây dé, cây bodé* and *cây buddé*, these confirming the correctness of this reduction.

Ficus simplicissima Lour. Fl. Cochinch. 667. 1790, ed. Willd. 821. 1793, Anamese *com nguoi chia lá*.

> *Ficus palmatiloba* Merr. in Philip. Journ. Sci. 21: 340. 1922; Gagnep. in Lecomte Fl. Gén. Indo-Chine 5: 791. 1928.

" Habitat in sylvis Cochinchinae." Loureiro's short description applies closely to the Hainan species described by me in 1922 as *Ficus palmatiloba*. Gagnepain recognized this as a valid species, citing various collections from Indo-China and giving *Ficus simplicissima* Lour. as a doubtful synonym. I accept Loureiro's name with confidence that this is the species he had in hand.

Humulus Linnaeus

Humulus scandens (Lour.) comb. nov.

> *Antidesma scandens* Lour. Fl. Cochinch. 617. 1790, ed. Willd. 757. 1793, Chinese *ù chaò lûm*.
> *Humulus japonicus* Sieb. & Zucc. in Abh. Akad. Muench. 4(3): 213. 1846 (Fl. Jap. Fam. Nat. 2: 89. 1846); Gagnep. in Lecomte Fl. Gén. Indo-Chine 5: 691. 1928.

" Habitat spontanea prope Cantonem Sinarum." Loureiro's poor description, based on a nearly glabrous form of which he saw only staminate specimens, applies unmistakably to the species currently known as *Humulus japonicus* Sieb. & Zucc. No other known Kwangtung species in any family remotely conforms to the characters indicated by Loureiro. Pax & Hoffmann (Pflanzenreich 81(IV–147–XV): 167. 1922), in excluding *Antidesma scandens* Lour. from the genus, merely state: " Quid ? Certe non *Antidesma*."

Cannabis (Tournefort) Linnaeus

Cannabis sativa Linn. Sp. Pl. 1027. 1753; Lour. Fl. Cochinch. 616. 1790, ed. Willd. 756. 1793, Chinese *mâ fuên, chú tsao*.

" Habitat culta in variis locis imperii Sinensi." The Linnaean species was correctly interpreted by Loureiro. *Cannabis indica* Rumph. (Herb. Amb. 5: 208. *pl. 77. f. 1*), cited as a synonym, is correctly placed.

URTICACEAE

Fleurya Gaudichaud

Fleurya interrupta (Linn.) Gaudich. Bot. Freyc. Voy. 497. 1826; Gagnep. in Lecomte Fl.
Gén. Indo-Chine **5**: 861. 1929.

Urtica interrupta Linn. Sp. Pl. 985. 1753; Lour. Fl. Cochinch. 557. 1790, ed. Willd.
682. 1793, Anamese *cây nanghai*, Chinese *tâm mâ*.

" Habitat agrestis in Cochinchina, & China." The description clearly applies to the
Linnaean species, a common and widely distributed weed in the Indo-Malaysian region.
The pre-Linnaean synonym *Urtica pilulifera* Burm. (Thes. Zeyl. 231. *pl. 110. f. 1.* 1737),
cited by Loureiro as a synonym, is correctly placed by both Linnaeus and by Loureiro,
but *Batti-Schorigenam* Rheede (Hort. Malabar. 2: 75. *pl. 40.* 1679), also cited as representing this species, is apparently some species of *Acalypha*.

Pellionia [64] Gaudichaud

Pellionia repens (Lour.) Merr. in Lingnan Sci. Journ. **6**: 276, 326. 1928 (based on *Polychroa repens* Lour.).

Polychroa repens Lour. Fl. Cochinch. 559. 1790, ed. Willd. 684. 1793, Anamese *beò taù*.

Begonia daveauana Carrière in Rev. Hort. **52**: 290. *1 pl.* 1880.

Elatostema repens Hall. f. in Ann. Jard. Bot. Buitenzorg **13**: 316. 1896 (based on *Polychroa repens* Lour.).

Pellionia daveauana N. E. Br. in Gard. Chron. n.s. **14**: 262. 1880; Gagnep. in Lecomte
Fl. Gén. Indo-Chine **5**: 902. 1929.

" Habitat inculta, coliturque ob pulchritudinem, late repens per scopulos fontanos in
China, & Cochinchina." Loureiro's description is distinctly good, the species being a very
characteristic one, which, when introduced into Europe as an ornamental, was promptly
described as a new species. It is not uncommon in cultivation in southern China, the
Philippines, and Malaysia, having apparently been disseminated by the Chinese. It is
now very generally cultivated in greenhouses in Europe and in America, and is native of
some parts of Indo-China. Loureiro's generic name long antedates that of Gaudichaud,
but *Pellionia* is the conserved one. A specimen from Loureiro listed as being among those
received from him has not been located in the herbarium of the British Museum.

Boehmeria Jacquin

Boehmeria nivea (Linn.) Gaudich. Bot. Freyc. Voy. 499. 1826; Gagnep. in Lecomte Fl.
Gén. Indo-Chine **5**: 845. 1929.

Urtica nivea Linn. Sp. Pl. 985. 1753; Lour. Fl. Cochinch. 558. 1790, ed. Willd. 683.
1793, Anamese *cây gai*, Chinese *pă má*.

" Habitat, & abundanter colitur in Cochinchina, & China." The Linnaean species
was correctly interpreted by Loureiro; it is the common China grass or ramie plant. A
specimen from Loureiro is preserved in the herbarium of the British Museum.

---

[64] *Pellionia* Gaudichaud (1826), conserved name, Cambridge Code. *Polychroa* Loureiro (1790) is the
oldest name for the genus.

Pouzolzia Gaudichaud

Pouzolzia zeylanica (Linn.) Benn. Pl. Jav. Rar. 67. 1838.

Parietaria zeylanica Linn. Sp. Pl. 1052. 1753.

Parietaria indica Linn. Mant. 1: 128. 1767.

Pouzolzia indica Gaudich. Bot. Freyc. Voy. 503. 1826; Gagnep. in Lecomte Fl. Gén. Indo-Chine 5: 848. 1929.

Parietaria cochinchinensis Lour. Fl. Cochinch. 654. 1790, ed. Willd. 804. 1793, Anamese *thuốc giòi*, Chinese *mau soi cŏt*.

Boehmeria cochinchinensis Spreng. Syst. 3: 844. 1826 (based on *Parietaria cochinchinensis* Lour.).

Pouzolzia cochinchinensis Blume Mus. Bot. Lugd.-Bat. 2: 245. 1856 (based on *Parietaria cochinchinensis* Lour.).

" Habitat in Cochinchina, & China, in hortis minus cultis." The description manifestly applies to a form of the common, polymorphous, and widely distributed *Pouzolzia zeylanica* (Linn.) Benn. (*P. indica* Gaudich.). Weddell (DC. Monog. Phan. 16(1): 221. 1869) placed it as a synonym of *P. indica* Gaudich. var. *microphylla* Wedd. Gagnepain fails to account for Loureiro's species or for either of the synonyms based upon it in his treatment of the Urticaceae of Indo-China (Lecomte Fl. Gén. Indo-Chine 5: 828–921. 1929).

PROTEACEAE

Helicia Loureiro

Helicia cochinchinensis Lour. Fl. Cochinch. 83. 1790, ed. Willd. 105. 1793, Anamese *cây côm vàng;* Lecomte Fl. Gén. Indo-Chine 4: 161. 1914.

Rhopala cochinchinensis R. Br. in Trans. Linn. Soc. 10: 192. 1811 (based on *Helicia cochinchinensis* Lour.).

" Habitat in sylvis Cochinchinae." Loureiro's type is preserved in the herbarium of the British Museum and R. Brown's amplified description of *Rhopala cochinchinensis* was based upon it. Hemsley (Journ. Linn. Soc. Bot. 26: 394. 1891) cites *Helicia lancifolia* Sieb. & Zucc. as a synonym, stating that Japanese and Hongkong specimens cannot be distinguished from Loureiro's species.

LORANTHACEAE

Loranthus [65] Linnaeus

Loranthus pentapetalus Roxb. Hort. Beng. 87. 1814, *nomen nudum*, Fl. Ind. 2: 211. 1824; Lecomte Fl. Gén. Indo-Chine 5: 196. 1915.

Helixanthera parasitica Lour. Fl. Cochinch. 142. 1790, ed. Willd. 176. 1793, Anamese *chanh coi do;* Danser in Bull. Jard. Bot. Buitenzorg III 10: 318. 1929, non *Loranthus parasiticus* Druce, nec Merr.

Helicia parasitica Pers. Syn. 1: 214. 1805 (based on *Helixanthera parasitica* Lour.).

" Habitat, & adrepit arboribus cultis in hortis Cochinchinae." Loureiro's description is definite and clearly applies to the widely distributed Indo-Malaysian *Loranthus penta-*

---

[65] *Loranthus* Linnaeus (1762), conserved name, Cambridge Code; *Scurrula* Linnaeus (1753) is older. Some authors restrict *Loranthus* to relatively few species and recognize *Scurrula* as a distinct genus.

petalus Linn. which is common in Indo-China; he was apparently misled but the characteristic, swollen, subglobose, lower part of the corolla and described it as a corolla-tube. The habit, habitat, long inflorescences, color of the flowers, and all other characters conform to Roxburgh's species. Loureiro's type is preserved in the herbarium of the British Museum. Danser retains *Helixanthera* Lour. as a valid genus citing 20 synonyms for *Helixanthera parasitica* Lour. Roemer & Schultes (Syst. **5**: X. 1819) spell the generic name as *Helicanthera* but formed no binomial under it.

Loranthus sp.

> *Pavetta parasitica* Lour. Fl. Cochinch. 73. 1790, ed. Willd. 93. 1793, Anamese *chanh coi tlòn lá.*
>
> *Ixora parasitica* Poir. in Lam. Encycl. Suppl. **3**: 209. 1813 (based on *Pavetta parasitica* Lour.).

" Nascitur frequens, & inhaeret arboribus in hortis Cochinchinae." The description clearly applies to some species of *Loranthus*, apparently in the group with *L. scurrula* Linn. Loureiro's specific name is invalidated in *Loranthus* by *L. parasiticus* Druce.

Elytranthe Blume

Elytranthe cochinchinensis (Lour.) G. Don Gen. Syst. **3**: 426. 1834 (based on *Loranthus cochinchinensis* Lour.).

> *Loranthus cochinchinensis* Lour. Fl. Cochinch. 195. 1790, ed. Willd. 241. 1793, Anamese *nhánh goi nhon lá.*
>
> *Loranthus ampullaceus* Roxb. Fl. Ind. **2**: 209. 1824.
>
> *Loranthus globosus* Roxb. *op. cit.* 206.
>
> *Elytranthe ampullacea* G. Don Gen. Syst. **3**: 425. 1834; Lecomte Fl. Gén. Indo-Chine **5**: 204. 1915.
>
> *Elytranthe globosa* G. Don *op. cit.* 426.
>
> *Macrosolen cochinchinensis* Van Tiegh. in Bull. Soc. Bot. France **41**: 122. 1894; Danser in Bull. Jard. Bot. Buitenzorg III **10**: 343. 1929 (based on *Loranthus cochinchinensis* Lour.).

" Habitat in ramis arborum hortensium in Cochinchina." Loureiro's description is very definite and applies unmistakably to the common, widely distributed, Indo-Malaysian species more generally known as *Loranthus ampullaceus* Roxb. It is manifest that Loureiro's specific name should be adopted in spite of Lecomte's opinion (Not. Syst. **3**: 98. 1915) that his description was insufficient. There are only four species of *Elytranthe* known from Indo-China and Loureiro's description definitely applies to the form Lecomte placed under *E. ampullacea* G. Don. Danser gives no less than forty synonyms for this much-named, relatively characteristic species.

ARISTOLOCHIACEAE

Asarum (Tournefort) Linnaeus

Asarum sieboldii Miq. Ann. Mus. Bot. Lugd.-Bat. **2**: 134. 1865.

> *Asarum virginicum* (non Linn.) Lour. Fl. Cochinch. 292. 1790, ed. Willd. 357. 1793, Anamese *tê tăn,* Chinese *si sĭn.*

" Habitat incultum in variis Sinarum provinciis." This reduction of Loureiro's species is based partly on his description, and partly on the Chinese name cited by him, which is a cognate form of *hsi hsin*, given by Hemsley (Journ. Linn. Soc. Bot. **26**: 359, 360. 1891) for both *Asarum himalayicum* Hook. f. and *A. sieboldii* Miq. The roots are much used by the Chinese in the practice of medicine. The internal evidence here is to the effect that inasmuch as Loureiro's Chinese experience was confined to Canton, he secured his material and information regarding this plant from herbalists.

Aristolochia (Tournefort) Linnaeus

Aristolochia tagala Cham. in Linnaea **7**: 207. *pl. 5. f.* 3. 1832.

> *Aristolochia roxburghiana* Klotzsch in Monatsschr. Akad. Berlin 596. 1859; Lecomte Fl. Gén. Indo-Chine **5**: 58. 1910.
> *Aristolochia indica* (non Linn.) Lour. Fl. Cochinch. 528. 1790, ed. Willd. 646. 1793, Anamese *cây khoai cà*.

" Habitat in sylvis Cochinchinae, maxime in montibus dictis *Nguon nhoung*." Loureiro doubtless interpreted the Linnaean species by *Radix puluronica* Rumph. (Herb. Amb. **5**: 476. *pl. 177*), cited by both Linnaeus and himself as a synonym; it represents a species very different from *Aristolochia tagala* Cham., one with non-cordate leaves. This Rumphian illustration typifies *Aristolochia rumphii* Kostel. (Allgem. Med.-Pharm. Fl. **2**: 465. 1831), a species of somewhat doubtful status but certainly not *A. tagala* Cham. The latter species, based on a Philippine specimen, is one of wide distribution in the Indo-Malaysian region, and Loureiro's description agrees with it.

Apama Lamarck

Apama racemosa (Lour.) O. Ktz. Rev. Gen. Pl. 563. 1891 (based on *Bragantia racemosa* Lour.).

> *Bragantia racemosa* Lour. Fl. Cochinch. 528. 1790, ed. Willd. 645. 1793, Anamese *hoa den mouc*.

" Habitat in montibus Cochinchinae." Duchartre (DC. Prodr. **15**: 429. 1864) gives a short description compiled from Loureiro's original one, quoting Bennett to the effect that Loureiro's type is preserved in the herbarium of the British Museum. It is thus all the more curious that Lecomte, in his treatment of the Aristolochiaceae of Indo-China (Fl. Gén. Indo-Chine **5**: 53-58. 1910) admits no representative of the genus. Rafinesque (Fl. Tellur. **4**: 99. 125. 1838) proposed the new generic name *Munnickia* for *Bragantia* Lour. (1790), non *Bragantia* Vandelli (1771) = *Gomphrena* Linnaeus, but published no binomial under it. In the Bentham and Hooker system *Bragantia* Lour. is recognized as a valid genus; in the Engler and Prantl system it is treated as a section of *Apama* Lam.

POLYGONACEAE

Rumex Linnaeus

Rumex trisetiferus Stokes Bot. Mat. Med. **2**: 305. 1812 (based on *Rumex crispus* Lour.).

> *Rumex chinensis* Campd. Monog. Rumex 63, 75. 1819; Courchet in Lecomte Fl. Gén. Indo-Chine **5**: 19. 1910.
> *Rumex loureirii* Campd. Monog. Rumex 142. 1819 (based on *Rumex crispus* Lour.).
> *Rumex loureirianus* Schult. & Schult. f. Syst. **7**: 1474. 1830 (based on *Rumex crispus* Lour.).

Rumex crispus (non Linn.) Lour. Fl. Cochinch. 216. 1790, ed. Willd. 269. 1793, Anamese *cây dieò hoang.*

"Habitat in Cochinchina prope flumina." I follow Courchet in this disposition of Loureiro's species. He retains *Rumex chinensis* Campd. as valid but other authors have reduced it to *Rumex maritimus* Linn.; it apparently is the species that appears in the literature of Indian and Chinese botany as *Rumex maritimus* Linn. Stokes' name is the oldest one.

Rumex sp.

Rheum barbarum Lour. Fl. Cochinch. 255. 1790 (error for *R. rhabarbarum*), Anamese *dai hoàng*, Chinese *tá hoâm, tay hoang.*

Rheum rhabarbarum (non Linn.) Lour. Fl. Cochinch. ed. Willd. 314. 1793.

Rheum cantoniense Lour.[66] ex Gomes in Mem. Acad. Sci. Lisb. Cl. Sci. Pol. Mor. Bel.- Let. n.s. **4**(1): 29. 1886.

"Habitat in multis locis imperii Chinensis: colitur etiam a curiosis in Cochinchina." A specimen from Loureiro in the herbarium of the Paris Museum was identified by Desvaux as *Rumex* sp. and Loureiro's poor description seems to apply to *Rumex* rather than to *Rheum.* The binomial *Rheum barbarum* as given in the original edition of 1790 is a manifest error for *rhabarbarum*, as Loureiro cites Linnaeus as the authority; there is no *Rheum barbarum* Linn.

Rheum Linnaeus

Rheum palmatum Linn. Sp. Pl. ed. 2, 531. 1762; Lour. Fl. Cochinch. 255. 1790, ed. Willd. 313. 1793, Anamese *dai hoàng*, Chinese *tà hoâm.*

"Habitat in provinciis borealibus imperii Chinensis, intra, & extra celebrem murum." The Linnaean species was probably interpreted by Loureiro correctly. Doubtless his material was secured from an herbalist.

Polygonum (Tournefort) Linnaeus

Polygonum barbatum Linn. Sp. Pl. 362. 1753; Lour. Fl. Cochinch. 241. 1790, ed. Willd. 296. 1793, Chinese *leào xí, hung hoang xeng.*

Pogalis barbata Raf. Fl. Tellur. **3**: 15. 1837.

"Habitat in China, Cantone, & alibi." It seems probable that Loureiro had a form of the widely distributed Linnaean species, in spite of his description of the leaves as glabrous.

Polygonum chinense Linn. Sp. Pl. 363. 1753; Lour. Fl. Cochinch. 241. 1790, ed. Willd. 297. 1793, Chinese *fŏ thân mû.*

Coccoloba asiatica Lour. Fl. Cochinch. 239. 1790, ed. Willd. 295. 1793, Anamese *cây muòng chuong.*

Polygonum asiaticum Jackson in Ind. Kew. **1**: 573. 1895, sub *Coccoloba asiatica* Lour. (based on *Coccoloba asiatica* Lour.).

Coccoloba cymosa Lour. Fl. Cochinch. 240. 1790, ed. Willd. 295. 1793, Anamese *mùòng chuong chum.*

Polygonum chinense Linn., "habitat prope Cantonem Sinarum," was correctly interpreted by Loureiro. At the same time he described the same species twice under *Cocco-*

[66] A Loureiro herbarium name here first published by Gomes.

loba, C. asiatica " habitat in sepibus, ac dumetis Cochinchinae," and *C. cymosa* " habitat similiter in sepibus Cochinchinae." It may be noted that this reduction of the last two species is in a measure verified by Courchet (Lecomte Fl. Gén. Indo-Chine **5**: 37. 1910) who cites the Anamese name *chuong chuong* for *Polygonum chinense* Linn. Courchet does not account for *Coccoloba asiatica* Lour., *Polygonum asiaticum* Jacks., and *Coccoloba cymosa* Lour. in his treatment of the Polygonaceae of Indo-China (Lecomte Fl. Gén. Indo-Chine **5**: 15–42. 1910).

Polygonum ciliatum Lour. Fl. Cochinch. 243. 1790, ed. Willd. 299. 1793, Chinese *hŏ xān kĭo.*
    Polygonum loureirii Poir. in Lam. Encycl. Suppl. **4**: 667. 1816 (based on *Polygonum ciliatum* Lour.).

" Habitat Cantone Sinarum." Hemsley (Journ. Linn. Soc. Bot. **26**: 336. 1891) states that this is an altogether obscure species. Steward (Contr. Gray Herb. **88**: 118. 1930) placed it among the unclassified and excluded ones. On the basis of Loureiro's description I have been unable to refer it to any of the species recorded from Kwangtung Province. There was no warrant for the publication of *Polygonum loureirii* Poir. as Loureiro's specific name was valid.

Polygonum fagopyrum Linn. Sp. Pl. 364. 1753.
    Fagopyrum esculentum Moench Meth. 290. 1794.
    Polygonum tataricum (non Linn.) Lour. Fl. Cochinch. 242. 1790, ed. Willd. 298. 1793, Chinese *tam cŏ mac.*
    Polygonum sinarum Desv. ex Meisn. Monog. Polygon. 62. 1826, in syn. (based on *Polygonum tataricum* Lour.).

" Habitat Cantone Sinarum." The form Loureiro had was apparently the common buckwheat rather than the more northern bitter buckwheat, *P. tataricum* Linn. Loureiro's specimen is preserved in the herbarium of the Paris Museum.

Polygonum glabrum Willd. Sp. Pl. **2**: 447. 1799; Courchet in Lecomte Fl. Gén. Indo-Chine **5**: 30. 1910.
    Polygonum hydropiper (non Linn.) Lour. Fl. Cochinch. 240. 1790, ed. Willd. 295. 1793, Anamese *rau ram nhà tlŏi,* Chinese *xūei leào.*

" Habitat in paludibus, & infra ripas fluminum, in Cochinchina, & China." This is reduced in Index Kewensis to *Polygonum serrulatum* Lag., with which species Loureiro's description does not well agree. Courchet placed it as a doubtful synonym of *P. flaccidum* Roxb., but the spikes are described by Loureiro as " congestae." On the whole Loureiro's description agrees very closely with the characters of *P. glabrum* Willd., which occurs in Indo-China and in Kwangtung Province, while the indicated habitat is that of Willdenow's species; in the latter the leaf-margins are not pilose as indicated by Loureiro.

Polygonum odoratum Lour. Fl. Cochinch. 243. 1790, ed. Willd. 299. 1793, Anamese *rau ram;* Meisn. in DC. Prodr. **14**: 106. 1857; Courchet in Lecomte Fl. Gén. Indo-Chine **5**: 29. 1910.

" Colitur in toto regno Cochinchinae: amat loca humida." This species is recorded only from Indo-China where it is used as a condiment. Courchet gives a detailed description of it based on modern collections. Steward (Contr. Gray Herb. **88**: 62. 1930) places it near *P. hydropiper* Linn., from which it differs in its smooth shiny achenes. A specimen from Loureiro listed as being among his plants in the herbarium of the British Museum has not been located.

Polygonum orientale Linn. Sp. Pl. 362. 1753; Meisn. in DC. Prodr. **14**: 123. 1837; Courchet in Lecomte Fl. Gén. Indo-Chine **5**: 37. 1910 (var. *pilosum* Meisn.); Steward in Contr. Gray Herb. **88**: 40. 1930.

Lagunea cochinchinensis Lour. Fl. Cochinch. 220. 1790, ed. Willd. 272. 1793, Anamese *cây ngai bà*, Chinese *pǎ niù*.

Polygonum cochinchinense Meisn. Monogr. Polygon. 55. 1826 (based on *Lagunea cochinchinensis* Lour.).

Polygonum subcordatum Miq. in Journ. Bot. Néerl. **1**: 95. 1861.

" Habitat frequenter prope fossas in Cochinchina, etiam in China." Specimens from Loureiro are preserved in the herbaria of the British and the Paris Museums, but even if authentic specimens were not extant, its identity is evident from the description, as the characters given by Loureiro apply unmistakably to the Linnaean species.

Polygonum perfoliatum Linn. Syst. ed. 10, 1006. 1759; Lour. Fl. Cochinch. 242. 1790, ed. Willd. 298. 1793, Anamese *rau sóung chua deei, bìm bìm gai;* Courchet in Lecomte Fl. Gén. Indo-Chine **5**: 38. 1910.

" Habitat in sepibus Cochinchinae." Loureiro correctly interpreted the Linnaean species, which is one of very wide distribution in Asia, extending to Japan, the Philippines and the Malay Archipelago. A specimen from Loureiro is preserved in the herbarium of the British Museum.

Polygonum persicaria Linn. Sp. Pl. 361. 1753; Lour. Fl. Cochinch. 240. 1790, ed. Willd. 296. 1793; Courchet in Lecomte Fl. Gén. Indo-Chine **5**: 32. 1910.

" Habitat incultum in hortis, & pratis humidis Cochinchinae." Courchet cites Loureiro as a collector in his consideration of *Polygonum persicaria* Linn. var. *agrestis* Meisn. and Loureiro's description apparently applies to a form of the Linnaean species.

Polygonum plebeium R. Br. Prodr. 420. 1810; Courchet in Lecomte Fl. Gén. Indo-Chine **5**: 24. 1910.

Polygonum aviculare (non Linn.) Lour. Fl. Cochinch. 241. 1790, ed. Willd. 297. 1793, Anamese *vien súc.*

" Habitat in China, & Cochinchina." Loureiro's description is very short but applies to *Polygonum plebeium* R. Br. rather than to *P. aviculare* Linn.

Polygonum strigosum R. Br. Prodr. 420. 1810; Courchet in Lecomte Fl. Gén. Indo-Chine **5**: 38. 1910.

? Rumex hostilis Lour. Fl. Cochinch. 217. 1790, ed. Willd. 269. 1793, Anamese *cây dieò gai.*

" Habitat agrestis in Cochinchina." The description is very short and imperfect. Courchet retains it under *Rumex* with the statement: " Plante très obscure, appartenant peut-être à une autre genre." If a *Polygonum*, Loureiro's description of the " petals " as 3, and the plant as dioecious, is incorrect, while the description of the stems as terete does not apply to *P. strigosum* R. Br. No *Rumex* has aculeate stems. The description does not remotely apply to any other polygonaceous plant known from southern and eastern

Asia. There is, of course, the chance that Loureiro had in hand a representative of some other family than the Polygonaceae.

Polygonum tinctorium Ait. Hort. Kew. 2: 31. 1789.

> *Polygonum tinctorium* Lour. Fl. Cochinch. 241. 1790, ed. Willd. 297. 1793, Chinese *hŏ lâm;* Meisn. in DC. Prodr. **14**: 102. 1857; Courchet in Lecomte Fl. Gén. Indo-Chine **5**: 25. 1910.
>
> *Pogalis tinctoria* Raf. Fl. Tellur. **3**: 15. 1836 (based on *Polygonum tinctorium* Lour. or *P. tinctorium* Ait.).

" Habitat Cantone Sinarum." Loureiro's type is preserved in the herbarium of the Paris Museum. The species is one occasionally cultivated in China, its status and relationships being well understood. While Loureiro is usually cited as the authority for the binomial, it was actually published one year earlier by Aiton, based on living specimens introduced into England from China in 1776; there is no doubt as to the specific identity of the two forms described independently under the same binomial.

CHENOPODIACEAE

Chenopodium (Tournefort) Linnaeus

Chenopodium album Linn. Sp. Pl. 219. 1753.

> *Chenopodium hybridum* (non Linn.) Lour. Fl. Cochinch. 174. 1790, ed. Willd. 217. 1793, Anamese *cây màn ri.*

" Habitat incultum in agris Cochinchinae." This reduction of Loureiro's species is based in part on the geographic distribution of *Chenopodium album* Linn. and its allies. *C. hybridum* Linn. is reported in China only from the northern provinces; nor does it occur in the warmer parts of British India. *C. album,* a variable species, occurs in the warmer parts of the Old World as well as in temperate regions. Courchet (Lecomte Fl. Gén. Indo-Chine **5**: 4. 1910) admits *C. hybridum* Linn. as an Indo-China plant solely on the basis of Loureiro's record. I am convinced that this species does not grow in Indo-China.

Beta (Tournefort) Linnaeus

Beta vulgaris Linn. Sp. Pl. 222. 1753; Lour. Fl. Cochinch. 174. 1790, ed. Willd. 217. 1793, Chinese *pǎ hung.*

" Habitat Cantone Sinarum, venditurque in foro cum aliis oleribus." Loureiro doubtless had a form of the common beet, cultivated for its edible leaves. He notes, however, that the roots were white and not edible.

Spinacia (Tournefort) Linnaeus

Spinacia oleracea Linn. Sp. Pl. 1027. 1753; Lour. Fl. Cochinch. 617. 1790, ed. Willd. 757. 1793.

" Habitat culta, & vulgaris Cantone Sinarum." The common spinach is still cultivated near Canton and Loureiro's description applies to it.

AMARANTHACEAE

Celosia Linnaeus

Celosia argentea Linn. Sp. Pl. 205. 1753; Lour. Fl. Cochinch. 163. 1790, ed. Willd. 203. 1793, Anamese *tanh thoung tu*, Chinese *tsīm sīam tsú*.

Celosia margaritacea Linn. *op. cit.* ed. 2, 297. 1762; Lour. Fl. Cochinch. 164. 1790, ed. Willd. 203. 1793, Anamese *ha khô thạo*, Chinese *hiá khù tsào*.

For the first Loureiro states: " in hortis, & agris," and for the second: " in agris," and for both: " Cochinchinae, & Chinae inculta." Both descriptions apply to the very common *Celosia argentea* Linn. of which *C. margaritacea* Linn. is a synonym. Specimens of both, as named by Loureiro, are in the herbarium of the British Museum.

Celosia cristata Linn. Sp. Pl. 205. 1753.

Celosia castrensis Linn. *op. cit.* ed. 2, 297. 1762; Lour. Fl. Cochinch. 163. 1790, ed. Willd. 202. 1793, Anamese *hoa moung gà*, Chinese *kī koán hōa*.

" Habitat passim culta in Cochinchina, & China." Loureiro's description applies unmistakably to *Celosia cristata* Linn., of which *C. castrensis* Linn. is a synonym. *Amaranthus vulgaris* Rumph. (Herb. Amb. **5**: 236. *pl. 84*), cited by Loureiro as a synonym, is correctly placed. *Celosia cristata* Linn. is apparently a cultigen derived from *C. argentea* Linn.

Amaranthus [67] Linnaeus

Amaranthus spinosus Linn. Sp. Pl. 991. 1753; Lour. Fl. Cochinch. 561. 1790, ed. Willd. 687. 1793, Anamese *rau gên gai*.

" Habitat spontaneus loca minus culta Cochinchinae." The description applies to the very common pantropic Linnaean species. *Blitum spinosum* Rumph. (Herb. Amb. **5**: 234. *pl. 83. f. 1*), cited by Loureiro as a synonym, after Linnaeus, is correctly placed.

Amaranthus oleraceus Linn. Sp. Pl. ed. 2, 1403. 1763; Lour. Fl. Cochinch. 561. 1790, ed. Willd. 686. 1793, Anamese *rau gên mùoi*.

" Habitat cultus, nec frequens in Cochinchina." Loureiro apparently had specimens representing *Amaranthus oleraceus* Linn., but the status of the Linnaean species is more or less doubtful. Hooker f. (Fl. Brit. Ind. **4**: 721. 1885) places it as a variety of *A. blitum* Linn.

Amaranthus tricolor Linn. Sp. Pl. 989. 1753; Lour. Fl. Cochinch. 560. 1790, ed. Willd. 685. 1793, Anamese *hoùng hien*, Chinese *hûm hién*.

Amaranthus cruentus (non Linn.) Lour. Fl. Cochinch. 561. 1790, ed. Willd. 687. 1793, Anamese *rau gên tiá*.

Amaranthus polygamus (non Linn.) Lour. Fl. Cochinch. 560. 1790, ed. Willd. 685. 1793, Anamese *rau gên tláng*, Chinese *pě hién*.

Amaranthus gangeticus Linn. Syst. ed. 10, 1268. 1759.

For the first two Loureiro states: " Habitat cultus in China, & Cochinchina "; and for *A. polygamus:* " Habitat in Cochinchina, & China tam cultus, quam spontaneus." I believe that all three descriptions appertain to color forms of the variable, widely distrib-

---

[67] Frequently spelled *Amarantus*, but the form selected by Linnaeus is here retained; see Sprague, Kew Bull. 287. 1928.

uted species commonly known as *Amaranthus gangeticus* Linn., but for which *A. tricolor* Linn. seems to be the oldest specific name.

Amaranthus viridis Linn. Sp. Pl. ed. 2, 1405. 1763.

> *Amaranthus tristis* (non Linn.) Lour. Fl. Cochinch. 560. 1790, ed. Willd. 686. 1793, Anamese *rau gen dat.*
>
> *Amaranthus gracilis* Desf. Tabl. Bot. 43. 1804; Standl. in N. Am. Fl. **21**: 117. 1917.
>
> " Habitat ubique incultus in hortis Cochinchinae, & Chinae, esculentus." This form is the one that occurs everywhere as a weed in gardens in southern China and is commonly cooked as a pot-herb. *Blitum terrestre* Rumph. (Herb. Amb. **5**: 232. *pl. 82. f. 2*), cited by Loureiro as representing his species, may represent a form of *A. gangeticus* Linn. This is *A. viridis* Linn. as currently interpreted in modern works on Asiatic botany, and as represented by the actual type in the Linnaean herbarium on which the Linnaean description, as far as his consideration of the species was a new one, was based. It is the form with very rugose utricles and is *A. gracilis* Desf. Thellung discusses *A. viridis* Linn. extensively in Mém. Soc. Nat. Cherbourg **38**: 212. 1912, but I see no reason whatever for considering *Amaranthus viridis* Linn. to other than as interpreted by practically all botanists who have considered the plants of the Old World tropics.

Achyranthes Linnaeus

Achyranthes aspera Linn. Sp. Pl. 204. 1753.

> *Cyathula geniculata* Lour. Fl. Cochinch. 102. 1790, ed. Willd. 124. 1793, Anamese *co xuóc, nguu tăt*, Chinese *niêu si;* Moore in Journ. Bot. **63**: 249. 1925.
>
> *Centrostachys aspera* Standl. in Journ. Washington Acad. Sci. **5**: 75. 1915.
>
> " Habitat in Cochinchina." Loureiro's type is preserved in the herbarium of the British Museum, and Hiern (Cat. Welw. Pl. 893. 1900) indicated that it is not the plant commonly but erroneously known as *Cyathula geniculata* Lour., but is the even more common *Achyranthes aspera* Linn. The current misinterpretation of *Cyathula* has largely been due to the fact that Loureiro's genus and species have in general been interpreted from an examination of *Auris canina I femina* Rumph. (Herb. Amb. **6**: 26. *pl. 11*) which is discussed by Loureiro in a note following his description, rather than from an examination of his extant type; the Rumphian illustration represents *Cyathula prostrata* Blume. Druce (Rept. Bot. Exch. Club Brit. Isles **4**: 618. 1917) erroneously adopts the binomial *Cyathula alternifolia* (Linn. f.) Druce, based on *Achyranthes alternifolia* Linn. f. (Suppl. 159. 1781) for the species commonly known as *Cyathula geniculata* Lour. In the first place *Achyranthes alternifolia* Linn. f. (1781) is invalidated by *A. alternifolia* Linn. (Mant. **1**: 50. 1767) = *Digera alternifolia* (Linn.) Aschers. (*D. arvensis* Forsk.), and secondly, *Achyranthes alternifolia* Linn. f. was in itself a mixture, not based on any botanical material, but on two pre-Linnaean references. The first *Achyranthes spicatus albus Lychnidis folio [Maderaspatensis]* Pluk. (Alm. 36 [26] *pl. 260. f. 1.* 1696) shows a plant with alternate leaves and terminal capitate inflorescences, undoubtedly a " *Cyathula* " and either *C. globosa* (Pers.) Moq. or *C. zeylanica* Hook. f. The second is *Amaranthus humilis, foliis oppositis flosculis in alis glomeratis* Burm. (Thes. Zeyl. 17. *pl. 4. f. 2.* 1737), a plant with opposite leaves and axillary inflorescences, which is clearly *Allmania nodiflora* (Linn.) R. Br. It is at least unfortunate that in the list of generic names conserved by the Cambridge Botanical Congress *Cyathula*

Blume (1825) has been retained as a valid generic name now that the earlier *Cyathula* Loureiro has been shown to be a synonym of *Achyranthes* Linn. The reason given was that no other generic name had been proposed for this particular group. However, *Polyscalis* Wallich (1832, *nomen nudum*), which was published by Moquin (1849) as a section of *Cyathula* with *Polyscalis capitata* Wall. retained in the section (as *C. capitata* Moq.), is congeneric with *Cyathula* Blume (non Loureiro). In my judgment this name should have been adopted rather than *Cyathula* Blume (non Loureiro).

Standley [68] calls attention to the fact that, strictly speaking, the *type* of the genus *Achyranthes* Linn. (Gen. Pl. 34. 1737, ed. 5, 96. 1794) is *Achyracantha* Dill. (Hort. Elth. 8. *pl. 7. f. 7.* 1732), which is an *Alternanthera;* Dillenius's illustration and description appertains to *Alternanthera achyrantha* R. Br. For *Achyranthes*, of all modern authors, he adopts *Centrostachys* Wall. in Roxb. Fl. Ind. **2**: 497. 1824. Turning to Linnaeus' Species Plantarum 204–205. 1753 we find that *Achyranthes* is there constituted as follows: *A. aspera* Linn. with two varieties *sicula* and *indica*, *A. lappacea* Linn. = *Pupalia lappacea* Moq., *A. lanata* Linn. = *Aerva lanata* Juss., *A. repens* Linn. = *Alternanthera achyrantha* R. Br., and *A. corymbosa* Linn. = *Polycarpaea corymbosa* Willd., representatives of five universally recognized genera in two families of plants. If Standley be followed, then all of the species he transferred to *Centrostachys* will again have to be transferred to *Cyathula* as Loureiro's genus has 58 years' priority over Wallich's and the two are congeneric. To avoid these changes and the additional resulting confusion I prefer to interpret *Achyranthes aspera* Linn., the first species in the genus, as the *standard species* of *Achyranthes*, thus conserving this name for the group of plants with which it has been associated by practically all botanists since 1753.

Aerva [69] Forskål

Aerva lanata (Linn.) Juss.[70] ex Schult. in Roem. & Schult. Syst. **5**: 564. 1819; Moq. in DC. Prod. **13**(2): 303. 1849.

Achyranthes lanata Linn. Sp. Pl. 204. 1753.

Illecebrum lanatum Linn. Mant. **2**: 344. 1771; Lour. Fl. Cochinch. 162. 1790, ed. Willd. 201. 1793, Anamese *rau chiéo*.

"Habitat incultum in agris, & hortis Cochinchinae." Loureiro was undoubtedly correct in his interpretation of this common and well known species.

Alternánthera Forskål

Alternanthera sessilis (Linn.) R. Br. ex Schult. in Roem. & Schult. Syst. **5**: 554. 1819.

Gomphrena sessilis Linn. Sp. Pl. 225. 1753.

Illecebrum sessile Linn. Sp. Pl. ed. 2, 300. 1762; Lour. Fl. Cochinch. 162. 1790, ed. Willd. 202. 1793, Chinese *fân kì kŏuc*.

[68] Standley, P. C. The application of the generic name Achyranthes. Journ. Washington Acad. Sci. **5**: 72–76. 1915.

[69] *Aerva* Forskål (1775), conserved name, Vienna Code; an older one is *Ouret* Adanson (1763, modified to *Uretia* O. Kuntze 1891). Sometimes spelled *Aerua* after Jussieu, but *Aerva* is the original form; see Sprague, Kew Bull. 342. 1928.

[70] Jussieu does not publish the binomial *Aerva* (*Aerua*) *lanata* in Ann. Mus. Hist. Nat. (Paris) **2**: 131. 1803. All that he says is: "Les feuilles nues sont alternes dans l'*amaranthus* le *celosia*, l'*Ærua* et le *Digera*, opposées dans l'*Iresine*, l'*Achyranthes*, le *Gomphrena* et l'*Illecebrum* réduit à un plus petit nombre d'espèces par la soustraction de celles à feuilles alternes (*Illecebrum lanatum, javanicum*) qui appertiennent à l'*Ærua*. . . ."

" Habitat in locis humidis prope Cantonem Sinarum." The Linnaean species is common in the vicinity of Canton and was correctly interpreted by Loureiro. *Olus squillarum* Rumph. (Herb. Amb. **6**: 37. *pl. 15. f. 1*) is correctly placed as a synonym. Robert Brown (Prodr. 417. 1810) does not publish the binomial *Alternanthera sessilis;* he there merely indicates that *Illecebrum sessile* Linn. belongs in *Alternanthera.*

Gomphrena Linnaeus

Gomphrena globosa Linn. Sp. Pl. 224. 1753; Lour. Fl. Cochinch. 175. 1790, ed. Willd. 218. 1793 (*Gomphraena*), Anamese *hoa nua ngài.*
" Habitat passim culta in Cochinchina, & China." The Linnaean species was correctly interpreted by Loureiro. *Flos globosus* Rumph. (Herb. Amb. **5**: 289. *pl. 100. f. 2*) is correctly placed as a synonym.

NYCTAGINACEAE

Mirabilis (Rivinius) Linnaeus

Mirabilis jalapa Linn. Sp. Pl. 177. 1753; Lour. Fl. Cochinch. 101. 1790, ed. Willd. 123. 1793, Anamese *hoa phân,* Chinese *jén chí hōa.*
" Habitat in Cochinchina, & China: vidi etiam in Africa." The Linnaean species, a common pantropic plant, was correctly interpreted by Loureiro.

Boerhavia [71] Linnaeus

Boerhavia diffusa Linn. Sp. Pl. 3. 1753; Lour. Fl. Cochinch. 15. 1790, ed. Willd. 20. 1793, Chinese *houng si sin.*
Boerhavia repens Linn. Sp. Pl. 3. 1753.
Axia cochinchinensis Lour. Fl. Cochinch. 36. 1790, ed. Willd. 44. 1793, Anamese *nhon sâm phu yen;* Moore in Journ. Bot. **63**: 247. 1925.
Regarding *Axia cochinchinensis* Loureiro states: " Frutex iste quem nullibi inveni, praeterquam in Cochinchina." The exact identity of Loureiro's genus and species was unknown until 1925, when Moore (Journ. Bot. **63**: 247) examined the type in the herbarium of the British Museum, although Bentham (Benth. & Hook. f. Gen. Pl. 2: 153. 1873) suggested that *Axia* was referable to *Boerhavia:* "*Axia,* Lour., est verisimiliter e charactere dato *Boerhaavia* inter *Nyctagineas,* bracteis pro calyce descriptis." Loureiro seriously misinterpreted certain floral characters as noted by Moore, which, from the description alone, would render definite placing of *Axia* almost impossible. Curiously Loureiro at the same time correctly interpreted *Boerhavia diffusa* Linn. from Canton material: " Habitat agros Cantonienses in China." *Boerhavia repens* Linn. is indistinguishable from *B. diffusa* Linn., a very common and widely distributed species occurring abundantly in waste places in or about towns throughout the Old World tropics.

Boerhavia africana Lour. Fl. Cochinch. 16. 1790, ed. Willd. 20. 1793; Choisy in DC. Prodr. 13(2): 456. 1849.
Boerhavia plumbaginea Cav. Ic. 2: 7. *pl. 112.* 1793; Baker & Wright in Thistelton-Dyer Fl. Trop. Afr. **6**: 6. 1909.

[71] Usually spelled *Boerhaavia,* but the original form is here retained; see Sprague, Kew Bull. 348. 1928.

" Habitat Mozambicci in Africa Orientali." Baker and Wright do not mention Loureiro's species in their treatment of the Nyctaginaceae of tropical Africa (Thistelton-Dyer Fl. Trop. Afr. **6**: 1–9. 1909). They record 6 species as occurring in the Mozambique District and of these Loureiro's description best conforms to the characters of *Boerhavia plumbaginea* Cav.

PHYTOLACCACEAE

Gisekia Linnaeus

Gisekia africana (Lour.) O. Ktz. Rev. Gen. Pl. **3**(2²): 108. 1898 (based on *Miltus africana* Lour.).

Miltus africana Lour. Fl. Cochinch. 302. 1790, ed. Willd. 370. 1793.

Glinus miltus Raeusch.[72] Nomencl. ed. 3, 141. 1797; Steud. Nomencl. 372. 1821.

Glinus mozambicensis Spreng. Syst. **2**: 467. 1825 (based on *Miltus africana* Lour.).

Gisekia miltus Fenzl in [Endl. & Fenzl] Nov. Stirp. Dec. Vind. **10**: 86. 1839 (based on *Miltus africanus* Lour.); Moq. in DC. Prodr. **13**(2): 28. 1849; Oliv. Fl. Trop. Afr. **2**: 594. 1871.

" Habitat in locis aridis insulae Mozambicci in Africa." A species definitely known from Lower Guinea and the Mozambique district, represented by various extant collections.

AIZOACEAE

Mollugo Linnaeus

Mollugo oppositifolia Linn. Sp. Pl. 89. 1753.

Pharnaceum mollugo Linn. Mant. **2**: 561. 1771; Lour. Fl. Cochinch. 185. 1790, ed. Willd. 230. 1793, Anamese *co dáng*.

" Habitat incultum in hortis Cochinchinae." Loureiro's description conforms closely with the characters of the common and widely distributed *Mollugo oppositifolia* Linn. of which *Pharnaceum mollugo* Linn. is a synonym. *Alsine erecta pentaphylla flore albo* Burm. (Thes. Zeyl. 13. *pl. 7*), cited by Loureiro, after Linnaeus, is an excellent illustration of the Linnaean species.

Mollugo pentaphylla Linn. Sp. Pl. 89. 1753.

Mollugo triphylla Lour. Fl. Cochinch. 62. 1790, ed. Willd. 79. 1793, Chinese *ha khim su*.

Pharnaceum triphyllum Spreng. Syst. **1**: 949. 1825 (based on *Mollugo triphylla* Lour.).

" Habitat Cantone Sinarum spontanea." Loureiro's description unmistakably applies to the Linnaean species. It is abundant in gardens and in recently disturbed soil in the vicinity of Canton and is a species of very wide geographic distribution.

PORTULACACEAE

Portulaca Linnaeus

Portulaca oleracea Linn. Sp. Pl. 445. 1753; Lour. Fl. Cochinch. 293. 1790, ed. Willd. 359. 1793, Anamese *rau sam*, Chinese *mà chi hién*.

" Habitat passim ad vias, & agros in Cochinchina, & China." Loureiro correctly interpreted the ubiquitous Linnaean species, the common purslane.

---

[72] This binomial here appears strictly as a *nomen nudum*, as Loureiro's binomial *Miltus africana* is not cited by Raeuschel; the latter was first definitely associated with *Glinus miltus* Raeusch. by Steudel.

BASELLACEAE

Basella (Rheede) Linnaeus

Basella rubra Linn. Sp. Pl. 272. 1753.

Basella nigra Lour. Fl. Cochinch. 183. 1790, ed. Willd. 229. 1793, Anamese *cây boung toi*, Chinese *lŏ quêi*.

Gandola nigra Raf. Sylva Tellur. 60. 1838 (based on *Basella nigra* Lour.).

"Habitat in Cochinchina & China, agrestis, cultaque, per sepes et crates hortorum implicata." *Basella rubra* Linn. is common and widely distributed in the Old World tropics, cultivated and semi-cultivated, and is very generally used for food as a pot herb. *Basella nigra* Lour. is manifestly the same as the Linnaean species, Willdenow making this reduction in 1793. *Gandola alba* Rumph. (Herb. Amb. **5**: 417. *pl. 154. f. 2*), cited by Loureiro as a synonym, is *Basella rubra* Linn. Rafinesque took the generic name *Gandola* from Rumphius; *Basella nigra* Lour. typifies this genus as described by him. Loureiro's specimen in the herbarium of the British Museum is *Basella rubra* Linn.

CARYOPHYLLACEAE

Stellaria Linnaeus

Stellaria uliginosa Murr. Prodr. Stirp. Gotting. 55. 1770; Gagnep. in Lecomte Fl. Gén. Indo-Chine **1**: 264. *f. 26, 1–5*. 1909.

Cerastium repens (non Linn.) Lour. Fl. Cochinch. 284. 1790, ed. Willd. 349. 1793, Chinese *a kīm tsao*.

"Habitat in pratis, & hortis Cochinchinae: etiam Cantone Sinarum." No representative of *Cerastium* is known from either Indo-China or from Kwangtung Province, China. Among the few possible species to which Loureiro's species may be referred, the description best agrees with *Stellaria uliginosa* Murr., the only species of the genus known from Indo-China and one of the three species known from Kwangtung Province.

Polycarpaea [73] Lamarck

Polycarpaea arenaria (Lour.) Gagnep. in Journ. de Bot. **21**: 280. 1908 (based on *Polia arenaria* Lour.), Bull. Soc. Bot. France **56**: 39. 1909, Lecomte Fl. Gén. Indo-Chine **1**: 269. 1909.

Polia arenaria Lour. Fl. Cochinch. 164. 1790, ed. Willd. 204. 1793, Anamese *sai hô nam*.

Polium arenarium Stokes Bot. Mat. Med. **1**: 477. 1812 (based on *Polia arenaria* Lour.).

"Habitat prope litora in Cochinchina." A species very similar to the widely distributed *Polycarpaea corymbosa* Lam., and separated from it only by minor characters. It is represented by *Clemens 3021, 3721*, from sand dunes at Tourane, near the classical locality Hue, and by *Squires 393* from Hue. *Polia* Lour. antedates *Polycarpaea* Lam., but the latter is conserved. Loureiro's type is preserved in the herbarium of the British Museum.

Lychnis Linnaeus

Lychnis coronata Thunb. Fl. Jap. 187. 1784.

Hedona sinensis Lour. Fl. Cochinch. 286. 1790, ed. Willd. 351. 1793, Chinese *yû mi*. *Lychnis grandiflora* Jacq. Ic. Pl. Rar. **1**: 9. *pl. 84*. 1786, Coll. **1**: 149. 1786.

[73] *Polycarpaea* Lamarck (1792), conserved name, Brussels Code; an older one is *Polia* Loureiro (1790).

" Colitur ob venustatem Cantone Sinarum." Willdenow in a footnote states: " Est *Lychnis coronata* Thunb. Jap. 187. et *Lychnis grandiflora* Jacquini." Loureiro's type preserved in the herbarium of the Paris Museum was identified by Desvaux as representing *Lychnis grandiflora* Jacq. Jacquin's specific name may or may not be older than Thunberg's. While Jacquin's *plate 84* may have been issued earlier than 1784, the date of Thunberg's publication, the descriptive text was not printed until 1786. The plates of Jacquin's Icones were prepared between 1781 and 1786. Thunberg's species was published in 1784.

Dianthus Linnaeus

Dianthus chinensis Linn. Sp. Pl. 411. 1753; Lour. Fl. Cochinch. 282. 1790, ed. Willd. 346.
1793, Anamese *cam trúoc hoa*.
" Habitat in viridariis Sinensibus, & Cochinchinensibus." Loureiro apparently had material representing a form of this very common Chinese species.

Dianthus caryophyllus Linn. Sp. Pl. 410. 1753; Lour. Fl. Cochinch. 281. 1790, ed. Willd.
345. 1793, Anamese *houng nhung hōa*.
" Colitur in China, unde in Cochinchinam creditur delatum." Loureiro described one of the garden forms with double flowers; it may or may not have been a form of the Linnaean species. Gagnepain (Lecomte Fl. Gén. Indo-Chine 1: 263. 1909) notes that *Dianthus plumaris* auct. occurs in Anam, expressing the opinion that it was introduced from Europe.

NYMPHAEACEAE

Nelumbium Jussieu

Nelumbium nelumbo (Linn.) Druce in Rept. Bot. Exch. Club Brit. Isles 3: 421. 1914; Merr.
Interpret. Herb. Amb. 218. 1917.
Nymphaea nelumbo Linn. Sp. Pl. 511. 1753; Lour. Fl. Cochinch. 340. 1790, ed. Willd.
416. 1793, Anamese *cây sen*, Chinese *liên hōa, heu xí hém*.
Nelumbium speciosum Willd. Sp. Pl. 2: 1258. 1799; Gagnep. in Lecomte Fl. Gén. Indo-Chine 1: 162. 1908.
" Habitat in paludibus coenosis Cochinchinae, & Chinae: colitur etiam in vasis amplis, & pretiosis in hortis, & atriis Magnatum." The Linnaean species, the common lotus, was correctly interpreted by Loureiro. *Nymphaea indica major* Rumph. (Herb. Amb. 6: 168. *pl. 73*) is correctly placed as a synonym.

RANUNCULACEAE

Paeonia (Tournefort) Linnaeus

Paeonia suffruticosa Andr. Bot. Repos. 6: *pl. 373*. 1804.
Paeonia moutan Sims in Curtis's Bot. Mag. 29: *pl. 1154*. 1808.
Paeonia officinalis (non Linn.) Lour. Fl. Cochinch. 343. 1790, ed. Willd. 419. 1793,
Anamese *thuoc duoc*, Chinese *xŏ yŏ*.
" Habitat culta, spontaneaque per totum imperium Sinense, maxime in provinciis Borealibus: in Cochinchinam inde translata." Loureiro apparently had garden forms of

the species commonly known as *Paeonia moutan* Sims = *P. suffruticosa* Andr., which is cultivated in China as far south as Canton.

Clematis (Dillenius) Linnaeus

Clematis chinensis Osbeck Dagbok Ostind. Resa 205. 1757; Merr. in Philip. Journ. Sci. Bot. **12**: 104. 1917; Rehd. in Journ. Arnold Arb. **14**: 200. 1933.

Clematis chinensis Retz. Obs. **2**: 18. *pl. 2.* 1781.

Clematis minor Lour. Fl. Cochinch. 345. 1790, ed. Willd. 422. 1793, Chinese *uei leng siēn.*

Clematis sinensis Lour. Fl. Cochinch. 345. 1790, ed. Willd. 422. 1793, Anamese *mouc thoung*, Chinese *mŭ tum.*

For *Clematis minor* Loureiro states: " Habitat in suburbiis Cantoniensibus Sinarum," and for *C. sinensis:* " Habitat agrestis in multis locis imperii Sinensis." I believe but one species is represented by the two descriptions and this the common plant in thickets near Canton, described by Osbeck in 1757 as *Clematis chinensis.* Loureiro's type is preserved in the herbarium of the Paris Museum and has been identified by Gagnepain as *Clematis chinensis* Retz. The flowers of *Clematis minor* are described as white, which is correct for *C. chinensis* Osbeck, those of *C. sinensis* Lour. are described as reddish-purple, which may be due to Loureiro's securing his data largely from dealers in drug plants. Willdenow in 1793 in a footnote to his edition of Loureiro definitely reduced *Clematis sinensis* Lour. to *C. chinensis* Retz. (= *C. chinensis* Osbeck). Rehder has verified the status of Osbeck's species by an actual examination of his type.

Clematis loureiriana DC. Syst. **1**: 144. 1818 (based on *Clematis dioica* Lour.).

Clematis dioica (non Linn.) Lour. Fl. Cochinch. 344. 1790, ed. Willd. 421. 1793, Anamese *thoung thao.*

Clematis smilacifolia Wall. in Asiat. Research. **13**: 402. 1820; Finet & Gagnep. in Lecomte Fl. Gén. Indo-Chine **1**: 3. 1907.

" Habitat in Cochinchina, arbores, & arundines scandens absque cirrhis." Loureiro's description applies unmistakably to the species currently known as *Clematis smilacifolia* Wall., which has been collected in the vicinity of Hue by Mrs. Clemens. De Candolle's specific name is valid, being two years older than Wallich's. While Finet and Gagnepain cite six synonyms of *Clematis smilacifolia* Wall., they fail to account for Loureiro's binomial and for that of de Candolle based upon it. Objection may be made to this interpretation of Loureiro's species on the basis of his description of the leaves as " 3-nata," i.e., 3-foliolate; but Finet and Gagnepain interpret *Clematis smilacifolia* Wall. as having simple and 3-foliolate leaves. The description of the species as dioecious is an error, Loureiro apparently being influenced by the Linnaean description which he assumed to apply to the material he had in hand.

Clematis meyeniana Walp. in Nov. Act. Acad. Nat. Cur. **19**: Suppl. **1**: 297. 1843; Finet & Gagnep. in Lecomte Fl. Gén. Indo-Chine **1**: 4. 1907.

Clematis virginiana (non Linn.) Lour. Fl. Cochinch. 345. 1790, ed. Willd. 422. 1793, Anamese *son mouc.*

" Habitat inter sepes Cochinchinae." De Candolle (Syst. **1**: 149. 1818) was certainly in error in referring Loureiro's species to *Clematis biternata* DC. The most probable identification of it is *Clematis meyeniana* Walp., which occurs in the vicinity of Hue to which Loureiro's imperfect description applies, as far as it goes. Hemsley (Journ. Linn. Soc. Bot. **23**: 2. 1886) and Rehder & Wilson (Sargent Pl. Wils. **1**: 336. 1913) reduce Loureiro's species to *Clematis apiifolia* DC., one unknown from Indo-China and one to which Loureiro's description does not apply. The statement in the diagnosis copied from Linnaeus, "foliolis . . . sublobato-angulatis," should not be considered as applying to the plant Loureiro actually described.

Ranunculus (Tournefort) Linnaeus

Ranunculus cantoniensis DC. Prodr. **1**: 43. 1824 (based on *Hecatonia pilosa* Lour.).
> *Hecatonia pilosa* Lour. Fl. Cochinch. 303. 1790, ed. Willd. 371. 1793, Chinese *khảo tsảo*, non *Ranunculus pilosus* HBK.
> *Ranunculus chinensis* Bunge in Mém. Acad. St. Pétersb. Sav. Étrang. **2**: 77. 1833 (Enum. Pl. Chin. Bor. 3. 1833).
> *Ranunculus pensylvanicus* Forbes & Hemsl. in Journ. Linn. Soc. Bot. **23**: 14. 1886; Finet & Gagnep. in Bull. Soc. Bot. France **51**: 309. 1904; Gagnep. in Lecomte Fl. Gén. Indo-Chine **1**: 10. *f. 3.* 1907, non Linn. f.
> *Ranunculus brachyrhynchus* Chien in Rhodora **18**: 189. 1916.
> *Ranunculus arcuans* Chien *op. cit.* 190.

"Habitat prope Cantonem Sinarum." While the eastern Asiatic form currently referred to *Ranunculus pensylvanicus* apparently does not represent the species the younger Linnaeus described, I suspect that both Chinese forms described by Chien in 1916 have valid names, if one wishes to recognize two species here. The more southern one, *R. brachyrhynchus* Chien, is safely the same as *R. cantoniensis* DC., while the more northern form, *R. arcuans* Chien, is apparently identical with *Ranunculus chinensis* Bunge. Handel-Mazzetti (Symb. Sin. **7**: 302, 305. 1931) places *Ranunculus arcuans* Chien as a doubtful synonym of *R. sieboldii* Miq. and *R. brachyrhynchus* Chien as a synonym of *R. langsdorffii* Spreng. (1825), retaining *R. chinensis* Bunge as a distinct species.

Ranunculus sceleratus Linn. Sp. Pl. 551. 1753.
> *Hecatonia palustris* Lour. Fl. Cochinch. 303. 1790, ed. Willd. 371. 1793, Chinese *chú lién*.

"Habitat in locis humidis prope Cantonem Sinarum." Loureiro's type is preserved in the herbarium of the Paris Museum and represents the Linnaean species. *Ranunculus sceleratus* Linn. is not uncommon in wet places in the vicinity of Canton.

MENISPERMACEAE

Pericampylus Miers

Pericampylus heterophyllus (Lour.) Diels in Pflanzenreich 46(IV-94): 220. 1910 (based on *Pselium heterophyllum* Lour.).
> *Pselium heterophyllum* Lour. Fl. Cochinch. 621. 1790, ed. Willd. 762. 1793, Anamese *deei môi*.
> *Pselium ambiguum* Miers Contrib. Bot. **3**: 123. *pl. 112.* 1871 (based on *Pselium heterophyllum* Lour., staminate specimen).

" Habitat in sylvis Cochinchinae." Diels examined Loureiro's type in the herbarium of the British Museum and his description is based solely on the type collection. Loureiro's generic name is much older than *Pericampylus* Miers but was not accepted by Diels for the reason that Loureiro's description of the fruits apparently appertains to *Stephania*. I suspect that the generic description was based on the *Pericampylus* as to the staminate flowers, and on a *Stephania* as to pistillate ones, as Loureiro described the staminate plant as having subcordate leaves, and the pistillate one as having peltate ones. Gagnepain does not mention Loureiro's genus and species in his treatment of the Menispermaceae of Indo-China (Lecomte Fl. Gén. Indo-Chine 1: 124–154. 1908). Miers' description and illustration of *Pselium ambiguum* were based on Loureiro's staminate specimen of *P. hetero-phyllum* in the herbarium of the British Museum, which Mr. Exell informs me has leaves slightly different from those of *Pericampylus glaucus* (Lam.) Merr., and calls attention to the differential stamen characters emphasized by Miers. No specimen of the pistillate plant described by Loureiro is extant.

Cocculus de Candolle [74]

Cocculus sarmentosus (Lour.) Diels in Pflanzenreich **46**(IV–94): 233. 1910 (based on *Ne-phroia sarmentosa* Lour.).

Nephroia sarmentosa Lour. Fl. Cochinch. 565. 1790, ed. Willd. 692. 1793, Anamese *deei xanh.*

Cocculus nephroia DC. Syst. 1: 531. 1818 (based on *Nephroia sarmentosa* Lour.).

Nephroica sarmentosa Miers Contrib. Bot. **3**: 261. 1871 (based on *Nephroia sarmen-tosa* Lour.).

Menispermum reniforme Spreng. Syst. **2**: 156. 1825 (based on *Nephroia sarmentosa* Lour.).

" Habitat in sylvis Cochinchinae." Loureiro's type is preserved in the herbarium of the British Museum and this was examined by Diels in connection with his monographic treatment of the family. The species is one of wide geographic distribution in southeastern Asia and Malaysia and often can scarcely be distinguished from *Cocculus trilobus* (Thunb.) DC. (*C. thunbergii* DC.); it is included in that species, and perhaps correctly so, by Gagnepain (Lecomte Fl. Gén. Indo-Chine 1: 142. 1908).

Stephania Loureiro

Stephania rotunda Lour. Fl. Cochinch. 608. 1790, ed. Willd. 747. 1793, Anamese *cu môt, tu nhien;* Miers Contrib. Bot. **3**: 215. 1871; Diels in Pflanzenreich **46**(IV–94): 275. 1910; Moore in Journ. Bot. **63**: 288. 1925.

Clypea rotunda Steud. Nomencl. ed. 2, 1: 387. 1840 (based on *Stephania rotunda* Lour.).

" Habitat in sylvis Cochinchinae." Miers, Moore, and apparently Diels, all examined Loureiro's type in the herbarium of the British Museum, the former basing his description upon it and describing the flowers, yet Moore, who gives additional data, cannot reconcile Diels' statement: " Cochinchina: o. n. O., mit abgefallenen Blüten (Loureiro-Original der

---

[74] *Cocculus* de Candolle (1818), conserved name, Vienna Code; older ones are *Cebatha* Forskål (1775), *Leaeba* Forskål (1775), *Epibaterium* Forskål (1776), *Nephroia* Loureiro (1790), *Baumgartia* Moench (1794), *Androphylax* Wendland (1798) and *Wendlandia* Willdenow (1799).

Art!) " with the British Museum specimen which has staminate flowers. Diels did not accept Gagnepain's interpretation of the species (Lecomte Fl. Gén. Indo-Chine 1: 148. *f. 14, 45–53*. 1908) (*S. rotunda* Gagnep. non Lour. = *S. pierrei* Diels), retaining *S. rotunda* Lour. as a valid species allied to *S. brachyandra* Diels. Moore, however, on the basis of the floral characters, states that the species is near *S. sinica* Diels. A critical comparison of the type with authentic material representing the several species involved is desirable.

Stephania longa Lour. Fl. Cochinch. 609. 1790, ed. Willd. 747. 1793, Anamese *deei môi tlon;* Diels in Pflanzenreich **46**(IV–94): 278. 1910.

 Clypea longa G. Don Gen. Syst. **1**: 113. 1831 (based on *Stephania longa* Lour.).

 " Habitat sepes arundinum in Cochinchina." This is retained by Diels as a valid species and redescribed on the basis of material from Kwangsi, Macao, Hongkong, Hainan, Anam, and Tonkin, with the reference to Loureiro cited thus: " Cochinchina, o. n. O. (Loureiro-Original der Art!)." I have found no record of Loureiro's type being extant in the herbaria of the two institutions known to contain specimens collected by him. Gagnepain does not mention the species in his treatment of the Menispermaceae of Indo-China (Lecomte Fl. Gén. Indo-Chine **1**: 124–154. 1908).

Fibraurea Loureiro

Fibraurea tinctoria Lour. Fl. Cochinch. 626. 1790, ed. Willd. 769. 1793, Anamese *cây vàng dăng*, Chinese *tiēn siēn tàn;* Miers Contrib. Bot. **3**: 41. 1871; Diels in Pflanzenreich **46**(IV–94): 122. 1910; Gagnep. in Lecomte Fl. Gén. Indo-Chine **1**: 135. 1908.

 Cocculus fibraurea DC. Syst. **1**: 525. 1818 (based on *Fibraurea tinctoria* Lour.).

 Menispermum tinctorium Spreng. Syst. **2**: 156. 1825 (based on *Fibraurea tinctoria* Lour.).

 Fibraurea recisa Pierre Fl. Forest. Cochinch. **2**: *pl. 111*. 1885.

 " Habitat in sylvis Cochinchinae, & Chinae." Loureiro's type, without flowers or fruit, in the herbarium of the British Museum, was examined by Miers and by Diels. The species is known only from Indo-China, Loureiro's statement that it occurred in China being probably due to misinformation secured from some herbalist. *Fibraurea tinctoria* of Indian botanists is *F. chloroleuca* Miers. Gagnepain maintains *Fibraurea recisa* Pierre as a valid species, his conception of *F. tinctoria* Lour. apparently having been based on descriptions and specimens of *F. chloroleuca* Miers which he cites as a synonym; Diels' note on Loureiro's type should be consulted.

Limacia Loureiro

Limacia scandens Lour. Fl. Cochinch. 620. 1790, ed. Willd. 761. 1793, Anamese *cây mê gà.*

 Cocculus limacia DC. Syst. **1**: 526. 1818 (based on *Limacia scandens* Lour.).

 Menispermum limacia Spreng. Syst. **2**: 155. 1825 (based on *Limacia scandens* Lour.).

 " Habitat in sylvis Cochinchinae." This species typifies the genus *Limacia* and an amplified description, based on Loureiro's type specimen in the herbarium of the British Museum, was published by Miers (Contrib. Bot. **3**: 109. 1871). It is therefore strange that the species was overlooked by Gagnepain in his treatment of the Menispermaceae on Indo-China (Lecomte Fl. Gén. Indo-Chine **1**: 124–154. 1908) although he admits three species of *Limacia* (two representing *Hypserpa* and one *Tiliacora* according to Diels' classification). I identified *Clemens 4332*, a fruiting specimen from Tourane, as representing *Limacia scandens* Lour., but this is clearly *Hypserpa laevifolia* Diels (*Limacia cuspidata* Gagnep.). Diels (Pflanzenreich 46(IV–94): 214. 1910) recognizes *Limacia scandens* Lour., citing Loureiro's type. He states that it is very near *L. oblonga* (Wall.) Miers and that probably the two species should be combined.

Tinospora Miers

Tinospora sinensis (Lour.) Merr. in Sunyatsenia 1: 193. 1934 (based on *Campylus sinensis* Lour.).

> *Campylus sinensis* Lour. Fl. Cochinch. 113. 1790, ed. Willd. 140. 1793, Chinese *xeng con thân*.
> *Cocculus tomentosus* Colebr. in Trans. Linn. Soc. 13: 59. 1822.
> *Menispermum malabaricum* Lam. Encycl. 4: 96. 1796.
> *Tinospora tomentosa* Miers in Ann. Nat. Hist. II 7: 38. 1851, Contrib. Bot. 3: 33. 1871.
> *Tinospora malabarica* Miers in Ann. Nat. Hist. II 7: 38. 1851; Diels in Pflanzenreich 46(IV–94): 142. 1910.

" Habitat in collibus nemorosis apud Sinas Cantonienses." A specimen from Loureiro in the herbarium of the Paris Museum has been identified by Gagnepain as *Tinospora tomentosa* Miers = *T. malabarica* (Lam.) Miers. I have a photograph of this specimen which conforms to Loureiro's species description, and which is well matched by three recent Kwangtung collections *Tso 21506, Liang 61590, 61729;* I cannot distinguish the Chinese from the Indian form. It should be noted that Loureiro's *generic* description calls for a plant with 5-merous, complete, perfect flowers, 5-lobed calyces, the lobes subulate, tubular 2-lipped corollas, 5-lobed stigma, and 5-celled, many-seeded capsules; these are not Menispermaceous characters. What manifestly happened here is that through some error Loureiro based his generic description on some entirely different plant than the single species he placed under *Campylus*. The species is clearly a *Tinospora*, but the genus *Campylus* represents some group remote from it.

MAGNOLIACEAE

Magnolia Linnaeus

Magnolia coco (Lour.) DC. Syst. 1: 459. 1818; Dandy in Lingnan Sci. Journ. 7: 141. 1931 (based on *Liriodendron coco* Lour.).

> *Liriodendron coco* Lour. Fl. Cochinch. 347. 1790, ed. Willd. 424. 1793, Anamese *hoa dea hap*, Portuguese (Macao) *fula coco.*
> *Magnolia pumila* Andr. Bot. Repos. 4: *pl. 226.* 1802.
> *Talauma pumila* Blume Fl. Jav. Magnol. 38. *pl. 12C.* 1828, pro parte.
> *Talauma coco* Merr. Sp. Blancoanae 12. 1918 (based on *Liriodendron coco* Lour.).

" Habitat in Cochinchina, Macai & Cantone, culta ob pulchritudinem et odorem floris." Loureiro's species is identical with the form commonly known as *Magnolia pumila* Andr. or *Talauma pumila* Blume; his specific name was derived from its Macao-Portuguese vernacular name, *fula coco.*

Manglietia Blume

Manglietia fordiana Oliv. in Hook. Ic. 20: *pl. 1953.* 1891.
? Magnolia ? inodora DC. Syst. 1: 459. 1818; Dandy in Lingnan Sci. Journ. 7: 142.
1931 (based on *Liriodendron liliifera* Lour.).
? Liriodendron liliifera (non Linn.) Lour. Fl. Cochinch. 346. 1790, ed. Willd. 424.
1793.

"Habitat agrestis prope Cantonem Sinarum." The exact identity of Loureiro's species is uncertain. It may be the same as *Manglietia fordiana* Oliv., which is the only representative of the family known to occur in a wild state in the vicinity of Canton, yet there are certain discrepancies between Loureiro's description and the characters of Oliver's species in the number of carpels, in the petals and in the anthers. Further field work in the vicinity of Canton is desirable before attempting a more definite placement of *Magnolia inodora* DC., which is a much older binomial than Oliver's. *Sampacca montana* Rumph. (Herb. Amb. 2: 204. *pl. 69*), cited by Loureiro as representing his species, must be excluded as it represents *Talauma rumphii* Blume and is the sole basis of *Liriodendron liliifera* Linn.

Michelia Linnaeus

Michelia champaca Linn. Sp. Pl. 536. 1753; Lour. Fl. Cochinch. 348. 1790, ed. Willd. 425.
1793, Anamese *hoa sú nam;* Finet & Gagnep. in Lecomte Fl. Gén. Indo-Chine 1:
38. 1907.
Michelia champava Lour. ex Gomes in Mem. Acad. Sci. Lisb. Cl. Sci. Mor. Pol. Bel.-
Let. n.s. 4(1): 27. 1868 (sphalm.).

"Habitat culta in hortis Cochinchinae, & Macai." Loureiro correctly interpreted the Linnaean species, which is much cultivated in the Indo-Malaysian region for its very fragrant flowers. *Sampacca* Rumph. (Herb. Amb. 2: 199. *pl. 67*), cited by Loureiro, is correctly placed, but *Sampacca silvestris* Rumph., also cited, appertains to an entirely different species and is an *Elmerillia* according to Mr. J. E. Dandy.

Michelia figo (Lour.) Spreng. Syst. 2: 643. 1825 (based on *Liriodendron figo* Lour.); Dandy
in Lingnan Sci. Journ. 7: 145. 1931.
Liriodendron figo Lour. Fl. Cochinch. 347. 1790, ed. Willd. 424. 1793, Portugese *fula
figo.*
Magnolia ? figo DC. Syst. 1: 460. 1818 (based on *Liriodendron figo* Lour.).
Magnolia fuscata Andr. Bot. Repos. 4: *pl. 229.* 1802.
Michelia fuscata Blume ex Wall. List no. 6495. 1832; Wight ex Steud. Nomencl. ed. 2,
2: 139. 1841.
Liriopsis fuscata Spach Hist. Veg. 7: 461. 1839.

"Habitat culta Macai, & Cantone Sinarum." This species is not uncommon in cultivation in southern China. The Cantonese name on recently collected material is *ham shiu.* Loureiro derived his specific name from the Macao-Portugese vernacular name, *fula figo.* Mr. J. E. Dandy called my attention to the fact that Blume did not publish the binomial *Michelia fuscata* (Fl. Jav. Magnol. 8. 1838) as indicated in Index Kewensis, but merely states that *Magnolia fuscata* is a *Michelia.*

Illicium Linnaeus

Illicium verum Hook. f. in Curtis's Bot. Mag. 114: *pl. 7005*. 1888; Finet & Gagnep. in Lecomte Fl. Gén. Indo-Chine 1: 30. *f. 6. 1–3*. 1907.

Illicium anisatum (non Linn.) Lour. Fl. Cochinch. 353. 1790, ed. Willd. 432. 1793, Anamese *dai hôi, bát giác hôi*, Chinese *pă có huèi hiàm*.

Badianifera officinarum O. Ktz. Rev. Gen. Pl. 6. 1891 (based on *Illicium anisatum* Lour.).

" Habitat agreste, cultumque in provinciis Sinensibus ad occasum Cantoniensis sitis." Loureiro's description conforms, at least in part, to the characters of *Illicium verum* Hook. f. It seems probable that it may have been based in part on some published description of *Illicium anisatum* Linn., at least as to the perianth segments and stamens, as these characters as given by him do not conform to those of *I. verum* Hook. f.

ANNONACEAE

Uvaria Linnaeus

Uvaria lurida Hook. f. & Th. Fl. Ind. 1: 101. 1855; Finet & Gagnep. in Lecomte Fl. Gén. Indo-Chine 1: 53. 1907.

? *Uvaria zeylanica* (non Linn.) Lour. Fl. Cochinch. 348. 1790, ed. Willd. 426. 1793, Anamese *cây mu tru*.

No locality is given but the Anamese name cited indicates an Indo-China specimen. In spite of certain discrepancies in Loureiro's description, notably the " petala . . . lato-lanceolata " and the fruits " dispositae in racemum simplicem," I am confident that he had a *Uvaria*, and in all probability the form interpreted by Finet and Gagnepain as *Uvaria lurida* Hook. f. & Th. *Funis musarius latifolius* Rumph. (Herb. 5: 78. *pl. 42*), cited by Loureiro as a synonym, must be excluded; it represents *Uvaria musaria* (Dunal) DC. of the Philippines and the Moluccas.

Desmos Loureiro

Desmos cochinchinensis Lour. Fl. Cochinch. 352. 1790, ed. Willd. 431. 1793, Anamese *cây châp chôi, cây cô chay*.

Unona desmos Dunal Monogr. Anon. 112. 1817 (based on *Desmos cochinchinensis* Lour.); Finet & Gagnep. in Lecomte Fl. Gén. Indo-Chine 1: 60. 1907.

Unona cochinchinensis DC. Syst. 1: 495. 1818 (based on *Desmos cochinchinensis* Lour.).

Desmos chinensis Lour. Fl. Cochinch. 352. 1790, ed. Willd. 431. 1793, Chinese *câu tsit fung*.

Unona discolor Vahl Symb. 2: 63. *pl. 36*. 1791; Finet & Gagnep. in Lecomte Fl. Gén. Indo-Chine 1: 63. 1907.

Unona chinensis DC. Syst. 1: 495. 1818 (based on *Desmos chinensis* Lour.).

For the first Loureiro states: " Habitat in dumetis Cochinchinae "; and for the second: " Habitat agrestis prope Cantonem Sinarum." A specimen from Loureiro, named *Arthroda*, is preserved in the herbarium of the Paris Museum and has been identified as *Desmos chinensis* Lour., and by Baillon as *Unona discolor* Vahl. There is a specimen of *Desmos cochinchinensis* Lour. among the Loureiro specimens in the herbarium of the British Museum. Finet and Gagnepain recognize two species here, but I can see no valid reasons for this procedure. Even Loureiro in describing *Desmos chinensis*, and in comparing it with his other species, states " unde fortasse illius varietas." The species is common about Canton and is apparently common in Indo-China. Safford (Bull. Torrey Bot. Club **39**: 501–508. 1912) definitely shows that the genus *Unona* was based wholly on American material and that the proper generic name for the so-called *Unona* species of the Old World is *Desmos*.

Polyalthia Blume

Polyalthia sp.

> *Melodorum fruticosum* Lour. Fl. Cochinch. 351. 1790, ed. Willd. 430. 1793, Anamese *cây bo gie*.
>
> *Unona dumetorum* Dunal Monogr. Anon. 116. 1817 (based on *Melodorum fruticosum* Lour.).

" Habitat in dumetis Cochinchinae." Loureiro's type, a leaf specimen with fragments of a broken flower, is preserved in the herbarium of the British Museum; this is a *Polyalthia*. Hooker f. & Thompson (Fl. Ind. **1**: 116. 1855) state that it does not represent *Melodorum* of Blume, but they considered it to be a species of doubtful affinity. It is not mentioned by Finet & Gagnepain in their treatment of the Annonaceae of Indo-China (Lecomte Fl. Gén. Indo-Chine **1**: 42–123. 1907). For the very numerous species erroneously placed in *Melodorum*, *Fissistigma* Griff., as I have shown, is the proper generic name (Philip. Journ. Sci. **15**: 125–137. 1919); *Melodorum* as interpreted by all modern authors has nothing to do with *Melodorum* Loureiro. I was apparently wrong in my synonymy of *Melodorum fruticosum* Lour. as given on page 128, and now am inclined to believe that *Unona dumetorum* Dunal is the only one of the 10 synonyms there cited that really belongs with Loureiro's species. It is clear that *Melodorum fruticosum* Loureiro represents a species of *Polyalthia* near *P. modesta* Finet & Gagnep. and *P. petelotii* Merr., as the leaves of his British Museum specimen are suspiciously like those of these two species; but here we are faced with a dilemma in that the floral characters of the genus *Melodorum* " petala 6, triangularia . . . *inflexo-clausa, bino ordine partes generationis occultantia*," do not apply to *Polyalthia*. Loureiro does not describe the floral details of either species, and he apparently derived these generic characters from the flowers of the second species, *M. arboreum* which I consider to represent a species of *Mitrephora*, and probably *M. thorelii* Pierre. Regarding its flowers Loureiro merely states: " Calyx, & corolla, ut in prima specie " [*M. fruticosum*].

Mitrephora [75] Hooker f. & Thomson

Mitrephora sp.

> *Melodorum arboreum* Lour. Fl. Cochinch. 351. 1790, ed. Willd. 430. 1793, Anamese *cây nhaoc*.
>
> *Unona sylvatica* Dunal Monog. Anon. 115. 1817 (based on *Melodorum arboreum* Lour.).

" Habitat in sylvis Cochinchinae." I suspect that *Mitrephora thorelii* Pierre (Fl. Forest Cochinch. **1**: *pl. 37.* 1881; Finet & Gagnepain in Lecomte Fl. Gén. Indo-Chine **1**: 91.

---

[75] *Melodorum* Loureiro (1790), if it be typified by *M. arboreum* Lour., is a much older generic name than *Mitrephora* Hooker f. & Thomson (1855); the flowers of *M. fruticosum* Lour. do not conform to the characters of the genus *Melodorum*, and this species apparently belongs in *Polyalthia*. It seems clear that the generic description must have been based on *Melodorum arboreum* Lour. not on *M. fruticosum* Lour.

1907) may prove to be the same as Loureiro's species; the latter authors do not mention *Melodorum arboreum* Lour. or Dunal's synonym based upon it. Loureiro's generic description of the flowers of *Melodorum* applies to *Mitrephora*.

Artabotrys R. Brown

Artabotrys uncinatus (Lam.) Merr. in Philip. Journ. Sci. Bot. **7**: 234. 1912; Enum. Philip. Fl. Pl. **2**: 173. 1923.

Annona uncinata Lam. Encycl. **2**: 127. 1786.

Unona hamata Dunal Monog. Anon. 106. 1817 (based on *Uvaria uncata* Lour.).

Artabotrys odoratissimus R. Br. in Bot. Reg. **5**: *pl. 423.* 1820; Finet & Gagnep. in Lecomte Fl. Gén. Indo-Chine **1**: 79. 1907.

Unona hamata DC. Syst. Nat. **1**: 491. 1818 (based on *Uvaria uncata* Lour.).

Uvaria uncata Lour. Fl. Cochinch. 349. 1790, ed. Willd. 426. 1793, Anamese *cây bút dieo*, Chinese *y͞m chào*.

" Habitat Cantone Sinarum: ubi ad vestiendos parietes pulchre aptatur, & absque violentia extenditur, & attolitur." Loureiro's description applies unmistakably to the species commonly known as *Artabotrys odoratissimus* R. Br. which is widely planted in the Indo-Malaysian region for its fragrant flowers and which is still not uncommon near Canton. *Funis uncatus* Rumph. (Herb. Amb. **5**: 65. *pl. 34*), discussed by Loureiro under his species, represents *Uncaria* of the Rubiaceae.

Annona [76] Linnaeus

Annona squamosa Linn. Sp. Pl. 537. 1753; Lour. Fl. Cochinch. 349. 1790, ed. Willd. 427. 1793, Anamese *cây mang câu;* Chinese *pû uôn xú.*

" Habitat in hortis Cochinchinae, & Chinae, & in multis aliis Asiae locis a me observata." Loureiro's description applies to the Linnaean species, which is widely cultivated in the Old World tropics. *Anona tuberosa* Rumph. (Herb. Amb. **1**: 138. *pl. 46*), cited by Loureiro as a synonym, is correctly placed.

Annona reticulata Linn. Sp. Pl. 537. 1753.

Annona asiatica (non Linn.) Lour. Fl. Cochinch. 350. 1790, ed. Willd. 428. 1793, Anamese *cây binh bat.*

" Habitat culta in hortis Cochinchinae, & alibi in India." Loureiro's description applies unmistakably to *Annona reticulata*, which, like *A. squamosa*, is widely cultivated in the Old World tropics. *Annona asiatica* Linn. is a synonym of *A. squamosa* Linn. *fide* Trimen (Fl. Ceyl. **1**: 32. 1893). Linnaeus used *Annona* in Hort. Cliff. 222. 1737, but *Anona* in Gen. Pl. 58. 1737, and in editions 2 to 4 of the latter work (1742–1752); he accepted *Annona* in Gen. Pl. ed. 5, 241. 1754, and in the succeeding editions of this work, as well as in Sp. Pl. 536. 1753, ed. 2, 756. 1763.

[76] Most modern botanists spell this generic name *Anona*. *Annona* is the correct form under the binomial system; see Sprague, Kew Bull. 344. 1928.

MYRISTICACEAE

Knema Loureiro

Knema corticosa Lour. Fl. Cochinch. 605. 1790, ed. Willd. 742. 1793, Anamese *cây máu chó;* Warb. in Nov. Act. Acad. Leop.-Carol. Nat. Cur. **68**: 593. *pl. 25. f. 1–4.* 1897; Lecomte Fl. Gén. Indo-Chine **5**: 105. 1914.

Knema bicolor Raf. Sylva Tellur. 137. 1838 (based on *Knema* [*corticosa*] Lour.).

Palala corticosa O. Ktz. Rev. Gen. Pl. 567. 1891 (based on *Knema corticosa* Lour.).

" Habitat in sylvis Cochinchinae." Loureiro's type, preserved in the herbarium of the British Museum, was examined by Warburg, who gives a detailed description of it. The species is known from Indo-China, Burma, Siam, Pulu Condor, and Hainan, and is the type of the genus.

LAURACEAE

Cinnamomum[77] (Tournefort) Sprengel

Cinnamomum camphora (Linn.) T. Nees & Eberm. Handb. Med.-Pharm. Bot. **2**: 430. 1831; Lecomte Fl. Gén. Indo-Chine **5**: 110. 1914.

Laurus camphora Linn. Sp. Pl. 369. 1753; Lour. Fl. Cochinch. 249. 1790, ed. Willd. 306. 1753, Anamese *laong nao,* Chinese *lùm nào hiàm.*

" Habitat frequens, & incultus non minus in China, quam in Japonia." The Linnaean species, the common camphor tree, was correctly interpreted by Loureiro.

Cinnamomum caryophyllus (Lour.) Moore in Journ. Bot. **63**: 255. 1925 (based on *Laurus caryophyllus* Lour.).

Laurus caryophyllus Lour. Fl. Cochinch. 250. 1790, ed. Willd. 307. 1793, Anamese *quê rành.*

Laurus caryophyllata Lour.[78] ex Gomes in Mem. Acad. Sci. Lisb. Cl. Sci. Pol. Mor. Bel.-Let. **4**(1): 27. 1868.

" Habitat in sylvis Cochinchinae." Meisner (DC. Prodr. **15**(1): 14. 1864) placed this as a doubtful synonym of *Cinnamomum culilawan* Blume var. *rubrum* Meisn., but Moore's examination of Loureiro's type in the herbarium of the British Museum shows that he was in error, and, furthermore, that the species is distinct from all of the 11 species described by Lecomte (Fl. Gén. Indo-Chine **5**: 109–117. 1914). Moore gives additional descriptive data based on Loureiro's type. The species is known only from the original collection.

Cinnamomum curvifolium (Lour.) Nees Syst. Laur. 80. 1836; Meisn. in DC. Prodr. **15**(1): 23. 1864 (based on *Laurus curvifolia* Lour.).

Laurus curvifolia Lour. Fl. Cochinch. 252. 1790, ed. Willd. 309. 1793, Anamese *miéng sanh caong lá.*

Cinnamomum albiflorum Nees in Wall. Pl. As. Rar. **2**: 75. 1831; Lecomte Fl. Gén. Indo-Chine **5**: 113. 1914.

[77] Farwell (Drugg. Circ. **72**: 535. 1918) accepted *Camphorina* Noronha (1790) as the oldest valid name for *Cinnamomum,* which I discussed briefly (Bot. Gaz. **70**: 84. 1920). Sprague (Kew Bull. 41. 1928) calls attention to the fact that *Camphorina* Noronha is a *nomen nudum* and that judging from the Sundanese name cited it pertains to some genus of the Annonaceae; and that *Septina* Noronha (1790) is also a *nomen nudum,* although it probably does pertain to *Cinnamomum;* neither was validly published.

[78] A Loureiro herbarium name here first published by Gomes.

Persea curvifolia Spreng. Syst. **2**: 268. 1825 (based on *Laurus curvifolia* Lour.).

" Habitat in sylvis montanis Cochinchinae." Meisner records Loureiro's species among those " non satis notae "; Lecomte does not mention it nor any of the synonyms based upon it. The description conforms closely with the characters of *Cinnamomum albiflorum* Nees, and I believe this, at least as interpreted by Lecomte, to be the same as *Cinnamomum curviflorum* (Lour.) Nees. Lecomte records Nees' species from Hue, the classical locality of Loureiro's species.

Cinnamomum litseaefolium Thw. Enum. Pl. Ceyl. 253. 1864; Lecomte Fl. Gén. Indo-Chine **5**: 113. *pl. 3.* 1913.

Laurus polyadelpha Lour. Fl. Cochinch. 251. 1790, ed. Willd. 309. 1793, Anamese *miéng sanh, bàng lá.*

" Habitat in montibus Cochinchinae." Meisner (DC. Prodr. **15**(1): 258. 1864) placed this among the " Lauraceae obscurae et quod ordinum dubiae," with the statement: " Fortasse Cinnamomi sp. quoad stamina incorrecte descripta." The peculiar statement in Loureiro's description in reference to the stamens, " omnia in tres fasciculos colligata," is apparently his attempt to describe the stamens and their basal glands. I believe the species to be the same as *Cinnamomum litseaefolium* Thw. as interpreted by Lecomte.

Cinnamomum loureirii Nees Syst. Laur. 65. 1836 (based on *Laurus cinnamomum* Lour.); Meisn. in DC. Prodr. **15**(1): 16. 1864; Lecomte in Not. Syst. **2**: 336. 1913.

Laurus cinnamomum (non Linn.) Lour. Fl. Cochinch. 249. 1790, ed. Willd. 305. 1793, Anamese *cây qûe,* Chinese *kúêi xú.*

" Habitat agrestris in altis montibus Cochinchinae, ad Occidentem, versus Laosios." Lecomte gives an ample description of *Cinnamomum loureirii* Nees as he understands it, based on specimens collected in Yunnan; he had no material from Indo-China. This may be *Camphorina saigonica* Farwell (*Cinnamomum saigonicum* Farwell Drugg. Circ. **62**: 535. 1918) which I have previously discussed (Bot. Gaz. **70**: 84–85. 1920); Farwell's two binomials are based on bark specimens from commercial sources. Chevalier, who states that the so-called Saigon cinnamon of commerce is the product of *Cinnamomum loureirii* Nees, also states that the bark is not a product of Indo-China but comes from China, Saigon being merely the place of export. It may be noted that the only species of the genus credited to Laos by Lecomte is *Cinnamomum cassia* Blume. There is no reason to believe that the Japanese plant actually described by Nees as *Cinnamomum loureirii* represents the species described by Loureiro. I believe that the binomial should be typified by the synonym on which it was based, not by the actual Japanese specimen Nees had.

Machilus Nees

Machilus odoratissimus Nees in Wall. Pl. As. Rar. **2**: 70. 1831; Meisn. in DC. Prodr. **15**(1): 40. 1864; Lecomte Fl. Gén. Indo-Chine **5**: 122. 1914.

Laurus indica (non Linn.) Lour. Fl. Cochinch. 253. 1790, ed. Willd. 311. 1793, Anamese *bâi loî dee.*

" Habitat frequenter in sylvis montanis Cochinchinae." This reduction follows Meisner, the only objection to it being Loureiro's description of the inflorescences as short. *Machilus* Rumph. (Herb. Amb. **3**: 68. *pl. 42*), placed by Loureiro as a synonym, certainly represents a different species and is perhaps a representative of some other genus than

Machilus of Nees. Loureiro's specimen in the herbarium of the British Museum is referred to *Machilus odoratissimus* Nees.

Actinodaphne Nees

Actinodaphne pilosa (Lour.) comb. nov.

> *Laurus pilosa* Lour. Fl. Cochinch. 253. 1790, ed. Willd. 311. 1793, Anamese *bâi loi loung, bâi loi vàng.*
>
> *Machilus pilosa* Nees Syst. Laur. 176. 1836; Meisn. in DC. Prodr. **15**(1): 43. 1864 (based on *Laurus pilosa* Lour.).
>
> *Tetranthera pilosa* Spreng. Syst. **2**: 267. 1825 (based on *Laurus pilosa* Lour.).
>
> *Actinodaphne cochinchinensis* Meisn. in DC. Prodr. **15**(1): 216. 1864; Lecomte in Not. Syst. **2**: 330. 1913, Fl. Gén. Indo-Chine **5**: 128. 1914.

" Habitat in sylvis montanis Cochinchinae." Here again Loureiro's description of the leaves as " enervia " merely indicates that they do not have the longitudinal nerves of *Cinnamomum*. Two specimens collected by de Pirey, *Chevalier 41226, 41330*, as *bai loi vang*, agree sufficiently well with Loureiro's description to warrant their reference to this species. Lecomte (Fl. Gén. Indo-Chine **5**: 107–158. 1914) does not account for Loureiro's binomial or the synonyms based upon it. He cites the local name *bai loi vang* for species of *Nothaphoebe* and *Litsea*. A specimen from Loureiro listed as being among his material in the herbarium of the British Museum has not been located.

Pseudosassafras Lecomte

Pseudosassafras tzumu (Hemsl.) Lecomte Not. Syst. **2**: 269. 1912.

> *Lindera tzumu* Hemsl. in Journ. Linn. Soc. Bot. **26**: 392. 1891.
>
> *Sassafras tzumu* Hemsl. in Kew Bull. 55. 1907, Hook. Ic. **29**: *pl. 2833*. 1907.
>
> *Litsea laxiflora* Hemsl. in Journ. Linn. Soc. Bot. **26**: 383. *pl. 8*. 1891.
>
> *?* *Laurus sassafras* (non Linn.) Lour. Fl. Cochinch. 254. 1790, ed. Willd. 312. 1793, Anamese *cáy vàng dee*, Chinese *hoám chăm.*
>
> *?* *Sassafras loureiri* Kostel. Allgem. Med.-Pharm. Fl. **2**: 481. 1831 (based on *Laurus sassafras* Lour.).
>
> *?* *Lindera loureiri* Blume Mus. Bot. Lugd.-Bat. **1**: 325. 1851; Meisn. in DC. Prodr. **15**(1): 246. 1864 (based on *Laurus sassafras* Lour.).

" Habitat in sylvis Cochinchinae, ad Boream, prope Tunkinum." There is little doubt that Loureiro had material representing the Chinese form commonly known as *Sassafras tzumu* Hemsl., which, although not known from Indo-China, grows in those Chinese provinces contiguous to Indo-China. Loureiro saw only leaf specimens and although he describes fruits he undoubtedly added these data from some previous description. His description of the younger leaves as entire and the older ones as 3-lobed applies to the Chinese *Sassafras* or *Pseudosassafras*, and again his description of the leaves as large indicates this genus rather than one of the species of *Lindera* with lobed or occasionally lobed leaves such as *L. cercidifolia* Hemsl.

Litsea [79] Lamarck

Litsea cubeba (Lour.) Pers. Syn. 2: 4. 1807; Merr. in Philip. Journ. Sci. **15**: 235. 1919 (based on *Laurus cubeba* Lour.).
Laurus cubeba Lour. Fl. Cochinch. 252. 1790, ed. Willd. 310. 1793, Anamese *cây mang tang*.
Litsea piperita Mirbel Hist. Nat. Pl. **11**: 150. 1804–05; Juss. in Pers. Syn. 2: 4. 1807, in syn. (based on *Laurus cubeba* Lour.).
Persea cubeba Spreng. Syst. 2: 269. 1825 (based on *Laurus cubeba* Lour.).
Daphnidium cubeba Nees Syst. Laur. 615. 1836 (based on *Laurus cubeba* Lour.).
Tetranthera cubeba Kostel. Allgem. Med.-Pharm. Fl. 2: 479. 1831; Meisn. in DC. Prodr. 15(1): 199. 1864 (based on *Laurus cubeba* Lour.).
Litsea citrata Blume Bijdr. 565. 1825; Lecomte Fl. Gén. Indo-Chine **5**: 138. 1914.
Malapoenna cubeba O. Ktz. Rev. Gen. Pl. 572. 1891 (based on *Laurus cubeba* Lour.).
" Habitat culta, nec rara in agris, & hortis Cochinchinae: puto, quod etiam in China."
Lecomte fails to account for Loureiro's species or any of the six synonyms based upon it. Hemsley (Journ. Linn. Soc. Bot. **26**: 380. 1891) states that he had seen only the fruit as it appears in commerce, although Loureiro's type is preserved in the herbarium of the British Museum. The description applies definitely to the common, well-known, and widely distributed species currently known as *Litsea citrata* Blume, which extends from eastern and central China to Burma and eastern India, and southward to Java. It is evident from Loureiro's description of some other species of *Laurus* that his statement that the leaves were " enervia " was merely intended to imply that they did not have the characteristic longitudinal nerves of *Cinnamomum*, for he placed all species of *Cinnamomum* that he described in *Laurus*.

Litsea glutinosa (Lour.) C. B. Rob. in Philip. Journ. Sci. Bot. **6**: 321. 1911 (based on *Sebifera glutinosa* Lour.); Merr. Enum. Philip. Fl. Pl. 2: 194. 1923.
Sebifera glutinosa Lour. Fl. Cochinch. 638. 1790, ed. Willd. 783. 1793, Anamese *bây loi nhót*, Chinese *cīen kām xú*.
Litsea sebifera Pers. Syn. 2: 4. 1807 (based on *Sebifera glutinosa* Lour.).
Litsea chinensis Lam. Encycl. **3**: 574. 1791.
Laurus involucrata Koenig in Retz. Obs. **6**: 27. 1791.
Tetranthera laurifolia Jacq. Pl. Rar. Hort. Schoenbr. 1: 59. *pl. 113.* 1797.
Litsea laurifolia Cordem. Fl. Ile Réunion 304. 1895.
" Habitat in sylvis Cochinchinae, & Chinae." A specimen from Loureiro is preserved in the herbarium of the Paris Museum, yet Lecomte (Fl. Gén. Indo-Chine **5**: 107–158. 1914) fails to account for Loureiro's binomial and curiously adds *Cylicodaphne sebifera* Blume, which is based on the totally different *Litsea sebifera* Blume, non Pers., as a synonym of *Litsea sebifera* Pers.; there is also a Loureiro specimen in the herbarium of the British Museum. Loureiro's description is imperfect and in some respects incorrect. He interpreted the involucre as the perianth, and the individual flowers as groups of stamens, his description reading: " Filamenta 100 circiter . . . distributa in 10 phalanges." The spe-

---

[79] *Litsea* Lamarck (1789), conserved name, Vienna Code; older ones are *Malapoenna* Adanson (1763) and *Tomex* Thunberg (1783). *Glabraria* Linnaeus (1771) currently placed here does not appertain to *Litsea*, but is a synonym of *Boschia* Korthals (1842), of the Bombacaceae.

cies is common and is one of wide distribution in the Indo-Malaysian region, having numerous synonyms; it is more generally known as *Litsea sebifera* Pers. *Glabraria tersa* Linn., which has been referred here by various authors, is no lauraceous plant but is a representative of the genus *Boschia* of the Bombacaceae. The type is in the Linnaean herbarium.

Litsea umbellata (Lour.) Merr. in Philip. Journ. Sci. **14**: 242. 1919 (based on *Hexanthus umbellatus* Lour.).

> *Hexanthus umbellatus* Lour. Fl. Cochinch. 196. 1790, ed. Willd. 242. 1793, Anamese *cây ngát;* Moore in Journ. Bot. **63**: 254. 1925.
> *Litsea hexantha* Juss. in Ann. Mus. Hist. Nat. (Paris) **6**: 212. 1805 (based on *Hexanthus umbellatus* Lour.).
> *Tetranthera ferruginea* R. Br. Prodr. 403. 1810, pro parte, quoad syn. Lour.
> *Litsea amara* Blume Bijdr. 563. 1825; Lecomte Fl. Gén. Indo-Chine **5**: 136. 1914.

" Habitat in montibus Cochinchinae." The above is the synonymy as I prepared it in my original manuscript of 1919, where I called attention to the fact that Loureiro in describing the genus observed filaments from which the anthers had fallen and described the three pairs of glands at the base of the inner three filaments as anthers. Moore's examination of Loureiro's type, preserved in the herbarium of the British Museum, confirms this identification of *Hexanthus umbellatus* Lour. with *Litsea amara* Blume. The species is common, extending from Indo-China to Sumatra and Java. Lecomte in his treatment of the Lauraceae of Indo-China (Fl. Gén. Indo-Chine **5**: 107–158. 1914) does not account for Loureiro's genus or species, or for *Litsea hexantha* Juss., which was based upon *Hexanthus umbellatus* Lour., although he does admit *Litsea amara* Blume as an Indo-China species.

Lindera [80] Thunberg

Lindera myrrha (Lour.) comb. nov.

> *Laurus myrrha* Lour. Fl. Cochinch. 251. 1790, ed. Willd. 308. 1793, Anamese *ô duoc, deau dáng,* Chinese *ū yŏ.*
> *Litsea trinervia* Pers. Syn. **2**: 4. 1807 (quoad syn. *Laurus myrrha* Lour.).
> *Tetranthera trinervia* Spreng. Syst. **2**: 266. 1825 (quoad syn. *Laurus myrrha* Lour.).
> *Litsea trinervia* Juss. in Ann. Mus. Hist. Nat. (Paris) **6**: 212. 1805 (based on *Laurus myrrha* Lour.).
> *Tetranthera myrrha* Kostel. Allgem. Med.-Pharm. Fl. **2**: 479. 1831 (based on *Laurus* [*myrrha*] Lour.).
> *Daphnidium myrrha* Nees Syst. Laur. 612. 1836, DC. Prodr. **15**(1): 230. 1864 (based on *Laurus myrrha* Lour.).
> *Lindera eberhardtii* Lecomte in Nouv. Arch. Mus. Hist. Nat. (Paris) V **5**: 115. 1913, Fl. Gén. Indo-Chine **5**: 156. 1914.

" Habitat frequens in dumetis Cochinchinae." R. Brown (Prodr. 403. 1810) compared Loureiro's species to *Laurus involucrata* Koen. = *Neolitsea zeylanica* (Nees) Merr., but he must have depended on Loureiro's description as Mr. Dandy informs me that there is no specimen from Loureiro in the herbarium of the British Museum, nor is the species checked in the Museum copy of the Flora Cochinchinensis as being among the plants received from Loureiro. Lecomte (Fl. Gén. Indo-Chine **5**: 107–158. 1914) fails to account for Loureiro's

[80] Conserved name, Cambridge Code; *Benzoin* Fabricius (1763) is older.

binomial or any of the synonyms based upon it. De Pirey's specimen of "*o duoc*" (*Chevalier 41202*) from Anam is sterile and represents a *Neolitsea*. Loureiro, however, describes his species as having 9 stamens; *Neolitsea* has six stamens. Lecomte's type of *Lindera eberhardtii* was from Hue, Loureiro's classical locality. Persoon (Syn. **2**: 4. 1807) placed Loureiro's species as a synonym of *Litsea trinervia* Pers. which in turn was based on *Laurus involucrata* Retz. *Litsea trinervia* Poir. (Lam. Encycl. Suppl. **3**: 480. 1813), cited by Nees as a synonym of *Daphnidium myrrha* Nees, was based on *Laurus involucrata* Koen.

Cassytha Linnaeus

Cassytha filiformis Linn. Sp. Pl. 35. 1753.

> *Calodium cochinchinense* Lour. Fl. Cochinch. 247. 1790, ed. Willd. 303. 1793, Anamese *to haong xanh*.

"Habitat ubique in sylvis Cochinchinae: puto, quod etiam in China quamvis ibi non viderim." Loureiro's species is manifestly identical with the very common, widely distributed, and well-known *Cassytha filiformis* Linn. *Cussuta s. cussutha indica* Rumph. (Herb. Amb. **5**: 491. *pl. 184. f. 4*), cited by Loureiro as a synonym, represents the Linnaean species. This reduction of *Calodium cochinchinense* was made by Willdenow in 1793. Loureiro's type is preserved in the herbarium of the British Museum.

PAPAVERACEAE

Chelidonium (Tournefort) Linnaeus

Chelidonium majus Linn. Sp. Pl. 505. 1753; Lour. Fl. Cochinch. 330. 1790, ed. Willd. 402. 1793, Anamese *hùynh lien*, Chinese *hoâm liên*.

> *Chelidonium sinense* DC. Syst. **2**: 100. 1821 (based on *Chelidonium majus* Lour.).
> *Chelidonium chinense* Kostel. Allgem. Med.-Pharm. Fl. **5**: 1605. 1836 (error for *C. sinense* DC.).

"Habitat tam culta, quam inculta in diversis provinciis Sinarum." Loureiro's description does not agree entirely with the characters of the Linnaean species, which is not uncommon in various parts of northern China. The slight discrepancies are probably due to the fact that Loureiro had only fragmentary material, secured from dealers in drug plants. There is no evidence that he personally saw the plant growing.

Argemone (Tournefort) Linnaeus

Argemone mexicana Linn. Sp. Pl. 508. 1753.

> *Echtrus trivialis* Lour. Fl. Cochinch. 344. 1790, ed. Willd. 421. 1793.

"Habitat in Benghala, & in ora Coromandelia, ubique in viis." This new genus is clearly identical with the older *Argemone mexicana* Linn., a pantropic weed of American origin. The reduction was suggested by Willdenow.

CRUCIFERAE

Lepidium Linnaeus

Lepidium ruderale Linn. Sp. Pl. 645. 1753.

> *? Lepidium petraeum* (non Linn.) Lour. Fl. Cochinch. 395. 1790, ed. Willd. 479. 1793, Anamese *dinh lich*, Chinese *tìm lì*.

? *Lepidium chinense* Raeusch. Nomencl. ed. 3, 184. 1797; Stokes Bot. Mat. Med. **3**: 429. 1812 (based on *Lepidium petraeum* Lour.).

? *Nasturtium sinense* DC. Syst. **2**: 699. 1821; Prodr. **1**: 236. 1824 (based on *Lepidium petraeum* Lour.).

? *Hutchinsia petraea* Desv. in Journ. Bot. **3**: 168. 1814 (based on *Lepidium petraeum* Lour.).

"Habitat agreste in China." Loureiro's description typifies all of the binomials cited above, but the species is one of more or less doubtful status. De Candolle (Prodr. **1**: 236. 1824) considered it to be an entirely doubtful species, even as to the genus. It is possible that Loureiro had dwarfed specimens of *Nasturtium globosum* Turcz., adding the fruit characters to make his description conform to the characters of the genus in which he placed his species. It may be noted that he describes the leaves both as " integerrima " and as "pinnata." It seems best to leave it for the present as a probable synonym of the Kwangtung form currently referred to *Lepidium ruderale* Linn.

Brassica (Tournefort) Linnaeus

Brassica chinensis Linn. Cent. Pl. **1**: 19. 1755, Amoen. Acad. **4**: 280. 1759; L. H. Bailey Gent. Herb. **1**: 99. *f. 46, 47.* 1922; **2**: 253. *f. 135, 136.* 1930.

Sinapis brassicata (non Linn.) Lour. Fl. Cochinch. 399. 1790, ed. Willd. 485. 1793, Anamese *cai bach thoi*, Chinese *pĕ kíai*.

"Habitat copiose culta in China, & Cochinchina." From Loureiro's description the form that he had with " petiolis longis, curvis, albissimis " is manifestly the one discussed and amply described by L. H. Bailey as *Brassica chinensis* Linn. Schulz (Pflanzenreich **70**(IV–105): 45. 1919) reduced *Brassica chinensis* Linn. to *Brassica napus* Linn. var. *chinensis* (Linn.) O. E. Schulz. The Linnaean species was based on plants grown at Upsala from seeds secured by Osbeck in Canton.

Brassica juncea (Linn.) Coss. in Bull. Soc. Bot. France **6**: 609. 1859; Gagnep. in Lecomte Fl. Gén. Indo-Chine **1**: 170. 1908; L. H. Bailey Gent. Herb. **1**: 91. *f. 38.* 1922, **2**: 258. *f. 141.* 1930.

Sinapis juncea Linn. Sp. Pl. 668. 1753.

Sinapis chinensis (non Linn.) Lour. Fl. Cochinch. 399. 1790, ed. Willd. 485. 1793, Anamese *cai cu, cai sen, cai mo*, Chinese *kíái tsái*.

"Habitat latissime culta in China, & Cochinchina." I follow L. H. Bailey in his interpretation of *Brassica juncea* (Linn.) Coss. and believe that *Sinapis chinensis* as described by Loureiro belongs here; as Bailey notes the type in the Linnaean species was material grown at Upsala from Chinese seeds. I do not think that the form described by Loureiro can be the same as *Brassica integrifolia* (West) O. E. Schulz (Pflanzenreich **70** (IV–105): 56. 1919).

Brassica oleracea Linn. Sp. Pl. 667. 1753; Lour. Fl. Cochinch. 396. 1790, ed. Willd. 481. 1793, Anamese *cai rô taù*.

"Habitat in Cochinchina, & China: dubito an indegena?" This is the common cabbage and Loureiro's short description unmistakably applies to it. It is commonly cultivated in both China and Cochinchina.

Brassica pekinensis (Lour.) Rupr. Fl. Ingr. 96. 1860; Gagnep. in Lecomte Fl. Gén. Indo-
Chine 1: 171. 1908, in nota; L. H. Bailey Gent. Herb. 1: 97. *f. 43.* 1922, 2: 250.
f. 132–134. 1930 (based on *Sinapis pekinensis* Lour.).
Sinapis pekinensis Lour. Fl. Cochinch. 400. 1790, ed. Willd. 485. 1793, Anamese *cai bén*, Chinese *pĕ tsái.*
Brassica pe-tsai L. H. Bailey in Cornell Agr. Exp. Sta. Bull. 67: 178, 190. 1894.
" Habitat Pekini culta, ubi omnium optima. In Cochinchina altius crescit, inferior qualitate." Without regard to its status in binomial nomenclature, *pe tsái* is a well-known cultivated plant. I here follow Doctor L. H. Bailey in retaining it as a species under Loureiro's specific name. Gagnepain (Lecomte Fl. Gén. Indo-Chine 1: 171. 1908) considers *pe tsái* to be but a variety of *Brassica campestris* Linn., a disposition of it which I cannot accept. Schulz (Pflanzenreich 70(IV–105): 59. 1919) reduced Loureiro's species to *Brassica cernua* (Thunb.) Forbes & Hemsl.; but Bailey (Gent. Herb. 1: 95. 1922) thinks that the type of *Sinapis cernua* Thunb. (Fl. Jap. 261. 1784), its name bringing synonym, apparently represents a species of *Raphanus.*

Brassica sp.
Brassica chinensis (non Linn.) Lour. Fl. Cochinch. 397. 1790, ed. Willd. 482. 1793, Anamese *cai rô*, Chinese *chāi lân tsái.*
" Habitat culta in Cochinchina, & China." The plant Loureiro described, clearly a *Brassica*, probably does not represent the Linnaean species, which, as I interpret it, was otherwise described by Loureiro as *Sinapis brassicata.* The exact status of Loureiro's concept of *Brassica chinensis* may possibly be determinable from the local name cited. The short description is too indefinite to warrant placing it this time as other than *Brassica* sp.

Raphanus [81] (Tournefort) Linnaeus

Raphanus sativus Linn. Sp. Pl. 669. 1753; Lour. Fl. Cochinch. 396. 1790, ed. Willd. 481.
1793, Anamese *la bac, cai cu*, Chinese *tsái fú kĕn.*
" Habitat cultus in Cochinchina, & China." The description applies to a form of the common radish which is very extensively cultivated in China.

Nasturtium [82] R. Brown

Nasturtium indicum (Linn.) DC. Syst. 2: 199. 1821, var. **apetalum** (Lour.) Gagnep. in
Lecomte Fl. Gén. Indo-Chine 1: 166. *f. 17.* 1908 (based on *Sisymbrium apetalum* Lour.).
Sisymbrium apetalum Lour. Fl. Cochinch. 400. 1790, ed. Willd. 486. 1793, Ánamese *cai hoang.*

[81] Sometimes spelled *Rhaphanus*, but the original form is here retained; see Sprague, Kew Bull. 360. 1928.

[82] *Nasturtium* R. Brown (1812), conserved name, Brussels Code; older ones are *Cardaminum* Moench (1794) and *Baeumerta* Gaertner (1800). Many botanists accept *Roripa* Scopoli (1772) in place of *Nasturtium* R. Br., while others retain *Nasturtium* for the white-flowered species, including the common water cress, and *Roripa* for the yellow-flowered species. *Nasturtium* has been discussed by Sprague (Journ. Bot. 62: 226. 1924), while (*op. cit.* 68: 219–220. 1930) he shows that the correct and original spelling of Scopoli's generic name is *Rorippa* (1760). He expresses the opinion that if but a single genus be recognized here, then *Rorippa* should be accepted. The International Code is not definite on this point.

Nasturtium apetalum DC. Syst. **2**: 200. 1821; A. Chev. Cat. Pl. Jard. Bot. Saigon 63. 1919 (based on *Sisymbrium apetalum* Lour.).

" Habitat spontaneum in hortis, & locis humidis Cochinchinae." Loureiro's description conforms closely with the characters of the Linnaean species, except in his statement that there are no petals. Gagnepain treats this apetalous form, which in all other respects is the same as the Linnaean species, as *N. indicum* (Linn.) DC. var. *apetalum* (Lour.) Gagnep.; this is clearly the form that Loureiro described.

Nasturtium microspermum DC. Syst. **2**: 199. 1821.

Ricotia cantoniensis Lour. Fl. Cochinch. 397. 1790, ed. Willd. 482. 1793, non *Nasturtium cantoniense* Hance, 1865.

Lunaria ? cantoniensis Desv. in Journ. Bot. **3**: 174. 1814 (based on *Ricotia cantoniensis* Lour.).

Lunaria ricotia Desv. ex Steud. Nomencl. ed. 2, **2**: 77. 1841 (based on *Ricotia cantoniensis* Lour.).

Roripa microsperma L. H. Bailey Gent. Herb. **1**: 25. 1920; Hand.-Maz. Symb. Sin. **7**: 358. 1931.

" Habitat inculta Cantone Sinarum." Loureiro's description applies definitely to *Nasturtium microspermum* DC., the type of which was from China. The species is fairly abundant in damp places near Canton, but the flowers are white rather than yellow as Loureiro describes them. It may be noted that Desvaux does not publish the binomial *Lunaria ricotia* as credited to him by Steudel and in Index Kewensis. At the place cited (Journ. Bot. **3**: 174. 1814) he does publish *Lunaria ? cantoniensis* definitely based on *Ricotia cantoniensis* Lour.

Cardamine (Tournefort) Linnaeus

Cardamine flexuosa With. Arrang. Brit. Pl. ed. 3, **3**: 578. 1796, subsp. **debilis** (D. Don) O. E. Schulz in Bot. Jahrb. **32**: 478. 1903.

Cardamine chelidonia (non Linn.) Lour. Fl. Cochinch. 398. 1790, ed. Willd. 484. 1793.

Cardamine debilis D. Don. Prodr. Fl. Nepal. 201. 1825.

" Habitat inculta Cantone Sinarum." This seems to be a form of the species occurring near Canton that is currently referred to *Cardamine hirsuta* Linn. but which, after O. E. Schulz, represents *C. flexuosa* With. subsp. *debilis* (D. Don) O. E. Schulz, rather than *C. hirsuta* Linn.

Capsella [83] Medikus

Capsella bursa-pastoris (Linn.) Medik. Pflanzengat. 85. 1792.

Thlaspi bursa-pastoris Linn. Sp. Pl. 647. 1753; Lour. Fl. Cochinch. 395. 1790, ed. Willd. 480. 1793, Chinese *hán san tsào*.

" Habitat incultum in China." Loureiro's description applies to the Linnaean species, the common shepherd's purse, which is abundant and widely distributed in China.

---

[83] *Capsella* Medikus (1792), conserved name, Vienna Code; older ones are *Bursa* (Siegesbeck) Weber (1780) and *Marsypocarpus* Necker (1790).

CAPPARIDACEAE

Gynandropsis [84] de Candolle

Gynandropsis gynandra (Linn.) Briq. in Ann. Conserv. Jard. Bot. Genève **17**: 382. 1914;
Merr. Enum. Philip. Fl. Pl. **2**: 209. 1923.
Cleome gynandra Linn. Sp. Pl. 671. 1753.
Cleome pentaphylla Linn. Sp. Pl. ed. 2, 938. 1763; Lour. Fl. Cochinch. 397. 1790, ed.
Willd. 482. 1793, Anamese *màn màn tía.*
Gynandropsis pentaphylla DC. Prodr. **1**: 238. 1824.
" Habitat inculta agros, & hortos Cochinchinae." The Linnaean species was correctly
interpreted by Loureiro. *Lagansa rubra* Rumph. (Herb. Amb. **5**: 280. *pl. 96. f. 3*), cited
by Loureiro as representing this species, is correctly placed. By error Loureiro reversed
the figures on Rumphius' plate as between this and the next species.

Polanisia Rafinesque

Polanisia icosandra (Linn.) W. & A. Prodr. 22. 1834.
Cleome icosandra Linn. Sp. Pl. 672. 1753; Lour. Fl. Cochinch. 398. 1790, ed. Willd.
483. 1793, Anamese *màn màn tláng.*
Cleome viscosa Linn. Sp. Pl. 672. 1753.
Polanisia viscosa DC. Prodr. **1**: 242. 1824.
Lagansa alba Raf. Sylva Tellur. 110. 1838 (based on *Cleome icosandra* Linn.; Lour.).
" Habitat inculta in hortis, & agris Cochinchinae." The Linnaean species was cor-
rectly interpreted by Loureiro; it is a very common and widely distributed weed. *Lagansa
alba* Rumph. (Herb. Amb. **5**: 280. *pl. 96. f. 2*) is correctly placed as a synonym.

Crataeva Linnaeus

Crataeva falcata (Lour.) DC. Prodr. **1**: 243. 1824 (based on *Capparis falcata* Lour.).
Capparis falcata Lour. Fl. Cochinch. 331. 1790, ed. Willd. 405. 1793.
Crataeva nurvala Ham. in Trans. Linn. Soc. 15: 121. 1827.
Crataeva religiosa Forst. var *nurvala* Hook. f. Fl. Brit. Ind. **1**: 172. 1872; Gagnep. in
Lecomte Fl. Gén. Indo-Chine **1**: 178. 1908.
Trichlanthera falcata Raf. Sylva Tellur. 108. 1838 (based on *Capparis falcata* Lour.).
" Habitat prope Cantonem Sinarum, inculta." Gagnepain (Lecomte Fl. Gén. Indo-
Chine **1**: 179. 1908) placed *Capparis falcata* Lour. as a doubtful synonym of *Crataeva
erythrocarpa* Gagnep. I have seen rather numerous specimens of the species from the
vicinity of Canton and they represent *Crataeva religiosa* Forst. f. as that species is currently
interpreted; Craib (Fl. Siam. Enum. **1**: 86. 1925), however, states that the Polynesian
Crataeva religiosa Forst. is totally different from the Asiatic plant currently referred to that
binomial. The lateral leaflets are *asymmetric* at the base, not falcate, Loureiro apparently
misapplying the latter term. The genus *Trichlanthera* Raf. was based on this and the next
species.

Crataeva magna (Lour.) DC. Prodr. **1**: 243. 1824 (based on *Capparis magna* Lour.).
Capparis magna Lour. Fl. Cochinch. 331. 1790, ed. Willd. 404. 1793, Anamese *cây bún.*

---

[84] *Gynandropsis* de Candolle (1824), conserved name, Vienna Code; an older one is *Pedicellaria* Schrank
(1790).

Trichlanthera corymbosa Raf. Sylva Tellur. 108. 1838 (based on *Capparis magna* Lour.).
Crataeva macrocarpa Kurz in Journ. Bot. **12**: 195. 1874; Gagnep. in Lecomte Fl. Gén.
  Indo-Chine **1**: 180. 1908.

" Habitat ripas fluminum im Cochinchina." Loureiro's type is preserved in the herbarium of the Paris Museum and is referred by Gagnepain, with doubt, to *Crataeva macrocarpa* Kurz. The species occurs in Indo-China extending to Hainan. I consider that there is no longer even reasonable doubt as to the identity of *Capparis magna* Lour. with *Crataeva macrocarpa* Kurz and accordingly accept the oldest binomial. The species is represented by *Clemens 3743, 4303* from Tourane near the classical locality Hue.

Capparis (Tournefort) Linnaeus

Capparis cantoniensis Lour. Fl. Cochinch. 331. 1790, ed. Willd. 404. 1793, Chinese *heang lăc phung*.
  Olofuton racemosum Raf. Sylva Tellur. 108. 1838 (based on *Capparis cantoniensis* Lour.).
  Capparis pumila Champ. in Hook. Kew Journ. Bot. **3**: 260. 1851.
  Capparis sciaphila Hance in Ann. Sci. Nat. V Bot. **5**: 206. 1866.

" Habitat agrestis prope Cantonem apud Sinas." This is the type of the genus *Olofuton* Raf. While Loureiro's description is imperfect and in some respects unsatisfactory, I believe *Capparis pumila* Champ., which is common in thickets near Canton, to be the species he attempted to describe, as it is the only *Capparis* in southern China that remotely agrees with the description. The flowers are in depauperate umbels which are racemosely arranged in terminal and lateral inflorescences. The fruits have but one or few seeds, while Loureiro's description calls for a many-seeded one.

Capparis sp.
  Capparis zeylanica (non Linn.) Lour. Fl. Cochinch. 330. 1790, ed. Willd. 403. 1793,
  Anamese *cây dùi tloúng*.

" Habitat sepes Cochinchinae." In my manuscript of 1919 I referred this to *Capparis sepiaria* Linn., but on restudying Loureiro's description I thought that it might possibly represent *C. micracantha* DC. (*Squires 352* from near Hue); however, Loureiro's description applies to neither of these species. I suspect that *C. zeylanica* Lour., non Linn., may be represented by *Squires 281* from near Hue, which I am unable to refer to any of the 22 species admitted by Gagnepain (Lecomte Fl. Gén. Indo-Chine **1**: 181–196. 1908). This specimen agrees very definitely with Loureiro's description in its paired, axillary pedicels described as " pedunculis monofloris, subbinatis, axillaribus." Loureiro is silent as to whether his plant was glabrous or pubescent; *Squires 281* is definitely pubescent.

Thylachium Loureiro

Thylachium (Thilachium) africanum Lour. Fl. Cochinch. 342. 1790, ed. Willd. 418. 1793;
  Oliv. Fl. Trop. Afr. **1**: 82. 1868.
  Thylachium ovalifolium Juss. in Ann. Mus. Hist. Nat. [Paris] **12**: 71. 1808 (based on
  Thilachium africanum Lour.).

" Habitat agreste in ora Africae orientali." The local name is given as *mangueiro*. De Candolle (Prodr. **1**: 254. 1824) accepted Jussieu's name *Thylachium ovalifolium* Juss.

Loureiro spelled the generic name *Thilachium*, but in the binomial, ed. 1, through typographical error it appears as *Thilakium;* Sprengel (Syst. 2: 606. 1825) gives it as *Thylacium*. It was corrected to *Thylachium* by Jussieu. Loureiro's type is preserved in the herbarium of the Paris Museum. The species is reported by Oliver only from the Mozambique District.

Stixis Loureiro

Stixis scandens Lour. Fl. Cochinch. 295. 1790, ed. Willd. 361. 1793, Anamese *cây cám;*
  Gagnep. in Lecomte Fl. Gén. Indo-Chine 1: 206. 1908.

" Habitat in sylvis Cochinchinae." Gagnepain (Lecomte Fl. Gén. Indo-Chine 1: 199–206. 1908) describes nine species of *Stixis*, his tenth being *S. scandens* Lour. which he considered to be one of doubtful status. He suggests that the latter may be near *S. elongata* Pierre. Among the fully described species of the genus it would seem to be identical with either *S. elongata* Pierre or *S. longiracemosa* A. DC. It is safely represented by *Clemens 3774, 4113*, the latter from Hue, the classical locality, the former from Mount Bana, near Tourane, a short distance from Hue; unfortunately both specimens are in fruit, and flowers are necessary to place this form in accordance with Gagnepain's arrangement of the species. *Atunus* Rumph. (Herb. Amb. 1: 171. *pl. 66*), which Loureiro thought might represent his genus, must be excluded as it represents the rosaceous *Parinarium glaberrimum* Hassk. A Loureiro specimen listed as being among those received from him has not been located in the herbarium of the British Museum.

RESEDACEAE

Reseda (Tournefort) Linnaeus

Reseda odorata Linn. Syst. Nat. ed. 10, 1046. 1759; Lour. Fl. Cochinch. 300. 1790, ed.
  Willd. 367. 1793; Muell.-Arg. in DC. Prodr. 16(2): 565. 1868.

" Observavi Cantone Sinarum ab Europa oriundam." No representative of the Resedaceae is known from China, either native or cultivated, yet unquestionably Loureiro had specimens of this Mediterranean species, which is sometimes cultivated in pots for ornamental or other purposes, as his excellent description applies strictly to *Reseda odorata* Linn.

MORINGACEAE

Moringa (Burman) Jussieu

Moringa oleifera Lam. Encycl. 1: 398. 1785.
  Guilandina moringa Linn. Sp. Pl. 381. 1753.
  Moringa pterygosperma Gaertn. Fruct. 2: 314. 1791.
  Anoma moringa Lour. Fl. Cochinch. 279. 1790, ed. Willd. 342. 1793, Bengalese *moringa*.
  Anoma morunga Lour. Fl. Cochinch. 279. 1790, ed. Willd. 343. 1793, Anamese *ba dàu
    dedi.*
  Moringa octogona Stokes Bot. Mat. Med. 2: 466. 1812 (in part based on *Anoma mo-
    ringa* Lour.).
  Moringa polygona DC. Prodr. 2: 478. 1825 (based on *Anoma moringa* Lour.).

Anoma was proposed and described by Loureiro as a new genus, with three species: *A. moringa* from Bengal published without reference to the earlier *Guilandina moringa*

Linn. = *Moringa oleifera* Lam. with which it is manifestly synonymous; *A. morunga* from Cochinchina and Mozambique which also cannot be distinguished from the latter species; and *A. cochinchinensis* apparently a representative of the Leguminosae-Caesalpinoideae.

NEPENTHACEAE

Nepenthes Linnaeus

Nepenthes mirabilis (Lour.) Druce in Rept. Bot. Exch. Club Brit. Isles **4**: 637. 1917 (July); Merr. Interpret. Herb. Amb. 242. 1917 (November) (based on *Phyllamphora mirabilis* Lour.).

Phyllamphora mirabilis Lour. Fl. Cochinch. 606. 1790, ed. Willd. 744. 1793, Anamese *cây nắp âm.*

Nepenthes phyllamphora Willd. Sp. Pl. **4**(2): 874. 1805 (based on *Phyllamphora mirabilis* Lour.); Lecomte Fl. Gén. Indo-Chine **5**: 52. 1910.

"Habitat loca humida, & agrestia Cochinchinae." A well-known and widely distributed species extending from southeastern China to the Malay Peninsula, Sumatra, Borneo, the Philippines, Moluccas, New Guinea and the Caroline Islands.

DROSERACEAE

Drosera Linnaeus

Drosera burmanni Vahl Symb. **3**: 50. 1794; Diels in Pflanzenreich **26**(IV–112): 75. *f. 27,* *E–G.* 1906; Gagnep. in Lecomte Fl. Gén. Indo-Chine **2**: 706. 1920.

Drosera rotundifolia (non Linn.) Lour. Fl. Cochinch. 186. 1790, ed. Willd. 232. 1793, Anamese *co tlôn gà.*

"Habitat in locis humidis Cochinchinae." Loureiro's description applies to Vahl's species, which is common in Indo-China and which occurs near Hue, *Squires 386, Clemens 4408.*

CRASSULACEAE

Sedum (Tournefort) Linnaeus

Sedum sarmentosum Bunge in Mém. Acad. St. Pétersb. Sav. Étrang. **2**: 104. 1833 (Enum. Pl. China Bor. 30).

? Sedum stellatum (non Linn.) Lour. Fl. Cochinch. 287. 1790, ed. Willd. 353. 1793, Cantonese *chêu lt.*

"Habitat Cantone Sinarum." Loureiro's species is reduced to *Sedum sarmentosum* Bunge with considerable confidence, partly because the latter is now known to occur near Canton. The description, however, does not conform to all characters of Bunge's species. It is possible that Loureiro had other than a *Sedum*, adding *Sedum* characters to his description.

Kalanchoe Adanson

Kalanchoe laciniata (Linn.) DC. Plant. Hist. Succul. *pl. 100.* 1801; Gagnep. in Lecomte Fl. Gén. Indo-Chine **2**: 702. 1920.

Cotyledon laciniata Linn. Sp. Pl. 430. 1753; Lour. Fl. Cochinch. 286. 1790, ed. Willd. 352. 1793, Anamese *truong sinh rách lá.*

" Habitat tam culta, quam inculta in Cochinchina." Loureiro's description applies unmistakably to the Linnaean species, which is widely distributed in cultivation in the Indo-Malaysian region.

Kalanchoe spathulata DC. Plant. Hist. Succul. *pl. 65.* 1801; Gagnep. in Lecomte Fl. Gén. Indo-Chine 2: 701. 1920.

? *Cotyledon serrata* (non Linn.) Lour. Fl. Cochinch. 287. 1790, ed. Willd. 352. 1793, Anamese *truong sinh lón lá.*

" Habitat in hortis Cochinchinae." Loureiro's description is short, indefinite, and very imperfect. He probably had a form of de Candolle's species; he certainly had no true *Cotyledon,* nor does his description or the figure cited (Hort. Elth. 113. *pl. 95. f. 112.* 1732) agree at all with *Bryophyllum pinnatum* (Lam.) Kurz (*B. calycinum* Salisb.) which Loureiro otherwise does not describe and which may have been as common in Indo-China in his time as it is to-day.

SAXIFRAGACEAE

Saxifraga (Tournefort) Linnaeus

Saxifraga sarmentosa Linn. f. Suppl. 240. 1781.

Saxifraga chinensis Lour. Fl. Cochinch. 281. 1790, ed. Willd. 345. 1793, Chinese *hŏ ngi tsao.*

" Habitat in agris, Cantone Sinarum." As noted by Hemsley (Journ. Linn. Soc. Bot. **23**: 266. 1887) this is the disposition of Loureiro's species made by Seringe and followed by Engler; but Hemsley, following Maximowicz, who thought, from the locality cited, that *S. chinensis* Lour. was probably distinct from *S. sarmentosa* Linn. f. and possibly the same as *S. cortusaefolia* S. & Z., retained *S. chinensis* Lour. as a distinct species. It may be noted that *S. sarmentosa* Linn. f. is the only species of the genus definitely known from Kwangtung Province, although it is doubtful if Loureiro saw other than specimens from cultivated plants. I can see no valid reason for considering that *Saxifraga chinensis* Lour. is other than a form of *S. sarmentosa* Linn. f. and believe Engler & Irmscher (Pflanzenreich **69**(IV–117–II): 652. 1919) to be correct in placing Loureiro's species as a synonym of the Linnaean one.

Dichroa Loureiro

Dichroa febrifuga Lour. Fl. Cochinch. 301. 1790, ed. Willd. 369. 1793, Anamese *cây thuong son,* Chinese *châm chān;* Gagnep. in Lecomte Fl. Gén. Indo-Chine 2: 688. *f. 68, 1–6.* 1920.

" Habitat loca montana in Cochinchina, & China." This is the type of the genus and is a well-understood species of southern China and Indo-China. It has by some authors been given a much wider geographic range but apparently several species are represented in what has generally been referred to *Dichroa febrifuga* Lour. Loureiro's type is preserved in the herbarium of the British Museum.

Hydrangea (Gronovius) Linnaeus

Hydrangea opuloides (Lam.) K. Koch Dendr. **1**: 353. 1869; Rehd. in Sargent Pl. Wils. **1**: 37. 1913 (var. *hortensia* Dippel).

Hortensia opuloides Lam. Encycl. **3**: 136. 1789.

Primula mutabilis Lour. Fl. Cochinch. 104. 1790, ed. Willd. 127. 1793, Chinese *sau cau hōa*.

Hydrangea hortensia Siebold in Nov. Act. Acad. Nat. Cur. **14**(2): 788. 1829.

" Habitat, & ob pulchritudinem collitur, Cantone Sinarum." The description applies unmistakably to the commonly cultivated garden hydrangea. Loureiro's type is preserved in the herbarium of the Paris Museum and Desvaux identified it as *Hydrangea hortensia* Siebold.

HAMAMELIDACEAE

Altingia Noronha

Altingia sp.

Amyris ambrosiaca (non Linn. f.) Lour. Fl. Cochinch. 230. 1790, ed. Willd. 283. 1793, Anamese *tô hap bình khang*.

" Habitat in montibus Cochinchinae, in provincia *Bình Khang*, circa 14 gradum latitudinis borealis." There is no proper description. Loureiro saw no flowers or fruits, but from statements made by the natives and the properties of the balsam examined by him he thought the species might be *Amyris ambrosiaca* Linn. f. Doctor A. Chevalier, however, reports the Indo-China species yielding the product described by Loureiro to be an *Altingia*. *Canarium odoriferum* Rumph., and *Nanarium minumum oleosum* Rumph., discussed by Loureiro under *Amyris ambrosiaca*, represent species of *Canarium* of the Burseraceae.

ROSACEAE

Spiraea Linnaeus

Spiraea cantoniensis Lour. Fl. Cochinch. 322. 1790, ed. Willd. 394. 1793, Chinese *tsi choŭc hōa*.

Spiraea sinensis Lour.[85] ex Gomes in Mem. Acad. Sci. Lisb. Cl. Sci. Pol. Mor. Bel.-Let. n.s. **4**(1): 29. 1868.

" Habitat Cantone Sinarum." A well-understood species, a native of southern China, and represented by numerous collections. Loureiro's type is preserved in the herbarium of the Paris Museum.

Chaenomeles Lindley

Chaenomeles lagenaria (Loisel.) Koidz. in Bot. Mag. Tokyo **23**: 173. 1909.

Cydonia lagenaria Loisel. in Nouv. Duhamel **6**: 255. *pl. 76*. (1813?).

Pyrus cydonia (non Linn.) Lour. Fl. Cochinch. 322. 1790, ed. Willd. 394. 1793, Anamese *muoc qua*, Chinese *mŭ qūa, mim xú*.

Chaenomeles japonica Spach Hist. Veg. **2**: 159. 1834.

" Habitat in China versus Septentrionem: Cantone raro." Hemsley (Journ. Linn. Soc. Bot. **23**: 256. 1887) treated this as *Pyrus cathayensis* Hemsl. which Rehder & Wilson (Sargent Pl. Wils. **2**: 297. 1915) consider to represent a variety of *Chaenomeles lagenaria* Koidz. It is widely cultivated in China and Japan.

[85] A Loureiro herbarium name here first published by Gomes.

Pyracantha M. Roemer

Pyracantha loureiri (Kostel.) comb. nov.

Mespilus loureiri Kostel. Allgem. Med.-Pharm. Fl. **4**: 1479. 1835 (based on *Mespilus pyracantha* Lour.).

Pyracantha chinensis M. Roem. Syn. **3**: 220. 1847 (based on *Mespilus pyracantha* Lour.).

Sportella atalantioides Hance in Journ. Bot. **15**: 207. 1877.

Pyracantha atalantioides Stapf in Curtis's Bot. Mag. **151**: *pl. 9099.* 1925; Hand.-Maz. Symb. Sin. **7**: 460. 1933.

Mespilus pyracantha (non Linn.) Lour. Fl. Cochinch. 320. 1790, ed. Willd. 392. 1793, Anamese *son tra, dang kieu tu,* Chinese *tàn kiêo tsù, xān chā.*

" Habitat in China tam agrestis, quam culta." Wilson (Sargent Pl. Wils. **1**: 177. 1912) reduced Hance's species to *Pyracantha crenulata* (Roxb.) M. Roem., to which it is manifestly allied, but recent authors seem to concur in the opinion that the Chinese form is distinct from the species originally characterized by Roxburgh. It is not at all common in those limited parts of Kwangtung that Loureiro visited, and it seems highly probable that he secured his material from herbalists, because of the medicinal uses he ascribes to it. No representative of the genus is known from Indo-China, yet Loureiro cites Anamese names, doubtlessly indicating an importation from China of material for medicinal use. I accept Kosteletzky's specific name which is much older than that proposed by Hance.

Pyrus [86] (Tournefort) Linnaeus

Pyrus calleryana Decne. Jard. Frut. **1**: sub *pl. 8.* 1872.

Pyrus communis (non Linn.) Lour. Fl. Cochinch. 321. 1790, ed. Willd. 393. 1793, Anamese *cây lê taù,* Chinese *lŷ tú.*

" Habitat frequens in hortis Cantoniensibus: in provinciis Sinensibus Borealibus succosiora, & sapidiora sunt pyra, non tamen Europaeorum aemula. In Cochinchina rara est haec arbor in hortis magnatum, nec ibi nunquam vidi fructificantem." The form Loureiro had was the common sand pear, now generally known as *sha lei* in Kwangtung where it is commonly planted; it is widely distributed in China. I strongly suspect that *Pyrus calleryana* Decne. will prove to be only a form of the older *P. sinensis* Lindl., as illustrated by Decaisne *l.c. pl. 5.* Cardot (Lecomte Fl. Gén. Indo-Chine **2**: 672. 1920), while not admitting *Pyrus calleryana* Decne. as occurring in Indo-China, mentions *P. ussuriensis* Max. or *P. serotina* Rehd. as cultivated in Tonkin.

Malus (Tournefort) Linnaeus

Malus baccata (Linn.) Borkh. Handb. Forstbot. **2**: 1280. 1803.

Pyrus baccata Linn. Mant. **1**: 75. 1767.

Pyrus malus (non Linn.) Lour. Fl. Cochinch. 321. 1790, ed. Willd. 393. 1793, Anamese *cây bình ba,* Chinese *pĭm pō.*

" Habitat in locis Borealibus imperii Sinensis." Loureiro's statement regarding the fruit, " aspectu jucundiora, quam gustu," seems clearly to indicate that he had other than the common apple, *Pyrus malus* Linn. = *Malus sylvestris* Mill. (*M. communis* Lam.). *Malus baccata* is doubtless the form he attempted to describe.

[86] Often spelled *Pirus,* but I follow Sprague, Kew Bull. 360. 1928 in retaining *Pyrus.*

Raphiolepis [87] Lindley

Raphiolepis indica (Linn.) Lindl. Bot. Reg. **6**: *pl. 468.* 1820; Cardot in Lecomte Fl. Gén. Indo-Chine **2**: 680. 1920.

Crataegus indica Linn. Sp. Pl. 477. 1753; Lour. Fl. Cochinch. 319. 1790, ed. Willd. 391. 1793, Anamese *cây boung vang tlái.*

Crataegus rubra Lour. Fl. Cochinch. 320. 1790, ed. Willd. 391. 1793, Chinese *û lý mŏ.*

Opa metrosideros Lour. Fl. Cochinch. 309. 1790, ed. Willd. 378. 1793, Anamese *cây boung váng.*

Mespilus rubra Stokes Bot. Mat. Med. **3**: 110. 1912 (based on *Crataegus rubra* Lour.).

Raphiolepis sinensis M. Roem. Syn. **3**: 114. 1847 (based on *Crataegus rubra* Lour.).

Raphiolepis loureiri Spreng. Syst. **2**: 508. 1825 (based on *Crataegus indica* Lour.).

Syzygium metrosideros DC. Prodr. **3**: 261. 1828 (based on *Opa metrosideros* Lour.).

Mespilus sinensis Poir. in Lam. Encycl. Suppl. **4**: 70. 1816 (based on *Crataegus rubra* Lour.).

Raphiolepis rubra Lindl. Collect. *pl. 3.* 1821 (cites *Crataegus rubra* Lour. as a doubtful synonym).

Crataegus sinensis Lour.[88] ex Gomes in Mem. Acad. Sci. Lisb. Cl. Sci. Pol. Mor. Bel.-Let. **4**(1): 29. 1868.

Eriobotrya metrosideros A. Chev.[89] Cat. Pl. Jard. Bot. Saigon 64. 1919 (based on *Opa metrosideros* Lour.).

It seems curious that Loureiro should have described the same species three times in two different genera, yet all three of his descriptions apparently apply to *Raphiolepis indica* Lindl. For *Opa metrosideros* he states: " Habitat in sylvis Cochinchinae," and cites as a doubtful synonym *Metrosideros vera parvifolia* Rumph. (Herb. Amb. **3**: 16. *pl. 7*) which is to be excluded as it represents the myrtaceous *Metrosideros vera* Roxb. Loureiro's type is preserved in the herbarium of the British Museum, where it was examined by Seemann (Journ. Bot. **1**: 280. 1863); it was identified by Spach as *Raphiolepis rubra* Lindl. Doctor Chevalier referred *Opa metrosideros* Lour. to *Eriobotrya* because a species of *Eriobotrya* (*E. benghalensis* Hook. f.) is common in the mountains of Anam and he had not seen *Raphiolepis* from that region; but *Clemens 3370* from Tourane is *Raphiolepis indica* Lindl. and Cardot cites various collections from Anam. Loureiro's description, as I understand it, scarcely applies to *Eriobotrya*. For *Crataegus rubra* Loureiro states: " Habitat agrestis prope Cantonem Sinarum "; Nakai (Journ. Arnold Arb. **5**: 66. 1924) retains *Raphiolepis rubra* (Lour.) Lindl. as a valid species closely allied to *R. indica* (Linn.) Lindl.; Loureiro's type is preserved in the herbarium of the British Museum. For *Crataegus indica* Loureiro states: " Habitat in sylvis Cochinchinae." I believe he had specimens of the Linnaean species, at least as *Raphiolepis indica* (Linn.) Lindl. is currently interpreted; Nakai (Journ. Arnold Arb. **5**: 66. 1924) refers it to *Raphiolepis indica* (Linn.) Lindl. var. *tashiroi* Hayata.

---

[87] *Raphiolepis* Lindley (1820), conserved name, Vienna Code (as *Rhaphiolepis*) over *Opa* Loureiro (1790). However *Opa* was based on two species *O. odorata* Lour., the first one described, which is a *Eugenia* (see p. 285), and Loureiro's generic description was apparently based on this, and *O. metrosideros* Lour. which is a *Raphiolepis.* Strictly *Opa* should, I believe, be considered a synonym of *Eugenia.*

[88] A Loureiro herbarium name here first published by Gomes.

[89] Doctor Chevalier cites *Eriobotrya benghalensis* Hook. f. as a synonym, which should be excluded.

Eriobotrya Lindley

Eriobotrya japonica (Thunb.) Lindl. in Trans. Linn. Soc. **13**: 102. 1821; Cardot in Lecomte Fl. Gén. Indo-Chine **2**: 678. 1920.

Mespilus japonicus Thunb. Fl. Jap. 206. 1784.

Crataegus bibas Lour. Fl. Cochinch. 319. 1790, ed. Willd. 391. 1793, Anamese *ti ba diep.* Chinese *pî pá xú.*

" Habitat abunde culta Macai, & Cantone Sinarum." Loureiro's description agrees in all respects with the characters of the widely cultivated loquat, *Eriobotrya japonica* (Thunb.) Lindl.

Photinia Lindley

Photinia sp.

Psidium caninum Lour. Fl. Cochinch. 310. 1790, ed. Willd. 379. 1793, Chinese *pǎ hōa.*

? *Photinia variabilis* Hemsl. in Journ. Linn. Soc. Bot. **23**: 263. 1887, pro parte.

" Habitat agreste prope Cantonem Sinarum." The subserrate and alternate leaves definitely exclude Loureiro's species from the Myrtaceae. The description, except in the single character " calyx . . . superus," applies fairly well to *Photinia*, but on the whole it agrees equally well with *Symplocos chinensis* (Lour.) Druce. The statement appertaining to its peculiar effect on dogs together with the local name may ultimately supply the clue to its identity.

Photinia sp.

Psidium nigrum Lour. Fl. Cochinch. 311. 1790, ed. Willd. 380. 1793, Anamese *cây nen.*

" Habitat in sylvis Cochinchinae." The description of the leaves as " leviter serratis " excludes this from the *Myrtaceae*, and of the fruits as " polysperma . . . semina nidulantia " from the genus *Eugenia*. *Photinia* may be the genus represented and possibly the species is that represented by *Squires 209* from Hue, with erroneous fruit characters added. Gagnepain does not mention Loureiro's species in his treatment of the Myrtaceae of Indo-China (Lecomte Fl. Gén. Indo-Chine **2**: 788–864. 1920–1921).

Rubus (Tournefort) Linnaeus

Rubus cochinchinensis Tratt. Rosa. Monog. **3**: 97. 1823 (based on *Rubus fruticosus* Lour.); Focke in Bibl. Bot. **17**(72): 49. 1910; Cardot in Lecomte Fl. Gén. Indo-Chine **2**: 632. 1920.

Rubus fruticosus (non Linn.) Lour. Fl. Cochinch. 325. 1790, ed. Willd. 398. 1793, Anamese *cây ngêi chia lá.*

Rubus playfairii Hemsl. in Journ. Linn. Soc. Bot. **23**: 235. 1887.

" Habitat sylvas, & sepes Cochinchinae." This is a well-understood species of the section *Malachobatus, Cochinchinenses,* amply described by Focke. It is represented by *Clemens 3881,* frequent in thickets at Tourane, *Kuntze 3607, 3608,* from Tourane, and *Clemens 3231,* from Mount Bana.

Rubus alceaefolius Poir. in Lam. Encycl. **6**: 247. 1804; Cardot in Lecomte Fl. Gén. Indo-Chine **2**: 635. 1920.

Rubus moluccanus (non Linn.) Lour. Fl. Cochinch. 324. 1790, ed. Willd. 397. 1793, Anamese *cây ngêi tlòn là.*

" Habitat in sylvis Cochinchinae." Poiret's species occurs in Indo-China, while *Rubus moluccanus* Linn. is unknown from that region. *Rubus latifolius* Rumph. (Herb. Amb. **5**: 88. *pl. 47. f. 2*), cited by Loureiro, after Linnaeus, typifies *Rubus moluccanus* Linn. Poiret's species was based on a specimen from Java; as interpreted by Cardot it is represented by *Clemens 3494* from thickets at Hue (*R. moluccanus* Lour.!).

Rubus parvifolius Linn. Sp. Pl. 1197. 1753; Lour. Fl. Cochinch. 324. 1790, ed. Willd. 398. 1793, Anamese *cây ngêi hoa tiá*.

" Habitat in sylvis Cochinchinae, & Chinae." Loureiro's description conforms to the characters of the Linnaean species, the type of which was an actual specimen collected by Osbeck near Canton; the species is very common there. *Rubus parvifolius* Rumph. (Herb. Amb. **5**: 88. *pl. 47. f. 1*), erroneously placed here by both Linnaeus and Loureiro, must be excluded, as it represents the very different *Rubus fraxinifolius* Poir.

Fragaria (Tournefort) Linnaeus

Fragaria elatior Ehrh. Beitr. **7**: 23. 1792; Hemsl. in Journ. Linn. Soc. Bot. **23**: 239. 1887. *Fragaria vesca* (non Linn.) Lour. Fl. Cochinch. 325. 1790, ed. Willd. 398. 1793, Anamese *phuc bôn tu*, Chinese *fú pûen tsù*.

" Habitat, & colitur in China." This reduction of Loureiro's species follows Hemsley, which is possibly the correct disposition of it.

Potentilla Linnaeus

Potentilla fruticosa Linn. Sp. Pl. 495. 1753; Lour. Fl. Cochinch. 326. 1790, ed. Willd. 399. 1793, Anamese *duong trinh daoc*, Chinese *yâm chi chŏ*.

" Habitat in provinciis Borealibus Sinensibus." Loureiro's description applies to *Potentilla fruticosa* Linn., a species not uncommon in northern China. The material Loureiro had was undoubtedly secured from an herbalist.

Rosa (Tournefort) Linnaeus

Rosa chinensis Jacq. Obs. Bot. **3**: 7. *pl. 55*. 1768; Rehder in Sargent Pl. Wils. **2**: 320. 1915. *Rosa nankinensis* Lour. Fl. Cochinch. 324. 1790, ed. Willd. 397. 1793, Anamese *hoa houng tieo*, Chinese *tsiaò mûi hōa*.

" Habitat Cantone Sinarum, & alibi, a Nankino oriunda." Loureiro's specimens were manifestly from a cultivated plant, very likely a hybrid. I follow Rehder in his reduction of *R. nankinensis* Lour. Hemsley (Journ. Linn. Soc. Bot. **23**: 250. 1887) placed it as a synonym of *Rosa indica* Linn., and Cardot (Lecomte Fl. Gén. Indo-Chine **2**: 664. 1920) places *Rosa chinensis* Jacq. as a synonym of the Linnaean species.

Rosa cochinchinensis G. Don Gen. Syst. **2**: 585. 1832 (based on *Rosa spinosissima* Lour.). *Rosa spinosissima* (non Linn.) Lour. Fl. Cochinch. 323. 1790, ed. Willd. 395. 1793, Anamese *hoa hoùng lot*.

" Habitat ubique in Cochinchina." Cardot does not mention this species in his treatment of the Rosaceae of Indo-China (Lecomte Fl. Gén. Indo-Chine **2**: 613–681. 1920) nor can I refer it to any of the 7 species of *Rosa* admitted by him, from Loureiro's very short and imperfect description. It is described as a simple-flowered form with reddish flowers. It is possible that Loureiro observed only cultivated plants, although he does not so state.

Rosa indica Linn. Sp. Pl. 492. 1753; ? Lour. Fl. Cochinch. 323. 1790, ed. Willd. 396. 1793, Anamese *hoa hoùng coung gai*, Chinese *tsiâm hõa*.

"Habitat in Cochinchina, & China communissima." The very short description, apparently taken from cultivated plants, is most unsatisfactory; it may or may not appertain to the Linnaean species.

Rosa laevigata Michx. Fl. Bor. Am. 1: 295. 1803; Cardot in Lecomte Fl. Gén. Indo-Chine 2: 660. 1920.

Rosa alba (non Linn.) Lour. Fl. Cochinch. 323. 1790, ed. Willd. 396. 1793, Anamese *kim anh tu, hoa houng tlăng*, Chinese *kĭn ȳm*.

"Habitat in China, & Cochinchina." Assuming that the Chinese and Indo-China forms placed by Loureiro under *Rosa alba* represent a single species, it is probable that *R. laevigata* Michx. is the species represented; this is now known in Canton as *kam ying*. The description does not agree with the characters of Michaux's species in all respects and Loureiro does not mention the spiny calyces, a very obvious character of *R. laevigata* Michx.

Rosa loureiriana G. Don Gen. Syst. 2: 585. 1832 (based on *Rosa cinnamomea* Lour.).

Rosa cinnamomea (non Linn.) Lour. Fl. Cochinch. 323. 1790, ed. Willd. 395. 1793, Anamese *hoa koe*, Chinese *mûi hõa*.

"Habitat ubique culta in Cochinchina, & China." A single-flowered form with red flowers, manifestly not the Linnaean species, but scarcely determinable from Loureiro's short description alone. It cannot be placed satisfactorily among the 7 Indo-China species admitted by Cardot.

Rosa sp.

Rosa centifolia (non Linn.) Lour. Fl. Cochinch. 323. 1790, ed. Willd. 396. 1793, Anamese *hoa hoùng taù*, Chinese *tá mûi hõa*.

"Habitat in China: aegre colitur in Cochinchina." As noted by Hemsley, Loureiro's description, which was based on a cultivated form with double flowers, does not conform to the characters of the Linnaean species. It seems impossible definitely to place it from the description alone.

Pygeum Gaertner

Pygeum sp.

Dodecadia agrestis Lour. Fl. Cochinch. 319. 1790, ed. Willd. 390. 1793, Anamese *cây chon dúng;* Moore in Journ. Bot. 63: 284. 1925.

"Habitat in sylvis Cochinchinae." *Dodecadia* has long been an enigma. Bentham & Hooker f. (Gen. Pl. 1: 1007. 1867) enumerate it as a genus of doubtful status, allied to *Homalium*, stating: "*Dodecadia* Lour. Fl. Cochinch. 318, genus a Reichenbachio Tiliaceis adscriptum, a Candollio, Endlicherio et Lindleyo ut videtur neglectum, est genus valde dubium forte *Homalio* affine." Engler & Prantl do not include it in the Natürlichen Pflanzenfamilien. De Dalla Torre & Harms admit it as a genus of doubtful status in the Flacourtiaceae. A specimen in the Paris Herbarium from Loureiro labeled "Dodecatria agrestis" is a *Grewia* near *G. microcos* Linn. but this cannot be the type as it does not remotely agree with Loureiro's description. Loureiro's specimen in the British Museum, which I examined in 1930, is definitely a *Pygeum*. Moore (Journ. Bot. 63: 284. 1925) published a long note on this specimen, thinking it might represent a *Diospyros*, noting however that there were numerous discrepancies in Loureiro's description. He apparently overlooked the characteristic basal glands on the leaves which point unmistakably to *Pygeum*, as do the 12-fid calyx, 30 stamens, and the style and stigma characters as given by Loureiro. Loureiro's description of the corolla as monopetalous must have been due to an error of observation. The description of the fruit as many-seeded may be ignored, as Loureiro further states: " Baccam maturam non vidi; ideo certo decernere non potui circa semina et loculos." In view of the fragmentary nature of the type specimen and the fact that it apparently represents a species different from any credited to Indo-China by Cardot (Lecomte Fl. Gén. Indo-Chine 2: 618–620. 1920) it seems best not to transfer Loureiro's specific name to *Pygeum* at this time. Loureiro's type has elliptic to elliptic-oblong, apparently obtuse, glabrous leaves, about 7 cm. long, and 3.5 cm. wide, with 5 nerves on each side of the midrib, the upper surface shining, the lower surface dull, dark-olivaceous, the petioles about 1.3 cm. long. The parts of the inflorescences remaining on the specimen are slightly pubescent; there are no flowers. The basal glands are black but not at all raised as is the case in some species of the genus.

Prunus (Tournefort) Linnaeus

Prunus amygdalus Stokes Bot. Mat. Med. **3**: 101. 1812.

Prunus communis Archangeli Fl. Ital. 209. 1882; Fritsch in Sitz. Ver. Akad. Wien **101**(1): 632. 1892, non Huds. 1778.

Amygdalus communis Linn. Sp. Pl. 473. 1752; ? Lour. Fl. Cochinch. 316. 1790, ed. Willd. 386. 1793, Anamese *hanh nhon*, Chinese *hîm ho gîn*.

" Habitat & colitur affatim in China, tam dulcis, quam amara. In Cochinchina puto, quod non sit: mihi certe non occurrit." It is highly probable that Loureiro had some material representing the true almond, perhaps merely the seeds from commercial sources; that he had the almond in mind is clear from his reference to the sweet and bitter forms. On the other hand there seems to be much doubt as to whether or not the true almond occurs in China, in spite of the fact that Bretschneider and Franchet record this species, the former as cultivated in the northern provinces, and the latter as occurring in Yunnan. Batalin states that the only almond that occurs in China is *Prunus tangutica* (Batalin) Koehne (*P. dehiscens* Koehne). Loureiro must have been thoroughly familiar with the true almond as it occurs in the Iberian Peninsula, and it is improbable that he would have misidentified an eastern Asiatic form, even if he had only seeds from commercial sources.

Prunus cochinchinensis (Lour.) Koehne in Bot. Jahrb. **52**: 300. 1915.

Amygdalus cochinchinensis Lour. Fl. Cochinch. 316. 1790, ed. Willd. 387. 1793, Anamese *cây giang cuóc*.

" Habitat in vastis sylvis Cochinchinae." *Amygdalus cochinchinensis* is known only from Loureiro's description; Cardot (Lecomte Fl. Gén. Indo-Chine 2: 621–629. 1920) does not mention it. Loureiro describes the fruits as being a half inch long, similar in form and odor to those of his *Amygdalus communis* = *Prunus amygdalus* Stokes. Koehne, from Loureiro's description, places it in *Sclerocraspedon* next to *Prunus marginata* Dunn, together with *P. spinulosa* Sieb. & Zucc., and *P. phaeosticta* Maxim. The characters as given by Loureiro do not agree with those of any of the species recorded from Indo-China.

Prunus persica (Linn.) Batsch Beytr. Entwick. Gesch. Naturr. 30. 1801; Stokes Bot. Mat. Med. **3**: 100. 1812.

Amygdalus persica Linn. Sp. Pl. 472. 1753; Lour. Fl. Cochinch. 315. 1790, ed. Willd. 386. 1793, Anamese *cây daò nhon*, Chinese *taô ho gîn.*

Amygdalus pumila (non Linn.) Lour. Fl. Cochinch. 316. 1790, ed. Willd. 387. 1793, Anamese *daò hoa houng.*

For *Amygdalus persica* Loureiro states: " Habitat culta in hortis Sinensibus, unde in Cochinchinam translata." The description applies to a form of the common peach. For *Amygdalus pumila* the statement is made: " Habitat in Cochinchina, non frequens, coliturque ob floris pulchritudinem." From the description it seems to be evident that Loureiro had in hand a double-flowered form of the common peach, as his description applies to this species, and particularly significant is his description of the flowers as " maiusculus " and the note on the fruit: " fructus Persico minor, forma similis, sapore acidus."

Prunus salicina Lindl. in Trans. Hort. Soc. London **7**: 239. 1830; Koehne in Sargent Pl. Wils. **1**: 580. 1913, **3**: 432. 1917.

Prunus triflora Roxb. Hort. Beng. 38. 1814, *nomen nudum*, Fl. Ind. ed. 2, **2**: 501. 1832 (err. *trifolia*); Cardot in Lecomte Fl. Gén. Indo-Chine **2**: 628. 1920.

Prunus domestica (non Linn.) Lour. Fl. Cochinch. 317. 1790, ed. Willd. 388. 1793, Anamese *cây môi*, Chinese *mûei xú.*

" Habitat in plerisque Sinarum provinciis: inde in Cochinchinam delata, ubi raro fructificat." Both Schneider and Koehne state that they saw no specimens of the true plum, *Prunus domestica* Linn., from China; hence its occurrence there must be considered as doubtful. The form Loureiro has was doubtless the species currently known as *Prunus triflora* Roxb. = *P. salicina* Lindl.

CONNARACEAE

Cnestis Jussieu

Cnestis palala (Lour.) Merr. in Journ. Straits Branch Roy. As. Soc. **85**: 201. 1922; Enum. Philip. Fl. Pl. **2**: 240. 1923 (based on *Thysanus palala* Lour.).

Thysanus palala Lour. Fl. Cochinch. 284. 1790, ed. Willd. 349. 1793 (excl. syn. Rumph.), Anamese *deei khe.*

Thysanus cochinchinensis DC. Prodr. **2**: 91. 1825 (based on *Thysanus palala* Lour.).

Cnestis ramiflora Griff. Not. **4**: 432. 1854; Lecomte Fl. Gén. Indo-Chine **2**: 44. 1908.

" Habitat in sylvis Cochinchinae." *Thysanus* Lour., long placed in the Connaraceae as a genus of doubtful status, is clearly the same as *Cnestis* Jussieu. *Thysanus palala* Lour. I believe to be identical with the rather widely distributed *Cnestis ramiflora* Griff. *Palala secunda* Rumph. (Herb. Amb. **2**: 26. *pl. 6*), cited by Loureiro as representing his species, and whence he derived his specific name, must be excluded as it represents the myristicaceous *Horsfieldia sylvestris* Warb. Loureiro's description, however, was based on actual Indo-China specimens, not on *Palala secunda* Rumph. His description of the ovary and styles is not good for *Cnestis*, but I suspect that his data were based on faulty observations. His specimen, listed as being in the herbarium of the British Museum, has not been located. The species is represented by *Clemens 3805* from thickets at Tourane near the classical locality. *Cnestis* Juss. has one year priority over *Thysanus* Lour.

Rourea Aublet

Rourea microphylla (Hook. & Arn.) Planch. in Linnaea **23**: 421. 1850; Gagnep. in Lecomte Fl. Gén. Indo-Chine **2**: 47. 1908.

Connarus microphyllus Hook. & Arn. Bot. Beechey's Voy. 179. 1833.

Santalodes microphyllum Schellenb. in Mitt. Bot. Mus. Univ. Zürich **50**: 53. 1910.

? Pterotum procumbens Lour. Fl. Cochinch. 293. 1790, ed. Willd. 358. 1793, Anamese *cây truong deei*.

"Habitat in sylvis Cochinchinae." De Dalla Torre and Harms (Gen. Siphon. 585. 1906) leave *Pterotum* along the *genera incertae sedis*. I have ventured to suggest its reduction to *Rourea* (*Santalodes*) although there are serious discrepancies between Loureiro's description and the characters of Aublet's genus. Moore (Journ. Bot. **63**: 245. 1925) failed to locate Loureiro's type which is supposed to be preserved in the herbarium of the British Museum, and Mr. A. H. G. Alston reported to me October, 1934, that it was not under *Rourea* nor *Tetracera*, possible reductions suggested by me after studying the original description. It is suspected that Loureiro's description is either erroneous, through the misinterpretation of certain morphological characters, or that it was based on material from two unrelated species.

LEGUMINOSAE
Mimosoideae

Pithecellobium [90] Martius

Pithecellobium clypearia (Jack) Benth. in Hook. Lond. Journ. Bot. **3**: 209. 1844 (*Pithecolobium*); Gagnep. in Lecomte Fl. Gén. Indo-Chine **2**: 106. 1913 (*Pithecolobium*).

Inga clypearia Jack in Malay. Miscel. 2(7): 78. 1822.

Mimosa vaga (non Linn.) Lour. Fl. Cochinch. 651. 1790, ed. Willd. 799. 1793, Anamese *cây tô dia*.

Pithecolobium vagum A. Chev. Cat. Pl. Jard. Bot. Saigon 64. 1919 (based on *Mimosa vaga* Lour.).

"Habitat in sylvis planis Cochinchinae." Loureiro's description conforms to the characters of *Pithecellobium clypearia* Benth. as interpreted by Gagnepain. It is of interest to note that the Anamese name cited by Gagnepain, *cây trô dia duc*, in a measure confirms the correctness of this reduction. Loureiro's specific name is invalid in this genus.

Pithecellobium clypearia (Jack) Benth. var. **acuminatum** Gagnep. in Lecomte Fl. Gén. Indo-Chine **2**: 107. 1913.

Mimosa nodosa (non Linn.) Lour. Fl. Cochinch. 649. 1790, ed. Willd. 798. 1793, Anamese *cây cô áo;* Moore in Journ. Bot. **63**: 290. 1925.

"Habitat agrestis in locis planis Cochinchinae." Moore's examination of Loureiro's fragmentary specimen in the herbarium of the British Museum definitely places this species which in my original manuscript of 1919 was left as *Pithecellobium* sp. De Pirey's specimen of *co ao* or *co uom, Chevalier 41188,* is *Pithecellobium lucidum* Benth.

[90] *Pithecellobium* Martius is the original spelling and is correctly formed from two Greek words meaning "monkey ear-ring" as explained by Martius. There is no warrant in changing the name to *Pithecolobium* as nearly all modern authors have done; see Merrill, E. D., in Journ. Washington Acad. Sci. **6**: 43. 1916; Sprague in Kew Bull. 243. 1929. The Brussels Code conserves *Pithecolobium* Martius (1837) over *Zygia* Boehmer (1760). The original and correct spelling is here accepted.

Albizzia Durazzini

Albizzia chinensis (Osbeck) Merr. in Am. Journ. Bot. **3**: 575. 1916.

Mimosa chinensis Osbeck Dagbok Ostind. Resa 233. 1757.

Mimosa marginata Lam. Encycl. **1**: 12. 1783.

Mimosa arborea (non Linn.) Lour. Fl. Cochinch. 651. 1790, ed. Willd. 800. 1793, Chinese *yâm mŏ*.

Mimosa stipulata Roxb. Hort. Beng. 40. 1814, *nomen nudum*.

Mimosa stipulacea Roxb. Fl. Ind. ed. 2, **2**: 549. 1832.

Albizzia stipulata Boiv. in Encycl. XIX Siècle **2**: 33. 1838; Benth. in Hook. Lond. Journ. Bot. **3**: 92. 1844; Gagnep. in Lecomte Fl. Gén. Indo-Chine **2**: 87. 1913.

Albizzia marginata Merr. in Philip. Journ. Sci. Bot. **5**: 23. 1910.

" Habitat agrestis prope Cantonem Sinarum." Osbeck's type was from the vicinity of Canton where the species is distinctly common. The species is allied to *Albizzia julibrissin* Durazz., and is sometimes confused with it. Loureiro's description of the fruits as subterete is not good for *Albizzia*, while the young branchlets usually bear few small spines, Loureiro describing the branches as unarmed; these discrepancies are almost certainly due to Loureiro having taken these data from the description of *Mimosa arborea* Thunb. (Fl. Jap. 229. 1784) which he erroneously cites as a synonym.

Albizzia corniculata (Lour.) Druce in Rept. Bot. Exch. Club Brit. Isles **4**: 603. 1917; Ricker in Journ. Washington Acad. Sci. **8**: 244. 1918; Merr. in Philip. Journ. Sci. Bot. **13**: 140. 1918 (based on *Mimosa corniculata* Lour.).

Mimosa corniculata Lour. Fl. Cochinch. 651. 1790, ed. Willd. 800. 1793, Chinese *hōai hōa*.

Albizzia millettii Benth. in Hook. Lond. Journ. Bat. **3**: 89. 1844; Gagnep. in Lecomte Fl. Gén. Indo-Chine **2**: 90. 1913.

" Habitat agrestis circa Cantonem Sinarum." Loureiro's description definitely applies to the Chinese species currently known as *Albizzia millettii* Benth. It is common in Kwangtung, still grows in the vicinity of Canton, and occurs in Indo-China.

Albizzia saponaria (Lour.) Blume ex Miq. Fl. Ind. Bat. **1**(1): 19. 1855; Gagnep. in Lecomte Fl. Gén. Indo-Chine **2**: 89. 1913 (based on *Mimosa saponaria* Lour.).

Mimosa saponaria Lour. Fl. Cochinch. 653. 1790, ed. Willd. 802. 1793, Anamese *cây chu blen*.

Inga saponaria Willd. Sp. Pl. **4**: 1008. 1805 (based on *Mimosa saponaria* Lour.).

" Habitat in sylvis Cochinchinae." Loureiro's description is poor and is distinctly inadequate, but as far as it goes, including the indicated uses, it agrees with the species as currently interpreted, which is one of wide geographic distribution in the Malaysian region. *Cortex saponarius* Rumph. (Herb. Amb. **4**: 131. *pl. 66*) is correctly placed as a synonym.

Acacia (Tournefort) Linnaeus

Acacia sinuata (Lour.) comb. nov.

Mimosa sinuata Lour. Fl. Cochinch. 653. 1790, ed. Willd. 802. 1793, Anamese *cây xoung tán*.

Mimosa concinna Willd. Sp. Pl. **4**: 1039. 1805.

Mimosa rugata Lam. Encycl. 1: 20. 1783, non *Acacia rugata* Ham.

Acacia concinna DC. Prodr. 2: 464. 1825; Gagnep. in Lecomte Fl. Gén. Indo-Chine 2: 81. 1913.

Acacia rugata Merr. in Philip. Journ. Sci. Bot. 5: 28. 1910, non Ham.

" Habitat in sylvis Cochinchina." Loureiro's description agrees closely with the characters of the species currently known as *Acacia concinna* DC. which is apparently common in Indo-China; the description of the flowers as 4-merous is perhaps an error on Loureiro's part. It may be noted that although the earlier *Mimosa rugata* Lam. probably presents the same species as *Acacia concinna* DC., that *Acacia rugata* Ham. was independently published and has nothing to do with Lamarck's species. Gagnepain (Lecomte Fl. Gén. Indo-Chine 2: 76–84. 1913) fails to account for Loureiro's species. G. Don (Gen. Syst. 2: 386. 1832) thought that it might be a species of *Entada*, but Loureiro's description does not apply to that genus.

Acacia farnesiana (Linn.) Willd. Sp. Pl. 4: 1083. 1805; Gagnep. in Lecomte Fl. Gén. Indo-Chine 2: 78. 1913.

Mimosa farnesiana Linn. Sp. Pl. 521. 1753; Lour. 652. 1790, ed. Willd. 801. 1793, Anamese *hoa xiem gai.*

" Habitat culta in hortis Cochinchinae, puto, quod non indigena." The description conforms with the characters of the Linnaean species, one of American origin but now common in most or all tropical countries.

Acacia pennata (Linn.) Willd. Sp. Pl. 4: 1090. 1805; Gagnep. in Lecomte Fl. Gén. Indo-Chine 2: 83. 1913.

Mimosa pennata Linn. Sp. Pl. 522. 1753; Lour. Fl. Cochinch. 652. 1790, ed. Willd. 802. 1793, Anamese *cây châm bia.*

" Habitat in sylvis Cochinchinae." Loureiro's description applies to the Linnaean species, one of wide geographical distribution in tropical Asia and apparently common in Indo-China.

Acacia sp. ?

Mimosa pilosa Lour. Fl. Cochinch. 650. 1790, ed. Willd. 798. 1793, Anamese *cây úp chén.*

" Habitat in sylvis Cochinchinae." Loureiro's description is definite, if correct, and may apply to some species of *Acacia*, yet I am unable to refer it to any of the species of the Mimosoideae admitted by Gagnepain in his treatment of the Leguminosae of Indo-China (Lecomte Fl. Gén. Indo-Chine 2: 57–110. 1913). In some respects the description conforms to the characters of *Acacia tomentosa* Willd., as interpreted by Gagnepain (*op. cit.* 79); but this species is recorded by Gagnepain only from Siam and Java. The statement " foliolis . . . inferioribus minoribus " suggests *Pithecellobium* and *Albizzia*, yet none of these have an inflorescence of the type described by Loureiro " flos terminalis, conglobatus in capitulum magnum." The description may have been based on material from two unrelated plants.

Neptunia Loureiro

Neptunia prostrata (Lam.) Baill. in Bull. Soc. Linn. Paris 1: 356. 1883; Macbr. in Contr. Gray Herb. 59: 15. 1919; Britt. & Rose N. Am. Fl. 23: 180. 1928.

Mimosa prostrata Lam. Encycl. 1: 10. 1783.

Neptunia oleracea Lour. Fl. Cochinch. 654. 1790, ed. Willd. 804. 1793, Anamese *rau nhút;* Gagnep. in Lecomte Fl. Gén. Indo-Chine **2**: 59. *f. 8.* 1913.

" Habitat culta in Cochinchina, fluctuans in stagnis & fluminibus lenti cursus." A well-known species, the type of the genus, occurring in the tropics of both hemispheres. Willdenow (Sp. Pl. **4**: 1044. 1805) reduced Loureiro's species to *Desmanthus natans* (Vahl) Willd., but both Loureiro's and Lamarck's names are older than Vahl's. Loureiro's type is preserved in the herbarium of the British Museum.

Entada [91] Adanson

Entada phaseoloides (Linn.) Merr. in Philip. Journ. Sci. Bot. **9**: 86. 1914, Interpret. Herb. Amb. 253. 1917, Enum. Philip. Fl. Pl. **2**: 252. 1923.

Lens phaseoloides Linn. in Stickman Herb. Amb. 18. 1754, Amoen. Acad. **4**: 128. 1759.

Mimosa scandens Linn. Sp. Pl. ed. 2, 1501. 1763; Lour. Fl. Cochinch. 650. 1790, ed. Willd. 798. 1793, Anamese *tlàm deei.*

Entada scandens Benth. in Hook. Lond. Journ. Bot. **4**: 332. 1842; Gagnep. in Lecomte Fl. Gén. Indo-Chine **2**: 62. 1913.

" Habitat in sylvis planis Cochinchinae." Loureiro's description applies to *Entada scandens* (Linn.) Benth. as currently interpreted. It also is definitely the form figured by Rumphius as *Faba marina major* (Herb. Amb. **5**: 5. *pl. 4*), cited by Loureiro as a synonym, which typifies *Entada phaseoloides* (Linn.) Merr., the oldest valid binomial for the species. A fruit from Loureiro is preserved in the herbarium of the British Museum.

Caesalpinoideae

Tamarindus Linnaeus

Tamarindus indica Linn. Sp. Pl. 34. 1753; Lour. Fl. Cochinch. 403. 1790, ed. Willd. 488. 1793, Anamese *cây me.*

" Habitat culta in hortis Cochinchinae." This is the common tamarind, the Linnaean species being correctly interpreted by Loureiro.

Bauhinia Linnaeus

Bauhinia coccinea (Lour.) DC. Prodr. **2**: 516. 1825 (based on *Phanera coccinea* Lour.).

Phanera coccinea Lour. Fl. Cochinch. 37. 1790, ed. Willd. 47. 1793, Anamese *rê quách;* Moore in Journ. Bot. **63**: 247. 1925.

" Habitat in sylvis Cochinchinae." This is the type of the genus *Phanera* and hence of the section generally recognized under this name. In Index Kewensis it is erroneously reduced to the very different Philippine *Bauhinia cumingiana* F.-Vill. *Folium linguae* Rumph. (Herb. Amb. **5**: 1. *pl. 1*), cited by Loureiro as representing his species, must be excluded as it typifies the equally different *Bauhinia lingua* DC. of the Moluccas. Gagnepain in his treatment of the species of *Bauhinia* of Indo-China (Lecomte Fl. Gén. Indo-Chine **2**: 119–151. 1913) does not mention Loureiro's species, and Moore, who gives additional descriptive data based on Gagnepain's type in the British Museum, failed to match it with any material then extant; nor can I, on the basis of very excellent material from Indo-China, *Clemens 4254* from Mount Bana, near Tourane, refer it to any of the 41 spe-

---

[91] *Entada* Adanson (1763), conserved name, Brussels Code; an older one is *Gigalobium* Boehmer (1760).

cies recognized by Gagnepain. The species must be a most conspicuous one in nature because of its numerous, large, red flowers.

Cassia (Tournefort) Linnaeus

Cassia fistula Linn. Sp. Pl. 377. 1753; Lour. Fl. Cochinch. 264. 1790, ed. Willd. 323. 1793, Anamese *tlái xiem*.

"Habitat in Cochinchina ad Austrum, prope Cambodiam." The description unmistakably applies to the well-known Linnaean species. *Cassia fistula* Rumph. (Herb. Amb. 2: 83. *pl. 21*), cited as a synonym, after Linnaeus, is correctly placed.

Cassia mimosoides Linn. Sp. Pl. 379. 1753; Gagnep. in Lecomte Fl. Gén. Indo-Chine 2: 162. 1913.

Cassia procumbens (non Linn.) Lour. Fl. Cochinch. 264. 1790, ed. Willd. 324. 1793, Anamese *cây me dât*.

"Habitat in Cochinchina per agros, & colles dispersa. Observata a me fuit in China pariter & in Africa, ubique constans, & sui similis." The description applies unmistakably to the common, widely distributed, and well-known *Cassia mimosoides* Linn.

Cassia obtusifolia Linn. Sp. Pl. 377. 1753; Lour. Fl. Cochinch. 263. 1790, ed. Willd. 323. 1793, Anamese *dâu ma*, Chinese *tsào kit lâm*.

"Habitat prope vias in Cochinchina, & in China." This species is very closely allied to *Cassia tora* Linn. Loureiro clearly indicates one of the fundamental differences between *C. obtusifolia* Linn. and *C. tora* Linn. in describing the pod of the former as terete, and that of the latter as somewhat 4-angled. Roxburgh noted the differences between the two, and Prain (Journ. As. Soc. Bengal 66(2): 159. 1897) has given detailed descriptions of both. He states that *C. obtusifolia* Linn. is fairly common in southeastern Asia. Gagnepain (Lecomte Fl. Gén. Indo-Chine 2: 163. 1913) does not recognize *C. obtusifolia* Linn. as occurring in Indo-China. *Gallinaria rotundifolia* Rumph. (Herb. Amb. 5: 283. *pl. 97. f. 2*), cited by Loureiro as a synonym, represents *Cassia tora* Linn.

Cassia sophera Linn. Sp. Pl. 379. 1753; Lour. Fl. Cochinch. 264. 1790, ed. Willd. 324. 1793, Anamese *thao kŭyet mình*, Chinese *xȳ tsi táu, kiue mim tsù*.

"Habitat in Sinarum locis Australibus." The description applies to the Linnaean species, which is common in southern China. Loureiro notes that he saw what he took to be the same species in Africa. *Gallinaria acutifolia* Rumph. (Herb. Amb. 5: 283. *pl. 97. f. 1*), cited by Loureiro as representing this species, is the allied *Cassia occidentalis* Linn.

Cassia tora Linn. Sp. Pl. 376. 1753; Lour. Fl. Cochinch. 263. 1790, ed. Willd. 322. 1793; Anamese *dâu muòng;* Gagnep. in Lecomte Fl. Gén. Indo-Chine 2: 163. 1913.

"Habitat inculta ubique in Cochinchina." The description clearly applies to the Linnaean species, which is one of the most common plants found in and about towns throughout the settled areas in the Old World tropics. A specimen from Loureiro is preserved in the herbarium of the British Museum.

Gleditsia [92] Clayton

Gleditsia fera (Lour.) Merr. in Philip. Journ. Sci. Bot. **13**: 141. 1918 (based on *Mimosa fera* Lour.).

Mimosa fera Lour. Fl. Cochinch. 652. 1790, ed. Willd. 801. 1793, Anamese *tao giác, nha tao*, Chinese *tsáo kiĕ*.

Gleditsia australis Hemsl. in Journ. Linn. Soc. Bot. **23**: 208. *pl. 5.* 1887.

Gleditsia thorelii Gagnep. in Not. Syst. **2**: 212. 1912, Lecomte Fl. Gén. Indo-Chine **2**: 114. *f. 13. 1–7.* 1913.

" Habitat agrestis in Cochinchina, & China; seritur quoque ad formandas sepes cuilibet animali horridas, & impenetrabiles." Kwangtung material shows great variation in the size of the fruit, which reaches a maximum length of at least 25 cm. Hemsley described *G. australis* as having fruits 4 to 5 inches long; Loureiro described them as 8 inches long. Doctor A. Chevalier in July, 1919, wrote me that he did not think that *G. thorelii* Gagnep. was distinct from *G. australis* Hemsl. as the differential characters indicated by Gagnepain are not constant and that the Indo-China form appeared to be a race found exclusively in the gardens of the natives.

Caesalpinia Linnaeus

Caesalpinia crista Linn. Sp. Pl. 380. 1753, pro majore parte; Merr. Interpret. Herb. Amb. 260. 1917; Enum. Philip. Fl. Pl. **2**: 266. 1923.

Guilandina bonducella Linn. Sp. Pl. ed. 2, 545. 1762.

Caesalpinia bonducella Flem. in Asiat. Res. **11**: 159. 1810; Gagnep. in Lecomte Fl. Gén. Indo-Chine **2**: 174. 1913; Petch in Ann. Bot. Gard. Peradeniya **9**: 299. 1925.

Guilandina gemina Lour. Fl. Cochinch. 265. 1790, ed. Willd. 325. 1793, Anamese *cây maóc meò;* Moore in Journ. Bot. **63**: 281. 1925.

" Habitat in sylvis planis Cochinchinae." Moore's examination of Loureiro's type in the herbarium of the British Museum (Journ. Bot. **63**: 281. 1925) verifies my reduction of *Guilandina gemina* Lour. to *Caesalpinia crista* Linn. (*C. bonducella* Flem.) made in 1919, the leaves being erroneously described by Loureiro as pinnate rather than bipinnate.

Caesalpinia minax Hance in Journ. Bot. **22**: 365. 1884.

Guilandina bonducella (non Linn.) Lour. Fl. Cochinch. 265. 1790, ed. Willd. 325. 1793, Chinese *nâm siĕ lâc.*

" Habitat in sylvis prope Cantonem Sinarum." In spite of Loureiro's description of the flowers as yellow, I believe that the form he had was *Caesalpinia minax* Hance, the type of which was from Kwangtung. The description of the leaflets as glabrous, and of the echinate pod having 6 or 7 oblong-ovate seeds, are characters of *C. minax* Hance, not of *G. bonducella* Linn. *Globuli majores* (i.e., *Frutex globulorum majorum*) Rumph. Herb. Amb. **5**: 92. *pl. 49. f. 1,* cited by Loureiro, is *Guilandina bonducella* Linn. = *Caesalpinia crista* Linn.

Caesalpinia nuga (Linn.) Ait. Hort. Kew. ed. 2, **3**: 32. 1811; Gagnep. in Lecomte Fl. Gén. Indo-Chine **2**: 181. 1913.

Guilandina nuga Linn. Sp. Pl. ed. 2, 546. 1762.

[92] Changed to *Gleditschia* and so used by many authors, but *Gleditsia* is the original and correct spelling; see Sprague, Kew Bull. 354. 1928.

Genista scandens Lour. Fl. Cochinch. 428. 1790, ed. Willd. 521. 1793, Anamese *cây gieng gieng.*

Butea loureirii Spreng. Syst. **3**: 186. 1826 (based on *Genista scandens* Lour.).

" Habitat in Cochinchina prope ripas fluminum." From the indicated habitat and the description there is no doubt that the widely distributed and abundant *Caesalpinia nuga* Ait. is the species intended, in spite of Loureiro's reference to the papilionaceous flowers; this term was no doubt used to make the description conform to the characters of *Genista* in which the species was erroneously placed. De Pirey's specimen of *ging ging,* *Chevalier 41206,* is *Caesalpinia nuga* Ait., further confirmation of the correctness of this reduction.

Caesalpinia pulcherrima (Linn.) Sw. Obs. 166. 1791; Gagnep. in Lecomte Fl. Gén. Indo-Chine **2**: 183. 1913.

Poinciana pulcherrima Linn. Sp. Pl. 380. 1753; Lour. Fl. Cochinch. 261. 1790, ed. Willd. 319. 1793, Anamese *hoa phung.*

Poinciana bijuga (non Linn.) Lour. Fl. Cochinch. 260. 1790, ed. Willd. 319. 1793.

Poinciana elata Lour. Fl. Cochinch. 261. 1790, ed. Willd. 320. 1793, Anamese *cây luc.*

For *Poinciana pulcherrima* Loureiro states: " Habitat in Cochinchina, & China, multisque Indiae locis." The description applies unmistakably to the Linnaean species. For *P. bijuga* he states: " Habitat agrestis in ora Africae orientali," and the description applies in all respects to *Caesalpinia pulcherrima* Sw. *Crista pavonis* Rumph. (Herb. Amb. **4**: 53. *pl. 20*), cited by Loureiro, after Linnaeus, as representing *P. bijuga,* is *Caesalpinia pulcherrima* Sw. Linnaeus' erroneous reduction of Rumphius' species misled Loureiro, who incidentally states under *P. pulcherrima* that the two supposed species did not appear to him to be distinct. For *P. elata* Loureiro states: " Habitat in sylvis Cochinchinae," from which one might refer that he was dealing with a native rather than with an introduced species. From the description alone I can see little reason for considering that he had other than an unarmed form of *Caesalpinia pulcherrima* Sw. Gagnepain is silent on this point as he does not mention Loureiro's species in his treatment of the Leguminosae of Indo-China (Lecomte Fl. Gén. Indo-Chine **2**: 57–613. 1913–20); under *Caesalpinia pulcherrima* Sw. he cites no local names that remotely resemble those given by Loureiro for the species here reduced:

Caesalpinia sappan Linn. Sp. Pl. 381. 1753; Lour. Fl. Cochinch. 262. 1790, ed. Willd. 320. 1793, Anamese *cây vang, tô mouc,* Chinese *sū fâm mŏ;* Gagnep. in Lecomte Fl. Gén. Indo-Chine **2**: 179. 1913.

" Habitat in altis montis Cochinchinae." Loureiro correctly interpreted the very common, widely distributed and well-known Linnaean species, although he erroneously described it as a large tree and it is not a species that grows on high mountains. *Lignum sappan* Rumph. (Herb. **4**: 56. *pl. 21*) is correctly placed, after Linnaeus, as a synonym.

Peltophorum [93] Walpers

Peltophorum sp.

Baryxylum rufum Lour. Fl. Cochinch. 266. 1790, ed. Willd. 326. 1793, Anamese *cày lim váng,* Chinese *tiĕ li mŭ;* Moore in Journ. Bot. **63**: 281. 1925.

[93] *Peltophorum* Walpers (1842), conserved name, Vienna Code; an older one is *Baryxylum* Loureiro, pro parte (1790).

" Habitat in altis montibus Cochinchinae ad Boream sitis." Loureiro's two specimens in the herbarium of the British Museum were examined by Moore who has given a long note regarding the problems raised. One specimen is a *Peltophorum*, but not *P. dasyrachis* Kurz to which Pierre tentatively referred it, nor do the leaves match any of the species of *Peltophorum* known from Indo-China. Another leaf specimen is apparently *Gymnocladus chinensis* Baill., while a fruit in the carpological collection represents the latter species. It is clear that the floral characters were drawn from a species of *Peltophorum*, and the fruit characters may have been taken from the *Gymnocladus*, as the fruit description is not at all that of *Peltophorum*. *Gymnocladus* is a plant of northern China and Loureiro doubtless secured this specimen from some source in China. *Peltophorum inerme* (Roxb.) Naves (*P. ferrugineum* Benth.) is normally a strand tree and does not grow in " altis montibus "; *P. dasyrachis* (Miq.) Kurz is also apparently a low altitude species. *Mitrosideros amboinensis* Rumph. (Herb. Amb. **3**: 21. *pl. 10*), cited by Loureiro as doubtfully representing his species, and from whence his fruit characters may have been taken, at least in part, must be excluded as it represents *Intsia bijuga* (Colebr.) O. Ktz. Gagnepain refers *Baryxylum rufum* Lour. in part to *Afzelia bijuga* A. Gray = *Intsia bijuga* O. Ktz. In Lecomte (Fl. Gén. Indo-Chine **2**: 189–192. 1913) he admits three species of *Peltophorum* and cites the local name *kim vang* for both *P. ferrugineum* Benth. = *P. inerme* (Roxb.) Naves, and *P. dasyrachis* (Miq.) Kurz. *Baryxylum* is, as to its floral characters. referable to *Peltophorum*, but in view of the manifest mixture of representatives of two or perhaps even three genera in the description, more definite placement of the species is impracticable. Here is a clear case where the binomial should be eliminated on the basis of the rule that a species based on a mixture of two or more different ones is invalid.

Cordyla Loureiro

Cordyla africana Lour. Fl. Cochinch. 412. 1790, ed. Willd. 500. 1793; Pers. Syn. **2**: 260, 1807 (*Cordylia*); Baker in Oliv. Fl. Trop. **2**: 257. 1871.

Calycandra pinnata A. Rich. in Guill. & Perr. Fl. Senegam. Tent. 30. *pl. 9*. 1832.

" Habitat ad oram Africae orientalem." A monotypic African genus for which Loureiro gives the local name *mutondo*. His type is preserved in the herbarium of the Paris Museum.

Papilionatae

Ormosia [94] Jackson

Ormosia pinnata (Lour.) Merr. in Lingnan Sci. Journ. **14**: 12. 1935 (based on *Cynometra pinnata* Lour.).

Cynometra pinnata Lour. Fl. Cochinch. 268. 1790, ed. Willd. 329. 1793, Anamese *cây rang;* Gagnep. in Lecomte Fl. Gén. Indo-Chine **2**: 155. 1913.

Ormosia hainanensis Gagnep. Not. Syst. **3**: 31. 1914, Lecomte Fl. Gén. Indo-Chine **2**: 511. 1920; Merr. in Lingnan Sci. Journ. **5**: 91. 1927.

" Habitat in sylvis Cochinchinae." Loureiro's description is definite, and *Clemens 4015*, from Tourane, which is identical with *Ormosia hainanensis* Gagnep., conforms to it. Gagnepain admits *Cynometra pinnata* Lour. with a description compiled from Loureiro, stating: " Espèce très douteuse, appartenant peut-être à un autre genre." He had only

[94] *Ormosia* Jackson (1811), conserved name, Vienna Code; an older one is *Toulichiba* Adanson (1763).

Hainan material representing *Ormosia hainanensis* Gagnep., admitting it as a species " à rechercher au Tonkin." The pods have from 1 to 3 or even 4 seeds. The species is now known from Anam, Hainan and Kwangtung.

Sophora Linnaeus

Sophora flavescens Ait. Hort. Kew. **2**: 43. 1789.

> *Robinia flava* Lour. Fl. Cochinch. 456. 1790, ed. Willd. 556. 1793, Anamese *hùynh câm*, Chinese *hôam khin*.
> *Caragana flava* Kostel. Allgem. Med.-Pharm. Fl. **4**: 1275. 1835 (based on *Robinia* [*flava*] Lour.).
> *? Robinia amara* Lour. Fl. Cochinch. 455. 1790, ed. Willd. 556. 1793, Anamese *khô sâm hoa tiá*, Chinese *khū sēm*.

For *Robinia flava* Loureiro states: " Habitat agrestis in provinciis Borealibus imperii Sinensis." In spite of certain discrepancies in Loureiro's description in comparison with *Sophora flavescens* Ait., notably the " pedunculis ternis 3-floris," it is probable that Aiton's species is the correct reduction of *Robinia flava* Lour. He apparently had only fragmentary material secured from an herbalist; he notes that he had not seen the living plant. This is Bretschneider's suggested reduction, quoted by Hemsley (Journ. Linn. Soc. Bot. **23**: 202. 1887). For *Robinia amara* Loureiro states: " Habitat inculta in Cochinchina, & China." Bretschneider, quoted by Hemsley, *l.c.*, states that the Chinese name quoted by Loureiro appertains to *Sophora angustifolia* Sieb. & Zucc. = *S. flavescens* Ait. Loureiro's description does not well agree with the characters of Aiton's species, as its flowers are not violet, and the species does not occur as far south as Indo-China. Here again Loureiro probably had fragmentary material secured from an herbalist.

Sophora japonica Linn. Mant. **1**: 68. 1767; Gagnep. in Lecomte Fl. Gén. Indo-Chine **2**: 504. 1916.

> *Anagyris foetida* Lour. Fl. Cochinch. 260. 1790, ed. Willd. 318. 1793, Chinese *pă pái*.
> *Macrotropis foetida* DC. Prodr. **2**: 99. 1825 (based on *Anagyris foetida* Lour.).
> *Anagyris chinensis* Spreng. Syst. **2**: 346. 1825 (based on *Anagyris foetida* Lour.).
> *Anagyris sinensis* Steud. Nomencl. ed. 2, **1**: 83. 1840 (based on *Anagyris foetida* Lour.).
> *? Robinia mitis* (non Linn.) Lour. Fl. Cochinch. 455. 1790, ed. Willd. 555. 1793, Anamese *khô sâm hoa vàng*, Chinese *khū sēm*.
> *? Pongamia chinensis* DC. Prodr. **2**: 416. 1825 (based on *Robinia mitis* Lour.).

For *Anagyris foetida* Loureiro states: " Habitat inculta prope Cantonem Sinarum." Hemsley (Journ. Linn. Soc. Bot. **23**: 204. 1887) merely states: " An obscure plant of this affinity," *i.e.*, *Ormosia* and *Sophora*. In spite of certain discrepancies in the description, I believe that what Loureiro intended to describe is the common *Sophora japonica* Linn. His definite description of the fruits as terete and many-seeded, while not too good for *Sophora*, would with greater definiteness exclude the species from *Ormosia*. I have one recorded local name from Canton for *Sophora japonica*, *pak wai far*, but none of my recorded names for *Ormosia* remotely resemble the one cited by Loureiro for *Anagyris foetida*. For *Robinia mitis* Loureiro states: " Habitat agrestis in China: raro in Cochinchina." The Linnaean species is a synonym of *Pongamia pinnata* (Linn.) Merr., one very different from the plant Loureiro described, if Loureiro's description may be relied upon; there are however serious discrepancies between the description and the characters of *Sophora japonica* Linn., yet at this time I am unable to suggest any other possible reduction of *Robinia mitis* Lour. De Candolle may have been influenced in his transfer of the species to *Pongamia* from an examination of Plukenet's figure cited by Loureiro although it does not represent the species the latter described.

Sophora sp.

? Anagyris inodora Lour. Fl. Cochinch. 260. 1790, ed. Willd. 318. 1793, Anamese *cây hay*.

? Macrotropis inodora DC. Prodr. 2: 99. 1825 (based on *Anagyris inodora* Lour.).

" Habitat in sylvis Cochinchinae." No definite determination of the proper place of this species seems to be possible from the original description and the inadequate data available. Gagnepain (Lecomte Fl. Gén. Indo-Chine 2: 57–613. 1913–20) makes no attempt to place it. It may be that a poorly and incorrectly described species of *Sophora* was considered; the local name *cây hay* is very similar to *cây hoè* cited by Gagnepain for *Sophora japonica* Linn., but that species occurs in Indo-China only as a cultivated plant, and Loureiro states that his specimens were from the forests; moreover, the fruit character, " legumen compressum," does not apply to *Sophora*. *Anygris foetida* Lour., however, seems definitely to be *Sophora japonica* Linn.

Crotalaria (Dillinius) Linnaeus

Crotalaria retusa Linn. Sp. Pl. 715. 1753.

Lupinus cochinchinensis Lour. Fl. Cochinch. 429. 1790, ed. Willd. 521. 1793, Anamese *cây luc lac*.

" Habitat agrestis in Cochinchina. Idem a me visus in Benghala." Loureiro's description applies unmistakably to the characteristic, widely distributed and well-known Linnaean species.

Crotalaria saltiana Andr. Bot. Repos. 10: *pl. 648*. 1811.

Crotalaria striata DC. Prodr. 2: 131. 1825; Baker in Oliver Fl. Trop. Afr. 2: 38. 1871.

? Lupinus africanus Lour. Fl. Cochinch. 429. 1790, ed. Willd. 522. 1793.

" Habitat agrestis in ora Orientali Africae." This seems to be the most probable identification of *Lupinus africanus* Lour. among the 16 species of *Crotalaria* with 3-foliolate leaflets credited by Baker (Oliver Fl. Trop. Afr. 2: 7–44. 1871) to the Mozambique District; Baker does not account for Loureiro's species. This suggested reduction does not explain Loureiro's term " calyx appendiculatus "; it may be that some other genus is represented. Loureiro's species is clearly not a *Lupinus* as it is from outside of the generic range of that group.

Crotalaria uncinella Lam. Encycl. 2: 200. 1786; Gagnep. in Lecomte Fl. Gén. Indo-Chine 2: 344. 1916.

Crotalaria elliptica Roxb. Fl. Ind. ed. 2, 3: 279. 1832.

Crotalaria vachellii Hook. & Arn. Bot. Beechey's Voy. 180. 1833.

? Lotus arabicus (non Linn.) Lour. Fl. Cochinch. 463. 1790, ed. Willd. 566. 1793, Anamese *dâu leo vàng*.

" Habitat agrestis in Cochinchina." This reduction is suggested merely as a possibility. Loureiro's description, however, does not agree particularly well with the characters of Lamarck's species.

Crotalaria sp. ?

Trifolium cuspidatum Lour. Fl. Cochinch. 445. 1790, ed. Willd. 542. 1793, Anamese *cây chia ba.*

" Habitat agreste in Cochinchina." The description suggests a species of *Crotalaria* of the section *Trifoliatae, Dispermae.* In 1919 I suggested *Crotalaria medicaginea* Lam. as a possibility; but Loureiro's description of the leaves as linear does not agree with the characters of Lamarck's species.

Medicago (Tournefort) Linnaeus

Medicago sp.

Medicago polymorpha (non Linn.) Lour. Fl. Cochinch. 453. 1790, ed. Willd. 553. 1793, Anamese *cô aó nho cây.*

" Habitat in agris Cochinchinae." The description seems clearly to apply to *Medicago*, yet no representative of the genus is as yet known from Indo-China. Loureiro's data agree closely with the characters of *Medicago orbicularis* All., but that species is scarcely to be considered as a possibility as its known range is remote from Indo-China.

Indigofera Linnaeus

Indigofera tinctoria Linn. Sp. Pl. 751. 1753; Lour. Fl. Cochinch. 458. 1790, ed. Willd. 560. 1793, Anamese *châm nho lá,* Chinese *lân tsâo, tá cīm;* Gagnep. in Lecomte Fl. Gén. Indo-Chine 2: 428. 1916.

" Habitat spontanea, coliturque vastissime in Cochinchina, & China." The description applies to the Linnaean species, one formerly of wide cultivation in tropical Asia as a source of indigo. *Indicum* Rumph. (Herb. Amb. **5**: 220. *pl. 80*), cited by Loureiro as illustrating the species, as to the plate, represents *Indigofera suffruticosa* Mill. (*I. anil* Linn.) although Rumphius' description for the most part applies to *Indigofera tinctoria* Linn.

Indigofera trifoliata Linn. Cent. Pl. 2: 29. 1756, Amoen. Acad. **4**: 327. 1759.

Indigofera coccinea Lour. Fl. Cochinch. 457. 1790, ed. Willd. 559. 1793, Chinese *louc hām tsào.*

Indigofera trita Hemsl. in Journ. Linn. Soc. Bot. **23**: 158. 1886, non Linn. f.

" Habitat agrestis circa Cantonem Sinarum." Loureiro's description applies closely to the not uncommon southern China form currently referred to *Indigofera trifoliata* Linn. Hemsley (Journ. Linn. Soc. Bot. **23**: 158. 1886) places *I. coccinea* Lour. as a synonym of *I. trita* Linn. f., his entire record of this as a Chinese plant being based on this reduction. He states: " Whether Loureiro's plant be the same is uncertain."

Tephrosia [95] Persoon

Tephrosia purpurea (Linn.) Pers. Syn. 2: 329. 1807; Gagnep. in Lecomte Fl. Gén. Indo-Chine 2: 270. *f. 28, 1-6.* 1916.

Cracca purpurea Linn. Sp. Pl. 752. 1753.

Galega purpurea Linn. Syst. ed. 10, 1172. 1759.

Hedysarum lineare Lour. Fl. Cochinch. 452. 1790, ed. Willd. 551. 1793, Anamese *cây ve ve cái.*

[95] *Tephrosia* Persoon (1807), conserved name, Vienna Code; older ones are *Cracca* Linnaeus (1747, 1753), *Colinil* Adanson (1763) and *Needhamia* Scopoli (1777).

" Habitat incultum Cochinchina." In my original manuscript of 1919 this was placed under *Uraria picta* (Jacq.) Desv. but a more critical examination of the description clearly indicates that Loureiro's species in its " legumen lineare, rectum, laeve, acuminatum, 6-spermum," cannot possibly represent a *Uraria*. The description does conform in all essentials with the characters of the very common and widely distributed *Tephrosia purpurea* Pers. as interpreted by Gagnepain, although the latter in his treatment of the Leguminosae of Indo-China (Lecomte Fl. Gén. Indo-Chine 2: 57–613. 1913–1920) fails to account for Loureiro's species. In proposing the binomial *Tephrosia purpurea*, Persoon actually cites " L., Burm. zeyl. 77. t. 33," *i.e.*, *Coronilla zeylanica, flore purpurascente* Burm. Thes. Zeyl. 77. *pl. 33*, the plate reference being a manifest error for *pl. 32*, because plate 33 is *Tephrosia villosa* Pers., while plate 32 is *T. purpurea* Pers. as currently interpreted; he doubtless intended to cite *Cracca purpurea* Linn. but did not do so.

Millettia Wight & Arnott

Millettia sp.

> *Crotalaria (Crotolaria) heptaphylla* Lour. Fl. Cochinch. 433. 1790, ed. Willd. 527. 1793.

" Habitat inculta in Cochinchina." The description apparently applies to one of the turgid-fruited species of *Millettia*. I cannot, from the data available, refer it to any of the 43 species of the genus admitted by Gagnepain in his treatment of the Leguminosae of Indo-China (Lecomte Fl. Gén. Indo-Chine 2: 361–396. 1916); Gagnepain does not mention Loureiro's species.

Millettia sp.

> *? Crotalaria (Crotolaria) scandens* Lour. Fl. Cochinch. 433. 1790, ed. Willd. 527. 1793,
> Anamese *cây hay*.

" Habitat in sylvis Cochinchinae." I had thought *Clemens 3735* from Mount Bana near Tourane, which apparently is a *Millettia*, might represent Loureiro's species, but its calyces are not glabrous. Loureiro does not indicate whether his species had simple or pinnate leaves, but one might infer simple ones from the fact that he does not state otherwise; if simple leaves, then no *Millettia* is represented. It is manifestly not a *Crotalaria*. Gagnepain does not mention Loureiro's species. The Anamese name is the same as that for *Anagyris inodora* Lour. = ? *Sophora* sp.

Millettia sp. ?

> *Psoralea rubescens* Lour. Fl. Cochinch. 444. 1790, ed. Willd. 541. 1793, Anamese *nênh hoa do*.

" Habitat in sylvis Cochinchinae." Gagnepain (Lecomte Fl. Gén. Indo-Chine 2: 307. 1916) mentions this in a footnote as apparently not belonging to *Psoralea*, and it cannot possibly be a *Psoralea* if Loureiro's description is correct. *Millettia* is suggested as a possibility. Loureiro's description of the calyx as " tuberculosus, glandulis binis, viscosis ad basim suffultus " may refer to the small bracteoles at the base of the calyx. It is suspected that the fruit characters were added to make the description conform to the generic characters of *Psoralea*.

Millettia sp. ?

Psoralea scutellata Lour. Fl. Cochinch. 443. 1790, ed. Willd. 540. 1793, Anamese *cây nênh hoa tím.*

Psoralea sentellata Gagnep. in Lecomte Fl. Gén. Indo-Chine 2: 307. 1916, in nota, sphalm.

" Habitat in sylvis Cochinchinae." This is clearly not a *Psoralea. Millettia* is suggested as a possibility, but Loureiro's description of the calyx as " crassus excrescentiis albis glandulosis conspersus " does not apply to any *Millettia* known to me, or for that matter to any leguminous tree with which I am familiar. In some respects the description suggests *Sophora japonica* Linn., but the discrepancies are as great as with *Millettia.* The statement that the fruit was 1- or 2-seeded was apparently added to make the description conform to the generic characters of *Psoralea.*

Sesbania [96] Scopoli

Sesbania cochinchinensis (Lour.) DC. Prodr. 2: 266. 1825 (based on *Coronilla cochinchinensis* Lour.).

Coronilla cochinchinensis Lour. Fl. Cochinch. 452. 1790, ed. Willd. 552. 1793, Anamese *dâu chi.*

Sesbania aculeata Gagnep. in Lecomte Fl. Gén. Indo-Chine 2: 411. 1916, non Pers.

" Habitat agrestis in Cochinchina." Loureiro's description is definite in the leaflets being about fifteen pairs, the inflorescences few-flowered (sub-trifloris), the pods torulose, the plant suffrutescent and about four feet high. Gagnepain (Lecomte Fl. Gén. Indo-Chine 2: 411. 1916) has given an amplified description of this form with torulose fruits under *Sesbania aculeata* Pers., citing *Coronilla cochinchinensis* Lour. as a synonym. He has, however, curiously misinterpreted *Sesbania aculeata* Pers., which is merely a new binomial based on *Aeschynomene bispinosa* Jacq. (Ic. Pl. Rar. 3: 13. *pl. 564.* 1793). Jacquin's most excellent colored plate represents a form with small leaflets, widely scattered small spines on the stems, small salmon-yellow flowers, and slender, cylindric, *non-torulose* pods, a form which cannot be distinguished from *Sesbania cannabina* (Retz.) Pers., amply described under this binomial by Gagnepain (*op. cit.* 410). *Sesbania cannabina* (Retz.) Pers. is apparently the correct binomial for this form with slender non-torulose pods, as it dates from 1789. *Aeschynomene bispinosa* Jacq. was published in 1793 (date of issue of the text), although the plate may have appeared before this date, as the illustrations were issued irregularly between 1786–1793.

Clianthus [97] Banks & Solander

Clianthus scandens (Lour.) Merr. in Journ. Bot. 66: 265. 1928 (based on *Sarcodum scandens* Lour.).

Sarcodum scandens Lour. Fl. Cochinch. 462. 1790, ed. Willd. 564. 1793, Anamese *cây mùòng deei;* Moore in Journ. Bot. 63: 286. *f. A–C.* 1925.

Sarcodium scandens Pers. Syn. 2: 352. 1807.

[96] *Sesbania* Scopoli (1777), conserved name, Brussels Code; older ones are *Sesban* Adanson (1763) and *Agati* Adanson (1763).

[97] *Clianthus* Banks & Solander (1832), conserved name, Vienna Code; an older one is *Sarcodum* Loureiro (1790); *Donia* G. Don, another synonym, was published in 1832.

Clianthus binnendyckianus Kurz in Journ. As. Soc. Bengal **40**(2): 51. 1871; Merr. Enum. Philip. Fl. Pl. **2**: 282. 1923.

" Habitat in sylvis Cochinchinae." This genus *Sarcodum* Lour. remained one of entirely doubtful status until 1927 when its identity became evident through excellent specimens collected near the classical locality by J. & M. S. *Clemens*, no. *3773*. The type in the herbarium of the British Museum is very fragmentary; Moore (Journ. Bot. **63**: 286–288. 1925) gives a photographic reproduction of it, with extensive additional descriptive data and an excellent discussion. From his data it was possible for me to recognize *Sarcodum scandens* Lour. as representing the Malaysian species currently known as *Clianthus binnendyckianus* Kurz when Indo-China material representing Loureiro's species was received in 1927. Loureiro's generic name *Sarcodum* is older than *Clianthus*, but the latter is retained in accordance with the approved list of *nomina generica conservanda*. Curiously *Clianthus* was unknown from continental Asia (except as to Loureiro's unrecognized genus *Sarcodum*) until the species was collected by Mrs. Clemens in Indo-China. Moore, on the basis of the very fragmentary type, failed to recognize *Clianthus* and thought that *Sarcodum* represented a distinct genus, inasmuch as nothing conforming to its characters is included by Gagnepain in his treatment of the Leguminosae of Indo-China (Lecomte Fl. Gén. Indo-Chine **2**: 57–613. 1913–20).

Glycyrrhiza (Tournefort) Linnaeus

Glycyrrhiza echinata Linn. Sp. Pl. 741. 1753; Lour. Fl. Cochinch. 445. 1790, ed. Willd:ˈ 543. 1793, Anamese *cám thao*, Chinese *fu chāu căn tsào*.

" Habitat in provinciis Borealibus imperii Chinensis." Loureiro was apparently correct in his interpretation of the Linnaean species, as this occurs in northern China (Hemsley Journ. Linn. Soc. Bot. **23**: 168. 1887). Loureiro probably secured his material from an herbalist.

Glycyrrhiza glabra Linn. Sp. Pl. 742. 1753; Lour. Fl. Cochinch. 446. 1790, ed. Willd. 543. 1793, Anamese *cam thao*, Chinese *fán chāu căn tsào*.

" Habitat agrestis cultaque in variis locis imperii Chinensis." It seems probable that Loureiro had material representing the Linnaean species and it is certain, if that be the case, that he secured this from commercial sources. The Linnaean species has never been collected in China, Hemsley (Journ. Linn. Soc. Bot. **23**: 168. 1887) admitting it solely on Bretschneider's interpretation of Loureiro's record.

Ormocarpum [98] Beauvois

Ormocarpum cochinchinense (Lour.) Merr. in Philip. Journ. Sci. **5**: Bot. 76. 1910 (based on *Diphaca cochinchinensis* Lour.).

Diphaca cochinchinensis Lour. Fl. Cochinch. 454. 1790, ed. Willd. 554. 1793, Anamese *kim phung*.

Hedysarum sennoides Willd. Sp. Pl. **3**: 1207. 1800.

Ormocarpum sennoides DC. Prodr. **2**: 315. 1825; Gagnep. in Lecomte Fl. Gén. Indo-Chine **2**: 560. 1920.

[98] *Ormocarpum* Beauvois (1806), conserved name, Vienna Code; an older one is *Diphaca* Loureiro (1790).

Solulus cochinchinensis O. Ktz. Rev. Gen. Pl. 205. 1891 (based on *Diphaca cochinchinensis* Lour.).

" Habitat culta in hortis Cochinchinae, & Chinae." *Solulus arbor* Rumph. (Herb. Amb. **3**: 200. *pl. 128*), cited by Loureiro under his species, represents *Ormocarpum orientale* (Spreng.) Merr. (*Ormocarpum glabrum* Teysm. & Binn.) which may well prove to be identical with *O. cochinchinensis* (Lour.) Merr.   Loureiro's type is preserved in the herbarium of the British Museum.

Aeschynomene Linnaeus

Aeschynomene aspera Linn. Sp. Pl. 713. 1753; Gagnep. in Lecomte Fl. Gén. Indo-Chine **2**: 558. 1920.

Aeschynomene lagenaria Lour. Fl. Cochinch. 446. 1790, ed. Willd. 544. 1793, Anamese *cây dien dien.*

" Habitat in paludibus, & locis caenosis, aquosisque in Cochinchina." Loureiro's description conforms closely to the characters of the Linnaean species, which is widely distributed in southern Asia.   Gagnepain does not account for Loureiro's species, although he does cite *Hedysarum lagenaria* Roxb. as a synonym of *A. aspera* Linn.   Loureiro's specimen, in the herbarium of the British Museum, represents *Aeschynomene aspera* Linn.

Aeschynomene indica Linn. Sp. Pl. 713. 1753; Gagnep. in Lecomte Fl. Gén. Indo-Chine **2**: 559. 1920.

Hedysarum alpinum (non Linn.) Lour. Fl. Cochinch. 451. 1790, ed. Willd. 551. 1793, Anamese *dâu nhút.*

Hedysarum virginicum (non Linn.) Lour. Fl. Cochinch. 451. 1790, ed. Willd. 551. 1793, Anamese *cây muóng dât.* ·

For *H. alpinum* Loureiro states: " Habitat agreste in Cochinchina," and for *H. virginicum:* " Habitat in agris Cochinchinae ad portum Turanum, vulgo *Hàn.*" If one compares the two descriptions critically, one notes few discrepancies between them. The Linnaean species is very common and is not otherwise described by Loureiro, while the flowers vary from yellow to purplish. The difference in the color of the flowers is one of the differential characters one notes in comparing Loureiro's descriptions. For the first Loureiro describes the fruits as " lineare, rectum, tenue, glabrum, articulatum, longum, pendulum," and for the second, " tenue, articulatum, glabrum, pedunculatum, erectum." It may be noted that the illustration cited under *H. alpinum* (Gmelin Fl. Sibir. **4**: 26. *pl. 10*) represents a true *Hedysarum* and a plant totally different from *Aeschynomene*; it is suspected that Loureiro's description was in part based on Gmelin's figure.

Arachis Linnaeus

Arachis hypogaea Linn. Sp. Pl. 741. 1753.

Arachis asiatica Lour. Fl. Cochinch. 430. 1790, ed. Willd. 522. 1793, Anamese *cây dâu phung.*

Arachis africana Lour. *l.cc.* 430, 523.

Arachis hypogaea Linn. var. *glabra* DC. Prodr. **2**: 474. 1825 (based on *Arachis asiatica* Lour.).

For *Arachis asiatica* Loureiro states: " Habitat culta abundantissime in Cochinchina, & in China," and for *A. africana:* " Habitat in variis locis Africae Orientalis "; both are manifestly only forms of the common peanut. *Chamaebalanus japonica* Rumph. (Herb. Amb. **5**: 426. *pl. 156. f. 2*), cited by Loureiro as a synonym of his *A. asiatica*, is *hypogaea* Linn.

Zornia Persoon

Zornia diphylla (Linn.) Pers. Syn. **2**: 318. 1807.

> *Hedysarum diphyllum* Linn. Sp. Pl. 747. 1753; Lour. Fl. Cochinch. 449. 1790, ed. Willd. 548. 1793.

" Habitat agreste prope Cantonem Sinarum." Loureiro's description conforms in all respects to the very common and widely distributed Linnaean species, which is abundant in open dry places near Canton.

Desmodium [99] Desvaux

Desmodium gangeticum (Linn.) DC. Prodr. **2**: 327. 1825; Gagnep. in Lecomte Fl. Gén. Indo-Chine **2**: 601. 1920.

> *Hedysarum gangeticum* Linn. Sp. Pl. 746. 1753; Lour. Fl. Cochinch. 448. 1790, ed. Willd. 547. 1793, Chinese *tsung loung thû*.

" Habitat agreste in suburbiis Cantoniensibus." Loureiro's description conforms to the characters of the Linnaean species. The species is common in the vicinity of Canton. Schindler (Fedde Repert. Beih. **49**: 6. 1928) is in agreement with this interpretation.

Desmodium heterocarpum (Linn.) DC. Prodr. **2**: 337. 1825.

> *Hedysarum heterocarpon* Linn. Sp. Pl. 747. 1753.
>
> *Hedysarum polycarpon* Poir. in Lam. Encycl. **6**: 413. 1804.
>
> *Desmodium polycarpum* DC. Prodr. **2**: 334. 1825; Gagnep. in Lecomte Fl. Gén. Indo-Chine **2**: 586. 1920.
>
> *Desmodium ? hippocrepis* DC. Prodr. **2**: 338. 1825 (based on *Hippocrepis barbata* Lour.).
>
> *Hippocrepis barbata* Lour. Fl. Cochinch. 453. 1790, ed. Willd. 553. 1793, Anamese *cây xuong nguc*.
>
> *Grona repens* Lour. Fl. Cochinch. 459. 1790, ed. Willd. 561. 1793, Anamese *dâu lan rùug;* Moore in Journ. Bot. **63**: 286. 1925.

For *Hippocrepis barbata* Loureiro states: " Habitat agrestis in Cochinchina." From the description this is clearly the species currently interpreted as *Desmodium heterocarpum* (Linn.) DC. (*D. polycarpum* DC.). Schindler (Fedde Repert. Beih. **49**: 6. 1928) states: " *Hippocrepis barbata* = ? *D. siliquosum* s. *heterocarpum* "; the description " legumen . . . barbatum " would seem to indicate the latter. For *Grona repens* Loureiro states: " Habitat in collibus agrestibus Cochinchinae." *Grona* is currently recognized as a valid genus; yet Loureiro's type, and the specimen agrees with his description, is *Desmodium heterocarpum* (Linn.) DC. Bentham (Bentham & Hooker f. Gen. Pl. **1**: 535. 1865) who, in recognizing *Grona* as a valid genus with three Indian species, states: " Character Loureireanus cum plantis nostris convenit. Specimen suum in Herb. Banks. servatum, errore quodam nomine *Gronae repentis* signatum, est *Desmodium polystachyum* DC."; " *Desmodium polystachyum* DC." does not exist and *D. polystachyum* Schlecht. is a Mexican species, this

---

[99] *Desmodium* Desvaux (1813), conserved name, Vienna Code; older ones are *Meibomia* Adanson (1763) and *Pleurolobus* J. St. Hiliare (1812); to this list should be added *Grona* Loureiro (1790).

being a *lapsus calami* on the part of Bentham, for *D. polycarpum* DC. = *D. heterocarpum* DC. Moore's examination of Loureiro's type verifies this reduction of *Grona repens* Lour. to *D. polycarpum* DC., yet Schindler (Fedde Repert. Beih. **49**: 6. 1928) states: " *Grona repens* = ! *D. siliquosum* Or. in H. Mus. Brit. Vergl. S. Moore in: Journ. Bot. (1925) 286." But Loureiro describes the fruits as hirsute, while Burman describes those of *Hedysarum siliquosum* as glabrous. According to Trimen (Fl. Ceyl. **2**: 53. 1894) the oldest valid name for the species is that supplied by *Hedysarum heterocarpon* Linn. (Sp. Pl. 757. 1753). Regarding *Hedysarum siliquosum* Burm. f. Fedde (Repert. **23**: 113. 1926) states: " In Burmann's Herbar sind sowohl diese Art wie auch *D. heterocarpum* unter dem Namen *H. siliquosum* vertreten. Da er aber nur das Exemplar " ekkor meong " zitiert, kann nur dieses als Or. angesehen werden." Burman describes the fruits as glabrous, which is not a character of *D. heterocarpum* (Linn.) DC. His description may have been based on the characters of both species mentioned by Schindler. It may be noted here that Gagnepain (Lecomte Fl. Gén. Indo-Chine **2**: 586. 588) retains *Desmodium polycarpum* DC. and *D. heterocarpum* DC. as distinct species, distinguished, however, on minor and perhaps not constant characters. In the original description of the species listed in the synonymy as above given, the fruits are all described as hirsute, hispid, bearded, etc., but *Hedysarum siliquosum* Burm. f., discussed above, has glabrous fruits. Gagnepain also (*op. cit. 407*) admits *Grona repens* Lour. with a very brief description compiled from Loureiro, as an obscure species not seen by him, but *Grona* Loureiro being a synonym of *Desmodium* Desv., the three Indian species erroneously placed under Loureiro's generic name need renaming. Accordingly a new generic name, an anagram of *Grona*, is proposed.[100]

Desmodium pseudotriquetrum DC. in Ann. Sci. Nat. **4**: 100. 1825, Prodr. **2**: 326. 1825; Prain in Journ. As. Soc. Bengal **66**(2): 390. 1897.

Pteroloma pseudotriquetrum Schindl. in Fedde Repert. **20**: 272. 1924.

Hedysarum triquetrum (non Linn.) Lour. Fl. Cochinch. 448. 1790, ed. Willd. 547. 1793, Anamese *cay do dot.*

" Habitat in collibus *Son cuong* in Cochinchina." Schindler (Fedde Repert. Beih. **49**: 6. 1928) considers that Loureiro correctly interpreted the Linnaean species. He states: " Merrill identifiziert die Art mit *Pt*[*eroloma*] *pseudotriquetrum*, das jedoch bisher aus Cochinchina nicht bekannt geworden ist," this being my identification of 1919. If Loureiro's description of the plant as prostrate is correct, the form he had cannot possibly represent *Desmodium triquetrum* (Linn.) DC., which is always strictly erect. Indo-China is well

---

[100] **Nogra** nom. nov.

(*Grona* Bentham, non Loureiro)

Nogra grahami (Wall.) comb. nov.

Glycine grahami Wall. List no. 5513. 1832, *nomen nudum;* Benth. in Miq. Pl. Jungh. 233. 1852.

Grona grahami Baker in Hook. f. Fl. Brit. Ind. **2**: 191. 1876; Gagnep. in Lecomte Fl. Gén. Indo-Chine **2**: 406. 1916.

Nogra filicaulis (Kurz) comb. nov.

Grona filicaulis Kurz in Journ. As. Soc. Bengal **42**(2): 232. 1873; Baker in Hook. f. Fl. Brit. Ind. **2**: 191. 1876.

Nogra dalzellii (Baker) comb. nov.

Grona dalzellii Baker in Hook. f. Fl. Brit. Ind. **2**: 191. 1876.

Galactia simplicifolia Dalz. in Hook. Journ. Bot. Kew Gard. Miscel. **3**: 209. 1851; Dalz. & Gibs. Bombay Fl. 69. 1861, non Spreng.

within the range of *Desmodium pseudotriquetrum* DC. although the species has not yet been reported from that country.

Desmodium rubrum (Lour.) DC. Prodr. **2**: 327. 1825 (based on *Ornithopus ruber* Lour.).

> *Ornithopus ruber* Lour. Fl. Cochinch. 452. 1790, ed. Willd. 552. 1793, Anamese *cây ve ve duc.*
>
> *Desmodium carlesii* Schindler in Bot. Jahrb. **54**: 56. 1916; Gagnep. in Lecomte Fl. Gén. Indo-Chine **2**: 606. 1920.
>
> *Meibomia rubra* O. Ktz. Rev. Gen. Pl. 198. 1891 (based on *Ornithopus ruber* Lour.).

" Habitat agrestis in Cochinchina." In my original manuscript of 1919 I placed this as a doubtful synonym of *Desmodium gangeticum* DC., and Schindler (Fedde Repert. Beih. **49**: 6. 1928) is correct in not accepting the suggested reduction. A more attentive examination of the description, together with a study of the *Desmodium* material available from the classical locality, clearly indicates that *Ornithopus ruber* Lour. is the same as *Desmodium carlesii* Schindl. It is represented by *Clemens 3257, 3296*, and *Squires 387* from Tourane and Hue.

Desmodium triflorum (Linn.) DC. Prodr. **2**: 334. 1825.

> *Hedysarum triflorum* Linn. Sp. Pl. 749. 1753; Lour. Fl. Cochinch. 450. 1790, ed. Willd. 549. 1793, Chinese *sié thòi tsào.*

" Habitat agreste prope Cantonem Sinarum." The description applies closely to the Linnaean species, which is very common near Canton, except that the flowers are purplish, not white as described by Loureiro.

Phyllodium Desvaux

Phyllodium elegans (Lour.) Desv. in Mém. Soc. Linn. Paris **4**: 324. 1826 (based on *Hedysarum elegans* Lour.); Schindl. in Fedde Repert. Beih. **49**: 6. 1928.

> *Hedysarum elegans* Lour. Fl. Cochinch. 450. 1790, ed. Willd. 549. 1793, Chinese *hàp chiong tsào.*
>
> *Desmodium elegans* Benth. Fl. Hongk. 83. 1861 (based on *Hedysarum elegans* Lour.), non DC. (1825), nec Schlecht. (1838).
>
> *Dicerma elegans* DC. Prodr. **2**: 339. 1925 (based on *Hedysarum elegans* Lour.).
>
> *Zornia elegans* Pers. Syn. **2**: 318. 1807 (based on *Hedysarum elegans* Lour.).
>
> *Aeschynomene heterophylla* Lour. Fl. Cochinch. 446. 1790, ed. Willd. 544. 1793, Anamese *cây bot múoi,* non *Desmodium heterophyllum* DC., nec Hook. & Arn.

For *Hedysarum elegans* Loureiro states: " Habitat incultum prope Cantonem Sinarum "; and for *Aeschynomene heterophylla* he states: " Habitat inculta in via *Nhà ho* Cochinchinae borealis." As to the exact status of *Hedysarum elegans* there is no doubt, as it represents a species closely allied to *Phyllodium pulchellum* (Linn.) Desv. with characteristic suborbicular, distichous leafy bracts. As to *Aeschynomene heterophylla,* Seemann (Bot. Voy. Herald 359. 1856) states under *Dicerma (Desmodium) elegans:* " to which may be referred, without doubt, the *Aeschynomene heterophylla* of Loureiro, hitherto undetermined." Schindler, however (Fedde Repert. Beih. **49**: 6. 1928) considered that Seemann was wrong and that Loureiro's species cannot represent a *Desmodium.* Were the characters assigned to it by Loureiro strictly correct, I would agree with Schindler, but it is clear that Loureiro misinterpreted bract characters and described the characteristic inflorescence as representing upper imparipinnate leaves.   It seems clear that Loureiro's lower 3-foliolate (3-nate) leaves are the normal ones of *Phyllodium elegans* and that his expression describing the upper leaves as "impari-pinnata, foliolis sub-rotundis sessilibus" applies to the characteristic inflorescences of the same species, the distichous, suborbicular, sessile bracts being misinterpreted as odd-pinnate leaves with sessile, suborbicular leaflets!   No leguminous plant known, or any other plant for that matter, has the peculiar combination of true leaf characters as described by Loureiro for this species.   It is all the more curious that he should so misinterpret the bracts and inflorescences as pinnate leaves, as in the same work he correctly interprets the similar organs of the closely allied *Phyllodium pulchellum* (Linn.) Desv. as well as those of his *Hedysarum elegans*.   I therefore cannot agree with Schindler's statement: "denn nach Loureiros Beschreibung kann keine *Desmodiine* gemeint sein."   Here is a clear case where Loureiro's description must be reasonably interpreted, not to be taken as botanically correct, for no plant exists that presents the vegetative characters assigned by Loureiro to *Aeschynomene heterophylla;* they include the true leaves, correctly described, and the inflorescences with their characteristic bracts erroneously described as pinnate leaves.

Phyllodium pulchellum (Linn.) Desv. in Mém. Soc. Linn. Paris **4**: 324. 1826; Schindl. in Fedde Repert. Beih. **49**: 6. 1928.

Hedysarum pulchellum Linn. Sp. Pl. 747. 1753; Lour. Fl. Cochinch. 449. 1790, ed. Willd. 548. 1793, Chinese *a phô sien.*

Desmodium pulchellum Benth. Fl. Hongk. 83. 1861.

"Habitat incultum prope Cantonem Sinarum."   The description clearly applies to the well-known, widely distributed, and characteristic Linnaean species.

Uraria Desvaux

Uraria crinita (Linn.) Desv. Journ. Bot. **1**: 123. 1813; Schindl. in Fedde Repert. Beih. **49**: 6. 1928.

Hedysarum crinitum Linn. Mant. **1**: 102. 1767; Lour. Fl. Cochinch. 451. 1790, ed. Willd. 550. 1793, Anamese *cây dây mâm.*

Hedysarum lagopodioides (non Linn.) Lour. Fl. Cochinch. 450. 1790, ed. Willd. 549. 1793, Chinese *tsui fúm tsào.*

For *Hedysarum crinitum* Loureiro states: "Habitat in collibus agrestibus Cochinchinae."   His description manifestly applies to the Linnaean species.   For *H. lagopodioides* he states: "Habitat incultum circa Cantonem Sinarum."   Judging from the description and from Loureiro's comment regarding the illustration cited (Burm. f. Fl. Ind. *pl. 53. f. 2*): "Quae tamen figura non satis convenit cum nostra observatione tam numero, quam forma foliorum," I take this to be *Uraria crinita* Desv. rather than *Uraria lagopodioides* Desv.   Schindler, however (Fedde Repert. Beih. **49**: 6. 1928), considers that Loureiro correctly interpreted *Hedysarum lagopodioides* Linn.   It is to be noted however that Loureiro's description, in its "folia 3–5 nata, ovato-lanceolata" and "pedunculis [pedicellis] . . . longis, inflexis," applies to *Uraria crinita* Desv. and not at all to *U. lagopodioides* Desv.   The latter species was apparently described by Loureiro as *Trifolium globosum* and *T. mel. indicum.*

Uraria lagopodioides (Linn.) Desv. in Mém. Soc. Linn. Paris **4**: 309. 1826 (*Urania*).

Hedysarum lagopodioides Linn. Sp. Pl. 1198. 1753.

Uraria lagopoides DC. Prodr. **2**: 324. 1825.

Trifolium mel. indicum (non Linn.) Lour. Fl. Cochinch. 444. 1790, ed. Willd. 541. 1793, Chinese *sam pa lim*.

Trifolium globosum (non Linn.) Lour. Fl. Cochinch. 444. 1790, ed. Willd. 542. 1793, Chinese *tsin li quong*.

Both species, as very imperfectly described by Loureiro, were from Canton, " in agris "; the type of *Hedysarum lagopodioides* Linn. was a specimen collected by Osbeck, near Canton. I place both of Loureiro's species as synonyms of *Uraria lagopodioides* Desv. with confidence, partly from the meager descriptive data, partly because of the fact that the Linnaean species is very common about Canton, and Loureiro, who would scarcely have overlooked it, otherwise does not describe it, if I am correct in my disposition of the species he did describe under *Hedysarum lagopodioides* Linn.; and partly because no other leguminous plants are known from Kwangtung Province whose characters even approxi· mate those given by Loureiro for the two forms he so inadequately described. Schindler (Fedde Repert. Beih. **49**: 5. 1928) in mentioning my suggested manuscript reductions, states regarding both: " die Bestimmung erscheint sehr zweifelhaft." In my mind there is no doubt as to the correctness of these reductions.

Lourea Necker

Lourea vespertilionis (Linn. f.) Desv. Journ. Bot. **1**: 122. *pl. 5. f. 18.* 1813; Gagnep. in Lecomte Fl. Gén. Indo-Chine **2**: 533. 1920; Schindl. in Fedde Repert. Beih. **49**: 6. 1928.

Hedysarum vespertilionis Linn. f. Suppl. 331. 1781; Lour. Fl. Cochinch. 447. 1790, ed. Willd. 546. 1793, Anamese *cây hô diep*.

" Habitat in Cochinchina, a Siamensi regno oriundum. Colitur ob pulchritudinem." Loureiro notes that he sent this to Europe in 1774, and this material no doubt yielded the specimen on which the original description of *Hedysarum vespertilionis* was based, the collection being cited by Linnaeus f. as " Habitat in Regno Cochin-China. Io. de Lou- rei." A specimen from Loureiro is preserved in the herbarium of the British Museum but there is no specimen in the Linnaean herbarium.

Lourea obcordata (Poir.) Desv. Journ. Bot. **1**: 122. 1813; DC. Prodr. **2**: 324. 1825; Gagnep. in Lecomte Fl. Gén. Indo-Chine **2**: 536. 1920 (incl. var. *reniformis* Gagnep. *l.c.* based on *Hedysarum reniforme* Lour.).

Hedysarum obcordatum Poir. in Lam. Encycl. **6**: 425. 1804.

Hedysarum reniforme (non Linn.) Lour. Fl. Cochinch. 447. 1790, ed. Willd. 545. 1793, Chinese *lô im tsào*.

Ploca humilis Lour.[1] ex Gomes in Mem. Acad. Sci. Lisb. Cl. Sci. Mor. Pol. Bel.-Let. n.s. **4**(1): 30. 1868.

Lourea reniformis DC. Prodr. **2**: 324. 1825; Schindl. in Fedde Repert. Beih. **49**: 5. 1928 (based on *Hedysarum reniforme* Lour.).

Hedysarum loureirii Spreng. Syst. Cur. Post. 292. 1827 (based on *Hedysarum reni- forme* Lour.).

---

[1] A Loureiro herbarium name here first published by Gomes.

" Habitat suburbia Cantonis Sinarum." A widely distributed and well-known Asiatic species. A specimen from Loureiro is preserved in the herbarium of the Paris Museum; Poiret's type was from Java. Even if the more northern form represents a variety, I do not see how Gagnepain's varietal name can stand, as its name-bringing synonym was invalid; consequently the binomial *Lourea reniformis* (Lour.) DC. is invalid.

Dalbergia [2] Linnaeus f.

Dalbergia pinnata (Lour.) Prain in Ann. Bot. Gard. Calcutta **10**(1): 48. 1904; Merr. in Philip. Journ. Sci. Bot. **5**: 96. 1910, Enum. Philip. Fl. Pl. **2**: 296. 1923.

Derris pinnata Lour. Fl. Cochinch. 432. 1790, ed. Willd. 526. 1793, Anamese *cham bia ăn tlâu.*

Dalbergia tamarindifolia Roxb. Hort. Beng. 53. 1814, *nomen nudum*, Fl. Ind. ed. 2, **3**: 233. 1832; Gagnep. in Lecomte Fl. Gén. Indo-Chine **2**: 485. 1916.

Amerimnon pinnatum O. Ktz. Rev. Gen. Pl. 159. 1891 (based on *Derris pinnata* Lour.).

" Habitat in sylvis Cochinchinae." Loureiro's type is preserved in the herbarium of the British Museum. The species is identical with the widely distributed Indo-Malaysian one currently known as *Dalbergia tamarindifolia* Roxb. On a strict interpretation this is the type of the genus *Derris*, not because the species is the first one listed under *Derris*, but because it agrees with the generic characters in having oblong, membranaceous, 1-seeded fruits, the generic name *Derris* being derived from the thin fruit character. Historically it is manifest that the original description of the genus was written in Indo-China and strictly on the basis of *Derris pinnata*. The description of the second species, *D. trifoliata*, was written during or after Loureiro's sojourn in Canton. He proceeded to Canton in 1778. To avoid a very great number of changes in nomenclature and much confusion, I prefer to consider the second species, *Derris trifoliata* Lour. (= *D. uliginosa* Roxb.), to represent the standard species of *Derris*, in spite of the fact that the genus *Derris* as actually defined by Loureiro is *Dalbergia*.

Pongamia [3] Ventenat

Pongamia pinnata (Linn.) Merr. Interpret. Herb. Amb. 271. 1917, Enum. Philip. Fl. Pl. **2**: 298. 1923.

Cytisus pinnatus Linn. Sp. Pl. 741. 1753.

Robinia mitis Linn. Sp. Pl. ed. 2, 1044. 1763.

Pongamia glabra Vent. Jard. Malm. **1**: 28. *pl. 28.* 1803; Gagnep. in Lecomte Fl. Gén. Indo-Chine **2**: 441. 1916.

Pterocarpus flavus Lour. Fl. Cochinch. 431. 1790, ed. Willd. 525. 1793, Anamese *hùynh bá,* Chinese *hoâm pě mŏ.*

" Habitat in sylvis Sinensibus." In spite of Loureiro's description of the flowers as yellow, his species is undoubtedly referable to *Pongamia pinnata* (Linn.) Merr., one not uncommon near Canton; the white to pink flowers turn yellowish in age. *Malaparius* Rumph. (Herb. Amb. **3**: 183. *pl. 117*), cited by Loureiro as representing his species, is the Linnaean species.

[2] *Dalbergia* Linnaeus f. (1781), conserved name, Vienna Code; older ones are *Amerimnon* P. Browne (1756), *Ecastaphyllum* P. Browne (1756) and ? *Acouroa* Aublet (1775).

[3] *Pongamia* Ventenat (1803), conserved name, Vienna Code; an older one is *Galedupa* Lamarck (1786), quoad descr.

Derris [4] Loureiro

Derris trifoliata Lour. Fl. Cochinch. 433. 1790, ed. Willd. 526. 1793, Chinese *săn leâo táu.*
  Robinia uliginosa Roxb. ex Willd. Sp. Pl. **3**: 1133. 1800.
  Derris uliginosa Benth. in Miq. Pl. Jungh. 252. 1852; Gagnep. in Lecomte Fl. Gén.
  Indo-Chine **2**: 453. 1916.
  Pterocarpus trifoliatus O. Ktz. Rev. Gen. Pl. 203. 1891 (based on *Derris trifoliata*
  Lour.).

  " Habitat in sylvis provinciae Cantoniensis Sinarum." De Candolle (Prodr. **2**: 415.
1825) states that he examined Loureiro's type in the herbarium of the Paris Museum, and
Doctor Gagnepain informs me that the specimen is identical with the species currently
known as *Derris uliginosa* Benth. Prain (Journ. As. Soc. Bengal **66**(2): 458. 1897) did not
accept Loureiro's binomial in place of Roxburgh's, as he thought that the description did
not apply sufficiently well to the latter. The species is very abundant, occurring within
the influence of salt or brackish water, and often gregarious along tidal streams throughout
the Indo-Malaysian region, and is common near Canton. I interpret this as the standard
species of the genus *Derris* in spite of the fact that it is the second species cited, and notwithstanding the fact that the generic name was derived from the fruit characters of the
first species described, *Derris pinnata* Lour. = *Dalbergia pinnata* (Lour.) Prain, while the
generic description of the fruits applies to the latter and not to the present species. See
Dalbergia pinnata (Lour.) Prain (p. 205).

Derris heptaphylla (Linn.) Merr. Interpret. Herb. Amb. 273. 1917, Enum. Philip. Fl. Pl.
  2: 299. 1923.
  Sophora heptaphylla Linn. Sp. Pl. 373. 1753, excl. syn. Plukenet.
  Pongamia sinuata Wall. List no. 5911. 1832, *nomen nudum.*
  Derris sinuata Thw. Enum. Pl. Zeyl. 93. 1859; Benth. in Journ. Linn. Soc. Bot. **4**:
  Suppl. 113. 1860.
  Aspalathus arborea Lour. Fl. Cochinch. 431. 1790, ed. Willd. 524. 1793, Anamese *cây
  kùa gà.*
  Semetor arborea Raf. Sylva Tellur. 69. 1838 (based on *Aspalathus arborea* Lour.).

  " Habitat in sylvis Cochinchinae." Loureiro's species is not accounted for by Gagnepain in his treatment of the Leguminosae of Indo-China (Lecomte Fl. Gén. Indo-Chine **2**:
57–613. 1913–20); nor have I found any suggested reduction of it by other authors. This
settles the status of the genus *Semetor* Rafinesque which was based wholly on the description of *Aspalathus arborea* Lour.; De Dalla Torre & Harms (Gen. Siphon. 225. 1901)
erroneously place *Semetor* Raf. as a synonym of *Aspalathus* Linnaeus. The identification
has been worked out in part by the process of elimination. The only possibilities that I
know of, assuming Loureiro's description to be reasonably accurate, are *Derris*, *Millettia*,
and *Pongamia*. In these genera the description best agrees with the characters of *Derris
heptaphylla* (Linn.) Merr., which incidentally must be distinctly common in the vicinity of
Hue, as Mrs. Clemens made six collections of it near there and Tourane. The description
does not agree in all particulars, the calyces are not 5-fid (but this character was apparently
repeated from the generic description of *Aspalathus* by Loureiro); the leaflets are not sessile;

---

[4] *Derris* Loureiro (1790), conserved name, Vienna Code; older ones are *Salken* Adanson (1763), *Solori*
Adanson (1763), *Deguelia* Aublet (1775) and *Cylizoma* Necker (1790).

and the plant is a large liana rather than a medium-sized tree; yet Loureiro describes the branches as " debilibus, reclinatis " indicating that he had in mind a semiscandent form. Mrs. Clemens records her number *4234* as a suberect shrub.   It is suspected that this form was included by Gagnepain (Lecomte Fl. Gén. Indo-Chine 2: 450. 1916) in *Derris thyrsiflora* Benth.

Derris sp.
> *? Anthyllis indica* Lour. Fl. Cochinch. 429. 1790, ed. Willd. 522. 1793, Anamese *cây káoc sát.*

" Habitat in montibus Cochinchinae."   Gagnepain makes no attempt to place Loureiro's species.   I merely suggest *Derris* as a possibility because of his description of the fruits as 2-seeded.   Some totally different genus may be represented.   In my original manuscript of 1919 I placed this as a possible *Millettia*, but the indicated seed characters do not conform.

Pisum Tournefort

Pisum sativum Linn. Sp. Pl. 727. 1753; Lour. Fl. Cochinch. 443. 1790, ed. Willd. 539. 1793, Anamese *dâu tlòn.*

" Habitat in China, & Cochinchina, non frequens, nec fortasse indigena."   The description applies to a form of the common pea, *Pisum sativum* Linn.

Vicia Tournefort

Vicia faba Linn. Sp. Pl. 737. 1753; Lour. Fl. Cochinch. 443. 1790, ed. Willd. 540. 1793, Anamese *tàm dâu*, Chinese *sàm têu.*

" Habitat culta in China, raro in Cochinchina."   The Linnaean species, which is widely cultivated in China, was correctly interpreted by Loureiro; its Cantonese name on recent collections appears as *chum tau.*

Abrus Linnaeus

Abrus precatorius Linn. Syst. ed. 12, 472. 1767; Lour. Fl. Cochinch. 428. 1790, ed. Willd. 520. 1793, Anamese *can thao do hôt.*
Glycine abrus Linn. Sp. Pl. 753. 1753.

" Habitat in dumetis, & sepibus Cochinchinae."   The description clearly applies to the very common and widely distributed Linnaean species.   *Abrus frutex* Rumph. (Herb. Amb. **5**: 57. *pl. 32*), cited by Loureiro, after Linnaeus, as a synonym, is correctly placed.

Clitoria Linnaeus

Clitoria ternatea Linn. Sp. Pl. 753. 1753; Lour. Fl. Cochinch. 454. 1790, ed. Willd. 555. 1793, Anamese *cây dâu biéc;* Gagnep. in Lecomte Fl. Gén. Indo-Chine 2:.310. 1916.

" Habitat agrestis, cultaque in Cochinchinae."   The Linnaean species, one of American origin but widely distributed in the Old World tropics, was correctly interpreted by Loureiro.   *Flos coeruleus* Rumph. (Herb. Amb. **5**: 56. *pl. 31*), cited by Loureiro as a synonym, after Linnaeus, is correctly placed.

Glycine Linnaeus

Glycine soja (Linn.) Sieb. & Zucc. in Abh. Akad. Muench. 4(2): 119. 1843; Gagnep. in Lecomte Fl. Gén. Indo-Chine 2: 398. 1916.

Dolichos soja Linn. Sp. Pl. 727. 1753; Lour. Fl. Cochinch. 441. 1790 (*D. soia*), ed. Willd. 537. 1793, Anamese *dâu nanh*, Chinese *hoâm têu*.

Phaseolus max Linn. Sp. Pl. 725. 1753.

Soja hispida Moench. Meth. 153. 1794.

Glycine hispida Max. in Bull. Acad. Sci. St. Pétersb. 18: 398. 1873.

Soja max Piper in Journ. Am. Soc. Agron. 6: 84. 1914.

Glycine max Merr. Interpret. Herb. Amb. 274. 1917.

" Habitat in Cochinchina, & in China frequenter cultus." Loureiro clearly described the soy bean. *Cadelium* Rumph. (Herb. Amb. **5**: 388. *pl. 140*) is correctly placed as a synonym. A specimen from Loureiro is preserved in the herbarium of the British Museum. Piper claims that the specific name *max* is the oldest valid one for the species whether considered under *Glycine* or *Soja*. However it has only page priority which is not recognized by the International Code of Botanical Nomenclature. Historically *Glycine* Linnaeus (Gen. 349. 1737) was based wholly on *Apios* Boerh. which is *Apios tuberosa* Moench. In the fifth edition of the Genera Plantarum (1754) the generic diagnosis remains unchanged, but in the Species Plantarum (1753) Linnaeus placed 8 species under *Glycine*. One of these, *Glycine javanica* Linn., is congeneric with *Glycine soja* (Linn.) S. & Z. and this I arbitrarily accept as the standard species to save *Glycine* in the sense in which most botanists have used it in the past 150 years, this in spite of the fact that *Glycine javanica* was certainly unknown to Linnaeus in 1737. The other species of *Glycine* as the genus was constituted by Linnaeus in 1753 represent seven different genera, *Apios*, *Wisteria*, *Krauhnia*, *Abrus*, *Rhynchosia*, *Amphicarpaea*, and *Fagelia*.

Erythrina Linnaeus

Erythrina variegata Linn. in Stickman Herb. Amb. 10. 1754, Amoen. Acad. **4**: 122. 1759, var. **orientalis** (Linn.) Merr. Interpret. Herb. Amb. 276. 1917, Enum. Philip. Fl. Pl. 2: 306. 1923.

Erythrina corallodendron Linn. var. *orientalis* Linn: Sp. Pl. 706. 1753.

Tetradapa javanorum Osbeck Dagbok Ostind. Resa 93. 1757.

Erythrina indica Lam. Encycl. 2: 391. 1788; Gagnep. in Lecomte Fl. Gén. Indo-Chine 2: 415. 1916.

Erythrina corallodendron (non Linn.) Lour. Fl. Cochinch. 427. 1790, ed. Willd. 519. 1793, Anamese *cây boung, thích doung bì*, Chinese *tum yĕ xú*.

Erythrina loureiri G. Don Gen. Syst. 2: 372. 1832 (based on *Erythrina corallodendron* Lour.).

" Habitat agrestis, cultaque in Cochinchina, & China Australi." Loureiro's description manifestly applies to the very common Indo-Malaysian species currently known as *Erythrina indica* Lam. I interpret the type of *Erythrina corallodendron* Linn. as American, *i.e.*, the var. *occidentalis* Linn. If variegated leaves be interpreted as a " monstrosity," then under the international code *Erythrina variegata* Linn. would have no standing, as the Rumphian illustration on which it is based is not the uncommon Malaysian form with variegated leaves.   Before invoking this rule I would first prefer to learn how many scores of binomials have been proposed on the basis of forms of plants with variegated leaves, many of which are universally accepted as valid.

Erythrina fusca Lour. Fl. Cochinch. 427. 1790, ed. Willd. 519. 1793, Anamese *cây son dong.*

  Erythrina ovalifolia Roxb. Hort. Beng. 53. 1814, *nomen nudum*, Fl. Ind. ed. 2, **3**: 254. 1832; Gagnep. in Lecomte Fl. Gén. Indo-Chine **2**: 417. 1916.

  Corallodendron fuscum O. Ktz. Rev. Gen. Pl. 172. 1891 (based on *Erythrina fusca* Lour.).

  "Habitat in Cochinchina ad ripas fluminum spontanea." This seems clearly to be identical with the widely distributed Indo-Malaysian species currently known as *Erythrina ovalifolia* Roxb. *Gelala aquatica* Rumph. (Herb. Amb. **2**: 235. *pl. 78*), cited by Loureiro as representing his species, is *Erythrina ovalifolia* Roxb. = *E. fusca* Lour.

Mucuna [5] Adanson

Mucuna cochinchinensis (Lour.) A. Cheval. in Bull. Agr. Inst. Sci. Saigon **1**: 91. 1919 (based on *Marcanthus cochinchinensis* Lour.); Merr. in Philip. Journ. Sci. **15**: 242. 1919, Enum. Philip. Fl. Pl. **2**: 307. 1923.

  Marcanthus cochinchinensis Lour. Fl. Cochinch. 461. 1790, ed. Willd. 563. 1793, Anamese *dâu mèo.*

  Carpopogon niveum Roxb. Hort. Beng. 54. 1814, *nomen nudum*, Fl. Ind. ed. 2, **3**: 285. 1832.

  Mucuna nivea DC. Prodr. **2**: 406. 1825.

  Mucuna utilis Wall. ex Wight Ic. **1**: *pl. 280.* 1838–39; Gagnep. in Lecomte Fl. Gén. Indo-Chine **2**: 321. 1916.

  Macranthus cochinchinensis Poir. in Lam. Encycl. Suppl. **3**: 569. 1813 (based on *Marcanthus cochinchinensis* Lour.).

  Stizolobium niveum O. Ktz. Rev. Gen. Pl. **1**: 208. 1891.

  "Habitat cultus in Cochinchina." Loureiro's specimen, in the herbarium of the British Museum, consists of leaves only. Piper, who examined it in 1912, states: "It might be any of the species allied to *Stizolobium niveum* but is probably *niveum.*" In 1919 Doctor Chevalier kindly sent me mature pods of the plant known at Hue, Indo-China, as *dâu mèo*, and a series of duplicate botanical specimens prepared from plants grown in Manila from these seeds were widely distributed to the larger herbaria of the world. This is *Marcanthus cochinchinensis* Lour., which in all respects is the species currently known as *Mucuna nivea* DC. Gagnepain in his treatment of the Leguminosae of Indo-China (Lecomte Fl. Gén. Indo-Chine **2**: 57–613. 1913–1920) does not mention Loureiro's genus and species.

Mucuna nigricans (Lour.) Steud. Nomencl. ed. 2, **2**: 163. 1841 (based on *Citta nigricans* Lour.); Merr. in Philip. Journ. Sci. Bot. **5**: 116. 1910, Enum. Philip. Fl. Pl. **2**: 309. 1923.

  Citta nigricans Lour. Fl. Cochinch. 456. 1790, ed. Willd. 557. 1793, Anamese *cây boung mât.*

  Carpopogon imbricatum Roxb. Hort. Beng. 54. 1814, *nomen nudum.*

  [5] *Mucuna* Adanson (1763), conserved name, Vienna Code; older ones are *Zoophthalmum* P. Browne (1756) and *Stizolobium* P. Browne (1756).

Mucuna imbricata DC. Prodr. 2: 406. 1825; Gagnep. in Lecomte Fl. Gén. Indo-Chine 2: 320. 1916.

Mucuna gigantea DC. var. *nigricans* DC. Prodr. 2: 405. 1825 (based on *Citta nigricans* Lour.).

Zoophthalmum nigricans Prain in Journ. Asiat. Soc. Bengal 66(2): 65. 1897, in syn. (based on *Citta nigricans* Lour.).

"Habitat inter sepes in Cochinchina." Loureiro's description is very definite and if it does not apply to *Mucuna imbricata* DC., then it represents a very closely allied species. It is to be noted that Gagnepain (Lecomte Fl. Gén. Indo-Chine 2: 315–324. 1916) admits *Mucuna imbricata* DC. on the basis of Loureiro's record, citing actual specimens only under the var. *bispicata* Gagnep. It would seem that more field work is desirable from which it will be possible more definitely to determine the exact status of Loureiro's species. *Lobus litoralis* Rumph. (Herb. Amb. 5: 10. *pl. 6*), cited by Loureiro, must be excluded as it represents *Mucuna gigantea* (Willd.) DC., a species with smooth fruit-valves, while Loureiro definitely describes his species as having plicate valves: "cortice exterius diviso in cellulas subquadratas, oblique formatas ex membranis verticalibus, ordinatis." A fragmentary specimen from Loureiro, with neither flowers nor fruits, is preserved in the herbarium of the British Museum.

Mucuna pruriens (Linn.) DC. Prodr. 2: 405. 1825; Merr. Interpret. Herb. Amb. 277. 1917.

Dolichos pruriens Linn. in Stickman Herb. Amb. 23. 1754, Amoen. Acad. 4: 132. 1759, Syst. ed. 10, 1162. 1759; Lour. Fl. Cochinch. 438. 1790, ed. Willd. 533. 1793, Anamese *dâu ngúa;* Gagnep. in Lecomte Fl. Gén. Indo-Chine 2: 323. 1916.

"Habitat ad ripas fluminum in Cochinchina, omnibus odiosus." The description manifestly applies to the common Indo-Malaysian species typified by *Cacara pruritus* Rumph. (Herb. Amb. 5: 393. *pl. 142*) which Loureiro cites, after Linnaeus, as representing the species.

Pueraria de Candolle

Pueraria montana (Lour.) comb. nov.

Dolichos montanus Lour. Fl. Cochinch. 440. 1790, ed. Willd. 536. 1793, Anamese *săn rùng.*

Pachyrhizus montanus DC. Prodr. 2: 402. 1825 (based on *Dolichos montanus* Lour.).

Stizolobium montanum Spreng. Syst. 3: 252. 1826 (based on *Dolichos montanus* Lour.).

Pueraria tonkinensis Gagnep. in Not. Syst. 3: 202. 1916, Lecomte Fl. Gén. Indo-Chine 2: 250. 1916.

Zeydora agrestis Lour.[6] ex Gomes in Mem. Acad. Sci. Lisb. Cl. Sci. Pol. Mor. Bel.-Let. n.s. 4(1): 27. 1868.

"Habitat in sylvis montanis Cochinchinae." Loureiro's species has been referred to *Pueraria phaseoloides* Benth. but his description does not apply to Bentham's species; it does, however, apply to the recently described *Pueraria tonkinensis* Gagnep. Gagnepain does not account for *Dolichos montanus* Lour., or any of the synonyms based upon it, in his treatment of the Leguminosae of Indo-China (Lecomte Fl. Gén. Indo-Chine 2: 57–613. 1913–20).

---

[6] A Loureiro herbarium name here first published by Gomes.

Pueraria thomsoni Benth. in Journ. Linn. Soc. Bot. **9**: 122. 1867; Gagnep. in Lecomte
Fl. Gén. Indo-Chine **2**: 251. *f. 25. 1–7.* 1916.

Dolichos spicatus Grah. in Wall. List no. 5557. 1832, *nomen nudum*, non Koenig.
Pachyrhizus trilobus DC. Prodr. **2**: 402. 1825 (based on *Dolichos trilobus* Lour.).
Dolichos grandifolius Grah. in Wall. List no. 5556. 1832, *nomen nudum*.
Pueraria triloba Mak. Iinuma Somoku-Dzusetsu, ed. 3, **3**: 954. *pl. 22.* 1912 (quoad
syn. Lour.).
Dolichos trilobus (non Linn.) Lour. Fl. Cochinch. 439. 1790, ed. Willd. 535. 1793,
Anamese *săn deai cu, cat căn*, Chinese *kēn cŏ*.

"Habitat cultus in Cochinchina, & China." This form has currently been referred
to *Pueraria thunbergiana* (Sieb. & Zucc.) Benth. (*Dolichos hirtus* Thunb., non *Pueraria
hirta* Kurz), a species originally described from Japan, and very closely allied to *P. thomsoni*
Benth. The geographic range clearly indicates the latter species for the form that Loureiro
described. I interpret Loureiro's statement "stipulis bicornibus" to mean a single lobe
of the stipule below and above its insertion, not 2-lobed below the insertion as in the Chinese
species *P. calycina* Franch., *P. bicalcarata* Gagnep. and *P. edulis* Pamh. *Pueraria triloba*
Makino, non Kurz, is an invalid name and was apparently intended for the Japanese form
currently known as *Pueraria thunbergiana* Benth. or for a segregate from that species;
I have not seen the work in which Makino's binomial is published. The reductions of
Dolichos spicatus Grah. and *D. grandifolius* Grah. were made by Baker.

Canavalia [7] de Candolle

Canavalia ensiformis (Linn.) DC. Prodr. **2**: 404. 1825; Gagnep. in Lecomte Fl. Gén. Indo-
Chine **2**: 260. 1916.

Dolichos ensiformis Linn. Sp. Pl. 725. 1753; Lour. Fl. Cochinch. 437. 1790, ed. Willd.
531. 1793, Anamese *dâu rua*, Chinese *tāo téu.*
Dolichos gladiatus Jacq. Coll. **2**: 276. 1788.
Canavalia loureiri G. Don Gen. Syst. **2**: 363. 1832 (based on *Dolichos ensiformis* Lour.).

"Habitat in Cochinchina, & China: coliturque pro umbraculo contra solis calorem:
cibo minus aptus." Loureiro's description applies unmistakably to the occasionally
cultivated form with very large pods, clearly described by Sloane as *Phaseolus maximus
perennis* etc. (Nat. Hist. Jamaica **1**: 177. *pl. 115.* 1707) as having pods a foot and a half
long and an inch broad. In view of the fact that no specimen exists in the Linnaean
herbarium, I interpret the Linnaean species from Sloane's description and illustration.
Dolichos gladiatus Jacq. = *Canavalia gladiata* DC. is manifestly a synonym. *Lobus
machaeroides* Rumph. (Herb. Amb. **5**: 376. *pl. 135. f. 1*), cited by Loureiro as a synonym,
represents the same species.

Cajanus [8] de Candolle

Cajanus cajan (Linn.) Millsp. in Field Columb. Mus. Bot. Ser. **2**: 53. 1900, 436. 1914;
Druce in Rept. Bot. Exch. Club Brit. Isles **4**: 611. 1917; Merr. Interpret. Herb.
Amb. 282. 1917, Enum. Philip. Fl. Pl. **2**: 314. 1923.

[7] *Canavalia* de Candolle (1825), conserved name, Brussels Code; older ones are *Canavali* Adanson
(1763) and *Clementea* Cavanilles (1804).

[8] *Cajanus* de Candolle (1813), conserved name, Brussels Code; an older one is *Cajan* Adanson (1763).

Cytisus cajan Linn. Sp. Pl. 739. 1753; Lour. Fl. Cochinch. 462. 1790, ed. Willd. 565. 1793, Anamese *dâu sắng*, Chinese *xān téu kēn, sần táu can.*

Cajanus indicus Spreng. Syst. **3**: 248. 1826; Gagnep. in Lecomte Fl. Gén. Indo-Chine **2**: 278. 1916.

Cajanum thora Raf. Sylva Tellur. 25. 1838.

Cytisus pseudocajan Jacq. Hort. Vind. **2**: 54. *pl. 119.* 1772.

Cajan inodorum Medic. in Vorles. Churpf. Phys. Ges. **2**: 363. 1787.

" Habitat incultus, cultusque ad ordinandas sepes in Cochinchina, & in China." Loureiro's specimen in the herbarium of the Paris Museum represents the Linnaean species and his ample description applies to it. *Phaseolus balicus* Rumph. (Herb. Amb. **5**: 377. *pl. 135. f. 2*) is correctly placed as a synonym.

Dunbaria Wight & Arnott

Dunbaria rotundifolia (Lour.) Merr. in Philip. Journ. Sci. **15**: 242. 1919 (based on *Indigofera rotundifolia* Lour.).

Indigofera rotundifolia Lour. Fl. Cochinch. 458. 1790, ed. Willd. 559. 1793, Chinese *ô tam sin.*

Dolichos conspersus Grah. in Wall. List no. 5542. 1832, *nomen nudum.*

Dunbaria conspersa Benth. in Miq. Pl. Jungh. 241. 1852.

Dolichos punctatus Wight & Arn. Prodr. 247. 1834.

Dunbaria punctata Benth. in Miq. Pl. Jungh. 242. 1852.

" Habitat agrestis circa Cantonem Sinarum." The Chinese species currently known as *Dunbaria conspersa* Benth. is not uncommon near Canton and is the only Kwangtung species that remotely agrees with Loureiro's description; the only discrepancy is Loureiro's description of the pods as 2-seeded. The Cantonese name on recently collected material appears as *chin tang.*

Dunbaria sp.?

Indigofera bufalina Lour. Fl. Cochinch. 458. 1790, ed. Willd. 559. 1793, Anamese *dâu tlâu.*

" Habitat in dumetis Cochinchinae." The description manifestly applies to no *Indigofera*, but apparently a species of *Dunbaria* or perhaps *Atylosia (Cantharospermum)* is represented by it. I am unable to suggest its further identification from the data and material at present available. Gagnepain does not mention Loureiro's species in his consideration of the Leguminosae on Indo-China. But for Loureiro's description of the leaflets as glabrous, I should be inclined to refer his species to *Atylosia crassa* Grah. as interpreted by Gagnepain (Lecomte Fl. Gén. Indo-Chine **2**: 280. 1916).

Rhynchosia [9] Loureiro

Rhynchosia volubilis Lour. Fl. Cochinch. 460. 1790, ed. Willd. 562. 1793, Chinese *chio táu.*

" Habitat inculta prope Cantonem Sinarum." A well-known species, abundant in thickets near Canton. Loureiro's type is preserved in the herbarium of the Paris Museum.

[9] *Rhynchosia* Loureiro (1790), conserved name, Vienna Code; an older one is *Dolicholus* Medikus (1787).

Rhynchosia densiflora (Roth) DC. Prodr. 2: 386. 1825; Baker in Oliv. Fl. Trop. Afr. 2: 222. 1871.

Glycine densiflora Roth Nov. Pl. Sp. 348. 1821.

Trifolium volubile Lour. Fl. Cochinch. 445. 1790, ed. Willd. 542. 1793, non *Rhynchosia volubilis* Lour.

" Habitat ad litora Africae orientalis." Among all the Leguminosae recorded from tropical East Africa, the only species that agrees at all with Loureiro's characters is *Rhynchosia densiflora* DC. and I believe *Trifolium volubile* Lour. to be safely reduced here. Baker does not account for Loureiro's binomial in his treatment of the Leguminosae of tropical Africa (Oliver Fl. Trop. Afr. 2: 1–364. 1871) nor has any author, as far as I know, suggested any reduction of it other than Willdenow, who thought that it might be near the form described by Walter as Anonymos no. 294.

Eriosema de Candolle

Eriosema chinense Vogel in Nov. Act. Acad. Nat. Cur. 19: Suppl. 1: 31. 1843.

Dolichos biflorus (non Linn.) Lour. Fl. Cochinch. 441. 1790 (err. typ. *biblorus*), ed. Willd. 537. 1793, Chinese *sān cŭ*.

" Habitat agrestis prope Cantonem Sinarum." Loureiro's description does not conform at all with the characters of the Linnaean species, yet Hemsley (Journ. Linn. Soc. Bot. 23: 194. 1887) solely on Loureiro's authority, admits the Linnaean species as a Chinese one. *Dolichos biflorus* as described by Loureiro is manifestly *Eriosema chinense* Vog., which is not uncommon in the vicinity of Canton. *Dolichos biflorus* Linn. is a totally different plant (Baker in Hook. f. Fl. Brit. Ind. 2: 210. 1876). Earlier descriptions than that of Vogel are supplied by *Crotalaria tuberosa* Ham. (D. Don Prodr. Fl. Nepal. 241. 1825) and *Pyrrhotrichia tuberosa* W. & A. (Prodr. 238. 1834) but the specific name *tuberosa* is preoccupied in *Eriosema* by *E. tuberosum* Richard and *E. tuberosum* Hochst., of Africa. Desvaux (Ann. Sci. Nat. 9: 421. 1826), by error, gives the generic name as *Euriosma*.

Flemingia Roxburgh

Flemingia macrophylla (Willd.) O. Ktz. ex Prain in Journ. As. Soc. Bengal. 66(2): 440. 1897, *in nota;* Merr. in Philip. Journ. Sci. Bot. 5: 130. 1910.

Crotalaria macrophylla Willd. Sp. Pl. 3: 982. 1800.

Flemingia congesta Roxb. ex Ait. Hort. Kew. ed. 2, 4: 349. 1812; Gagnep. in Lecomte Fl. Gén. Indo-Chine 2: 302. 1916.

Ervum hirsutum (non Linn.) Lour. Fl. Cochinch. 461. 1790, ed. Willd. 563. 1793, Anamese *cây deái chôn.*

Ervum cochinchinense Pers. Syn. 2: 309. 1807 (based on *Ervum hirsutum* Lour.).

" Habitat incultum, per agros dispersum in Cochinchina." Loureiro's description applies closely to the common and widely distributed *Flemingia macrophylla* (Willd.) O. Ktz. currently known as *F. congesta* Roxb.

Phaseolus (Tournefort) Linnaeus

Phaseolus aureus Roxb. Hort. Beng. 55. 1814, *nomen nudum,* Fl. Ind. ed. 2, 3: 297. 1832; Piper in U. S. Dept. Agr. Bull. 119: 16. *pl. 4.* 1914; Merr. Enum. Philip. Fl. Pl. 2: 318. 1923.

Phaseolus mungo Gagnep. in Lecomte Fl. Gén. Indo-Chine 2: 231. 1916, non Linn.

Phaseolus radiatus (non Linn.) Lour. Fl. Cochinch. 435. 1790, ed. Willd. 529. 1793, Anamese *dâu xanh, luc dâu,* Chinese *liu téu.*

" Habitat, & colitur abundanter in Cochinchina, & China." Loureiro's description applies to the green or golden gram, a species often but erroneously known as *Phaseolus mungo* Linn. *Phaseolus minimus* Rumph. (Herb. Amb. **5**: 386. *pl. 139. f. 2*) is correctly placed as a synonym.

Phaseolus lunatus Linn. Sp. Pl. 724. 1753; Lour. Fl. Cochinch. 434. 1790, ed. Willd. 528. 1793, Anamese *dâu dai;* Gagnep. in Lecomte Fl. Gén. Indo-Chine 2: 227. 1916.

Phaseolus tunkinensis Lour. Fl. Cochinch. 435. 1790, ed. Willd. 529. 1793, Anamese *dâu ke bác.*

For *Phaseolus lunatus* Loureiro states: " Habitat cultus in Cochinchina." The description applies to a form of the common lima bean, *Phaseolus lunatus* Linn. For *P. tunkinensis* he states: " Habitat cultus in Cochinchina, ab Tunkino oriundus," and the description manifestly applies to a form or variety of the polymorphous *P. lunatus* Linn.

Phaseolus radiatus Linn. Sp. Pl. 725. 1753.

Phaseolus sublobatus Roxb. Fl. Ind. ed. 2, **3**: 288. 1832; Gagnep. in Lecomte Fl. Gén. Indo-Chine **2**: 232. 1916.

? *Phaseolus mungo* (non Linn.) Lour. Fl. Cochinch. 435. 1790, ed. Willd. 530. 1793, Anamese *dâû muòng ăn,* Chinese *siâo téu, tsiám téu.*

" Habitat in Cochinchina, & China in agris cultis." The form that Loureiro described as *Phaseolus mungo* may represent the true *P. radiatus* Linn., the type of which was a specimen that was grown at Upsala from seeds secured in Canton. This Linnaean species, however, is supposed to be the same as *P. sublobatus* Roxb. (See Piper in U. S. Dept. Agr. Bur. Pl. Ind. Bull. **119**: 17. 1914.)

Phaseolus vulgaris Linn. Sp. Pl. 723. 1753; Lour. Fl. Cochinch. 434. 1790, ed. Willd. 527. 1793, Anamese *dâu tláng tàu;* Gagnep. in Lecomte Fl. Gén. Indo-Chine **2**: 225. 1916.

" Habitat raro in Cochinchina, non indigena." The description applies well to a scandent form of the polymorphous Linnaean species.

Voandzeia Thouars

Voandzeia subterranea (Linn.) Thouars Gen. Nov. Madagasc. 23. 1806.

Glycine subterranea Linn. Sp. Pl. ed. 2, 1023. 1763; Lour. Fl. Cochinch. 457. 1790, ed. Willd. 558. 1793.

" Habitat prope Sofalam, & in variis locis Africae Orientalis." Loureiro's description was apparently based on specimens grown by him in Portugal, as he states that he succeeded in growing the plant from seeds for two years. He correctly interpreted the Linnaean species.

Vigna Savi

Vigna sinensis (Linn.) Savi ex Hassk. Cat. Hort. Bogor. 279. 1844; Endl. ex Hassk. Pl. Jav. Rar. 386. 1848; Merr. Enum. Philip. Fl. Pl. **2**: 320. 1923.

Dolichos sinensis Linn. Cent. Pl. **2**: 28. 1756, Amoen. Acad. **4**: 132, 326. 1759; Lour. Fl. Cochinch. 436. 1790, ed. Willd. 530. 1793, Anamese *dâu dua,* Chinese *táu cŏ, téu cŏ.*

Dolichos catjang Burm. f. Fl. Ind. 161. 1768; Linn. Mant. **2**: 269. 1771; Lour. Fl. Cochinch. 442. 1790, ed. Willd. 538. 1793, Anamese *dâu den, dâu bac, dâu deà,* Chinese *hĕ téu, min téu, siào hûm téu;* Gagnep. in Lecomte Fl. Gén. Indo-Chine **2**: 243. 1916.

Dolichos hastatus Lour. Fl. Cochinch. 442. 1790, ed. Willd. 539. 1793.

Dolichos unguiculatus (non Linn.?) Lour. Fl. Cochinch. 436. 1790, ed. Willd. 531. 1793.

Phaseolus cylindricus Linn. Amoen. Acad. **4**: 132. 1759.

Vigna cylindrica Skeels in U. S. Dept. Agr. Bur. Pl. Ind. Bull. **282**: 32. 1913.

Dolichos sesquipedalis Linn. Sp. Pl. ed. 2, 1019. 1763.

Vigna catjang Walp. in Linnaea **13**: 533. 1839.

Dolichos hastifolius Stokes Bot. Mat. Med. **4**: 27. 1812 (based on *Dolichos hastatus* Lour.).

For *Dolichos sinensis* Loureiro states: " Habitat cultus ubique in Cochinchina, & China "; for *D. catjang:* " Habitat in agris culti in Cochinchina, & China "; for *D. unguiculatus:* " Habitat in China, unde in Lusitaniam translatus "; and for *D. hastatus:* " Habitat cultus in ora Africae Orientali." All descriptions apply to cultivated forms or varieties of a single variable species, *D. sinensis* being the form with pods from one to two feet in length, the asparagus bean; *D. catjang* a form with short firm pods, three to five inches in length, the catjang; *D. unguiculatus,* the cow pea, a form with firm pods from eight inches to a foot in length, and *D. hastatus* apparently one of the cultivated forms in East Africa with short pods and slightly lobed leaves somewhat as in *D. tranquebaricus* Jacq. It seems highly probable that the oldest binomial for this collective species is *Dolichos unguiculatus* Linn. (Sp. Pl. 725. 1753) but no type is extant and the description is too inadequate to be certain as to its identity. Piper (Torreya **12**: 190. 1912) identified *Dolichos unguiculatus* Linn. as *Phaseolus unguiculatus* (Linn.) Piper (*P. antillanus* Urb.) on the basis of an examination of " Linnaeus' original specimen " in the herbarium of the Linnaean Society, but there must be some error here, as Jackson (Proc. Linn. Soc. **124**: 69. 1912) in his Index to the Linnaean Herbarium lists *D. unguiculatus* as not being represented by any specimen so named by Linnaeus and checked as being in the herbarium in 1753, 1755, or 1767. Baker, in his treatment of the Leguminosae (Oliver Fl. Trop. Africa **2**: 1–364. 1871), does not account for *Dolichos hastatus* Lour. but does not credit *Vigna sinensis* to the Mozambique District.

Pachyrhizus [10] Richard

Pachyrhizus erosus (Linn.) Urban Symb. Antill. **4**: 311. 1905.

Dolichos erosus Linn. Sp. Pl. 726. 1753.

Dolichos bulbosus Linn. Sp. Pl. ed. 2, 1021. 1763; Lour. Fl. Cochinch. 439. 1790, ed. Willd. 534. 1793, Anamese *săn rúong.*

Pachyrhizus angulatus Rich. in DC. Prodr. **2**: 402. 1825; Gagnep. in Lecomte Fl. Gén. Indo-Chine **2**: 256. 1916.

" Habitat cultus in Cochinchina." The description clearly applies to the yam bean. *Cacara bulbosa* Rumph. (Herb. Amb. **5**: 373. *pl. 132. f. 2*) cited by Loureiro, after Linnaeus, as a synonym, is correctly placed.

[10] *Pachyrhizus* Richard (1825), conserved name, Vienna Code; an older one is *Cacara* (Rumphius) Thouin (1805).

Dolichos Linnaeus

Dolichos lablab Linn. Sp. Pl. 725. 1753; Freem. in Bot. Gaz. **66**: 513. *f. 1–2.* 1918.

Dolichos purpureus Linn. Sp. Pl. ed. 2, 1021. 1763; Lour. Fl. Cochinch. 438. 1790, ed. Willd. 534. 1793, Anamese *dâu bán tiá,* Chinese *tsù piēn téu.*

Dolichos albus Lour. Fl. Cochinch. 439. 1790, ed. Willd. 534. 1793, Anamese *dâu bán tláng,* Chinese *pĕ piēn téu.*

Dolichos altissimus (non Jacq.) Lour. Fl. Cochinch. 438. 1790, ed. Willd. 533. 1793, Anamese *dâu bán phu yen.*

Lablab vulgaris Savi Osserv. Phaseolus et Dolichos 19. 1822; Gagnep. in Lecomte Fl. Gén. Indo-Chine **2**: 235. 1916.

Lablab perennans DC. Prodr. **2**: 402. 1825 (based on *Dolichos albus* Lour.).

For *Dolichos purpureus* and *D. albus* Loureiro states: " Habitat cultus in Cochinchina, & China "; and for *D. altissimus:* " Habitat incultus, cultusque in Cochinchina." The first species was correctly interpreted but it is a synonym of *Dolichos lablab* Linn. The second is merely a white-flowered form of the same species. The third I believe also is to be reduced to *D. lablab* Linn. as *Cacara perennis* Rumph. (Herb. Amb. **5**: 378. *pl. 136*), cited as a synonym, is the common hyacinth bean, *Dolichos lablab* Linn. It is curious that some authors continue to maintain *Lablab* as a genus distinct from *Dolichos.* I interpret *D. lablab* Linn., the first species cited under *Dolichos* by Linnaeus, as the standard species of the genus; it is also the type of *Lablab.* With its removal from *Dolichos* to *Lablab* it may be noted that not a single one of the 12 species originally proposed by Linnaeus in 1753 would remain in *Dolichos.*

Psophocarpus [11] Necker

Psophocarpus tetragonolobus (Linn.) DC. Prodr. **2**: 403. 1825; Gagnep. in Lecomte Fl. Gén. Indo-Chine **2**: 259. 1916.

Dolichos tetragonolobus Linn. in Stickman Herb. Amb. 23. 1754, Amoen. Acad. **4**: 132. 1759, Syst. ed. 10, 1162. 1759, Sp. Pl. ed. 2, 1020. 1763; Lour. Fl. Cochinch. 437. 1790, ed. Willd. 532. 1793, Anamese *dâu roùng.*

" Habitat, non frequens, in Cochinchina, & China." Loureiro clearly describes the Linnaean species which is widely cultivated in the Old World tropics. *Lobus quadrangularis* Rumph. (Herb. Amb. **5**: 374. *pl. 133*) cited by Loureiro, after Linnaeus, is the actual basis of the Linnaean species. -

Leguminosae of Doubtful Generic Status

Phaseolus tuberosus Lour. Fl. Cochinch. 434. 1790, ed. Willd. 528. 1793, Anamese *dâu săn rùng.*

" Habitat agrestis in Cochinchina." From the description one would suspect a *Pueraria* or perhaps *Pachyrhizus,* except for Loureiro's description of the flowers as "integrè luteis." It is suspected that he had flowering material of some species of *Phaseolus* or *Vigna,* ascribing to his species erroneous root characters, " radix tuberosa, fasciculata, magna, esculenta," due to a mixture of material.

---

[11] *Psophocarpus* Necker (1790), conserved name, Vienna Code; an older one is *Botor* Adanson (1763).

Anoma cochinchinensis Lour. Fl. Cochinch. 280. 1790, ed. Willd. 343. 1793, Anamese *ba dàu nho lá*.

"Habitat in sylvis Cochinchinae." The plant described is suspected of being a representative of the *Leguminosae-Caesalpinoideae;* certainly from the description it is not a representative of the *Moringaceae.* It is not accounted for in Lecomte's Flore Générale de Indo-Chine. A Loureiro specimen listed as being among his plants in the herbarium of the British Museum has not been located.

OXALIDACEAE

Oxalis Linnaeus

Oxalis repens Thunb. Diss. Oxalis 16. 1781; B. L. Rob. in Journ. Bot. **44**: 391. 1906.

Oxalis corniculata (non Linn.) Lour. Fl. Cochinch. 285. 1790, ed. Willd. 350. 1793, Anamese *chua me ba chia,* Chinese *tsŏ tsiān tsao;* Guillaumin in Lecomte Fl. Gén. Indo-Chine **1**: 610. 1911.

Oxalis corniculata Linn. var. *repens* Zucc. in Abh. Akad. Muench. **1**: 230. 1829–30; Knuth in Pflanzenreich **95** (IV–130): 150. 1930.

"Habitat sparsa per vias, & hortos in Cochinchina, & China." The common form in the oriental tropics is currently referred to *Oxalis corniculata* Linn. but I follow Doctor Robinson in accepting Thunberg's specific name for this species. The true *O. corniculata* Linn. is the more northern form.

Biophytum de Candolle

Biophytum sensitivum (Linn.) DC. Prodr. **1**: 690. 1824; Guillaumin in Lecomte Fl. Gén. Indo-Chine **1**: 608. 1911.

Oxalis sensitiva Linn. Sp. Pl. 434. 1753; Lour. Fl. Cochinch. 285. 1790, ed. Willd. 350. 1793, Anamese *chua me lá me,* Chinese *chăn tsú.*

"Habitat Cantone Sinarum, & in Cochinchina per hortos, & prata." A common, widely distributed and well understood species is represented here; Loureiro was correct in his interpretation of the Linnaean species.

Averrhoa Linnaeus

Averrhoa bilimbi Linn. Sp. Pl. 428. 1753; Lour. Fl. Cochinch. 289. 1790, ed. Willd. 355. 1793.

"Habitat Goae, & in multis locis Indiae intra, & extra Gangem." Loureiro correctly interpreted the Linnaean species, which is widely cultivated in the tropics of both hemispheres. *Blimbingum teres* Rumph. (Herb. Amb. **1**: 118. *pl. 36*), cited by Loureiro as a synonym, is correctly placed.

Averrhoa carambola Linn. Sp. Pl. 428. 1753; Lour. Fl. Cochinch. 288. 1790, ed. Willd. 354. 1793, Anamese *cây khê,* Chinese *yâm tao.*

"Habitat ubique culta per totam Cochinchinam, & in provinciis Australibus imperii Sinensis: vidi etiam in Malabaria, & in aliis Indiae locis." This species, like the preceding, is widely cultivated in the tropics of both hemispheres. *Prunum stellatum* Rumph. (Herb. Amb. **1**: 115. *pl. 35*), cited by Loureiro as a synonym, is correctly placed.

ZYGOPHYLLACEAE

Tribulus (Tournefort) Linnaeus

Tribulus terrestris Linn. Sp. Pl. 387. 1753; Lour. Fl. Cochinch. 270. 1790, ed. Willd. 331.
1793, Anamese *gai ma vuong, bach tât lê*, Chinese *ciĕ li tsù*.
" Habitat in Cochinchina, & China ad maris litora. Vidi etiam in ora Coromandelia,
prope Pondechery." The description conforms to the characters of the widely distributed
Linnaean species, which is represented by *Clemens 3071* from sand dunes on the coast at
Tourane. No representative of the family is admitted for Indo-China in Lecomte Fl.
Gén. Indo-Chine 1: 1–1070. 1907–12.

RUTACEAE

Xanthoxylum Linnaeus

Xanthoxylum avicennae (Lam.) DC. Prodr. 1: 726. 1824; Guillaumin in Lecomte Fl. Gén.
  Indo-Chine 1: 638. 1911.
  Fagara avicennae Lam. Encycl. 2: 445. 1788.
  Xanthoxylum clavaherculis (non Linn.) Lour. Fl. Cochinch. 659. 1790, ed. Willd. 810.
  1793, Anamese *cây múong troúng*, Chinese *sŏ*.
" Habitat in sylvis Cochinchinae, & Chinae." Loureiro's description agrees with the
characters of the common *Xanthoxylum avicennae* DC., the type of which was from China
and probably from Kwangtung Province. Two specimens from Loureiro are preserved in
the herbarium of the British Museum. Linnaeus published the generic name as *Zanthoxylum*. It was corrected by J. F. Gmelin to *Xanthoxylum* and the latter form was
included in the additions to the list of conserved generic names by the Cambridge Botanical
Congress.

Xanthoxylum nitidum (Roxb.) DC. Prodr. 1: 727. 1824; Guillaumin in Lecomte Fl. Gén.
  Indo-Chine 1: 641. 1911.
  Fagara nitida Roxb. Fl. Ind. 1: 439. 1820.
  Fagara piperita (non Linn.) Lour. Fl. Cochinch. 80. 1790, ed. Willd. 101. 1793, Anamese
  cây song, hùynh luc, Chinese *hôam liu*.
  Piper pinnatum Lour. Fl. Cochinch. 31. 1790, ed. Willd. 38. 1793, Anamese *thuc tieo*,
  Chinese *xú tsiào*, non *Zanthoxylum pinnatum* Druce.
For *Fagara piperita* Loureiro states: " Habitat in sepibus, & fruticetis Cochinchinae."
The form he describes can scarcely be other than the common *Xanthoxylum nitidum*
(Roxb.) DC., the type of which was a specimen from a plant cultivated at Calcutta, and
introduced from Canton. It is very common in Kwangtung Province and apparently
also in Indo-China. No locality is cited for *Piper pinnatum* and from the short description
and indicated medicinal uses it is suspected that Loureiro had only fragmentary material
secured from an herbalist. It is indeed curious that he should have described this as a
Piper, being misled by the very peppery taste of the fruits. Guillaumin (Lecomte Fl.
Gén. Indo-Chine 1: 644. 1911) retains both of Loureiro's species as forms of doubtful
status under *Xanthoxylum*.

Euodia [12] J. R. & G. Forster

Euodia trichotoma (Lour.) Pierre Fl. Forest. Cochinch. **3**: *pl. 287.* 1893 (based on *Tetradium trichotomum* Lour.); Rehd. & Wils. in Sargent Pl. Wils. **2**: 132. 1914.
Tetradium trichotomum Lour. Fl. Cochinch. 91. 1790, ed. Willd. 115. 1793, Anamese *cây dâu deâú.*
Brucea trichotoma Spreng. Syst. **1**: 441. 1825 (based on *Tetradium trichotomum* Lour.).
Ampacus trichotoma O. Ktz. Rev. Gen. Pl. 98. 1891 (based on *Tetradium trichotomum* Lour.).
Euodia viridans Drake in Journ. de Bot. **6**: 273. 1892; Guillaumin in Lecomte Fl. Gén. Indo-Chine **1**: 634. 1911.
Euodia colorata Dunn in Kew Bull. 2. 1906.
" Habitat in montibus Cochinchinae." Pierre was unquestionably correct in his interpretation of Loureiro's species, and the latter's specific name should be retained. The species extends from Kwangtung to Yunnan and Indo-China. Loureiro's type is preserved in the herbarium of the British Museum; for a note on this see Bennett (Pl. Jav. Rar. 199. 1844).

Euodia lepta (Spreng.) comb. nov.
Ilex lepta Spreng. Syst. **1**: 496. 1825 (based on *Lepta triphylla* Lour.).
Zanthoxylum pteleaefolium Champ. ex Benth. in Hook. Kew Journ. Bot. **3**: 330. 1851, excl. *Cuming 1819.*
Euodia triphylla Guillaumin in Lecomte Fl. Gen. Indo-Chine **1**: 632. 1911, non DC.
Brucea triphylla Dryander ex Britten ex Moore in Journ. Bot. **63**: 248. 1925 (based on *Lepta triphylla* Lour.).
Euodia pteleaefolia Merr. in Philip. Journ. Sci. Bot. **7**: 377. 1912.
Lepta triphylla Lour. Fl. Cochinch. 82. 1790, ed. Willd. 104. 1793, Anamese *cây mat.*
" Habitat in sylvis Cochinchinae." Britten, quoted by Moore (Journ. Bot. **63**: 248–249. 1925), briefly discusses Loureiro's type which is preserved in the herbarium of the British Museum, calling attention to Bennett's long discussion of it (Pl. Jav. Rar. 199. 1844). Bennett notes that *Lepta* had been " most singularly bandied about by various authors." Guillaumin, in spite of the fact that Sonnerat's specimen, the true basis of *Zanthoxylon triphyllum* Lam. (= *Euodia triphylla* DC. = *Melicope triphylla* Merr. in Philip. Journ. Sci. Bot. **7**: 375. 1912), is preserved in the Lamarck herbarium in the Paris Museum, follows current but erroneous usage (Lecomte Fl. Gén. Indo-Chine **1**: 632. 1911) in admitting what is really an endemic Philippine species of *Melicope*, as a widespread *Euodia*. I have examined Indo-China specimens, erroneously named by Guillaumin as *Euodia triphylla*, and find them to be identical with Kwangtung material representing typical *Euodia pteleaefolia* (Champ.) Merr. Attention is called to the fact that *Lepta triphylla* Lour. has no connection whatever with the binomial *Euodia triphylla* DC., which was based on *Zanthoxylum triphyllum* Lam., and which is *Melicope triphylla* (Lam.) Merr. The oldest valid specific name for this much-named plant is supplied by *Ilex lepta* Spreng.

[12] Sprague, Kew Bull. 353. 1928, calls attention to the fact that the original, and an orthographically correct form of this name, as published by J. R. & G. Forster, is *Euodia*, and that this form should be accepted under the provisions of the International Code of Botanical Nomenclature, unless *Evodia* be included in some future list of *nomina conservanda.*

Ruta (Tournefort) Linnaeus

Ruta chalepensis Linn. Mant. 1: 69. 1767; Lour. Fl. Cochinch. 269. 1790, ed. Willd. 330.
1793, Anamese *kuu li huong,* Chinese *sāo tsào;* Guillaumin in Lecomte Fl. Gén.
Indo-Chine 1: 645. 1911.

" Habitat inculta in hortis Cochinchinae, & Chinae." Guillaumin cites cultivated
specimens collected by Bon and by Thorel in Indo-China. It is possible that this cultivated
plant in China and Indo-China may prove to be but a form or variety of *Ruta graveolens*
Linn.

Acronychia [13] J. R. & G. Forster

Acronychia pedunculata (Linn.) Miq. Fl. Ind. Bat. Suppl. 532. 1861; Merr. Enum. Philip.
Fl. Pl. 2: 333. 1923.
Jambolifera pedunculata Linn. Sp. Pl. 349. 1753, pro parte.
Jambolifera rezinosa Lour. Fl. Cochinch. 231. 1790, ed. Willd. 285. 1793, Anamese
săn cây.
Cyminosma resinosa DC. Prodr. 1: 722. 1824 (based on *Jambolifera rezinosa* Lour.).
Gela lanceolata Lour. Fl. Cochinch. 232. 1790, ed. Willd. 285. 1793, Anamese *cây bái
bái.*
Ximenia lanceolata DC. Prodr. 1: 533. 1824 (based on *Gela lanceolata* Lour.).
Selas lanceolatum Spreng. Syst. 2: 216. 1825 (based on *Gela lanceolata* Lour.).
Acronychia laurifolia Blume Cat. Gew. Buitenzorg 27, 63. 1823; Guillaumin in Lecomte
Fl. Gén. Indo-Chine 1: 646. 1911.
Cyminosma pedunculata DC. Prodr. 1: 722. 1824.

For *Jambolifera rezinosa* Loureiro states: " Habitat in dumetis Cochinchinae," and
his description applies to the species commonly known as *Acronychia laurifolia* Blume.
For *Gela lanceolata* he states: " Habitat agrestis in Cochinchina," and the description also
applies to Blume's species; Loureiro's type is preserved in the herbarium of the British
Museum. The synonymy is curiously complicated. The original description of *Jambolifera* Linn. (Gen. Pl. 165. 1754) applies unmistakably to *Acronychia,* but the sole species,
J. pedunculata Linn. (Sp. Pl. 349. 1753), is typified by *Jambolifera* Linn. (Fl. Zeyl. 58. 1747)
and the description in the Flora Zeylanica applies to the species currently known as *Eugenia
jambolana* Lam. = *E. cumini* (Linn.) Druce. Trimen (Journ. Linn. Soc. Bot. 24: 140.
1888) notes that the specimens 139 and 185 of the Flora Zeylanica with their native names
were transposed; the specimen under 139 is the *Acronychia,* the one under 185 is the *Eugenia.*
Fl. Zeyl. 139 is the type of *Jambolifera pedunculata* as actually published by Linnaeus.
Jambolones Bauhin (Pinax 466), the second reference given by Linnaeus, is doubtless
Eugenia cumini (Linn.) Druce, but the Linnaean generic description does not apply to
this species. Miquel based the binomial *Acronychia pedunculata* on " *Cuminosma* [*Cyminosma*] DC. Prodr. I. p. 722. W. et Arn. Prodr. I. p. 146, excl. quib. syn.," but most of the
few synonyms cited are apparently the species as here interpreted. In accordance with
this interpretation of *Jambolifera* Linnaeus this generic name is much older than *Acronychia*
Forster. Gagnepain (Not. Syst. 3: 331. 1918) discusses *Jambolifera rezinosa* Lour. under
Eugenia resinosa Gagnep., calling attention to the fact that *cay san* and *cay san huyen,*

---

[13] *Acronychia* J. R. & G. Forster (1776), conserved name, Vienna Code; older ones are *Cunto* Adanson
(1763) and *Jambolana* Adanson (1763). To this list should be added *Jambolifera* Linnaeus (1753).

the Anamese names of *Eugenia resinosa* Gagnep., correspond to *săn cây*, the Anamese name of *Jambolifera rezinosa* Lour. His binomial, however, is not based on Loureiro's earlier one and Loureiro's description " stamina 8 " does not apply to *Eugenia*.

Acronychia sp.
> *Jambolifera odorata* Lour. Fl. Cochinch. 231. 1790, ed. Willd. 284. 1793, Anamese *rau ton.*
> *Cyminosma odorata* DC. Prodr. 1: 722. 1824 (based on *Jambolifera odorata* Lour.).

" Habitat in hortis Cochinchinae." The description seems to apply to *Acronychia* but I do not know of any cultivated representatives of this genus. The description of the leaves as " basi oblique truncata . . . inferius albicantia " hardly applies to the common *Acronychia pedunculata* Miq. Guillaumin does not account for this species in his treatment of the Rutaceae of Indo-China (Lecomte Fl. Gén. Indo-Chine 1: 629–687. 1911); it may prove to be merely a form of *Acronychia pedunculata* Miq.

Glycosmis Correa

Glycosmis cochinchinensis (Lour.) Pierre ex Engler in Engler & Prantl Nat. Pflanzenfam. 3(4): 185. *f. 106.* 1895 (based on *Toluifera cochinchinensis* Lour.); Guillaumin in Lecomte Fl. Gén. Indo-Chine 1: 653. 1911, pro parte.
> *Loureira cochinchinensis* Meisn. Gen. Comm. 53. 1837 (based on *Toluifera cochinchinensis* Lour.).
> *Toluifera cochinchinensis* Lour. Fl. Cochinch. 262. 1790, ed. Willd. 321. 1793, Anamese *cây cam ruu.*

" Habitat inculta in locis planis Cochinchinae." Although Engler in taking up Pierre's transfer of Loureiro's specific name intended it to replace *G. pentaphylla* Corr. as a collective species, it is clear that the form Loureiro described with simple leaves, 3 inches long, is not the same as *Limonia pentaphylla* Retz (Obs. **5**: 24. 1789) = *Glycosmis pentaphylla* Corr. Loureiro's species is represented by *Clemens 3363, 4448,* from thickets at Hue and Tourane, and de Pirey's specimen of *cam ruou, Chevalier 41186.* Guillaumin's description of *Glycosmis cochinchinensis* (Lour.) Pierre applies only in small part to Loureiro's species, as he treated it as a collective one, citing 14 synonyms, most of which will have to be excluded with the restriction of specific limits to the form actually described by Loureiro. True *Glycosmis pentaphylla* (Retz.) Corr. does not occur in Indo-China.

Micromelum [14] Blume

Micromelum falcatum (Lour.) Tanaka in Journ. Bot. **68**: 225. 1930 (based on *Aulacia falcata* Lour.).
> *Aulacia falcata* Lour. Fl. Cochinch. 273. 1790, ed. Willd. 335. 1793, Anamese *cây cham tlâu tláng;* Moore in Journ. Bot. **63**: 282. 1925.
> *Aulacia falcifolia* Stokes Bot. Mat. Med. **2**: 481. 1812 (based on *Aulacia falcata* Lour.).
> *Cookia falcata* DC. Prodr. **1**: 537. 1824 (based on *Aulacia falcata* Lour.).
> *Micromelum octandrum* Turcz. in Bull. Soc. Nat. Mosc. **36**(1): 578. 1863.

---

[14] *Micromelum* Blume (1825), conserved name, Cambridge Code. *Aulacia* Loureiro (1790) is the oldest name for the genus.

" Habitat in sylvis Cochinchinae." In my preliminary manuscript of 1919 I concluded that *Aulacia falcata* was the same as *Micromelum pubescens* Blume, and Moore confirmed this by an examination of Loureiro's type in the herbarium of the British Museum, i.e., as *M. pubescens* has been currently interpreted. Tanaka, however, who later examined Loureiro's type, considered that the species is distinct from the Javan form Blume described, and that it represents the species later described as *Micromelum octandrum* Turcz. The species extends from Indo-China to Burma, Tenasserim, and the Andaman Islands. *Aulacia falcata* Lour., in spite of its extant type, is not accounted for by Guillaumin in his treatment of the Rutaceae of Indo-China (Lecomte Fl. Gén. Indo-Chine 1: 629–687. 1911). The genus *Aulacia* is currently but erroneously reduced to *Clausena*.

Murraya [15] (*Murraea*) Koenig

Murraya paniculata (Linn.) Jack in Malay. Misc. 1: 31. 1820; Merr. Interpret. Herb. Amb. 292. 1917.

> *Chalcas paniculata* Linn. Mant. 1: 68. 1767; Lour. Fl. Cochinch. 270. 1790, ed. Willd. 331. 1793, Anamese *cây nguyet qúi*, Chinese *caō li yòng.*
> *Chalcas japonensis* Lour. Fl. Cochinch. 271. 1790, ed. Willd. 332. 1793, Anamese *nguyet qúi taù.*
> *Chalcas camuneng* Burm. f. Fl. Ind. 104. 1768.
> *Murraea exotica* Linn. Mant. 2: 563. 1771.

For *Chalcas paniculata* Loureiro states: " Habitat agrestis in Cochinchina, & China: colitur etiam frequenter ob florum fragrantiam "; and for *C. japonensis:* " Colitur in Cochinchina, & China." The Linnaean species was correctly interpreted by Loureiro, and *Camunium vulgare* Rumph. (Herb. Amb. 5: 26. *pl. 17*) is correctly placed as a synonym. *C. japonensis* Lour. is merely a cultivated form of *M. paniculata* Jack with small leaves, well represented by *Camunium japonicum* Rumph. (Herb. Amb. 5: 29. *pl. 18. f. 2*), cited by Loureiro as a synonym. A specimen of *Chalcas paniculata* from Loureiro is preserved in the herbarium of the British Museum.

Clausena [16] Burman f.

Clausena lansium (Lour.) Skeels in U. S. Dept. Agr. Bur. Plant Ind. Bull. **168**: 31. 1909 (based on *Quinaria lansium* Lour.).

> *Quinaria lansium* Lour. Fl. Cochinch. 272. 1790, ed. Willd. 334. 1793, Chinese *uan pî chū.*
> *Cookia punctata* Sonn. Voy. Ind. Or. 2: 231. *pl. 130.* 1782, non *Clausena punctata* W. & A.
> *Clausena punctata* Rehd. & Wils. in Sargent Pl. Wils. 2: 140. 1914, non W. & A.
> *Cookia wampi* Blanco Fl. Filip. 358. 1837.
> *Clausena wampi* Oliv. in Journ. Linn. Soc. Bot. 5: Suppl. 2: 34. 1861; Guillaumin in Lecomte Fl. Gén. Indo-Chine 1: 664. *f. 70. 1–3* (*vampi*). 1911.

" Colitur Cantone Sinarum, ubi fructus prostant in foro venales." Loureiro's type is preserved in the herbarium of the Paris Museum; a specimen listed as being among his

---

[15] *Murraya* (*Murraea*) Koenig (1771) conserved name, Brussels Code; older ones are *Camunium* Adanson (1763), and *Chalcas* Linnaeus (1767), while *Bergera* Koenig was published in 1771.

[16] *Claucena* in Burman's text, changed to *Clausena* in his index [p. 1].

plants in the herbarium of the British Museum has not been located. The species is currently known as *Clausena wampi* (Blanco) Oliv. It is very commonly cultivated near Canton, local names appearing on recently collected material being *wong pei, wong pa* and *wong poi*. Swingle and Tanaka adopt Burman's original spelling *Claucena*, but this is a manifest typographical error—the name was derived from the proper name Clausen and appears in the index to Burman's work in its correct form *Clausena*, rather clearly indicating his intention to use the latter spelling.

Clausena excavata Burm. f. Fl. Ind. 87. *pl. 29. f. 2.* 1768; Guillaumin in Lecomte Fl. Gén. Indo-Chine 1: 661. 1911.

> *Lawsonia falcata* Lour. Fl. Cochinch. 229. 1790, ed. Willd. 282. 1793, Anamese *cây chàm tlâu, cây méo.*
> *Lausonia falcifolia* Stokes Bot. Mat. Med. 2: 364. 1812 (based on *Lawsonia falcata* Lour.).

"Habitat ubique in dumetis Cochinchinae." Koehne (Pflanzenreich 17 (IV–216): 272. 1903), in excluding Loureiro's species from the *Lythraceae*, referred it to *Solanum* sp., which is manifestly wrong. He probably was influenced by Loureiro's erroneous reference of *Adulterina* Rumph. (Herb. Amb. 6: 58. *pl. 25. f. 1*) to *Lawsonia falcata;* the Rumphian illustration represents *Solanum verbascifolium* Linn., but Loureiro's description is of some plant totally different from this. *Lawsonia falcata* Lour., as described, in spite of the fact that pinnate leaves are not mentioned, surely represents *Clausena excavata* Burm. f., which is common at Hue, widely distributed in Indo-China, and which is not otherwise described by Loureiro.

Triphasia Loureiro

Triphasia trifolia (Burm. f.) P. Wils. in Torreya 9: 33. 1909.

> *Limonia trifolia* Burm. f. Fl. Ind. 103. 1768.
> *Limonia trifoliata* Linn. Mant. 2: 237. 1771.
> *Triphasia trifoliata* DC. Prodr. 1: 536. 1824.
> *Triphasia aurantiola* Lour. Fl. Cochinch. 153. 1790, ed. Willd. 189. 1793, Anamese *kim kúit;* Guillaumin in Lecomte Fl. Gén. Indo-Chine 1: 650. 1911.

"Habitat in Cochinchina, & China: coliturque in viridariis magnatum ob odorem, & formae elegantiam." A common, widely distributed and well-known species in the Indo-Malaysian region. Loureiro's type is preserved in the herbarium of the British Museum

Atalantia [17] Correa

Atalantia buxifolia (Poir.) Oliv. in Journ. Linn. Soc. Bot. 5: Suppl. 2: 26. 1861.

> *Citrus buxifolia* Poir. in Lam. Encycl. 4: 580. 1798.
> *Limonia monophylla* Linn. Mant. 1: 237. 1767; Lour. Fl. Cochinch. 271. 1790, ed. Willd. 333. 1793, Anamese *cây cam dàng,* Chinese *săo peng lâc, xăc may lâc,* non *Atalantia monophylla* (Roxb.) DC. Prodr. 1: 535. 1824.
> *Atalantia loureiriana* M. Roem. Syn. 1: 37. 1846 (based on *Limonia monophylla* Lour.).
> *Severinia buxifolia* Tenore Ind. Sem. Hort. Neap. 3. 1840; Swingle in Journ. Washington Acad. Sci. 6: 655. *f. 1–2.* 1916.

[17] *Atalantia* Correa (1805), conserved name, Vienna Code; an older one is *Malnaregam* Adanson (1763).

Limonia bilocularis Roxb. Fl. Ind. ed. 2, **2**: 377. 1832; Guillaumin in Lecomte Fl. Gén. Indo-Chine **1**: 672. 1911.

Dumula sinensis Lour.[18] ex Gomes in Mem. Acad. Sci. Lisb. Cl. Sci. Pol. Mor. Bel.-Let. n.s. **4**(1): 29. 1868.

Severinia monophylla Tanaka in Journ. Bot. **68**: 232. 1930, Bull. Mus. Hist. Nat. [Paris] II **2**: 163. 1930.

" Habitat in dumetis Cochinchinae, & Chinae." Loureiro was apparently correct in his interpretation of the Linnaean species, whereas most modern authors may have misinterpreted it. Tanaka states that there are three specimens in the Linnaean herbarium and that these represent *Limonia monophylla* Linn., citing *Severinia buxifolia* Ten. as a synonym. Jackson, however (Proc. Linn. Soc. **124**: Suppl. 97. 1912), indicates that these specimens were not named by Linnaeus. Specimens from Loureiro preserved in the herbaria of the British and the Paris Museums apparently represent the same form. The species is very common in southeastern China and is also apparently abundant in Indo-China. Tanaka (Journ. Bot. **68**: 232. 1930), in discussing the specimens in the Linnaean herbarium, erroneously cites the publication of *Limonia monophylla* as Linnaeus Sp. Pl. **1**: 237. 1753; it was actually published in Mantissa **1**: 237. 1767. Some authors credit the binomial *Atalantia monophylla* to Correa (Ann. Mus. Hist. Nat. [Paris] **6**: 383. 1805), but Correa in this paper merely gives the generic name *Atalantia* and proposed no new binomial under it. It should be noted that *Atalantia monophylla* DC. (1824) was based wholly on *Limonia monophylla* Roxb. Pl. Coromandel **1**: 59. *pl. 83.* 1795, a species distinct from *Limonia monophylla* Linn.

Citrus Linnaeus

Citrus aurantifolia (Christm.) Swingle in Journ. Washington Acad. Sci. **3**: 465. 1913.

Limonia aurantifolia Christm. Pflanzensyst. **1**: 618. 1777.

Citrus lima Lunan Hort. Jamaic. **1**: 451. 1814.

Citrus acida Roxb. Fl. Ind. ed. 2, **3**: 390. 1832.

Citrus medica Linn. var. 2. *Citrus limon* Lour. Fl. Cochinch. 465. 1790, ed. Willd. 568. 1793, Anamese *cây canh*, Chinese *tsĭm pi xú*.

The description applies to a form of the common lime. *Limonellus* Rumph. (Herb. Amb. **2**: 107. *pl. 29*), cited by Loureiro as a synonym, is correctly placed.

Citrus aurantium Linn. Sp. Pl. 782. 1753.

Citrus fusca Lour. Fl. Cochinch. 467. 1790, ed. Willd. 571. 1793, Anamese *cây baòng, chì xác*, Chinese *chì kéu*.

" Habitat latissime in Cochinchina, rarius in China." Loureiro's description apparently applies to a form of the sour orange.

Citrus maxima (Burm.) Merr. Interpret. Herb. Amb. 296. 1917.

Aurantium maximum Burm. ex Rumph. Herb. Amb. Auctuarium Ind. Univ. [16]. 1755.

Citrus grandis Osbeck Dagbok Ostind. Resa 98. 1757.

Citrus decumana Linn. Syst. Nat. ed. 12, **2**: 508. 1767 (*decumanus*); Lour. Fl. Cochinch. 467. 1790, ed. Willd. 571. 1793, Anamese *cây buoi*, Chinese *yéu xú*.

" Habitat frequenter in Cochinchina & China." The description applies to the common pomelo.

[18] A Loureiro herbarium name here first published by Gomes.

Citrus medica Linn. Sp. Pl. 782. 1753; Lour. Fl. Cochinch. 465. 1790, ed. Willd. 568. 1793, Anamese *cây tanh yen*.

" Habitat culta in Cochinchina, & China." Loureiro's description definitely applies to the citron, *Citrus medica* Linn. The form discussed in the note immediately following the description for which Loureiro gives the name *phat thu*, and for which he gives the incidental name " Citrus digitata, seu Chirocarpus," is the form known as Buddha's fingers, *Citrus medica* Linn. var. *sarcodactylis* (Nooten) Swingle in Sargent Pl. Wils. 2: 141. 1914.

Citrus nobilis Lour. Fl. Cochinch. 466. 1790, ed. Willd. 569. 1793, Anamese *cam sanh*, Chinese *tsem căn*.

" Habitat abundanter in Cochinchina: etiam in China, quamvis Cantone illam non viderim." This is a form of the orange with a thick, juicy, sweet, edible pericarp, noted by Loureiro to be the best of all citrus fruits. It is generally admitted as a valid species, but is possibly a cultigen derived from *C. aurantium* Linn.

Citrus sinensis Osbeck Dagbok Ostind. Resa 41. 1757, *nomen nudum*, Reise Ostind. China 250. 1765; Swingle in Sargent Pl. Wils. 2: 148. 1914.

  Citrus aurantium (non Linn.) Lour. Fl. Cochinch. 466. 1790, ed. Willd. 569. 1793, Anamese *cây cam*, Chinese *căn xú*.

" Habitat culta, incultaque in Cochinchina, & China." This is manifestly the sweet orange, claimed by Swingle to represent a species distinct from the sour orange, *C. aurantium* Linn.; presumably it is a cultigen derived from the latter.

Fortunella Swingle

Fortunella margarita (Lour.) Swingle in Journ. Washington Acad. Sci. **5**: 170. *f. 2*. 1915 (based on *Citrus margarita* Lour.).

  Citrus margarita Lour. Fl. Cochinch. 467. 1790, ed. Willd. 570. 1793, Anamese *chŭ tsù*, Chinese *châu tu*.

" Habitat Cantone Sinarum, nec rara: in Cochinchina mihi non visa." Doctor Swingle retains this, the oval kumquat, as specifically distinct from the round-fruited form, *Fortunella japonica* (Thunb.) Swingle (*Citrus japonica* Thunb.).

Fortunella japonica (Thunb.) Swingle in Journ. Washington Acad. Sci. **5**: 171. *f. 3*. 1915.

  Citrus japonica Thunb. in Nov. Act. Upsal. 3: 199. 1780; Fl. Jap. 292. 1784.

  Citrus madurensis Lour. Fl. Cochinch. 467. 1790, ed. Willd. 570. 1793, Anamese *kīm kúit*, Chinese *kīn, kúit xú*.

" Habitat in Cochinchina, & China: coliturque ob pulchritudinem." Loureiro's description applies unmistakably to the round or common kumquat, whether this is specifically distinct from the oval kumquat, *Fortunella margarita* (Lour.) Swingle, as Swingle claims, or not. *Limonellus madurensis* Rumph. (Herb. Amb. **2**: 110. *pl. 31*) is correctly placed by Loureiro as a synonym.

SIMARUBACEAE
Eurycoma Jack

Eurycoma longifolia Jack in Malay. Miscel. 2(7): 45. 1822; Lecomte Fl. Gén. Indo-Chine 1: 695. 1911.

  Crassula pinnata (non Linn. f.) Lour. Fl. Cochinch. 185. 1790, ed. Willd. 231. 1793, Anamese *cây bap benh*.

" Habitat agrestis in locis arenosis Cochinchinae." In spite of certain discrepancies in Loureiro's description, *Crassula pinnata* as described by him is manifestly referable to *Eurycoma longifolia* Jack. He notes that the entire plant is very bitter, a simarubaceous character, while Lecomte, who recognizes three varieties in Indo-China, gives the Anamese name *bâ binh*, a cognate form of *cây bap benh;* the latter does not account for Loureiro's binomial.

Brucea [19] J. F. Miller

Brucea javanica (Linn.) Merr. in Journ. Arnold Arb. **9**: 3. *pl. 10.* 1928.

> *Rhus javanica* Linn. Sp. Pl. 265. 1753.
> *Gonus amarissimus* Lour. Fl. Cochinch. 658. 1790, ed. Willd. 809. 1793, Anamese *sâu dâu rùng*, Chinese *a tàm tsào*.
> *Brucea sumatrana* Roxb. Hort. Beng. 12. 1814, *nomen nudum*, Fl. Ind. ed. 2, **1**: 449. 1832; Spreng. Syst. **1**: 441. 1825; Lecomte Fl. Gén. Indo-Chine **1**: 698. 1911.
> *Brucea amarissima* Desv. ex Gomes in Mem. Acad. Sci. Lisb. Cl. Sci. Pol. Mor. Bel.- Let. n.s. 4(1): 30. 1868; Merr. in Philip. Journ. Sci. Bot. **10**: 18. 1915, Interpret. Herb. Amb. 299. 1917, Enum. Philip. Fl. Pl. **2**: 347. 1923; Druce in Rept. Bot. Exch. Club Brit. Isles **4**: 611. 1917 (based on *Gonus amarissimus* Lour.).
> *Lussa amarissima* O. Ktz. Rev. Gen. Pl. 104. 1891 (based on *Gonus amarissimus* Lour.).

" Habitat in sylvis Cochinchinae, & Chinae." Loureiro's type is preserved in the herbarium of the Paris Museum. The species is a common, widely distributed, and well-known one in the Indo-Malaysian region. It is abundant in thickets near Canton and also in Indo-China. Lecomte does not account for *Gonus amarissimus* Lour. although it was based on Indo-China material. The illustration cited above under *Brucea javanica* is a photographic reproduction of Linnaeus' type specimen. Britten (Journ. Bot. **38**: 315–316. 1900) retains *Rhus javanica* Linn. as a true *Rhus;* his note should be consulted. My different conclusion is expressed in Journ. Arnold Arb. **9**: 1–4. 1928.

BURSERACEAE

Canarium (Rumphius) Linnaeus

Canarium album (Lour.) Raeusch. Nomencl. ed. 3, 287. 1797; Hance in Journ. Bot. **9**: 39. 1871; Guillaumin in Bull. Soc. Bot. France **55**: 617. *pl. 19. f. 1.* 1908; Lecomte Fl. Gén. Indo-Chine **1**: 714. 1911 (based on *Pimela alba* Lour.).

> *Pimela alba* Lour. Fl. Cochinch. 408. 1790, ed. Willd. 495. 1793, Anamese *cây ca na*, Chinese *pă lâm*.

" Habitat in sylvis Cochinchinae, & Chinae." This species is commonly cultivated for its edible fruits in Kwangtung Province, where it is generally known as *paak lam*. Engler (DC. Monog. Phan. **4**: 149. 1883) placed it as a species of doubtful status, although Hance, twelve years earlier, had clearly indicated the differences between *Canarium album* Raeusch. and *C. pimela* Koenig. The fruits of these two species are the so-called " Chinese olives " which are extensively utilized by the Chinese for food. Guillaumin has given a very extended description of the species.

[19] *Brucea* J. F. Miller (1780), conserved name, Vienna Code; an older one is *Lussa* Rumphius (1755 O. Kuntze 1891).

Canarium pimela Koenig in Koenig & Sims Ann. Bot. 1: 361. *pl. 7. f. 1.* 1805; Hance in
    Journ. Bot. **9**: 39. 1871; Engler in DC. Monog. Phan. **4**: 122. 1883 (based on *Pimela*
    nigra Lour.).
    Canarium nigrum Engler in Engler & Prantl Nat. Pflanzenfam. 3(4): 240. 1896;
    Guillaumin in Lecomte Fl. Gén. Indo-Chine 1: 710. 1911 (based on *Pimela nigra*
    Lour.), non Roxb.
    Pimela nigra Lour. Fl. Cochinch. 407. 1790, ed. Willd. 495. 1793, Anamese *cây bùi*,
    Chinese *õ lâm.*
    Lipara nigra Lour.[20] ex Gomes in Mem. Acad. Sci. Lisb. Cl. Sci. Pol. Mor. Bel.-Let.
    n.s. 4(1): 30. 1868.
    " Habitat in sylvis Cochinchinae, & Chinae." This species is commonly cultivated
in Kwangtung Province for its edible fruits, and is currently known as *oo lam* and *moo lam.*
Loureiro cites *Canarium nanarium* Rumph. (Herb. Amb. **2**: 155. *pl. 49*) as doubtfully
representing his species; it must be excluded as it represents *Canarium sylvestre* Gaertn.
Specimens from Loureiro are preserved in the herbaria of the Paris and the British Museums.
Koenig's amplified description was based on Loureiro's British Museum specimen.

Canarium sp.
    Pimela oleosa Lour. Fl. Cochinch. 408. 1790, ed. Willd. 496. 1793, Anamese *cây deâu*
    rái.
    " Habitat frequens in sylvis Cochinchinae, maxime versus Cambodiam." For the
reason that Loureiro erroneously cited *Nanarium oleosum* Rumph. (Herb. Amb. **2**: 162.
pl. 54) as representing his species, the latter based on an actual Indo-Chinese specimen,
Pimela oleosa Lour. has been placed by various authors as a synonym of *Canarium oleosum*
(Lam.) Engl. (*Amyris oleosa* Lam., *Canarium microcarpum* Engl.). Guillaumin (Lecomte
Fl. Gén. Indo-Chine 1: 710. 1911), following current usage, retained *Pimela oleosa* Lour.
as a synonym of *Canarium oleosum* (Lam.) Engl., admitting the latter species as one
occurring in Indo-China solely on the basis of Loureiro's synonym; he gives the other
range as the Sunda Islands and Timor; yet Lamarck's species was based on Rumphius'
description and illustration, which in turn were based on Amboina material. *Robinson Pl.*
Rumph. Amb. 376, represents Lamarck's species (Merrill, Interpret. Herb. Amb. 303.
1917). *Pimela oleosa* Lour. has little in common with *Canarium oleosum* (Lam.) Engl.,
but is probably a synonym of some one of the other species described by Guillaumin.
In any case *Canarium oleosum* (Lam.) Engl. must be excluded from the Indo-China flora.

MELIACEAE

Melia Linnaeus

Melia azedarach Linn. Sp. Pl. 384. 1753; Lour. Fl. Cochinch. 269. 1790, ed. Willd. 329.
    1753, Anamese *cây sâu dâu, xuyen luyen,* Chinese *xún lién;* Pellegr. in Lecomte Fl.
    Gén. Indo-Chine 1: 727. 1911.
    Melia cochinchinensis M. Roem. Syn. 1: 95. 1846 (based on *Melia azedarach* Lour.).
    " Habitat inculta, cultaque in Cochinchina." The Linnaean species was correctly
interpreted by Loureiro. This may or may not be the form characterized by Pierre as

---

[20] A Loureiro herbarium name here first published by Gomes.

Melia azedarach L. var. *cochinchinensis* Pierre (Fl. Forest. Cochinch. **4**: sub *pl. 356*. 1897) which was published without reference to *Melia cochinchinensis* M. Roem.

Dysoxylum Blume

Dysoxylum loureiri Pierre Fl. Forest. Cochinch. **4**: *pl. 352*. 1896; Pellegr. in Lecomte Fl. Gén. Indo-Chine **1**: 742. 1911.

> *Epicharis loureiri* Pierre in Bull. Soc. Linn. Paris **1**: 291. 1881.
> *Santalum album* (non Linn.) Lour. Fl. Cochinch. 87. 1790, ed. Willd. 109. 1793, Anamese *cây hùynh dàn, bach dàn*, Chinese *tân yàm*.

" Vidi autem saepe viventes arbores in *Doung nai*, provincia Australiori Cochinchinae." Loureiro included in his discussion the uses of the true sandalwood, *Santalum album* Linn. The plant with pinnate leaves which he describes and of which he saw no flowers is *Dysoxylum loureiri* Pierre. See Pierre, L. Sur deux espèces d'*Epicharis* produisant les bois dits: Sandal citrin et Sandal rouge (Bull. Soc. Linn. Paris **1**: 289–292. 1881).

Aglaia Loureiro

Aglaia odorata Lour. Fl. Cochinch. 173. 1790, ed. Willd. 216. 1793, Anamese *cây ngâu*.

> *Opilia odorata* Spreng. Syst. **1**: 766. 1825 (based on *Aglaia odorata* Lour.).

" Habitat in Cochinchina, & China tam agrestis, quam culta in hortis magnatum." This is a very well-known species, the type of the genus. The Cantonese name is *mai yai lun* or *mai chai lun*. *Camunium sinense* Rumph. (Herb. Amb. **5**: 26. *pl. 18. f. 1*), cited by Loureiro as a synonym, is correctly placed.

POLYGALACEAE

Polygala (Tournefort) Linnaeus

Polygala chinensis Linn. Sp. Pl. 704. 1753, et herb. Linn.

> *Polygala glomerata* Lour. Fl. Cochinch. 426. 1790, ed. Willd. 518. 1793, Chinese *tāi kām;* Gagnep. in Lecomte Fl. Gén. Indo-Chine **1**: 257. 1909.
> *Polygala densiflora* Blume Bijdr. 59. 1825; Chodat in Mém. Soc. Phys. Hist. Nat. Genève **31**(2): 380. 1893.

" Habitat inculta prope Cantonem Sinarum." Loureiro's species is a well-known one extending from the Himalayan region to southern China, Java, Borneo, and the Philippines. Var. β of *Polygala thea* Burm. f. Fl. Ind. 154. 1768 is a synonym, but Burman's species itself (= *P. theezans* Linn.) is a synonym of *Ionidium suffruticosum* (Linn.) Ging. Loureiro's type is preserved in the herbarium of the Paris Museum. The relationship of Loureiro's species to *Polygala chinensis* Linn. has not previously been brought out. In comparing a photograph of Loureiro's type with a photograph of the type of the Linnaean species no differences could be detected. Hemsley (Journ. Linn. Soc. Bot. **23**: 59. 1886) does not admit *P. chinensis* Linn. as a Chinese species, stating that it was only recorded as an Indian plant by its author; there is nothing on the sheet in the Linnaean herbarium to indicate whether the specimen came from China or from India, but the specific name would seem to be evidence that it came from the former country; it beautifully matches a series of Kwangtung specimens, and Loureiro's type also came from this same region.

Bennett, however (Hook. f. Fl. Brit. Ind. 1: 204. 1872), retained *P. chinensis* Linn. and *P. glomerata* Lour. as distinct, and clearly examined the Linnaean type.

Polygala japonica Houtt. Nat. Hist. II **10**: 89. *pl. 62. f. 1.* 1779; Chodat in Mém. Soc. Phys. Hist. Nat. Genève **31**(2): 353. *pl. 28. f. 18–20.* 1893.

Polygala sibirica (non Linn.) Lour. Fl. Cochinch. 426. 1790, ed. Willd. 517. 1793.

Polygala loureirii Gard. & Champ. in Hook. Journ. Bot. Kew Miscel. **1**: 242. 1849 (based in part, as to synonymy, on *Polygala sibirica* Lour.).

"Habitat inculta Cantone Sinarum." Many botanists, including Hemsley (Journ. Linn. Soc. Bot. **23**: 62. 1886) reduce *Polygala japonica* Houtt. to *P. sibirica* Linn., and with this wider interpretation Loureiro would be correct in his use of the Linnaean specific name. Chodat, however, limits *P. sibirica* Linn. to Russia and northern and central Asia, with a southward range in China to Shantung and Shansi, giving *P. japonica* Houtt. a range of from Japan to Celebes and Timor. The Kwangtung form described by Loureiro is safely *P. japonica* Houtt. as interpreted by Chodat.

Salomonia Loureiro

Salomonia cantoniensis Lour. Fl. Cochinch. 14. 1790, ed. Willd. 18. 1793, Chinese *siau lam teng.*

"Habitat inculta Cantone Sinarum." Loureiro's type is preserved in the herbarium of the Paris Museum. This is a well-known species, the type of the genus, extending from Assam to southern China and Malaysia.

Xanthophyllum [21] Roxburgh

Xanthophyllum sylvestre (Lour.) Moore in Journ. Bot. **63**: 254. 1925 (based on *Eystathes sylvestris* Lour.).

Eystathes sylvestris Lour. Fl. Cochinch. 235. 1790, ed. Willd. 289. 1793, Anamese *cây tlám.*

Valentinia sylvestris Raeusch. Nomencl. ed. 3, 109. 1797 (based on *Eystathes sylvestris* Lour.).

" Habitat in altis montibus Cochinchinae." Loureiro's genus *Eystathes* remained one of entirely doubtful status until 1925 when Moore made a critical examination of the type in the herbarium of the British Museum, finding, as Robert Brown had indicated on the specimen, that *Xanthophyllum* is the genus represented. He could not place the species among the five admitted by Gagnepain (Lecomte Fl. Gén. Indo-Chine **1**: 242–247. 1909) and gives additional descriptive data based on the type. *Eystathes* Loureiro has twenty-four years priority over *Xanthophyllum* Roxb.

EUPHORBIACEAE

Phyllanthus Linnaeus

Phyllanthus cochinchinensis Spreng. Syst. **3**: 21. 1826 (based on *Cathetus fasciculata* Lour.).

Cathetus fasciculata Lour. Fl. Cochinch. 608. 1790, ed. Willd. 746. 1793, Anamese *chôi duc.*

[21] *Xanthophyllum* Roxburgh (1814), conserved name, Brussels Code; older ones are *Palae* Adanson (1763) and *Eystathes* Loureiro (1790).

Phyllanthus fasciculatus Muell.-Arg. in DC. Prodr. **15**(2): 350. 1866; Beille in Lecomte
Fl. Gén. Indo-Chine **5**: 579. 1927 (based on *Cathetus fasciculata* Lour.), non *P.
fasciculatus* Poir. (1804).
Phyllanthus cinerascens Hook. & Arn. Bot. Beechey's Voy. 211. 1836.
Diasperus fasciculatus O. Ktz. Rev. Gen. Pl. 599. 1891 (based on *Cathetus fasciculatus*
Lour.).
"Habitat agrestis in collibus Cochinchinae." Loureiro's type is preserved in the
herbarium of the British Museum and its status was first determined by Seemann (Bonplandia 7: 48. 1859); Mueller-Arg. also examined the type and gives an ample description
of it (DC. Prodr. **15**(2): 350. 1866). It is represented by *Squires 406* and *Clemens 3183*
from Tourane and Hue, Anam, and is very common on dry hills near Canton.

Phyllanthus emblica Linn. Sp. Pl. 982. 1753; Lour. Fl. Cochinch. 553. 1790, ed. Willd.
677. 1793, Anamese *cây boung ngot*, Chinese *hac mîn săn;* Moore in Journ. Bot.
63: 288. 1925.
"Habitat agrestis in Cochinchina, & China." In my original manuscript of 1919
this was placed as certainly representing the Linnaean species. Moore, however, states
that the unsatisfactory Loureiro specimen in the herbarium of the British Museum is
quite different in foliage from *Phyllanthus emblica* Linn., although a specimen of the
section *Emblica* may be represented by it. *Phyllanthus emblica* Linn., as currently interpreted, is distinctly variable in its vegetative characters, is common in Kwangtung Province,
and is very common in Indo-China. The Anamese names given by Beille (Lecomte Fl.
Gén. Indo-Chine **5**: 580. 1927) *kam lam* and *kam lan ko* suggest nothing in common with
the one cited by Loureiro, yet I judge that Loureiro had a form of the common and somewhat variable Linnaean species.

Phyllanthus niruri Linn. Sp. Pl. 981. 1753; Beille in Lecomte Fl. Gén. Indo-Chine **5**:
577. 1927.
Nymphanthus niruri Lour. Fl. Cochinch. 545. 1790, ed. Willd. 665. 1793, Anamese
cây chó de.
"Habitat inculta in hortis, & agris Cochinchinae." Loureiro's binomial was based
on the Linnaean species and should be interpreted by this. A Loureiro specimen in the
herbarium of the Paris Museum labeled merely "*Nympanthus*" has been identified as
Nymphanthus niruri Lour., but Mueller-Arg. has identified it as *Phyllanthus urinaria* Linn.;
there is also a specimen from Loureiro in the herbarium of the British Museum. *Herba
moeroris alba* Rumph. (Herb. Amb. **6**: 41. *pl. 17, f. 1*) is *Phyllanthus niruri* Linn.

Phyllanthus ruber (Lour.) Spreng. Syst. **3**: 22. 1826; Muell.-Arg. in DC. Prodr. **15**(2): 419.
1866; Beille in Lecomte Fl. Gén. Indo-Chine **5**: 607. 1927 (based on *Nymphanthus
rubra* Lour.).
Nymphanthus rubra Lour. Fl. Cochinch. 544. 1790, ed. Willd. 665. 1793, Anamese
cây do dot.
Diasperus ruber O. Ktz. Rev. Gen. Pl. 600. 1891 (based on *Nymphanthus rubra* Lour.).
"Habitat in sylvis Cochinchinae." Mueller's amplified description was based on
Loureiro's type which is preserved in the herbarium of the British Museum; Beille's description is based on several modern collections. The species is also represented by *Clemens
3142, 4394,* and *Squires 405* from Tourane and Hue and it also occurs in Hainan.

Phyllanthus squamifolius (Lour.) Stokes Bot. Mat. Med. **4**: 364. 1812; Spreng. Syst. **3**: 21. 1826; Muell.-Arg. in DC. Prodr. **15**(2): 433. 1866 (based on *Nymphanthus squamifolius* Lour.).

    Nymphanthus squamifolius Lour. Fl. Cochinch. 544. 1790, ed. Willd. 663. 1793, Anamese *cây bay oúc.*

    Diasperus squamifolius O. Ktz. Rev. Gen. Pl. 601. 1891 (based on *Nymphanthus squamifolius* Lour.).

"Habitat in sylvis montanis Cochinchinae." This is known only from Loureiro's short description and I am unable to place it among the 42 species of the genus admitted by Beille (Lecomte Fl. Gén. Indo-Chine **5**: 571–608. 1927) who does not account for it. Mueller-Arg. (DC. Prodr. **15**(2): 433. 1866) leaves it among the species "non satis notae," with the query "an a vulgari *P. urinaria* diversus?"; but Loureiro describes it as a large tree, the wood suitable for house construction. If a *Phyllanthus*, Loureiro interpreted the branchlets as pinnate leaves. A Loureiro specimen listed as being among those received from him in the herbarium of the British Museum, has not been located.

Phyllanthus urinaria Linn. Sp. Pl. 982. 1753; Lour. Fl. Cochinch. 554. 1790, ed. Willd. 677. 1793, Anamese *co sua*, Chinese *fí yòng tsào;* Muell.-Arg. in DC. Prodr. **15**(2): 364. 1866; Beille in Lecomte Fl. Gén. Indo-Chine **5**: 586. 1927.

"Habitat, & a me observata in Cochinchina, Cantone Sinarum, & in Africa orientali." Loureiro's description applies to the Linnaean species, a common and widely distributed weed in the warmer parts of the Old World.

Phyllanthus sp.

    Tricarium cochinchinense Lour. Fl. Cochinch. 557. 1790, ed. Willd. 681. 1793, Anamese *cây trâm ung.*

    Phyllanthus cochinchinensis Muell.-Arg. in DC. Prodr. **15**(2): 417. 1866 (based on *Tricarium cochinchinense* Lour.), non Stokes.

    Diasperus cochinchinensis O. Ktz. Rev. Gen. Pl. 599. 1891 (based on *Tricarium cochinchinense* Lour.).

"Habitat in sylvis Cochinchinae." Mueller-Arg. placed this in the section *Prosorus* next to *Phyllanthus acidissimus* (Blanco) Muell.-Arg. = *Cicca disticha* Linn. = *Cicca acida* (Linn.) Merr. It may prove to be the same as *Cicca acida* (Linn.) Merr. which Loureiro described on the preceding page as *Cicca racemosa* Lour. although, if Loureiro was correct in the habitat cited, this is unlikely. Beille, in his treatment of *Phyllanthus* (Lecomte Fl. Gén. Indo-Chine **5**: 571–608. 1927) does not mention Loureiro's species. A specimen listed as being among those received from Loureiro has not been located in the herbarium of the British Museum.

Cicca Linnaeus

Cicca acida (Linn.) Merr. Interpret. Herb. Amb. 314. 1917.

    Averrhoa acida Linn. Sp. Pl. 428. 1753.

    Cicca disticha Linn. Mant. **1**: 124. 1767.

    Phyllanthus distichus Muell.-Arg. in DC. Prodr. **15**(2): 413. 1866; Beille in Lecomte Fl. Gén. Indo-Chine **5**: 594. 1927.

    Phyllanthus acidus Skeels in U. S. Dept. Agr. Bur. Pl. Ind. Bull. **148**: 17. 1909.

Cicca racemosa Lour. Fl. Cochinch. 556. 1790, ed. Willd. 680. 1793, Anamese *cây tam buot.*

" Habitat frequens in regno Champáva: colitur raro in metropoli Cochinchinae." Loureiro's description applies unmistakably to the Linnaean species; his type is preserved in the herbarium of the British Museum.   Beille, however, does not account for Loureiro's species, yet for *Phyllanthus distichus* Muell.-Arg. he cites a cognate form of the same Anamese name that Loureiro gives, *cây tam mot.*   Mueller-Arg. (DC. Prodr. 15(2): 417. 1866) considered that Loureiro's species was probably identical with *Cicca acidissima* Blanco which is a synonym of *Cicca acida* (Linn.) Merr.   The species is commonly planted in the Old World tropics for its acid edible fruit.

Glochidion J. R. & G. Forster

Glochidion pilosum (Lour.) comb. nov.
> *Nymphanthus pilosus* Lour. Fl. Cochinch. 544. 1790, ed. Willd. 664. 1793, Anamese *cây bot múoi.*
> *Emblica pilosa* Spreng. Syst. 3: 20. 1826 (based on *Nymphanthus pilosus* Lour.).
> *Phyllanthus pilosus* Muell.-Arg. in DC. Prodr. 15(2): 432. 1866 (based on *Nymphanthus pilosus* Lour.).
> *Diasperus pilosus* O. Ktz. Rev. Gen. Pl. 600. 1891 (based on *Nymphanthus pilosus* Lour.).
> *Glochidion annamense* Beille in Lecomte Fl. Gén. Indo-Chine 5: 627. *f. 74, 18–20.* 1927.

" Habitat in sylvis Cochinchinae."   Loureiro's species has hitherto not been placed. Mueller-Arg. left it among the " species non satis notae," suggesting that *Andrachne* may be the genus described.   I believe it to be the species described by Beille in 1927 as *Glochidion annamense*, and to be represented by *Clemens 4183, 4239,* from Tourane; Loureiro misinterpreted the branchlets with their distichous leaflets as pinnate leaves. Beille does not account for Loureiro's binomial or for the synonyms based upon it.   A specimen from Loureiro listed as being among his plants in the herbarium of the British Museum has not been located.

Glochidion puberum (Linn.) Hutchinson in Sargent Pl. Wils. 2: 518. 1916; Druce in Rept. Bot. Exch. Club Brit. Isles 4: 624. 1917.
> *Agyneia pubera* Linn. Mant. 2: 296. 1771.
> *Nymphanthus chinensis* Lour. Fl. Cochinch. 544. 1790, ed. Willd. 664. 1793, Chinese *siong chu tsao.*
> *Phyllanthus villosus* Poir. in Lam. Encycl. 5: 297. 1804.

" Habitat agrestis prope Cantonem Sinarum."   I believe this to be the correct reduction of Loureiro's species, the indicated fruit characters perhaps having been added to make the description conform to the generic characters of *Nymphanthus*.   Mueller-Arg. followed Sprengel in placing *Nymphanthus chinensis* Lour. as a synonym of *Phyllanthus villosus* Poir.; from the description I take the latter to be a form of *Glochidion puberum* Hutch.

Baccaurea Loureiro

Baccaurea ramiflora Lour. Fl. Cochinch. 661. 1790, ed. Willd. 813. 1793, Anamese *giâu tien;* Muell.-Arg. in DC. Prodr. 15(2): 458. 1866; Pax & Hoffm. in Pflanzenreich 81 (IV–147–XV): 71. 1922; Gagnep. in Lecomte Fl. Gén. Indo-Chine 5: 551. 1927.

Baccaurea cauliflora Lour. Fl. Cochinch. 661. 1790, ed. Willd. 813. 1793, Anamese
giâu dât; Muell.-Arg. in DC. Prodr. **15**(2): 458. 1866; Pax & Hoffm. in Pflanzenreich
81 (IV–147–XV): 70. 1922; Gagnep. in Lecomte Fl. Gén. Indo-Chine **5**: 551. 1927.
Pierardia sapida Roxb. Fl. Ind. ed. 2, **2**: 254. 1832.
Baccaurea sapida Muell.-Arg. in DC. Prodr. **15**(2): 459. 1866; Pax & Hoffm. in
Pflanzenreich **81** (IV–147–XV): 52. 1922; Gagnep. in Lecomte Fl. Gén. Indo-Chine
5: 548. 1927.

Baccaurea ramiflora Lour. is the type of the genus: " Habitat frequens in hortis Cochinchinae." For *B. cauliflora*, which Loureiro states differed from *B. ramiflora* chiefly
or only by its infructescences being borne on the trunk rather than on the branches, he
states: " Habitat in hortis Cochinchinae minus frequens culta"; I do not hesitate in reducing this to *B. ramiflora*. With even less hesitancy I reduce *Baccaurea sapida* Muell.-
Arg. to *B. ramiflora* Lour., as, according to Gagnepain, this species is frequent in Indo-China
and is commonly cultivated; in fact it is the only cultivated *Baccaurea* recorded from
Indo-China of the four that Gagnepain fully describes. Pax and Hoffmann state regarding
both of Loureiro's species: " vix recognoscenda est." In *Baccaurea sapida* Muell.-Arg.
the inflorescences are borne both on the trunks and on the branches. In all the citations
given above under Loureiro's two binomials, the descriptions are compiled from his originals.
Nani hua Rumph. (Herb. Amb. **3**: 21. *pl. 9*), cited by Loureiro as representing *B. ramiflora*,
must be excluded as it represents the very different *B. nanihua* Merr. of Amboina.

Baccaurea sylvestris Lour. Fl. Cochinch. 662. 1790, ed. Willd. 813. 1793, Anamese *cây
lon bon;* Muell.-Arg. in DC. Prodr. **15**(2): 457. 1866; Pax & Hoffm. in Pflanzenreich
81 (IV–147–XV): 61. 1922; Gagnep. in Lecomte Fl. Gén. Indo-Chine **5**: 552. 1927.
Baccaurea annamensis Gagnep. in Bull. Soc. Bot. France **70**: 235. 1923; Gagnep. in
Lecomte Fl. Gén. Indo-Chine **5**: 550. 1927.

" Habitat in sylvis montanis Cochinchinae." In all the references above cited the
descriptions are merely compiled from Loureiro's original. Mueller-Arg. placed the species
in the section *Hedycarpus* with *Baccaurea lanceolata* Muell.-Arg. Moore (Journ. Bot. **63**:
290. 1925) states that the type specimen in the herbarium of the British Museum is a
pistillate one so that its true position within the genus cannot be determined. Chevalier
(Cat. Pl. Jard Bot. Saigon 64. 1919) is in error in reducing *Baccaurea sylvestris* Lour. to
the meliaceous *Lansium domesticum* Correa. Loureiro's extant type, which I examined
in 1930, represents a true *Baccaurea* and I have no hesitation in reducing *Baccaurea
annamensis* Gagnep. to *B. sylvestris* Lour.

Antidesma (Burman) Linnaeus

Antidesma fruticosum (Lour.) Muell.-Arg. in DC. Prodr. **15**(2): 259. 1866; Pax & Hoffm.
in Pflanzenreich **81** (IV–147–XV): 122. 1922 (based on *Rhytis fruticosa* Lour.).
Rhytis fruticosa Lour. Fl. Cochinch. 660. 1790, ed. Willd. 812. 1793, Anamese *cây
chòi mòi;* Moore in Journ. Bot. **63**: 290. 1925.

" Habitat in sylvis Cochinchinae." Mueller's short description was based on Loureiro's type in the herbarium of the British Museum, while Pax and Hoffmann, who place
the species in the section *Roxburghiana*, give an even shorter description; Moore (Journ.
Bot. **63**: 290. 1925) has published a full description based on the type specimen. Gagnepain

places both binomials as doubtful synonyms of *Antidesma japonicum* Sieb. & Zucc. (Lecomte Fl. Gén. Indo-Chine 5: 518. 1927) which is a manifest error. Loureiro's description of the fruit as 3-seeded is an error on his part, and, as Moore notes, he mistook the pistillode for the ovary and erroneously interpreted some of the flowers as perfect. Loureiro's species is represented by *Clemens 4199, 4012,* from Hue, the classical locality, and from Tourane.

Bridelia Willdenow

Bridelia monoica (Lour.) Merr. in Philip. Journ. Sci. Bot. **13**: 142. 1918 (based on *Clutia monoica* Lour.).

Clutia monoica Lour. Fl. Cochinch. 638. 1790, ed. Willd. 784. 1793, Chinese *xún ti fūm.*

Cleistanthus monoicus Muell.-Arg. in DC. Prodr. **15**(2): 508. 1866 (based on *Clutia monoica* Lour.); Jabl. in Pflanzenreich **65** (IV–147–VIII): 53. 1915.

Bridelia loureiri Hook. & Arn. Bot. Beechey's Voy. 211. 1836 (based on *Clutia monoica* Lour.).

Kaluhaburunghos monoicus O. Ktz. Rev. Gen. Pl. 607. 1891 (based on *Clutia monoica* Lour.).

Bridelia tomentosa Blume Bijdr. 597. 1825; Jabl. in Pflanzenreich **65** (IV–147–VIII): 58. 1915, cum syn.

" Habitat spontanea prope Cantonem, Sinarum." This species is common in thickets in the vicinity of Canton but no *Cleistanthus* occurs in the vicinity of that city. Loureiro's description applies absolutely to the species currently known as *Bridelia tomentosa* Blume, one of very wide geographic distribution, except in the description of the fruit as a 3-celled capsule. It is evident that here Loureiro described his species from flowering specimens and added fruit characters to make his description conform to the characters of *Clutia* in which genus he erroneously placed his species.

Croton Linnaeus

Croton cascarilloides Raeusch. Nomencl. ed. 3, 280. 1797 (based on *Croton punctatus* Lour.).

Croton punctatus (non Jacq.) Lour. Fl. Cochinch. 581. 1790, ed. Willd. 712. 1793, Anamese *bac thau rùng;* Muell.-Arg. in DC. Prodr. **15**(2): 565. 1866; Gagnep. in Lecomte Fl. Gén. Indo-Chine 5: 290. 1925.

Rottlera punctata A. Juss. ex Spreng. Syst. **3**: 877. 1826 (based on *Croton punctatus* Lour.).

Croton cumingii Muell.-Arg. in Linnaea **34**: 101. 1865, DC. Prodr. **15**(2): 566. 1866; Gagnep. in Lecomte Fl. Gén. Indo-Chine **5**: 264. 1925.

Oxydectes punctata O. Ktz. Rev. Gen. Pl. 612. 1891 (based on *Croton punctatus* Lour.).

" Habitat in sylvis Cochinchinae." Loureiro's type is preserved in the herbarium of the British Museum and Mueller's amplified description was based upon it. Mueller-Arg. apparently failed to realize that this was no other than the species he also described as *Croton cumingii,* based on a Philippine specimen. In any case, Loureiro's specific name is invalidated by *Croton punctatus* Jacq. (Coll. **1**: 166. 1786) which is apparently the oldest valid name for the American species currently known as *Croton argyracanthus* Michx. Gagnepain retained *Croton punctatus* Lour. as an imperfectly known species. It is represented by *Clemens 3929* from Hue, the classical locality. The species extends from the

Malay Peninsula and Indo-China to Borneo, Hainan, Formosa, the Riu Kiu Islands and the Philippines. In Index Kewensis *Croton cascarilloides* Raeusch., which is based solely on *Croton punctatus* Lour., is erroneously reduced to *Mallotus philippensis* (Lam.) Muell.-Arg. *Croton cascarilloides* Geiseler (Crot. Monog. 8. 1807; Muell.-Arg. in DC. Prodr. 15(2): 555. 1866) based on *Croton cascarilla* Lam., non Linn., of Santo Domingo, is an entirely different species for which a new binomial is needed; *Oxydectes cascarilloides* O. Ktz. (Rev. Gen. Pl. 611. 1891) was based on this, not on *C. cascarilloides* Raeusch.

Croton crassifolius Geisel. Crot. Monogr. 19. 1807.

> *Tridesmis tomentosa* Lour. Fl. Cochinch. 576. 1790, ed. Willd. 707. 1793, Chinese *ca xí mà*, non *Croton tomentosus* Link.
>
> *Tridesmis hispida* Lour. Fl. Cochinch. 576. 1790, ed. Willd. 706. 1793, Chinese *kí quăt yòng*, non *Croton hispidus* H.B.K.
>
> *Croton tomentosus* Muell.-Arg. in Linnaea **34**: 107. 1865, DC. Prodr. 15(2): 588. 1866; Gagnep. in Lecomte Fl. Gén. Indo-Chine **5**: 262. 1925 (based on *Tridesmis tomentosa* Lour.), non Link.
>
> *Oxydectes tomentosa* O. Ktz. Rev. Gen. Pl. 613, 614. 1891 (based on *Tridesmis tomentosa* Lour.).

For *Tridesmis tomentosa* Loureiro states: " Habitat agrestis circa Cantonem Sinarum." The species is still not uncommon there. Loureiro's type is preserved in the herbarium of the Paris Museum where it was examined by Mueller-Arg. For *Tridesmis hispida* Loureiro states: " Habitat in dumetis circa Cantonem Sinarum." This I believe to be only a depauperate form of *Croton crassifolius* Geisel., as such forms still occur in dry sterile places near Canton. It may be noted that although Loureiro describes the leaves as hispid in the diagnosis, yet in the short description that follows he describes them as pilose. The specific name is invalidated in *Croton* by *C. hispidus* H.B.K. Geiseler's species was based on a specimen collected by Dahl in Hainan.

Croton lasianthus Pers. Syn. **2**: 586. 1807; Muell.-Arg. in DC. Prodr. 15(2): 602. 1866; Gagnep. in Lecomte Fl. Gén. Indo-Chine **5**: 290. 1925 (based on *Croton lanatus* Lour.).

> *Croton lanatus* (non Lam.) Lour. Fl. Cochinch. 581. 1790, ed. Willd. 713. 1793, Anamese *cây tlai*.
>
> *Croton erioanthemum* Sm. in Rees Cyclop. **10**: no. 21. 1808 [22] (based on *Croton lanatus* Lour.).
>
> *Triplandra lanata* Raf. Sylva Tellur. 63. 1838 (based on *Croton lanatus* Lour.).
>
> *Oxydectes lasiantha* O. Ktz. Rev. Gen. Pl. 612. 1891 (based on *Croton lanatus* Lour.).

" Habitat in sylvis montanis Cochinchinae." Loureiro's type is preserved in the herbarium of the British Museum and Mueller's amplified description was based upon it; the latter notes that it is similar to *C. laevifolius* Blume differing in the form of its calyces and in the number of stamens. Gagnepain placed the species among the imperfectly known ones: " Voisin de *C. laevifolius* Bl.? "; from the data available I cannot place the species with certainty among the 40 representatives of the genus admitted by him (Lecomte Fl. Gén. Indo-Chine **5**: 256–290. 1925). It may be represented by *Clemens 4233, 4386,*

---

[22] The title page for the volume is dated 1819. It was issued in 1808 *fide* Jackson in Journ. Bot. **34**: 310. 1896.

4397, from Tourane, a common shrub in thickets, but in all these specimens the leaves are finely toothed. *Croton erioanthemum* Sm. appears in Index Kewensis as *C. erianthum* Sm.

Croton tiglium Linn. Sp. Pl. 1004. 1753; Lour. Fl. Cochinch. 582. 1790, ed. Willd. 714.
1793, Anamese *ba dâu taù*, Chinese *pă téu;* Gagnep. in Lecomte Fl. Gén. Indo-Chine **5**: 285. 1925.

" Habitat incultum in Cochinchina, & in China." The common Linnaean species was correctly interpreted by Loureiro and the two pre-Linnaean synonyms cited by him are correctly placed.

Croton sp.

Croton lacciferus (non Linn.) Lour. Fl. Cochinch. 582. 1790, ed. Willd. 714. 1793, Anamese *cây kánh kién.*

" Habitat agreste in provinciis Australibus Cochinchinae, & in Cambodia." The description is clearly that of a *Croton* and doubtless appertains to one of the 40 species of the genus recognized by Gagnepain (Lecomte Fl. Gén. Indo-Chine **5**: 256–290. 1925) as occurring in Indo-China; I am, however, unable to suggest which. *Halecus terrestris* Rumph. (Herb. Amb. **3**: 197. *pl. 127*), cited by Loureiro as representing the species, is *Macaranga involucrata* (Roxb.) Baillon and *Ricinoides aromatica* Burm. Zeyl. 201. *pl. 91*, represents *Croton aromaticus* Linn. var. *lacciferus* (Linn.) Trimen.

Claoxylon A. L. de Jussieu

Claoxylon hainanense Pax & Hoffm. in Pflanzenreich **63** (IV–147–VII): 128. 1914; Gagnep. in Lecomte Fl. Gén. Indo-Chine **5**: 421. 1926.

Mercurialis indica (non Linn.) Lour. Fl. Cochinch. 628. 1790, ed. Willd. 771. 1793, Anamese *rau mai, luc mai.*

" Habitat tam culta, quam spontanea in Cochinchina." Mueller-Arg. (DC. Prodr. **15**(2): 798. 1866) correctly excluded this from *Mercurialis* with the comment: " Forte Claoxyli species, sed forma antherarum, si bene observata, obstat. An potius species Malloti? "; Pax and Hoffmann (Pflanzenreich **63** (IV–147–VII): 281. 1914) query: " An species *Malloti?* ". Loureiro's species is safely the same as *Claoxylon hainanense* Pax & Hoffm., with the characters of which his description agrees, a species that occurs near Hue, and for which Gagnepain, independent of any suggestion from Loureiro, cites an Anamese name *cây luc mâi.*

Mallotus Loureiro

Mallotus anamiticus O. Ktz. Rev. Gen. Pl. 608. 1891; Pax & Hoffm. in Pflanzenreich **63** (IV–147–VII): 204. 1914.

Coelodiscus anamiticus Gagnep. in Lecomte Fl. Gén. Indo-Chine **5**: 375. 1926.

Ricinus tanarius (non Linn.) Lour. Fl. Cochinch. 584. 1790, ed. Willd. 717. 1793, Anamese *bach dàn nam.*

" Habitat in sylvis Cochinchinae." Mueller-Arg., who examined Loureiro's type preserved in the herbarium of the British Museum, referred it to *Mallotus floribundus* (Blume) Muell.-Arg. It is, however, clearly *M. anamiticus* O. Ktz., a species very closely allied to *M. floribundus*, but one that Pax and Hoffmann left among the " species dubiae vel incertae." Kuntze's type, from Tourane, is preserved in the herbarium of the New York Botanical Garden; the species is also represented by *Squires 175* and *Clemens 3130*,

4099, from Mount Bana and Tourane. *Tanarius minor* Rumph. (Herb. Amb. **3**: 190. *pl. 121*), cited by Loureiro as representing the species he described, typifies *Macaranga tanarius* (L.) Muell.-Arg., a species very different from *Mallotus anamiticus* O. Ktz.

Mallotus apelta (Lour.) Muell.-Arg. in Linnaea **34**: 189. 1865, DC. Prodr. **15**(2): 963. 1866; Pax & Hoffm. in Pflanzenreich **63** (IV–147–VII): 171. 1914; Gagnep. in Lecomte Fl. Gén. Indo-Chine **5**: 354. 1925 (based on *Ricinus apelta* Lour.).

  Ricinus apelta Lour. Fl. Cochinch. 585. 1790, ed. Willd. 718. 1793, Chinese *xān pĕ xú.*

  Croton chinensis Geisel. Crot. Monog. 24. 1807.

  Rottlera chinensis Juss. Euph. Tent. 33. 1824.

  Rottlera cantoniensis Spreng. Syst. **3**: 878. 1826 (based on *Ricinus apelta* Lour.).

  " Habitat agrestis circa Cantonem Sinarum." A well-known species common in eastern and southern China and fairly common in thickets near Canton. It is the form indicated by Pax as *Mallotus apelta* var. *chinensis* (Geisel.) Pax & Hoffm. (Pflanzenreich **63** (IV–147–VII): 171. 1914) but there seems to be no reason for distinguishing the variety. Loureiro's type is preserved in the herbarium of the Paris Museum where it was examined by Mueller-Arg.

Mallotus paniculatus (Lam.) Muell.-Arg. in Linnaea **34**: 189. 1865, DC. Prodr. **15**(2): 965. 1866.

  Croton paniculatus Lam. Encycl. **2**: 207. 1786.

  Mallotus cochinchinensis Lour. Fl. Cochinch. 635. 1790, ed. Willd. 781. 1793, Anamese *cây bét;* Pax & Hoffm. in Pflanzenreich **63** (IV–147–VII): 166. 1914; Gagnep. in Lecomte Fl. Gén. Indo-Chine **5**: 355. 1925.

  Echinus trisulcus Lour. Fl. Cochinch. 633. 1790, ed. Willd. 778. 1793, Anamese *cây hón.*

  Mappa cochinchinensis Spreng. Syst. **3**: 878. 1826 (based on *Echinus trisulcus* Lour.).

  Lasipana tricuspis Raf. Sylva Tellur. 22. 1838 (based on *Echinus trisulcus* Lour.).

  For *Mallotus cochinchinensis* Loureiro states: " Habitat sepes, & hortos minus cultos Cochinchinae, & Chinae." This species is the type of the genus and is a common, widely distributed, well understood one; Loureiro's type is preserved in the herbarium of the British Museum. For *Echinus trisulcus* Loureiro states: " Habitat agrestis loca plana Cochinchinae." The generic description shows some discrepancies when critically compared with *Mallotus*, but these were apparently due to errors of observation on the part of Loureiro. The genus and species seem safely to be the same as *Mallotus paniculatus* (Lam.) Muell.-Arg. *Ulassium mas* Rumph. (Herb. Amb. **3**: 42. *pl. 23*) cited by Loureiro as doubtfully representing *Echinus trisulcus*, must be excluded as it represents the totally different rubiaceous *Adina fagifolia* (Teysm. & Binn.) Valeton. A specimen from Loureiro listed as being among his plants in the herbarium of the British Museum has not been located.

Alchornea Swartz

Alchornea rugosa (Lour.) Muell.-Arg. in Linnaea **34**: 170. 1865, DC. Prodr. **15**(2): 905. 1866; Pax & Hoffm. in Pflanzenreich **63** (IV–147–VII): 243. 1914; Merr. in Philip. Journ. Sci. **15**: 244. 1919 (based on *Cladodes rugosa* Lour.).

  Cladodes rugosa Lour. Fl. Cochinch. 574. 1790, ed. Willd. 704. 1793, Anamese *cây mót.*

  Alchornea hainanensis Pax & Hoffm. in Pflanzenreich **63** (IV–147–VII): 242. 1914.

" Habitat in sylvis Cochinchinae." Loureiro's species is a widely distributed, well-known one extending from Burma through Malaysia and the Philippines to New Guinea. His type is preserved in the herbarium of the British Museum.

Alchornea trewioides (Champ.) Muell.-Arg. in Linnaea **34**: 168. 1865, DC. Prodr. **15**(2): 901. 1866.

Stipellaria trewioides Champ. in Hook. Journ. Bot. Kew Gard. Miscel. **6**: 3. 1854.

Croton aromaticum (non Linn.) Lour. Fl. Cochinch. 583. 1790, ed. Willd. 715. 1793, Chinese *pa táu yòng*.

" Habitat incultum circa Cantonem Sinarum." The description in general applies closely to *Alchornea trewioides* Muell.-Arg. which is common in Kwangtung Province, except in the statement that the styles (or stigmas) are branched. *Halecus littorea* Rumph. (Herb. Amb. **3**: 196. *pl. 126*), cited by Loureiro as representing his concept of *Croton aromaticum*, must be excluded as it represents *Mallotus tiliifolius* (Blume) Muell.-Arg.

Acalypha Linnaeus

Acalypha australis Linn. Sp. Pl. 1004. 1753; Gagnep. in Lecomte Fl. Gén. Indo-Chine **5**: 338. 1925.

Urtica gemina Lour. Fl. Cochinch. 558. 1790, ed. Willd. 682. 1793, Anamese *nang hai tlon lá*.

Acalypha gemina Spreng. Syst. **3**: 880. 1826 (based on *Urtica gemina* Lour.).

" Habitat agrestis in Cochinchina." Loureiro's description agrees closely with the characters of the common and widely distributed *Acalypha australis* Linn., which is certainly the correct disposition of the species.

Acalypha boehmerioides Miq. Fl. Ind. Bat. Suppl. 459. 1861; Gagnep. in Lecomte Fl. Gén. Indo-Chine **5**: 337. 1925.

Urtica pilosa Lour. Fl. Cochinch. 558. 1790, ed. Willd. 682. 1793, Anamese *nang hai loung*, non *Acalypha pilosa* Cav.

" Habitat agrestis in Cochinchina." The description agrees closely with the characters of Miquel's species, but the indicated habitat is better for the even more common *Acalypha indica* Linn.; both species occur in the vicinity of Hue and Tourane, but *A. boehmerioides* Miq. is apparently the more common plant there. It is represented by *Bauche & Couderc*, cited by Gagnepain, and *Clemens 3928*. Loureiro interpreted the characteristic lobed bracts subtending the pistillate flowers as 12-fid calyces. Mueller-Arg. (DC. Prodr. **15**(2): 880. 1866) erred in reducing *Urtica pilosa* Lour. to the American *Acalypha poiretii* Spreng., and he also erred in citing Loureiro's binomial as *Urtica hispida* rather than *U. pilosa*. Gagnepain does not account for Loureiro's species in his treatment of the Euphorbiaceae or the Urticaceae of Indo-China (Lecomte Fl. Gén. Indo-Chine **5**: 229–673, 828–921. 1925–29).

Ricinus (Tournefort) Linnaeus

Ricinus communis Linn. Sp. Pl. 1007. 1753; Lour. Fl. Cochinch. 584. 1790, ed. Willd. 716. 1793, Anamese *cây du du deau*, Chinese *pi mâ, hô mâ;* Gagnep. in Lecomte Fl. Gén. Indo-Chine **5**: 327. 1925.

" Habitat incultus, cultusque in Cochinchina, & China." The Linnaean species, the very common castor oil plant, was correctly interpreted by Loureiro.

Homonoia Loureiro

Homonoia riparia Lour. Fl. Cochinch. 637. 1790, ed. Willd. 783. 1793, Anamese *rì rì bò foung;* Gagnep. in Lecomte Fl. Gén. Indo-Chine **5**: 330. *f. 38. 5–8.* 1925.

"Habitat spontanea ripas fluminum in Cochinchina." It is the type of the genus and a very well-known species of wide distribution in the Indo-Malaysian region, found only along and in the beds of swiftly flowing streams. Loureiro's type is preserved in the herbarium of the British Museum.

Aleurites J. R. & G. Forster

Aleurites moluccana (Linn.) Willd. Sp. Pl. **4**: 590. 1805; Gagnep. in Lecomte Fl. Gén. Indo-Chine **5**: 291. 1925.

Jatropha moluccana Linn. Sp. Pl. 1006. 1753.

Juglans camirium Lour. Fl. Cochinch. 573. 1790, ed. Willd. 702. 1793, Anamese *dêau lai.*

"Habitat agrestis, cultaque in Cochinchina." Loureiro's description applies to the common candlenut. He took his specific name from *Camirium* Rumph. (Herb. Amb. 2: 180. *pl. 58*) which is a synonym of the Linnaean species.

Aleurites montana (Lour.) Wilson in Bull. Imper. Inst. **11**: 460. 1913; Pax & Hoffm. in Pflanzenreich 68 (IV–147–XIV): 8. 1919 (based on *Vernicia montana* Lour.).

Vernicia montana Lour. Fl. Cochinch. 587. 1790, ed. Willd. 721. 1793, Anamese *cây dêau son,* Chinese *tong xú.*

Dryandra vernicia Correa in Ann. Mus. Hist. Nat. (Paris) **8**: 69. *pl. 32.* 1806 (based on *Vernicia montana* Lour.).

Elaeococca vernicia A. Juss. ex Spreng. Syst. **3**: 884. 1826 (based on *Vernicia montana* Lour.).

Aleurites vernicia Hassk. in Flora **25**: Beibl. 2: 40. 1842; Gagnep. in Lecomte Fl. Gén. Indo-Chine **5**: 1093. 1931 (based on *Elaeococca vernicia* A. Juss., *Vernicia montana* Lour.).

Aleurites cordata Gagnep. in Lecomte Fl. Gén. Indo-Chine **5**: 294. 1925, non R. Br.

"Habitat in sylvis montanis Cochinchinae, simul & in China." Loureiro's type is preserved in the herbarium of the British Museum. The species has long been confused with *Aleurites cordata* (Thunb.) R. Br. of Japan and northern China. Gagnepain in the *errata* (Lecomte Fl. Gén. Indo-Chine **5**: 1093. 1931) adopts *Aleurites vernicia* Hassk. and adds the two other synonyms based on Loureiro's binomial, which he did not account for in his original consideration of the species; here I accept the oldest valid specific name. Correa's figure was based on Loureiro's specimen in the Banksian herbarium now at the British Museum.

Jatropha Linnaeus

Jatropha curcas Linn. Sp. Pl. 1006. 1753; Gagnep. in Lecomte Fl. Gén. Indo-Chine **5**: 324. 1925.

Croton moluccanus (non Linn.) Lour. Fl. Cochinch. 583. 1790, ed. Willd. 716. 1793, Anamese *ba dâu nam.*

"Habitat cultum in hortis Cochinchinae." Loureiro's description manifestly does not apply to *Croton moluccanus* Linn. but does conform to the characters of the common and widely planted *Jatropha curcas* Linn., except as to the statement that the leaves were tomentose beneath; this was manifestly taken from the Linnaean diagnosis cited. Plukenet's figure cited by Loureiro does not represent the species the latter described. Gagnepain gives, as one of the Anamese names for *Jatropha curcas* Linn., *cây bâ dâu*, which confirms the correctness of this reduction. The conventional interpretation of *Croton moluccanus* Linn. is *Mallotus moluccanus* Muell.-Arg. = *Melanolepis moluccana* Pax & Hoffm.; the correct interpretation is in part *Aleurites moluccana* Willd., and in part *Givotia rottleriformis* Griff. The oldest valid name for the species long masquerading under the Linnaean specific name is *Melanolepis multiglandulosa* (Reinw.) Reichb. f. & Zoll. (Merrill, Interpret. Herb. Amb. 318. 1917).

Manihot (Tournefort) Adanson

Manihot esculenta Crantz Inst. 1: 167. 1766.

> *Jatropha manihot* Linn. Sp. Pl. 1007. 1753.
>
> *Jatropha janipha* (non Linn.) Lour. Fl. Cochinch. 585. 1790, ed. Willd. 718. 1793, Anamese *bach phu tu*, Chinese *pĕ fú tsù*.
>
> *Manihot utilissima* Pohl Pl. Bras. Ic. Descr. 1: 32. *pl. 24.* 1827; Pax in Pflanzenreich **44** (IV–147–II): 67. *f. 24.* 1910.

"Habitat inculta apud Sinas." The description clearly applies to the common manioc, *Manihot esculenta* Crantz. The reference to Jacquin (Amer. *pl. 162. f. 1*) is to be excluded as it represents a species of *Jatropha*. What is apparently the oldest valid specific name is here adopted for this well-known species.

Codiaeum [23] (Rumphius) A. L. de Jussieu

Codiaeum variegatum (Linn.) Blume Bijdr. 606. 1825; Gagnep. in Lecomte Fl. Gén. Indo-Chine **5**: 408. 1926.

> *Croton variegatus* Linn. Sp. Pl. 1199. 1753.
>
> *Phyllaurea codiaeum* Lour. Fl. Cochinch. 575. 1790, ed. Willd. 705. 1793, Anamese *hùynh bá lá.*
>
> *Chrozophylla elliptica* Raf. Sylva Tellur. 64. 1838.

"Habitat culta in Cochinchina, & China: puto, quod etiam agrestis." *Phyllaurea* was proposed as a new generic name for *Croton variegatus* Linn., the specific name *codiaeum* being taken from *Codiaeum medium chrysosticon* Rumph. (Herb. Amb. 4: 65. *pl. 25*) which typifies the Linnaean species. Loureiro's generic name is much older than that of Jussieu, but the latter is the conserved one. It is certain that Loureiro saw only cultivated plants as this species is not native of Indo-China. A specimen from Loureiro is preserved in the herbarium of the British Museum. *Chrozophylla* Raf. was a new generic name for the group including *Phyllaurea* Lour.

Excoecaria Linnaeus

Excoecaria agallocha Linn. Syst. ed. 10, 1288. 1759; Pax & Hoffm. in Pflanzenreich **52** (IV–147–V): 165. *f. 30.* 1912.

> *Commia cochinchinensis* Lour. Fl. Cochinch. 606. 1790, ed. Willd. 743. 1793, Anamese *cây son giá.*

[23] *Codiaeum* (Rumphius) A. Jussieu (1824), conserved name, Vienna Code; an older one is *Phyllaurea* Loureiro (1790).

" Habitat agrestis prope litora in Cochinchina." Loureiro's type is preserved in the herbarium of the British Museum. Seemann who examined it (Journ. Bot. 1: 281. 1863) notes that the apparent structural differences between *Commia* and *Excoecaria* resolve themselves into errors of description on Loureiro's part. The species is very common along the seashore throughout the Indo-Malaysian region.

Excoecaria cochinchinensis Lour. Fl. Cochinch. 612. 1790, ed. Willd. 750. 1793, Anamese *cây lieo do;* Moore in Journ. Bot. **63**: 289. 1925.

Excoecaria bicolor Hassk. in Nat. Tijdschr. Nederl. Ind. **10**: 158. 1855 (Retzia **1**: 158); Retzia ed. alt. 31. 1858; Lecomte Fl. Gén. Indo-Chine **5**: 404. *f. 47. 1-7.* 1926.

" Habitat in Cochinchina, & China, ubi colitur propter foliorum rubrorum pulchritudinem." Loureiro's type is preserved in the herbarium of the British Museum, as noted by Mueller-Arg. (DC. Prod. **15**(2): 1215. 1866) who erroneously considered it to be conspecific with the Indian *E. crenulata* Wight. Pax & Hoffmann (Pflanzenreich **52** (IV–147– V): 160. 1912) state: " Verisimiliter cum *E. bicolore* convenit *Excoecaria cochinchinensis* Lour. Fl. Cochinch. II (1790) 612, sed bracteae pluriflorae describuntur. Cum *E. crenulata* vel *oppositifolia* species Loureiroana autem non quadrat. Loureiro speciem suam etiam cultam tantum vidit." *Excoecaria bicolor* Hassk. occurs in Indo-China, is still cultivated in Canton, and Pax and Hoffmann's surmise is verified by Moore who states, after examining Loureiro's type (herb. Brit. Mus.), that Loureiro's species is the same as *E. bicolor* Hassk.

Sapium (P. Browne) Jacquin

Sapium cochinchinense (Lour.) O. Ktz. Rev. Gen. Pl. **3**(2): 293. 1898; Pax & Hoffm. in Pflanzenreich **52** (IV–147–V): 252. 1912; Gagnep. in Lecomte Fl. Gén. Indo-Chine **5**: 401. 1926 (based on *Triadica cochinchinensis* Lour.).

Triadica cochinchinensis Lour. Fl. Cochinch. 610. 1790, ed. Willd. 749. 1793, Anamese *cây soi tiá, cây cha dam;* Moore in Journ. Bot. **63**: 288. 1925.

Stillingia cochinchinensis Baill. Adansonia **1**: 351. 1861 (based on *Triadica cochinchinensis* Lour.).

Excoecaria loureiroana Muell.-Arg. in DC. Prodr. **15**(2): 1217. 1866 (based on *Triadica cochinchinensis* Lour.).

" Habitat in sylvis Cochinchinae." Loureiro's type is preserved in the herbarium of the British Museum and Moore (Journ. Bot. **63**: 288. 1925) has given additional much needed descriptive data based upon it, as the description of Mueller-Arg., although based on Loureiro's type, is imperfect. Gagnepain merely includes the species with a short description as one of doubtful status. Pax and Hoffmann place it next to the African *S. ellipticum* (Hochst.) Pax, while Moore states that it seems to be allied to that species. Mr. A. W. Exell, who kindly re-examined Loureiro's type at my request, reports that it is not *Sapium baccatum* Roxb. which has sharply long-acuminate leaves, whereas Loureiro's specimen has obtuse or very shortly blunt-acuminate leaves.

Sapium sebiferum (Linn.) Roxb. Fl. Ind. ed. 2, **3**: 693. 1832; Gagnep. in Lecomte Fl. Gén. Indo-Chine **5**: 398. 1926.

Croton sebiferus Linn. Sp. Pl. 1004. 1753.

Triadica sinensis Lour. Fl. Cochinch. 610. 1790, ed. Willd. 749. 1793, Chinese *ū khau mŏ.*

Triadica chinensis Spreng. Syst. 1: 93. 1825 (based on *Triadica sinensis* Lour.).

Seborium chinense Raf. Sylva Tellur. 63. 1838 (based on *Croton sebiferus* Linn.).

" Habitat agrestis circa Cantonem Sinarum." Loureiro's description applies to the Linnaean species, the type of the latter being a specimen collected by Osbeck near Canton. The species is common near Canton and is locally known as *oo k'au shue*.

Euphorbia Linnaeus

Euphorbia antiquorum Linn. Sp. Pl. 450. 1753; Lour. Fl. Cochinch, 298. 1790, ed. Willd.
365. 1793, Anamese *cây xuong raong, thanh laong;* Gagnep. in Lecomte Fl. Gén.
Indo-Chine 5: 240. *f. 26. 1–7.* 1925.

" Habitat in sepibus Cochinchinae." Loureiro's description, although short and imperfect, conforms to the characters of the Linnaean species as far as it goes and there is little doubt that he had specimens representing it. The description agrees almost equally well with the characters of *Euphorbia trigona* Haworth, but the latter species is not recorded from Indo-China.

Euphorbia helioscopia Linn. Sp. Pl. 459. 1753.

Sedum anacampseros (non Linn.) Lour. Fl. Cochinch. 287. 1790, ed. Willd. 353. 1793,
Anamese *truong sinh tlòn lá,* Chinese *pă touc sān.*

" Habitat in Cochinchina, & China. Hic florentem observavi non illic, quamvis non raro occurrat." As far as the description goes a *Sedum* seems to be indicated, but no *Sedum* is known from low altitudes in Indo-China, and I know of no Chinese species to which Loureiro's description applies. It is probable that Loureiro had sterile material from cultivated plants of *Euphorbia helioscopia* Linn., in spite of his description of the leaves as opposite, and added the brief flower and fruit characters to make his description conform to the characters of *Sedum;* or more likely his material came from an herbalist. This species is cultivated in southeastern China and also occurs there in waste places; it is not recorded from Indo-China.

Euphorbia neriifolia Linn. Sp. Pl. 451. 1753; Lour. Fl. Cochinch. 298. 1790, ed. Willd.
366. 1793, Anamese *xuong raong rao;* Gagnep. in Lecomte Fl. Gén. Indo-Chine 5:
239. 1925.

? Euphorbia edulis Lour. Fl. Cochinch. 298. 1790, ed. Willd. 365. 1793, Anamese
xuong raong lá.

For *Euphorbia neriifolia* Loureiro states: " Habitat sepes Cochinchinae "; it is evident that he correctly interpreted the Linnaean species. For *E. edulis* he states: " Habitat culta in hortis Cochinchinae, puto, quod etiam agrestis," and notes that the leaves, cooked with other pot herbs, are eaten; it seems doubtful if this is correct. Gagnepain does not mention Loureiro's species. I suspect it to be a form of *E. neriifolia* Linn. but this needs verification from field work. It is highly probable that the description was based in part on hear-say evidence.

Euphorbia tirucalli Linn. Sp. Pl. 452. 1753; Lour. Fl. Cochinch. 299. 1790, ed. Willd. 366.
1793, Anamese *cây san hô xanh;* Gagnep. in Lecomte Fl. Gén. Indo-Chine 5: 254.
1925.

" Habitat inter sepes in Cochinchina, uti & in Malabaria." The Linnaean species was correctly interpreted by Loureiro, although his description of the flowers is incorrect.

Ossifraga lactea Rumph. (Herb. Amb. Auct. 62. *pl. 29*) cited, after Linnaeus, as a synonym, is correctly placed.

BUXACEAE

Buxus Linnaeus

Buxus sp.

Buxus sempervirens (non Linn.) Lour. Fl. Cochinch. 554. 1790, ed. Willd. 678. 1793, Anamese *huỳnh duong*, Chinese *hoâm tuŏn*.

" Habitat in sylvis planis Cochinchinae, & Chinae." As *Buxus sempervirens* Linn. was interpreted by Hemsley (Journ. Linn. Soc. Bot. **26**: 418. 1894) the form Loureiro observed would doubtless be included as representing the Linnaean species. Gagnepain (Lecomte Fl. Gén. Indo-Chine **5**: 660–664. 1927) admits four species of *Buxus* for Indo-China, but *B. sempervirens* Linn. is not included. As the various Asiatic species are now interpreted, it seems to be doubtful if true *Buxus sempervirens* Linn. occurs in China. Loureiro probably included *B. harlandii* Hance in his concept of the Linnaean species.

ANACARDIACEAE

Anacardium Linnaeus

Anacardium occidentale Linn. Sp. Pl. 383. 1753; Lour. Fl. Cochinch. 248. 1790, ed. Willd. 304. 1793.

" Habitat in ora orientali Africae. Vidi quoque in Malabaria, & in Bengalia." Loureiro correctly interpreted the Linnaean species. *Cassuvium* Rumph. (Herb. Amb. **1**: 177. *pl. 69*), cited as a synonym, is correctly placed.

Mangifera Linnaeus

Mangifera foetida Lour. Fl. Cochinch. 160. 1790, ed. Willd. 199. 1793, Anamese *xoài hôi;* Lecomte Fl. Gén. Indo-Chine **2**: 15. 1908.

" Habitat inculta in locis planis, agrestibus." The species has probably been correctly interpreted by modern authors, although such interpretations may have been based more on *Manga foetida* Rumph. (Herb. Amb. **1**: 98. *pl. 28*), cited by Loureiro as a synonym, than on Loureiro's description.

Mangifera indica Linn. Sp. Pl. 200. 1753; Lour. Fl. Cochinch. 159. 1790, ed. Willd. 198. 1793, Anamese *cây xoài*, Chinese *căn xú*.

" Habitat in Cochinchina, & in omnibus fere locis utriusque Indiae frigoris vehementis expertibus." Loureiro correctly interpreted the Linnaean species. *Manga domestica* Rumph. (Herb. Amb. **1**: 93. *pl. 25*) is correctly placed as a synonym.

Gluta Linnaeus

Gluta nitida (Lour.) Merr. in Sunyatsenia **2**: 35. 1934 (based on *Penaea nitida* Lour.). *Penaea nitida* Lour. Fl. Cochinch. 72. 1790, ed. Willd. 91. 1793, Anamese *cây son*.

" Habitat agrestis in Cochinchina." Loureiro's species has not hitherto been placed. The character " stylo lateri, non apici, insidente " taken together with other data clearly indicates the Anacardiaceae and in this family the genus *Gluta*. Loureiro's species is represented by *Clemens 3270, 3504*, a small tree occurring in thickets at Tourane, the material agreeing in all respects with Loureiro's description except the "capsula ovata, polysperma," data manifestly added to make the species description conform to the characters of *Penaea;* Loureiro further notes regarding the fruits "maturam non vidi, ideo de numero loculorum incertus mansi." This species, the flowers 4- or 5-merous, is clearly allied to *Gluta elegans* (Wall.) Hook. f. of the Malay Peninsula and is distinct from both *G. cambodiana* Pierre and *G. coarctata* Hook. f., the only representatives of the genus recorded from Indo-China by Lecomte (Fl. Gén. Indo-Chine 2: 20–22. 1908). The local name *cây son* is recorded by Gagnepain for *Melanorrhoea laccifera* Pierre of the Anacardiaceae, and for *Gelonium cicerospermum* Gagnep. of the Euphorbiaceae.

Rhus (Tournefort) Linnaeus

Rhus chinensis Mill.[24] Gard. Dict. ed. 8, no. 7. 1768; Britten in Journ. Bot. **38**: 316. 1900, in nota.

> *Rhus semialata* Murr. in Comm. Götting. **6**: 27. *pl. 3.* 1784; Lecomte Fl. Gén. Indo-Chine **2**: 35. 1908; Merr. in Journ. Arnold Arb. **9**: 3. *pl. 11.* 1928.
>
> *Rhus javanica* (non Linn.) Lour. Fl. Cochinch. 183. 1790, ed. Willd. 228. 1793, Chinese *xiong tsät.*

"Habitat in sylvis Cantoniensibus Sinarum." Loureiro's description applies unmistakably to the species that many authors (including Britten in Journ. Bot. **38**: 315–316. 1900) interpret as *Rhus javanica* Linn. On a strict interpretation of types, however, *Rhus javanica* Linn. = *Brucea javanica* (Linn.) Merr.; see Merrill, E. D. "On the type of *Rhus javanica* Linn." (Journ. Arnold Arb. **9**: 1–4. *pl. 10–11.* 1928). The illustrations in the latter paper are photographic reproductions of the specimens in the Linnaean herbarium, one the type of *Rhus javanica* Linn. as named by Linnaeus = *Brucea javanica* (Linn.) Merr. (*B. sumatrana* Roxb.), the other a specimen not named by Linnaeus, which represents *Rhus semialata* Murr. var. *roxburghii* DC, the oldest specific name for which is *Rhus chinensis* Mill.

Rhus succedanea Linn. Mant. **2**: 221. 1771; Lecomte Fl. Gén. Indo-Chine **2**: 36. 1908.

> *Augia sinensis* Lour. Fl. Cochinch. 337. 1790, ed. Willd. 411. 1793, pro majore parte, Anamese *cây son,* Chinese *tsí xú, tsăt xú.*
>
> *Calophyllum augia* Steud. Nomencl. ed. 2, **1**: 261. 1840 (based on *Augia sinensis* Lour.).

"Habitat in sylvis Cochinchinae, Chinae, Cambodiae, & Siami." Loureiro's description is manifestly based on a mixture but it is essentially, except for certain floral characters (notably in having 100 stamens), that of *Rhus.* Loureiro intended to describe the varnish tree, and his description of the leaves and of the fruits appertains to *Rhus.* Pierre (Fl. Forest. Cochinch. **4**: sub. *pl. 367*) suggested that the mixture consisted of leaves and fruits of *Rhus succedanea* Linn., and the flowers of some species of *Melanorrhoea; cây son* is the Anamese name of *M. laccifera* Pierre. The "100 stamens" is no anacardiaceous character but was possibly taken from a *Calophyllum* flower and this character was what must have led Steudel to place the species in *Calophyllum.* De Dalla Torre &

---

[24] *Rhus chinensis* Osbeck Dagbok Ostind. Resa 232. 1757 is a *nomen nudum;* it is probably the same as Miller's species.

Harms (Gen. Siphon. 320. 1901) place *Augia* as a definite synonym of *Calophyllum* where it cannot possibly belong except as to the indicated number of stamens which was due to some kind of a mixture of material. The description for the most part appertains to *Rhus succedanea* Linn. and I believe Loureiro's genus and species should be placed as a synonym of the Linnaean species. Rehder & Wilson (Pl. Wils. **2**: 183. 1916) in a note appended to *Rhus succedanea* Linn., state that it is highly probable that *Augia sinensis* Lour. belongs here, *R. succedanea* Linn. being mistaken by some of the older authors for the true varnish tree, *R. verniciflua* Stokes.

AQUIFOLIACEAE

Ilex (Tournefort) Linnaeus

Ilex cochinchinensis (Lour.) Loesen. in Nov. Act. Akad. Naturf. **78**: 230. 1901; Pitard
in Lecomte Fl. Gén. Indo-Chine **1**: 853. 1912 (based on *Hexadica cochinchinensis*
Lour.).
Hexadica cochinchinensis Lour. Fl. Cochinch. 562. 1790, ed. Willd. 688. 1793,
Anamese *cây bùi den.*
Hexacadica corymbosa Raf. Sylva Tellur. 158. 1838 (based on *Hexadica cochinchinensis*
Lour.).

" Habitat in sylvis Cochinchinae." Loureiro's type is preserved in the herbarium of the British Museum and Loesener's greatly amplified description is based in part on this specimen, perhaps in part on *Bon 2172*, regarding which Loesener queries: " an eadem species? "; Pitard's description is apparently based on Bon's specimen, as this is the only one he cites other than Loureiro's, as apparently he did not see the latter. Loureiro's description of the fruit as a capsule is a manifest error. De Pirey's *tram bui, tram bui bui*, and *bui bui, Chevalier 41249, 41250, 41251*, all represent a single species of *Ilex* and apparently the one Loureiro described.

HIPPOCRATEACEAE

Salacia Linnaeus

Salacia cochinchinensis Lour. Fl. Cochinch. 526. 1790, ed. Willd. 642. 1793, Anamese
cây tráoc mau; Pierre Fl. Forest. Cochinch. **3**: *pl. 298*. 1893; Pitard in Lecomte Fl.
Gén. Indo-Chine **1**: 904. 1912.
Salacia saigonensis Baill. Adansonia **11**: 272. 1874.

" Habitat in dumetis Cochinchinae." Loureiro's type is preserved in the herbarium of the British Museum. The species is apparently represented by *Clemens 3508* from Tourane, near the classical locality Hue. *Tonsella chinensis* Spreng. (Syst. **1**: 177. 1825) is essentially based on *Salacia chinensis* Linn. (Mant. **2**: 293. 1771) the first synonym cited by Sprengel, although he also adds *Salacia cochinchinensis* Lour. Hemsley (Journ. Linn. Soc. Bot. **23**: 125. 1886) states that the Linnaean species is an obscure one but that it is certainly not the same as *Salacia cochinchinensis* Lour.

STAPHYLEACEAE

Turpinia [25] Ventenat

Turpinia sp.

Triceros cochinchinensis Lour. Fl. Cochinch. 184. 1790, ed. Willd. 230. 1793, Anamese *cây áu rùng.*

Maurocenia cochinchinensis O. Ktz. Rev. Gen. Pl. 150. 1891 (based on *Triceros cochinchinensis* Lour.).

"Habitat inculta in montibus Cochinchinae." *Triceros* was long considered to represent a genus of uncertain status in the Anacardiaceae, and in my preliminary study of 1919 I left it as a genus of doubtful status, perhaps allied to *Euodia* of the Rutaceae. The flowers are described as 5-merous, the styles as 3, and the fruit as 3-celled, 2-seeded, with three horns at the apex, while in the diagnostic sentence the leaves are described as "quinatis" and in the description as "impari-bipinnata, 2-juga"; that is simply pinnate, with 5 leaflets. I am convinced that O. Kuntze was correct in referring *Triceros* to *Maurocenia = Turpinia*. The limits of the species, as between *Turpinia nepalensis* Wall., *T. latifolia* Wall., and *T. sphaerocarpa* Hassk., are not clearly defined. *T. pomifera* (Roxb.) DC. has been by some authors treated more or less as a collective species and currently many forms are referred to it that do not belong in its alliance; the species is characterized by its unusually large fruits and does not occur in Indo-China or in Malaysia. Loureiro's specimen, listed as being among the material received from him, has not been located in the herbarium of the British Museum. *Turpinia* was overlooked by the authors of the Flora générale de l'Indo-Chine; the Clemens collection contains representatives of two species of the genus from Indo-China.

SAPINDACEAE ·

Cardiospermum Linnaeus

Cardiospermum halicacabum Linn. Sp. Pl. 366. 1753; Lour. Fl. Cochinch. 239. 1790, ed. Willd. 294. 1793, Anamese *cây tam phoung.*

Rhodiola biternata Lour. Fl. Cochinch. 627. 1790, ed. Willd. 770. 1793.

For the first Loureiro states: "Habitat incultum in Cochinchina"; for the second "Habitat in hortis minus cultis Cochinchinae, non frequens." Both descriptions apply to the common, well-known, and very widely distributed *Cardiospermum halicacabum* Linn.

Cardiospermum corindum Linn. Sp. Pl. ed. 2, 526. 1762; Lour. Fl. Cochinch. 239. 1790, ed. Willd. 294. 1793; Radlk. in Pflanzenreich 98 (IV–165): 397. 1932.

"Habitat incultum in suburbis Cantoniensibus Sinarum." I had placed this under *C. halicacabum* Linn. but I defer to Radlkofer's judgment (Pflanzenreich 98 (IV–165): 397. 1932) who retains *C. corindum* Linn. as a distinct species and considers that Loureiro correctly interpreted it.

Allophylus Linnaeus

Allophylus racemosus (Linn.) Radlk. in Engler & Prantl Nat. Pflanzenfam. 3(5): 313. 1895; Lecomte Fl. Gén. Indo-Chine 1: 1013. 1912; Radlk. in Pflanzenreich 98 (IV–165): 568. 1932.

[25] *Triceros* Loureiro (1790) is older than *Turpinia* Ventenat (1803); the latter should be included in some future list of *nomina conservanda.*

Schmidelia racemosa Linn. Mant. **1**: 67. 1767.

Allophylus ternatus Lour. Fl. Cochinch. 232. 1790, ed. Willd. 286. 1793, Anamese *cây chánh ba.*

Gemella trifolia Lour. Fl. Cochinch. 649. 1790, ed. Willd. 796. 1793, Anamese *cây nhánh ba.*

Schmidelia cochinchinensis DC. Prodr. **1**: 611. 1824 (based on *Allophylus ternatus* Lour.).

Aporetica gemella DC. Prodr. **1**: 610. 1824 (based on *Gemella trifolia* Lour.).

Schmidelia gemella Cambess. in Mém. Mus. Hist. Nat. [Paris] **18**: 24. 1829 (based on *Gemella trifolia* Lour.).

For *Allophylus ternatus* Loureiro states: " Habitat ad ripas fluminum in Cochinchina." For *Gemella trifolia*, which was described as a new genus, no habitat is given although manifestly the type came from Indo-China as indicated by the local name cited. The native names cited for the two species are suspiciously alike and I can detect nothing in the two descriptions by which two species can be recognized. The descriptions are sufficiently definite so that it is evident that in both cases either *Allophylus racemosus* (Linn.) Radlk. or the closely allied *A. glaber* (Roxb.) Radlk. was intended. As between these two species, both of which occur in Indo-China, it is impossible to determine to which Loureiro's names belong, and I have rather arbitrarily referred them and the several synonyms based upon them to *Allophylus racemosus* Radlk., as this entails no changes in nomenclature or in currently used specific names. Loureiro's type of *Gemella trifolia* is preserved in the herbarium of the British Museum.

Sapindus (Tournefort) Linnaeus

Sapindus mukorossi Gaertn. Fruct. **1**: 342. *pl. 70.* 1788; Lecomte Fl. Gén. Indo-Chine **1**: 1041. 1912; Radlk. in Pflanzenreich 98 (IV–165): 652. 1932.

Sapindus abruptus Lour. Fl. Cochinch. 238. 1790, ed. Willd. 293. 1793, Chinese *mu hôan xú.*

Sapindus saponaria (non Linn.) Lour. Fl. Cochinch. 238. 1790, ed. Willd. 293. 1793, Anamese *cây bòn hòn.*

For *Sapindus abruptus* Loureiro states: " Habitat Cantone Sinarum "; it can scarcely be other than *Sapindus mukorossi* Gaertn., although described as having 4 petals and sepals instead of 5; this species is now currently known in Canton as *muk wah;* Radlkofer leaves it as a doubtful synonym of Gaertner's species. For *S. saponaria* he states: " Habitat agrestis, cultusque in Cochinchina." This I believe also to be a form of *S. mukorossi* Gaertn. *Saponaria* Rumph. (Herb. Amb. **2**: 134), cited by Loureiro as a synonym, is *Sapindus rarak* DC. (*Dittelasma rarak* Hook. f.).

Schleichera [26] Willdenow

Schleichera oleosa (Lour.) Merr. Interpret. Herb. Amb. 337. 1917.

Pistacia oleosa Lour. Fl. Cochinch. 615. 1790, ed. Willd. 755. 1793, Anamese *cây deâu truòng.*

[26] *Schleichera* Willdenow (1805), conserved name, Vienna Code; older ones are *Cussambium* (Rumphius) Lamarck (1786) and *Koon* Gaertner (1791).

Schleichera trijuga Willd. Sp. Pl. **4**: 1096. 1805; Lecomte Fl. Gén. Indo-Chine **1**: 1034. 1912; Radlk. in Pflanzenreich **98** (IV–165): 874. 1932.

Cussambium oleosum O. Ktz. Rev. Gen. Pl. 143. 1891 (based on *Pistacia oleosa* Lour.).

" Habitat agrestis, cultaque in Cochinchina." Loureiro's description unmistakably applies to the species currently known as *Schleichera trijuga* Willd. *Cussambium* Rumph. (Herb. Amb. **1**: 154. *pl. 57*), cited by Loureiro as a synonym, represents the same species. As partial confirmation of this interpretation of Loureiro's species, it may be noted that Lecomte, although not accounting for Loureiro's binomial, cites an Anamese name, *cây dean trûong*, apparently a misprint for *cây deau trûong*, the local name given by Loureiro.

Euphoria Commerson

Euphoria longan (Lour.) Steud. Nomencl. 328. 1821 (based on *Dimocarpus longan* Lour.).

Dimocarpus longan Lour. Fl. Cochinch. 233. 1790, ed. Willd. 288. 1793, Anamese *cây nhon, laong nhan*, Chinese *lûm yèn*.

Euphoria longana Lam. Encycl. **3**: 574. 1791; Lecomte Fl. Gén. Indo-Chine **1**: 1046. 1912; Radlk. in Pflanzenreich **98** (IV–165): 898. 1932.

Nephelium longana Cambess. in Mém. Mus. Hist. Nat. [Paris] **18**: 30. 1829.

" Habitat culta in China, & Cochinchina." This is the well-known *longan* or *lungan*, extensively cultivated for its fruit in southern China, and in tropical Asia generally. Radlkofer gives the date of publication of Lamarck's species as 1789; pages 1–360 of the Encyclopédie were published in 1789; 361–755 in 1791. Loureiro's specific name thus has clear priority over that of Lamarck.

Litchi Sonnerat

Litchi chinensis Sonn. Voy. Ind. **2**: 230. *pl. 129*. 1782; Lecomte Fl. Gén. Indo-China **1**: 1047. 1912; Radlk. in Pflanzenreich **98** (IV–165): 917. 1932.

Dimocarpus litchi Lour. Fl. Cochinch. 233. 1790, ed. Willd. 287. 1793, Anamese *cây bai*, Chinese *ly chi*.

Nephelium litchi Cambess. in Mém. Mus. Hist. Nat. (Paris) **18**: 30. 1829.

Euphoria litchi Desf. Tabl. 135. 1804.[27]

Sapindus edulis Ait. Hort. Kew. **2**: 36. 1789.

Euphoria sinensis Gmel. Syst. 611. 1791.

Scytalia chinensis Gaertn. Fruct. **1**: 197. *pl. 42*. 1788.

Nephelium chinense Druce in Rept. Bot. Exch. Club. Brit. Isles **4**: 637. 1917.

Litchi litchi Britton Fl. Bermud. 226. 1918.

"Colitur abundantissime in provinciis Australibus imperii Chinensis, & in borealibus regni Cochinchinae." A well-known species, extensively cultivated, with a very large number of horticultural varieties; the most highly prized fruit produced in southern China. Willdenow in his edition of the Flora Cochinchinensis in a footnote calls attention to the fact that Loureiro's species is the same as that named and illustrated by Sonnerat in 1782. Loureiro's type is preserved in the herbarium of the British Museum.

[27] Index Kewensis credits this binomial to Jussieu Gen. 248. 1789. It is not there published, appearing thus: "Huc referuntur Lit-chi & Lon-gan Sinensium fructus exquisiti."

Nephelium Linnaeus

Nephelium lappaceum Linn. Mant. **1**: 125. 1767; Lecomte Fl. Gén. Indo-Chine **1**: 1051.
    f. 131, 4. 1912; Radlk. in Pflanzenreich **98** (IV–165): 957. 1932.
    Dimocarpus crinita Lour. Fl. Cochinch. 234. 1790, ed. Willd. 288. 1793, Anamese
    cây chôm chôm.

" Habitat in sylvis Cochinchinae, & Javae." The Linnaean species occurs in Indo-China and there seems to be no reason for considering that Loureiro had other than specimens of it. The reference to Java was apparently taken from Bontius.

Nephelium informe (Lour.) Cambess. in Mém. Mus. Hist. Nat. [Paris] **18**: 30. 1829 (based
    on *Dimocarpus informis* Lour.).
    Dimocarpus informis Lour. Fl. Cochinch. 234. 1790, ed. Willd. 288. 1793, Anamese
    nhón cút deê.
    Euphoria informis Poir. in Lam. Encycl. Suppl. **3**: 478. 1814 (based on *Dimocarpus*
    informis Lour.).

" Habitat in sylvis Cochinchinae." Loureiro's description is distinctly definite and in my opinion it applies to *Nephelium* or to one of the closely allied genera. Pierre (Fl. Forest. Cochinch. **4**: sub *pl. 318*. 1894) thought that it might not appertain to the Sapindaceae. Lecomte does not mention it nor any of the synonyms based upon it in his treatment of the Sapindaceae of Indo-China (Fl. Gén. Indo-Chine **1**: 1001–1053. 1912). In the British Museum list of Loureiro's species under *Dimocarpus litchi* this statement appears: " There is another sheet endorsed simply Cochinchina, J. de Loureiro by Dryander which may be *Dimocarpus informis* which is queried in Fl. Coch." It may be that *Dimocarpus informis* Lour. is but a form of *Litchi chinensis* Sonn. Radlkofer (Pflanzenr. **98** (IV–165): 945. 1932) leaves it as a doubtful synonym of *Xerospermum microcarpum* Pierre.

Mischocarpus [28] Blume

Mischocarpus flexuosus (Lour.) comb. nov.
    Vateria flexuosa Lour. Fl. Cochinch. 334. 1790, ed. Willd. 407. 1793, Anamese *cây*
    truòng; Moore in Journ. Bot. **63**: 285. 1925.
    Mischocarpus fuscescens Blume Rumphia **3**: 169. 1847; Lecomte Fl. Gén. Indo-Chine
    1: 1028. 1912; Radlk. in Pflanzenreich **98**(IV–165): 1294. 1933.
    Pedicellia fuscescens Hu in Bull. Fan Mem. Inst. Biol. **1**: 31. 1929.

" Habitat in sylvis Cochinchinae." A very fragmentary specimen, detached fruits only, is among the Loureiro material in the herbarium of the British Museum and Moore (Journ. Bot. **63**: 285. 1925) has published a short note regarding it, citing Brandis' reference of the material to *Mischocarpus*. Assuming Loureiro's description to be correct, the species cannot, as Moore indicates, be *Mischocarpus sundaicus* Blume which is apetalous, but rather the petaliferous *M. fuscescens* Blume is represented, a species that occurs at Hue (Harmand, cited by Lecomte). It is curious that in neither of the descriptions which I here refer to *Mischocarpus* did Loureiro mention the pinnate leaves; from the descriptions one would infer that the plant he had in mind had simple leaves.

[28] *Mischocarpus* Blume (1825), conserved name, Cambridge Code; *Pedicellia* Loureiro (1790) is the oldest name for the genus.

Mischocarpus oppositifolius (Lour.) Merr. in Lingnan Sci. Journ. **7**: 313. 1931 (based on *Pedicellia oppositifolia* Lour.).

Pedicellia oppositifolia Lour. Fl. Cochinch. 655. 1790, ed. Willd. 806. 1793, Anamese *cây truòng truòng.*

Mischocarpus sundaicus Blume Bijdr. 238. 1825; Lecomte Fl. Gén. Indo-Chine **1**: 1029. 1912; Radlk. in Pflanzenreich **98**(IV–165): 1299. 1933.

Pedicellia loureiri Pierre Fl. Forest. Cochinch. **4**: *pl. 323A.* 1895 (nomenclaturally based on *Pedicellia oppositifolia* Lour.).

" Habitat in sylvis Cochinchinae." Loureiro's description applies to *Mischocarpus* and this is the only apetalous representative of the genus known from Indo-China. He neglected to state whether the leaves were simple or pinnate. In reference to the genus *Pedicellia*, Radlkofer (Sitz. Math.-Phys. Kl. Acad. Wissensch. München **9**: 648. 1879) states under *Mischocarpus:* " Hujus forsan generis species (suadente Blumeo) *Pedicellia oppositifolia* (Cochinchina)." The species extends from Indo-China to the Andaman Islands, through the Philippines and Malaysia to New Guinea. A specimen from Loureiro listed as being among the plants received from him has not been located in the herbarium of the British Museum. *Truong truong* collected by de Pirey for Doctor Chevalier represents both this species and *Arytera littoralis* Blume.

SAPINDACEAE OF UNCERTAIN GENERIC STATUS

Acer pinnatum Lour. Fl. Cochinch. 649. 1790, ed. Willd. 797. 1793, Anamese *nhon cút deê, troung khê.*

Negundo cochinchinensis DC. Prodr. **1**: 596. 1824 (based on *Acer pinnatum* Lour.).

Atalaya cochinchinensis Blume Rumphia **3**: 186. 1847 (based on *Acer pinnatum* Lour.).

" Habitat in sylvis Cochinchinae." In proposing the binomial *Negundo cochinchinensis*, which was based on *Acer pinnatum* Lour., de Candolle queried: " an potius Sapindacearum genus? " The description is rather short and indefinite. It is suspected that the description of the fruits was added without Loureiro actually having seen them, merely to make the description conform to the characters of *Acer* in which he erroneously placed his species. It is not accounted for in Lecomte's treatment of the Sapindaceae of Indo-China. (Fl. Gén. Indo-Chine **1**: 1001–1053. 1912.) Radlkofer (Pflanzenreich **98**(IV–165): 612. 1932) leaves it among the species " omnino dubiae."

BALSAMINACEAE

Impatiens (Rivinius) Linnaeus

Impatiens balsamina Linn. Sp. Pl. 938. 1753; Lour. Fl. Cochinch. 512. 1790, ed. Willd. 626. 1793, Anamese *maóng tay co.*

Impatiens cornuta Linn. Sp. Pl. 937. 1753; Lour. Fl. Cochinch. 511. 1790, ed. Willd. 626. 1793, Anamese *cây nác ne.*

Impatiens mutila Lour. Fl. Cochinch. 512. 1790, ed. Willd. 626. 1793, Anamese *maóng tay tàu.*

Balsamina mutila DC. Prodr. **1**: 686. 1824 (based on *Impatiens mutila* Lour.).

Impatiens balsamina Linn., " agrestis, cultaque in Cochinchina," was correctly interpreted by Loureiro. *I. cornuta* Linn., " ubique spontanea in hortis Cochinchinae," while probably not strictly the Linnaean species, was undoubtedly a form of the exceedingly variable *I. balsamina* Linn.; true *I. cornuta* Linn. is the wild Ceylon form, supposedly the one from which the cultivated forms have been derived. *I. mutila* Lour., " culta in Cochinchina, puto, quod a Sinis translata," is also apparently but a garden form of the common balsam.

Impatiens chinensis Linn. Sp. Pl. 937. 1753; Lour. Fl. Cochinch. 511. 1790, ed. Willd. 625. 1793, Chinese *hûm thâu kiŏ.*

> *Impatiens cochleata* Lour. Fl. Cochinch. 512. 1790, ed. Willd. 627. 1793, Chinese *tsīen chí hûm.*
>
> *Balsamina cochleata* DC. Prodr. 1: 686. 1824 (based on *Impatiens balsamina* Linn.).

For *Impatiens chinensis* Loureiro states: " Habitat spontanea cultaque Cantone Sinarum." The Linnaean species is common in the vicinity of Canton, and Loureiro's description seems to apply to it. For *I. cochleata* he states: " Habitat culta Cantone Sinarum." Hemsley (Journ. Linn. Soc. Bot. **23**: 101. 1886) noted Loureiro's description of the spur as " nectario magno compresso in spiram convoluto," stating that he had seen no Chinese specimens presenting this character. The spur of *Impatiens chinensis* Linn. is very strongly curved, sometimes even subspiral, and I suspect that Loureiro had a form of the Linnaean species. However, I have no information that would lead me to believe that *Impatiens chinensis* Linn. is ever cultivated in Canton; various forms of *Impatiens balsamina* Linn. are there cultivated, but neither of Loureiro's descriptions apply to that species.

RHAMNACEAE

Paliurus (Tournefort) Miller

Paliurus ramosissimus (Lour.) Poir. in Lam. Encycl. Suppl. **4**: 262. 1816 (based on *Aubletia ramosissima* Lour.).

> *Aubletia ramosissima* Lour. Fl. Cochinch. 283. 1790, ed. Willd. 348. 1793, Chinese *an păt pouc.*
>
> *Paliurus aubletia* Schult. in Roem. & Schult. Syst. **5**: 343. 1819 (based on *Aubletia ramosissima* Lour.).
>
> *Zizyphus ramosissima* Spreng. Syst. **1**: 771. 1825 (based on *Aubletia ramosissima* Lour.).

" Habitat Cantone Sinarum inculta." The species is a well-known one, widely distributed in eastern Asia. Loureiro's type is preserved in the herbarium of the Paris Museum.

Zizyphus (Tournefort) Linnaeus

Zizyphus jujuba Mill. Gard. Dict. ed. 8, no. 1. 1768.

> *Rhamnus zizyphus* Linn. Sp. Pl. 194. 1753; Lour. Fl. Cochinch. 158. 1790, ed. Willd. 195. 1793, Anamese *hoùng táo,* Chinese *hûm tsáo.*
>
> *Zizyphus sativa* Gaertn. Fruct. **1**: 202. 1788.
>
> *Zizyphus vulgaris* Lam. Encycl. **3**: 316. 1789.
>
> *Rhamnus soporifer* Lour. Fl. Cochinch. 158. 1790, ed. Willd. 196. 1793, Anamese *toan táo,* Chinese *soăn tsáo.*

Zizyphus soporifer Schult. in Roem. & Schult. Syst. **5**: 340. 1819 (based on *Rhamnus soporifer* Lour.).

For *Rhamnus zizyphus* Loureiro states: " Habitat in Cochinchina, & Cantone Sinarum, ubi drupae sapore acridulces prostant in foro venales tempore autumnali." This is the common jujube or Chinese date, the Linnaean species, but for which Miller's binomial is the oldest valid one in *Zizyphus;* it was published independently by Miller who cites only " Zizyphus, Dod. p. 807." For *Rhamnus soporifer* Loureiro states: " Habitat in provinciis Borealibus Sinarum." He must have had only fragmentary material of this, probably secured from an herbalist. The description is short and imperfect and in its entire, nerveless leaves does not agree with *Zizyphus jujuba* Mill., yet undoubtedly Loureiro had a form of this distinctly variable species. The species commonly known as *Zizyphus jujuba* Lam. (1789) is *Z. mauritiana* Lam., one not closely allied to *Z. jujuba* Mill.

Zizyphus mauritiana Lam. Encycl. **3**: 319. 1789.

> *Rhamnus jujuba* Linn. Sp. Pl. 194. 1753; Lour. Fl. Cochinch. 157. 1790, ed. Willd. 195. 1793, Anamese *dai táo*, Chinese *tá tsáo*.
>
> *Zizyphus jujuba* Lam. Encycl. **3**: 318. 1789; Pitard in Lecomte Fl. Gén. Indo-Chine **1**: 918. 1912, non Mill.

" Habitat culta in China, & Cochinchina." The Linnaean species was correctly interpreted by Loureiro; his specific name, however, is invalidated in *Zizyphus* by *Z. jujuba* Mill., the correct name for the species commonly known as *Z. sativa* Gaertn. or as *Z. vulgaris* Lam.

Zizyphus agrestis Schult. in Roem. & Schult. Syst. **5**: 341. 1819 (based on *Rhamnus agrestis* Lour.).

> *Rhamnus agrestis* Lour. Fl. Cochinch. 158. 1790, ed. Willd. 197. 1793, Anamese *cây na*.

No locality is cited, but the Anamese name clearly indicates an Indo-China specimen. The short description suggests *Zizyphus* except in the statement " corolla nulla " and " stamina 5, villis multis circumdata." It is possible that some non-rhamnaceous plant is represented, but I have no suggestion as to what it might be. Pitard, in his treatment of the Rhamnaceae of Indo-China (Lecomte Fl. Gén. Indo-Chine **1**: 908–934. 1912) does not mention either of the binomials given above.

Berchemia Necker

Berchemia lineata (Linn.) DC. Prodr. **2**: 23. 1825; Pitard in Lecomte Fl. Gén. Indo-Chine **1**: 923. 1812.

> *Rhamnus lineata* Linn. Cent. Pl. **2**: 11. 1756, Amoen. Acad. **4**: 308. 1759; Lour. Fl. Cochinch. 159. 1790, ed. Willd. 197. 1793, Anamese *cây ráo ráo*, Chinese *chè lûm*.
>
> *Berchemia loureiriana* DC. Prodr. **2**: 23. 1825 (based on *Rhamnus lineata* Lour.).
>
> *Zizyphus loureiriana* Dietr. Syn. Pl. **1**: 811. 1839 (based on *Rhamnus lineata* Lour.).

" Habitat inter dumeta, & arundinum sepes in Cochinchina, & China." Specimens from Loureiro are preserved in the herbaria of the British and the Paris Museums; the Linnaean species, type from near Canton, was correctly interpreted by him. The species is common in the vicinity of Canton and is apparently also common in some parts of Indo-China.

Colubrina [29] L. C. Richard

Colubrina asiatica (Linn.) Brongn. in Ann. Sci. Nat. **10**: 369. 1827; Pitard in Lecomte Fl. Gén. Indo-Chine **1**: 930. 1912.

Ceanothus asiaticus Linn. Sp. Pl. 196. 1753.

Tralliana scandens Lour. Fl. Cochinch. 157. 1790, ed. Willd. 195. 1793, Anamese *cây rác*.

Rhamnus scandens Spreng. Syst. **1**: 768. 1825 (based on *Tralliana scandens* Lour.).

"Habitat, & scandit quascunque arbores obvias in Cochinchina." *Tralliana* Lour. has long been an enigmatic genus. I believe the genus and species described by Loureiro to be nothing but the very common and widely distributed *Colubrina asiatica* (Linn.) Brongn., which Loureiro otherwise does not describe, the slight discrepancies between Loureiro's description and the characters of the Linnaean species being due to faulty observation on Loureiro's part.

VITACEAE

Vitis (Tournefort) Linnaeus

Vitis pentagona Diels & Gilg in Bot. Jahrb. **29**: 460. 1900; Gagnep. in Lecomte Fl. Gén. Indo-Chine **1**: 998. 1912.

Vitis indica (non Linn.) Lour. Fl. Cochinch. 155. 1790, ed. Willd. 192. 1793, Anamese *nho rùng, nhon lá*.

"Habitat inculta, scandens per sepes in Cochinchina." This is probably the correct disposition of the form that Loureiro so inadequately described. The description agrees fairly well with the characters of *Vitis pentagona* as far as it goes.

Vitis vinifera Linn. Sp. Pl. 202. 1753; ? Lour. Fl. Cochinch. 155. 1790, ed. Willd. 192. 1793, Anamese *cây nho tàu, bô dao*, Chinese *pû tâo*.

"Habitat in China: inde delata colitur raro in Cochinchina." Loureiro's description is very short and imperfect and may apply to the Linnaean species; doubtless some forms of the European grape occurred in cultivation in China in Loureiro's time. But for his definite statement that the Indo-China form observed by him was cultivated and introduced from China, I should be inclined to think that what he had was, at least in part, *Vitis balanseana* Planch. This occurs at Tourane, near Hue, and Gagnepain, who records its Anamese name as *cây nho*, notes that its fruits are edible; this species also occurs in Kwangtung Province. There is no record, however, that it is cultivated. It seems improbable that Loureiro would have misinterpreted a species with which he must have been thoroughly familiar in Portugal.

Ampelocissus [30] Planchon

Ampelocissus africanus (Lour.) comb. nov.

Botria africana Lour. Fl. Cochinch. 154. 1790, ed. Willd. 191. 1793.

Ampelopsis botria DC. Prodr. **1**: 633. 1824 (based on *Botria africana* Lour.).

[29] *Colubrina* L. C. Richard (1827), conserved name, Vienna Code; older ones are *Marcorella* Necker (1790), *Tralliana* Loureiro (1790), and *Tubanthera* Commelin (1825).

[30] *Ampelocissus* Planchon (1887), conserved name, Cambridge Code. *Botria* Loureiro (1790) is the oldest name for the genus.

Ampelopsis botrya Kostel. Allgem. Med.-Pharm. Fl. **4**: 1198. 1835 (based on *Botria africana* Lour.).

Vitis africana Spreng. Syst. **1**: 778. 1825 (based on *Botria africana* Lour.).

Vitis mossambicensis Klotzsch in Peters Reise Mossam. Bot. **1**: 180. 1862–64; Baker in Oliv. Fl. Trop. Afr. **1**: 397. 1868.

Ampelocissus mossambicensis Planch. in Journ. Vigne Am. **9**: 49. 1885, DC. Monog. Phan. **5**: 392. 1887.

"Habitat agrestis prope litora orae Zanguebariae in Africa." Loureiro's description is ample and I believe the species he described to be identical with *Ampelocissus mossambicensis* Planch. His type is preserved in the herbarium of the Paris Museum and this specimen has been identified by Baillon as *Vitis* sp. Loureiro cites the local name *muzarrúba*, and states that the plant was known to the Portuguese as *parreira brava*. *Botria* has generally been reduced to *Cissus*, but I do not hesitate in placing it under *Ampelocissus*.

Ampelocissus martini Planch. in DC. Monog. Phan. **5**: 373. 1887; Gagnep. in Lecomte Fl. Gén. Indo-Chine **1**: 992. *f. 122.* 1912.

Vitis labrusca (non Linn.) Lour. Fl. Cochinch. 155. 1790, ed. Willd. 193. 1793, Anamese *nho rùng, chia lá*.

"Habitat in sylvis provinciarum Australium Cochinchinae, prope Cambodiam." Loureiro's description agrees very closely with the characters of Planchon's species. The species is closely allied to *Ampelocissus arachnoidea* (Hassk.) Planch., the latter being represented by *Labrusca molucca* Rumph. (Herb. Amb. **5**: 452. *pl. 167*); the Rumphian synonym is cited by Loureiro as representing *Vitis labrusca* as described by him. Gagnepain, without citing Loureiro's synonym, gives the Anamese name *cay nho rung* for *Ampelocissus martini* Planch.

Cissus Linnaeus

Cissus quadrangularis Linn. Mant. **1**: 39. 1767; Lour. Fl. Cochinch. 84. 1790, ed. Willd. 106. 1793, Anamese *deei xanh vuong*.

"Habitat, & scandit arbores sylvestres in India. Vidi etiam prope Mozambiccum in Africa." Loureiro's description applies unmistakably to the very characteristic Linnaean species. *Funis quadrangularis* Rumph. (Herb. Amb. **5**: 83. *pl. 44. f. 2*) is correctly placed as a synonym.

Cissus triloba (Lour.) comb. nov.

Callicarpa triloba Lour. Fl. Cochinch. 70. 1790, ed. Willd. 89. 1793, Anamese *rát chia ba*, Chinese *ca fú thây*.

Cissus vitiginea (non Linn.) Lour. Fl. Cochinch. 83. 1790, ed. Willd. 105. 1793, Anamese *cây dau xuong*.

Cissus modeccoides Planch. in DC. Monog. Phan. **5**: 503. 1887; Gagnep. in Lecomte Fl. Gén. Indo-Chine **1**: 971. 1912.

For *Cissus vitiginea* no locality is cited but the Anamese name clearly indicates an Indo-China specimen. For *Callicarpa triloba* Loureiro states: "Habitat in China, & Cochinchina spontanea." I believe both descriptions apply to *Cissus modeccoides* Planch., although Loureiro possibly confused some other Chinese form of the genus in his *Callicarpa triloba*, as Planchon's species does not occur near Canton. *Callicarpa triloba* Lour. is manifestly not a verbenaceous species.

Cayratia [31] A. L. de Jussieu

Cayratia pedata (Lour.) Juss. ex Gagnep. in Not. Syst. **1**: 346. 1910; Lecomte Fl. Gén. Indo-Chine **1**: 979. 1912 (based on *Columella pedata* Lour.).

Columella pedata Lour. Fl. Cochinch. 86. 1790, ed. Willd. 108. 1793, Anamese *cây rát loung;* Moore in Journ. Bot. **63**: 249. 1925.

Cissus pedata Lam. Encycl. **1**: 31. 1783.

Vitis pedata Vahl ex Wall. List no. 6027. 1831 (based on various herbarium specimens, not on any previously published binomial).

Cissus cochinchinensis Spreng. Syst. **1**: 450. 1825 (based on *Columella pedata* Lour.).

Lagenula pedata Lour Fl. Cochinch. 88. 1790, ed. Willd. 111. 1793, Anamese *cây rát nho lá.*

Fusanus pedatus Spreng. Syst. **1**: 490. 1825 (based on *Lagenula pedata* Lour.).

Pedastis indica Raf. Sylva Tellur. 87. 1838 (based on *Cissus pedata* "auct.," prob. incl. Lour.).

For *Columella pedata* Loureiro states: "Habitat agrestis in Cochinchina." Regarding Loureiro's type in the herbarium of the British Museum, Moore (Journ. Bot. **63**: 249. 1925) states: "This is accepted as equivalent to *Cissus pedata* Lam. and perhaps correctly, but the specimen is extremely poor." *Columella pedata* Lour. was described as a new genus and species, independently of the earlier *Cissus pedata* Lam. *Columella* Lour. as a generic name long antedates *Cayratia* Jussieu, but the latter is retained in the list of *nomina generica conservanda* as approved by the Cambridge Botanical Congress because of the confusion that would result in adopting Loureiro's earlier name; *Columellia* Ruíz & Pavon (1794) typifies the family Columelliaceae. It is of interest to note that Jussieu proposed only the new generic name *Cayratia* (Dict. Class. Hist. Nat. **4**: 346 [not 146 as cited by Gagnepain] 1823) deriving this from the Anamese name *cây rát;* he there published no new binomial, merely mentioning *Columella pedata* Lour. as the basis of his new generic name. The binomial *Cayratia pedata* Juss. dates from Gagnepain's use of it in 1910. *Cissus pedata* Lam. and *Vitis pedata* Vahl were published independently and without reference to *Columella pedata* Lour. It is only an inference that the binomial *Cayratia pedata* Juss., as actually published by Gagnepain, may be considered to have been based on *Columella pedata* Lour. rather than on the earlier *Cissus pedata* Lam., both of which are cited by Gagnepain as synonyms. In proposing the generic name *Cayratia* Jussieu considered only *Columella pedata* Lour., and if the independently published *Cissus pedata* Lam. and *Vitis pedata* Vahl are really synonyms of it, it is merely a coincidence that the several authors selected the same specific name. For *Lagenula pedata* Loureiro states: "Habitat in montibus Cochinchinae." This genus has been referred by all authors, since Sprengel, to the santalaceous *Fusanus*, with which it has nothing in common. Loureiro merely misinterpreted certain perianth characters; his description is essentially that of *Columella* = *Cayratia* and the local name cited by him, *cây rát nho lá*, confirms this reduction of *Lagenula*. I believe the species to be merely a form of *Cayratia pedata* Juss.; it is however possible that *C. pellita* Gagnep. is the species represented. Two overlooked or forgotten generic names of Rafinesque are synonyms of *Cayratia* Juss., i.e., *Pedastis* Raf. (Sylva

[31] *Cayratia* A. L. de Jussieu (1818), conserved name, Cambridge Code, *Columella* Loureiro and *Lagenula* Loureiro (1790) are both older.

Tellur. 87. 1838) based on *Cissus pedata* " auct." = *Cayratia pedata* (Lour.) Juss., and *Causonia* Raf. l.c. based on *Vitis japonica* Thunb. = *Cayratia japonica* (Thunb.) Gagnep.

Cayratia geniculata (Blume) Gagnep. in Not. Syst. 1: 345. 1911; Lecomte Fl. Gén. Indo-Chine 1: 976. 1912.

> *Cissus geniculata* Blume Bijdr. 184. 1825.
> *Columella geniculata* Merr. in Philip. Journ. Sci. Bot. 11: 132. 1916.
> *Cissus trifoliata* (non Linn.) Lour. Fl. Cochinch. 83. 1790, ed. Willd. 105. 1793, Anamese *cây rát*.

No locality is given by Loureiro but it is evident from the native name cited, *cây rát*, that he had specimens from Indo-China. His description applies to *Cayratia geniculata* (Blume) Gagnep., not to *Cissus trifoliata* Linn., which is an American species based entirely on Jamaican references.

ELAEOCARPACEAE

Elaeocarpus (Burman) Linnaeus

Elaeocarpus sylvestris (Lour.) Poir. in Lam. Encycl. Suppl. 2: 704. 1812; Moore in Journ. Bot. 63: 282. 1925 (based on *Adenodus sylvestris* Lour.).

> *Adenodus sylvestris* Lour. Fl. Cochinch. 294. 1790, ed. Willd. 361. 1793, Anamese *cây côm tláng*.
> *Elaeocarpus henryi* Hance in Journ. Bot. 23: 322. 1885.
> *Elaeocarpus decipiens* Hemsl. in Journ. Linn. Soc. Bot. 23: 94. 1886.
> *Elaeocarpus glabripetalus* Merr. in Philip. Journ. Sci. 21: 501. 1922.
> *Elaeocarpus kwangtungensis* Hu in Journ. Arnold Arb. 5: 229. 1924.
> *Elaeocarpus subsessilis* Hand.-Maz. Symb. Sin. 7: 614. 1933.

" Habitat in sylvis Cochinchinae." In my original manuscript of 1919 I failed to place this species among those admitted by Gagnepain (Lecomte Fl. Gén. Indo-Chine 1: 564–582. 1911) thinking that it might represent a species near *E. bonii* Gagnep. Moore (Journ. Bot. 63: 282. 1925) gives an amplified description based on Loureiro's type in the herbarium of the British Museum; he failed to match the species. I examined the type in 1930 and also Hemsley's Kwangtung specimens representing the very inadequately described *E. decipiens* Hemsl., and can see no reason for considering that more than one species is represented. I also place *E. henryi* Hance here. *Elaeocarpus glabripetalus* Merr. is certainly the same as *E. decipiens* Hemsl., and the type *E. kwangtungensis* Hu is a fruiting specimen of the same species. This is a common species in Kwangtung Province, China, but is apparently known from Indo-China only from Loureiro's original collection.

Elaeocarpus tectorius (Lour.) Poir. in Lam. Encycl. Suppl. 2: 704. 1812; Seem. Fl. Vit. 28. 1865, in nota (based on *Craspedum tectorium* Lour.).

> *Craspedum tectorium* Lour. Fl. Cochinch. 336. 1790, ed. Willd. 411. 1793, Anamese *lá măt căt*.
> *Elaeocarpus robustus* Roxb. Hort. Beng. 42. 1812, *nomen nudum*, Fl. Ind. ed. 2, 2: 597. 1832; Mast. in Hook. f. Fl. Brit. Ind. 1: 402. 1874; Gagnep. in Lecomte Fl. Gén. Indo-Chine 1: 577. 1911.

Elaeocarpus leptostachyus Wall. List. no. 2672. 1831, *nomen nudum*, K. Muell. Anot.
Fam. Elaeocarp. 23. 1849.

Elaeocarpus ovalifolius Wall. List. no. 2665. 1831, *nomen nudum;* K. Muell. Anot.
Fam. Elaeocarp. 21. 1849.

" Habitat in altis sylvis Cochinchinae." Loureiro's type is preserved in the herbarium
of the British Museum. Seemann who examined it (Fl. Vit. 28. 1865) states that both
Elaeocarpus leptostachyus Wall. and *E. ovalifolius* Wall. are synonyms of Loureiro's species;
the latter is identical with *Elaeocarpus robustus* Roxb., and King (Journ. As. Soc. Bengal
60(2): 127. 1891) placed the former as a variety of Roxburgh's species. Loureiro's species
is not mentioned by Gagnepain in his treatment of the genus *Elaeocarpus* in Indo-China
(Lecomte Fl. Gén. Indo-Chine 1: 564–582. 1911) although he admits *Elaeocarpus robustus*
Roxb. as occurring in Indo-China. Loureiro's description of the fruit must be excluded as
these data do not apply to *Elaeocarpus*. On the basis that Loureiro's generic description
was based on two unrelated species, some authors would perhaps consider the binomial
Elaeocarpus tectorius to be an invalid one. I interpret the genus and the species from the
flowering specimen described; the fruits are not mentioned in the description of the species.

TILIACEAE

Corchorus (Tournefort) Linnaeus

Corchorus capsularis Linn. Sp. Pl. 529. 1753; Lour. Fl. Cochinch. 334. 1790, ed. Willd.
408. 1793, Chinese *san lim mâ.*

Rhizanota cannabina Lour.[32] ex Gomes in Mem. Acad. Sci. Lisb. Cl. Sci. Pol. Mor.
Bel.-Let. n.3. 4(1): 29. 1838.

" Habitat agrestis, coliturque in provincia Cantoniensi Sinarum." Loureiro's description applies to the Linnaean species; it is the common jute plant. The specimen
from Loureiro, preserved in the herbarium of the Paris Museum, represents the Linnaean
species.

Microcos (Burman) Linnaeus

Microcos paniculata Linn. Sp. Pl. 514. 1753; Burret in Notizbl. Bot. Gart. Berlin 9: 773.
1926.

Grewia microcos Linn. Syst. ed. 12. 2: 602. 1767; Gagnepain in Lecomte Fl. Gén. Indo-
Chine 1: 543. 1911.

Grewia microcos Linn. var. *rugosa* (Lour.) Mast. in Hook. Fl. Brit. Ind. 1: 393. 1874
(based on *Arsis rugosa* Lour.).

Fallopia nervosa Lour. Fl. Cochinch. 336. 1790, ed. Willd. 410. 1793, Chinese *hai pú ip.*

Arsis rugosa Lour. Fl. Cochinch. 335. 1790, ed. Willd. 409. 1793, Anamese *cây chua ke;*
Moore in Journ. Bot. 63: 285. 1925.

Both *Fallopia* and *Arsis* were monotypic genera proposed by Loureiro, yet apparently
but a single species is represented by the two descriptions. For *Arsis* he states: " Habitat
in sylvis Cochinchinae"; for *Fallopia:* " Habitat inculta prope Cantonem Sinarum." The
type of *Arsis rugosa* Lour. is preserved in the herbarium of the British Museum, and
Moore, who examined it, states (Journ. Bot. 63: 285. 1925) that it agrees closely with

---

[32] A Loureiro herbarium name here first published by Gomes.

Wallich 1098b referred by Masters to *Grewia microcos* Linn. var. *rugosa* (Lour.) Mast. in Hook. f. Fl. Brit. Ind. **1**: 393. 1874. The first published reference to Loureiro's type that I have seen is Trimen's comment appended to Hance's article on *Fallopia* (Journ. Bot. **9**: 240. 1871). *Fallopia nervosa* Lour. was there correctly interpreted by Hance, who reduced it to *Grewia microcos* Linn. = *Microcos paniculata* Linn., on the basis of material and data secured by him from herbalists in Canton, particularly with reference to the Chinese name and the economic uses cited by Loureiro. The species is rather common in Kwangtung and still occurs in the immediate vicinity of Canton.

Triumfetta (Plumier) Linnaeus

Triumfetta bartramia Linn. Syst. ed. 10, 1044. 1759; Merr. Interpret. Herb. Amb. 354. 1917.

Triumfetta rhomboidea Jacq. Enum. Pl. Carib. 22. 1760.

Urena polyflora Lour. Fl. Cochinch. 417. 1790, ed. Willd. 508. 1793, Chinese *xie thâu fŏ*.

Malachra ? urena DC. Prodr. **1**: 441. 1824 (based on *Urena polyflora* Lour.).

" Habitat agrestis prope Cantonem Sinarum." Gagnepain (Lecomte Fl. Gén. Indo-Chine **1**: 493. 1911) reduced Loureiro's species to *Helicteres lanceolata* DC., a species unknown from China and one to which Loureiro's description does not remotely apply. This reduction was apparently based on the assumption that Loureiro's specimen on "*Mopex sinensis*" in the herbarium of the Paris Museum represents *Urena polyflora* Lour., an assumption that seems to be unwarranted from Loureiro's description. Except for certain floral details which may have been based on faulty observations, Loureiro's description agrees fairly well with the characters of the ubiquitous *Triumfetta rhomboidea* Jacq. = *T. bartramia* Linn.

Triumfetta grandidens Hance in Journ. Bot. **15**: 329. 1877; Gagnep. in Lecomte Fl. Gén. Indo-Chine **1**: 553. 1911.

Urena procumbens (non Linn.) Lour. Fl. Cochinch. 417. 1790, ed. Willd. 507. 1793, Anamese *cày báy cát*.

" Habitat loca arenosa in Cochinchina, & China." Loureiro's description agrees well with the characters of Hance's species, and I believe this disposition of *Urena procumbens* Lour. to be correct. The type of *Urena procumbens* Linn. was a specimen collected by Osbeck near Canton and judging from a sketch of this specimen in the Linnaean herbarium made by the late B. Daydon Jackson it is merely a form of the ubiquitous *Urena lobata* Linn.

MALVACEAE

Abutilon (Tournefort) Adanson

Abutilon indicum (Linn.) Sweet Hort. Brit. 54. 1826; Gagnep. in Lecomte Fl. Gén. Indo-Chine **1**: 409. 1912.

Sida indica Linn. Cent. Pl. **2**: 26. 1756; Lour. Fl. Cochinch. 414. 1790, ed. Willd. 503. 1793, Anamese *cây kôi xay*.

" Habitat inculta in hortis, & agris Cochinchinae." Loureiro's description conforms to the characters of the Linnaean species, which is very common and widely distributed in the Indo-Malaysian region. *Abutilon laeve* Rumph. (Herb. Amb. **4**: 31. *pl. 11*), cited by Loureiro as illustrating his species, is correctly placed as a synonym.

Malva (Tournefort) Linnaeus

Malva verticillata Linn. Sp. Pl. 689. 1753; Lour. Fl. Cochinch. 422. 1790, ed. Willd. 514.
 1793, Anamese *doung qui tu*, Chinese *tūng quéi tsŭ*.
 "*Habitat culta Cantone Sinarum. In Cochinchina raro.*" The Linnaean species
occurs naturally in northern and central China, and is apparently naturalized in Hongkong.
Loureiro presumably saw only cultivated specimens, and the medicinal use of the plant
would account for its cultivation. I can see no reason for considering that he had other
than specimens of the Linnaean species.

Sida Linnaeus

Sida acuta Burm. f. Fl. Ind. 147. 1768; Gagnep. in Lecomte Fl. Gén. Indo-Chine **1**: 402.
 1910.
 Sida carpinifolia Linn. f. Suppl. 307. 1781.
 Sida scoparia Lour. Fl. Cochinch. 414. 1790, ed. Willd. 504. 1793, Anamese *cây báy chôi*.
 "*Habitat in loca agrestia in Cochinchina.*" Loureiro cites *Sida acuta* Burm. f. (Fl.
Ind. 147. 1768) as a synonym of his species, and also *Sigalurium longifolium* Rumph.
(Herb. Amb. **6**: 45. *pl. 18. f. 2*) which represents Burman's species. His description applies
in all respects to this very common and widely distributed plant.

Sida cordifolia Linn. Sp. Pl. 684. 1753; Lour. Fl. Cochinch. 414. 1790, ed. Willd. 503. 1793;
 Gagnep. in Lecomte Fl. Gén. Indo-Chine **1**: 400. 1910.
 Malva tomentosa Linn. Sp. Pl. 687. 1753; Lour. Fl. Cochinch. 422. 1790, ed. Willd.
 514. 1793, Anamese *cây bái thi*.
 For *Sida cordifolia* Loureiro states: "*Habitat in agris Cochinchinae,*" and his description apparently applies to the Linnaean species which is very common and widely
distributed in the Indo-Malaysian region. For *Malva tomentosa* he states: "*Habitat
inculta in Cochinchina,*" and apparently this description applies to the Linnaean species,
which, however, according to Trimen (Fl. Ceyl. **1**: 143. 1893) is a synonym of *Sida cordifolia*
Linn.

Sida mysorensis W. & A. Prodr. 59. 1834; Gagnep. in Lecomte Fl. Gén. Indo-Chine **1**:
 403. 1910.
 Sida viscosa (non Linn.) Lour. Fl. Cochinch. 413. 1790, ed. Willd. 502. 1793, Anamese
 cây báy xôi.
 "*Habitat in agris Cochinchinae.*" Loureiro's description agrees fairly well with the
characters of *Sida mysorensis* W. & A., a viscid species of wide distribution in the Old
World tropics.

Sida rhombifolia Linn. Sp. Pl. 684. 1753; Gagnep. in Lecomte Fl. Gén. Indo-Chine **1**: 405.
 1910.
 Sida alnifolia (non Linn.) Lour. Fl. Cochinch. 413. 1790, ed. Willd. 502. 1793, Anamese
 cây báy doùng tièn.
 "*Habitat in agris Cochinchinae.*" Loureiro's description applies to the very common
Linnaean species as it is currently interpreted. According to Trimen (Fl. Ceyl. **1**: 142.
1893) *Sida alnifolia* Linn. is in part *S. rhombifolia* Linn. and in part *S. cordifolia* Linn.

Urena (Dillenius) Linnaeus

Urena lobata Linn. Sp. Pl. 692. 1753; Lour. Fl. Cochinch. 416. 1790, ed. Willd. 507. 1793,
Anamese *cây báy loung*, Chinese *siê thâu fŏ*.
Urena sinuata Linn. Sp. Pl. 692. 1753; Lour. Fl. Cochinch. 417. 1690, ed. Willd. 507.
1793, Anamese *cây báy oúc*.
Urena monopetala Lour. Fl. Cochinch. 418. 1790, ed. Willd. 508. 1793, Anamese *cây
báy chíeo*.

For the first Loureiro states: " Habitat spontanea in Cochinchina, & China "; for the
second: " Habitat agrestis in Cochinchina "; and for the third: " Habitat inculta in Cochinchina." The three descriptions apparently apply to forms of the very common,
widely distributed and variable *Urena lobata* Linn.; *U. sinuata* Linn. has more deeply
lobed leaves than *U. lobata* and by some authors is considered as a variety of the latter,
but all variations of leaf shape are sometimes found on the same plant, and there are no
distinguishing fruit or flower characters. *U. monopetala* Lour. is a form with narrower,
scarcely lobed or angled, tomentose leaves, and is apparently a form of *Urena lobata* Linn.
var. *scabriuscula* A. Gray.

Hibiscus Linnaeus

Hibiscus mutabilis Linn. Sp. Pl. 694. 1753; Lour. Fl. Cochinch. 419. 1790, ed. Willd. 511.
1793, Anamese *phú duong*, Chinese *fū yung;* Gagnep. in Lecomte Fl. Gén. Indo-
Chine 1: 428. 1910.
" Habitat in hortis Cochinchinae, & Chinae." The Linnaean species was correctly
interpreted by Loureiro. *Flos horarius* Rumph. (Herb. Amb. **4**: 27. *pl. 9*) is correctly
placed as a synonym.

Hibiscus rosa-sinensis Linn. Sp. Pl. 694. 1753; Lour. Fl. Cochinch. 419. 1790, ed. Willd.
510. 1793, Anamese *hoùng kân;* Gagnep. Lecomte Fl. Gén. Indo-Chine 1: 429. 1910.
" Habitat tam culta, quam spontanea in Cochinchina, & China." Loureiro correctly
interpreted the Linnaean species, but it is exceedingly doubtful if he saw any wild form of it.
This is one of the most commonly cultivated shrubs in the Indo-Malaysian region. *Flos
festalis* Rumph. (Herb. Amb. **4**: 24. *pl. 8*) is correctly placed as a synonym.

Hibiscus surattensis Linn. Sp. Pl. 696. 1793; Lour. Fl. Cochinch. 420. 1790, ed. Willd.
512. 1793, Anamese *cây soung chua;* Gagnep. in Lecomte Fl. Gén. Indo-Chine 1:
423. 1910.
" Habitat inter sepes in Cochinchinae." Loureiro's description applies unmistakably
to the Linnaean species, except in his statement that it is a " frutex 6 pedalis," which,
however, he modified by adding " scandens." The stems do attain this length, but are
suffrutescent, prostrate, or more or less clambering in thickets. The acid leaves are
commonly eaten as noted by Loureiro.

Hibiscus syriacus Linn. Sp. Pl. 695. 1753; Lour. Fl. Cochinch. 420. 1790, ed. Willd. 511.
1793, Anamese *houng kan biéc;* Gagnep. Lecomte Fl. Gén. Indo-Chine 1: 428. 1910.
" Habitat in hortis Cochinchinae: in China mihi non visus." Loureiro's description
applies to the commonly cultivated, double-flowered form of the Linnaean species, which
is widely distributed in the warmer parts of Asia in cultivation.

Hibiscus tiliaceus Linn. Sp. Pl. 694. 1753; Lour. Fl. Cochinch. 418. 1790, ed. Willd. 509.
1793, Anamese *cày tla làm chíeo;* Gagnep. in Lecomte Fl. Gén. Indo-Chine 1: 431.
1910.

   Paritium tiliaceum St. Hil. Fl. Bras. Merid. 1: 256. 1825.

   " Habitat agrestis in Cochinchina, maxime ad ripas fluminum, & maris litora: similiter in China." Loureiro's description applies unmistakably to the common and widely
distributed Linnaean species. *Novella daun* Rumph. (Herb. Amb. 2: 218. *pl. 73*) is correctly placed as a synonym.

Abelmoschus Medikus

Abelmoschus esculentus (Linn.) Moench Meth. 617. 1794.

   Hibiscus esculentus Linn. Sp. Pl. 696. 1753; Lour. Fl. Cochinch. 421. 1790, ed. Willd.
   512. 1793, Anamese *cây boung vàng, chia lá,* Chinese *hoàng sóuc qúei;* Gagnep. in
   Lecomte Fl. Gén. Indo-Chine 1: 433. 1910.

   " Habitat in hortis Cochinchinae, & Chinae." The Linnaean species was correctly
interpreted by Loureiro; it is the common okra.

Thespesia Solander

Thespesia populnea (Linn.) Soland. ex Corr. in Ann. Mus. Hist. Nat. (Paris) 9: 290. *pl. 8.
   f. 1.* 1807; Gagnep. in Lecomte Fl. Gén. Indo-Chine 1: 436. 1910.

   Hibiscus populneus Linn. Sp. Pl. 694. 1753; Lour. Fl. Cochinch. 418. 1790, ed. Willd.
   509. 1793, Anamese *cây tla.*

   " Habitat agrestis in Cochinchina. Etiam longa serie plantatus in urbe Gallorum
Pondichery, ubi jucundum umbraculum praebet viatoribus." The description definitely
applies to the common and widely distributed littoral Linnaean species. *Novella litorea*
Rumph. (Herb. Amb. 2: 224. *pl. 74*) is correctly placed as a synonym.

Gossypium Linnaeus

Gossypium barbadense Linn. Sp. Pl. 693. 1753; Mast. in Oliv. Fl. Trop. Afr. 1: 210. 1868;
   Watt Wild and Cult. Cotton Pl. 265. *f. 46–48.* 1907.

   Gossypium arboreum (non Linn.) Lour. Fl. Cochinch. 416. 1790, ed. Willd. 506. 1793.

   " Habitat in ora Africae Orientali." Loureiro describes the flowers as " integre
luteus, " while the typical form of *Gossypium arboreum* Linn. has purple flowers. There is
little doubt that the form Loureiro had was *G. barbadense* Linn., at least as interpreted
by Masters.

Gossypium herbaceum Linn. Sp. Pl. 693. 1753; ? Lour. Fl. Cochinch. 415. 1790, ed. Willd.
   505. 1793, Anamese *cây boung,* Chinese *miên fú.*

   " Colitur in tota Cochinchina, China, & plerisque Asiae locis non nimis frigidis."
Loureiro apparently included more than the Linnaean species in his conception of *Gossypium herbaceum,* but it is impossible to determine the true status of the form or forms that
he described under the Linnaean binomial. It is suspected that the Chinese form was
Gossypium nanking Meyen.

Kosteletzkya Presl

Kosteletzkya adoensis (Hochst.) Mast. in Oliv. Fl. Trop. Afr. 1: 194. 1868.

   Hibiscus adoensis Hochst. in Flora 24 (2): Intelligenzbl. 29. 1841, *nomen nudum;*
   Walp. Ann. 2: 143. 1851.

Hibiscus terniflorus Garcke in Bot. Zeit. **7**: 833. 1849.

? *Althaea africana* DC. Prodr. **1**: 437. 1824 (based on *Alcea africana* Lour.).

? *Alcea africana* Lour. Fl. Cochinch. 421. 1790, ed. Willd. 513. 1793.

" Habitat ad litora Africae orientalis." Willdenow in a footnote states: " An *Althaea sinensis* Cavanilles diss. II. pag. 92. tab. 29. fig. 3.? " The description " arillis 5, monospermis, laevibus " excludes *Althaea* as a possibility; moreover, no representative of that genus is known from tropical Africa. *Kosteletzkya* seems to be a possibility, although Loureiro's description does not conform entirely with the characters of Hochstetter's species which is recorded from the Mozambique District. Masters does not mention Loureiro's species in his treatment of the Malvaceae (Oliver Fl. Trop. Afr. **1**: 175–214. 1868). It is possible that a species of *Pavonia* is represented. In view of the present uncertainty as to the application of Loureiro's specific name, no change is here made. Garcke's specific name is older, as a properly published one, than is Hochstetter's.

BOMBACACEAE

Adansonia Linnaeus

Adansonia digitata Linn. Sp. Pl. 1190. 1753; Mast. in Oliv. Fl. Trop. Afr. **1**: 212. 1868.

Ophelus sitularius Lour. Fl. Cochinch. 412. 1790, ed. Willd. 501. 1793.

Adansonia situla Spreng. Syst. **3**: 124. 1826 (based on *Ophelus sitularius* Lour.).

Adansonia integrifolia Raf. Sylva Tellur. 149. 1838 (based on *Ophelus* [*sitularius*] Lour.).

" Habitat agrestis ad litora Africae orientalis." The description unmistakably applies to the common boabab. Sprengel recognized *A. situla* as a distinct species on the assumption that Loureiro's species had simple leaves; from Loureiro's description the inference is that the leaves were simple, although he does not definitely so state. It is suspected that he based his description on fragmentary material. He cites the local name *mulambeira*. Masters does not account for Loureiro's species or Sprengel's synonym based upon it. He states that specimens from the Mozambique district differ from those of other parts of Africa, in having rather narrow leaflets which are sometimes smooth on the lower surfaces; Loureiro describes the leaves as glabrous.

Ceiba Medikus

Ceiba pentandra (Linn.) Gaertn. Fruct. **2**: 244. *pl. 133.* 1791.

Bombax pentandrum Linn. Sp. Pl. 511. 1753; Lour. Fl. Cochinch. 415. 1790, ed. Willd. 504. 1793, Anamese *cây gòn*, Chinese *mo miên hōa, uēn xú*.

Eriodendron anfractuosum DC. Prodr. **1**: 479. 1824.

" Habitat agrestis in Cochinchina, & China, copiosius autem in Cambodia." Loureiro clearly describes the kapok or silk cotton tree *Ceiba pentandra* Gaertn. *Eriophorus javana* Rumph. (Herb. Amb. **1**: 194. *pl. 80*), cited by Loureiro as representing the species, is correctly placed.

STERCULIACEAE

Pentapetes Linnaeus

Pentapetes phoenicea Linn. Sp. Pl. 698. 1753; Lour. Fl. Cochinch. 409. 1790, ed. Willd. 497. 1793, Anamese *hoa ti ngo;* Gagnep. in Lecomte Fl. Gén. Indo-Chine **1**: 511. *f. 49, 1–5.* 1911.

"Habitat culta in hortis Cochinchinae, & Chinae." The description applies to the Linnaean species, which is widely distributed in the Indo-Malaysian region. *Flos impius* Rumph. (Herb. Amb. **5**: 288. *pl. 100. f. 1*) cited by Loureiro as a synonym, after Linnaeus, is correctly placed. A specimen from Loureiro listed as being among the plants received from him has not been located in the herbarium of the British Museum.

Sterculia Linnaeus

Sterculia africana (Lour.) comb. nov.

Triphaca africana Lour. Fl. Cochinch. 577. 1790, ed. Willd. 708. 1793.

Triplobus cordata Raf. Sylva Tellur. 111. 1838 (based on *Triphaca africana* Lour.).

Sterculia triphaca R. Br. in Benn. Pl. Jav. Rar. 228. 1844 (based on *Triphaca africana* Lour.); Mast. in Oliv. Fl. Trop. Afr. **1**: 216. 1868.

Sterculia ipomoeaefolia Garcke in Peters Reise Mossamb. **2**: Bot. 130. 1862–64.

Clompanus africana O. Ktz. Rev. Gen. Pl. 77. 1891 (based on *Triphaca africana* Lour.).

"Habitat agrestis in continenti Orientali Africae, contra Mocambiccum." Loureiro gives the local name as *mutonha*. His type is preserved in the herbarium of the Paris Museum. De Candolle (Prodr. **1**: 483. 1824) queries: "An Sterculiae species?", and this suggested reduction was verified by R. Brown and by Baillon. *Triplobus* was proposed as a new generic name by Rafinesque on the basis that *Triphaca* was an erroneously constructed one.

Sterculia foetida Linn. Sp. Pl. 1008. 1753; Lour. Fl. Cochinch. 586. 1790, ed. Willd. 719. 1793, Anamese *chim chim rùng;* Gagnep. in Lecomte Fl. Gén. Indo-Chine **1**: 461. 1911.

"Habitat in sylvis Cochinchinae." Loureiro's description applies to the common and widely distributed Linnaean species. *Clompanus major* Rumph. (Herb. Amb. **3**: 168. *pl. 107*), cited as a synonym, is correctly placed.

Sterculia lanceolata Cav. Diss. **5**: 287. *pl. 143. f. 1.* 1788; Lindl. Bot. Reg. **15**: *pl. 1256.* 1829; Gagnep. in Lecomte Fl. Gén. Indo-Chine **1**: 470. 1911.

Helicteres undulata Lour. Fl. Cochinch. 531. 1790, ed. Willd. 649. 1793, Anamese *cây uói deăi lá.*

Camaion undulata Raf. Sylva Tellur. 75. 1838 (based on *Helicteres undulata* Lour.).

Sterculia balansae Aug. DC. in Bull. Herb. Boiss. II. **3**: 369. 1903.

"Habitat in sylvis Cochinchinae." Loureiro's description clearly applies to *Sterculia*. I believe the indicated reduction in Index Kewensis to *Sterculia lanceolata* Cav. to be correct. Cavanilles' original description was based on a fruiting specimen from China and all descriptions that I have seen, except that of *Sterculia balansae* A.DC., do not indicate the number of stamens. Aug. de Candolle gives the number as 10 for his species, which conforms with Loureiro's description. Gagnepain, in his treatment of the Sterculiaceae of Indo-China (Lecomte Fl. Gén. Indo-Chine **1**: 454–522. 1911) fails to account for Loureiro's species or for Rafinesque's synonym based upon it. It is represented by *Clemens 3620, 3747,* from Tourane near Hue.

Sterculia sp.

Helicteres paniculata Lour. Fl. Cochinch. 531. 1790, ed. Willd. 649. 1793, Anamese *cây uói tlòn lá,* non *Sterculia paniculata* Wall.

Icosinia paniculata Raf. Sylva Tellur. 75. 1838 (based on *Helicteres paniculata* Lour.).
" Habitat in sylvis Cochinchinae." The description definitely applies to *Sterculia*.
In Index Kewensis this is erroneously reduced to *Sterculia grandiflora* Vent., a species inadequately characterized, one of very uncertain status, but probably a true *Sterculia* and not *Cola acuminata* Schott & Endl. to which it has been reduced. Loureiro's species probably belongs in the group with *Sterculia lanceolata* Cav., differing in its ovate (not lanceolate) leaves, its lax, not congested panicles, and its stamens 20 or more, not 10 as in Cavanilles' species. Gagnepain does not mention Loureiro's species in his treatment of the Sterculiaceae of Indo-China (Lecomte Fl. Gén. Indo-Chine 1: 454–522. 1911). In any case Loureiro's specific name is invalidated in *Sterculia* by *S. paniculata* Wall.

Melochia (Dillenius) Linnaeus

Melochia corchorifolia Linn. Sp. Pl. 675. 1753; Lour. Fl. Cochinch. 407. 1790, ed. Willd. 494. 1793, Anamese *cây bái giéi*.
Melochia concatenata Linn. l.c.
" Habitat in agris Cochinchinae, inculta." The Linnaean species, a very common and widely distributed one in the Old World tropics, was correctly interpreted by Loureiro. *Melochia concatenata* Linn. is the same species, but has merely place priority, hence the commonly used name is here accepted.

Helicteres (Plukenet) Linnaeus

Helicteres angustifolia Linn. Sp. Pl. 963. 1753; Lour. Fl. Cochinch. 530. 1790, ed. Willd. 647. 1793, Chinese *sān chí mà*.
" Habitat agrestis circa Cantonem Sinarum." The type of the Linnaean species is a specimen collected near Canton by Osbeck, and Loureiro's description conforms to its characters. The species is abundant in open dry grassy places in the vicinity of Canton and is one of wide geographical distribution in tropical Asia and Malaysia.

Helicteres hirsuta Lour. Fl. Cochinch. 530. 1790, ed. Willd. 648. 1793, Anamese *cây dúoi chôn;* Gagnep. in Lecomte Fl. Gén. Indo-Chine 1: 490. 1911.
Camaion hirsuta Raf. Sylva Tellur. 75. 1838 (based on *Helicteres hirsuta* Lour.).
Helicteres spicata Colebr. ex Roxb. Hort. Beng. 97. 1814, *nomen nudum;* Mast. in Hook. f. Fl. Brit. Ind. 1: 366. 1874.
" Habitat in sylvis planis Cochinchinae." As currently interpreted, this is a species of wide geographic distribution in the Indo-Malaysian region. There are three sheets from Loureiro in the herbarium of the British Museum.

DILLENIACEAE

Tetracera Linnaeus

Tetracera scandens (Linn.) Merr. Interpret. Herb. Amb. 365. 1917.
Tragia scandens Linn. in Stickman Herb. Amb. 18. 1754, Amoen. Acad. 4: 128. 1759.
Delima sarmentosa Linn. Syst. ed. 10, 1076. 1759.
Tetracera sarmentosa Vahl Symb. 3: 70. 1794; Finet & Gagnep. in Lecomte Fl. Gén. Indo-Chine 1: 15. 1907.

Actaea aspera Lour. Fl. Cochinch. 332. 1790, ed. Willd. 405. 1793, Chinese *tsia ip.*
Calligonum asperum Lour. Fl. Cochinch. 342. 1790, ed. Willd. 418. 1793, Anamese
    deei chio tlái.
Seguieria asiatica Lour. Fl. Cochinch. 341. 1790, ed. Willd. 417. 1793, Anamese *deei chio.*
Tetracera aspera Raeusch. Nomencl. ed. 3, 147. 1797 (based on *Calligonum asperum*
    Lour.).
Trachytella actaea DC. Syst. 1: 410. 1818 (based on *Actaea aspera* Lour.).
Trachytella calligonum DC. Syst. 1: 410. 1818 (based on *Calligonum asperum* Lour.).
Traxilisa aspera Raf. Sylva Tellur. 162. 1838 (based on *Calligonum asperum* Lour.).
Tetracera volubilis Merr. Sp. Blancoanae 262. 1918, sphalm. (*T. scandens* intended).

For *Actaea aspera* Loureiro states: " Habitat inculta prope Cantonem Sinarum, " and
for *Calligonum asperum* " Habitat sylvas Cochinchinae." The descriptions of both
indicate *Tetracera* and unquestionably both appertain to the widely distributed *Tetracera
scandens* (Linn.) Merr. (*T. sarmentosa* Vahl) which is common both in the vicinity of
Canton and in Anam. For *Seguieria asiatica* Loureiro states: " Habitat in sylvis Cochinchinae." The rather poor, and apparently inaccurate description seems manifestly
to appertain to *Tetracera*, and apparently to a form of *T. scandens* (Linn.) Merr. In
partial confirmation of the correctness of this reduction, at least to the genus, the Anamese
name *deei chio* indicates a plant in the alliance with *Calligonum asperum* Lour., for which
the Anamese name *deei chio tlái* is cited.

OCHNACEAE

Ochna Linnaeus

Ochna integerrima (Lour.) comb. nov.
    Elaeocarpus integerrimus Lour. Fl. Cochinch. 338. 1790, ed. Willd. 412. 1793, Anamese
    cây mai boung vang.
    Discladium harmandii Van Tiegh. in Ann. Sci. Nat. VIII Bot. 16: 351. 1902.
    Ochna harmandii Lecomte Fl. Gén. Indo-Chine 1: 706. *f. 75.* 1911.

" Habitat agrestis in Cochinchina; colitur etiam in hortis ob odorem, & pulchritudinem
florum. . . ." Loureiro failed to add the conventional sign † used by him to indicate
his new species. This was originally placed by me as perhaps an *Elaeocarpus* near *E.
hainanensis* Oliv., in spite of serious discrepancies between the characters of that species
and the one Loureiro attempted to describe. A specimen collected by de Pirey under the
local name *bong mai vang* (*Chevalier 41165*) at Long Quang Tri, Anam, supplied the clue
to the present interpretation. The leaves are *not* " integerrima " but are minutely toothed;
the petals are *not* " lacera " but are entire. The description otherwise is absolutely that
of *Ochna*. To be noted particularly is the description of inflorescence and the flowers and
fruits, particularly the: " Germina 10, sub-rotunda, minuta, ad basim styli circumposita.
Drupa 1 maturescens solitaria, reliquis germinibus omnibus suffocatis "; these are *Ochna*,
not *Elaeocarpus*, characters. It is manifest that Loureiro's inaccurate description was not
based on a mixture of material, but that he depended on his memory and added the petal
character of *Elaeocarpus* from the generic description; significant of this is his statement
at the end of the fruit description " nec amplius memini "; this interpretation is then of
Elaeocarpus integerrimus Lour. *emended.* Lecomte admits four varieties of this species,

reducing as synonyms 10 binomials proposed by Van Tieghem in 1902 and 1907 who, however, did not provide satisfactory descriptions; these binomials are in the genera *Discladium, Diporidium, Polythecium* and *Polythecanthum*.

THEACEAE

Thea Linnaeus

Thea oleosa Lour. Fl. Cochinch. 339. 1790, ed. Willd. 414. 1793, Anamese *chè deâu*, Chinese *yêu châ*.

Camellia drupifera Lour. Fl. Cochinch. 411. 1790, ed. Willd. 499. 1793, Anamese *cây deâù so*.

Mesua bracteata Spreng. Syst. 3: 127. 1826 (based on *Camellia drupifera* Lour.).

Theaphylla oleifera Raf. Sylva Tellur. 139. 1838 (based on *Thea oleosa* Lour.).

Drupifera oleosa Raf. Sylva Tellur. 140. 1838 (based on *Camellia drupifera* Lour.).

Thea olearia Lour.[33] ex Gomes in Mem. Acad. Sci. Lisb. Cl. Sci. Pol. Mor. Bel.-Let. n.s. 4(1): 29. 1868.

Thea drupifera Pierre Fl. Forest. Cochinch. 2: sub *pl. 119*, 1887 (based on *Camellia drupifera* Lour.).

Thea sasanqua Nois. var. *oleosa* Pierre Fl. Forest. Cochinch. 2: *pl. 116. f. B 1, 2*. 1887 (not based on *Thea oleosa* Lour.).

Thea sasanqua Nois. var. *loureiri* Pierre Fl. Forest. Cochinch. 2: *pl. 115*. 1887; Pitard in Lecomte Fl. Gén. Indo-Chine 1: 344. 1910 (not based on any Loureiro description).

For *Thea oleosa* Loureiro states: " Habitat agrestis circa Cantonem Sinarum "; and for *Camellia drupifera:* " Habitat inculta, cultaque in Cochinchina." Various authors have interpreted them differently, but I believe that both are but forms of a single species and one closely allied to and by no means always easily distinguishable from *Thea sasanqua* (Thunb.) Nois. Seemann (Bonplandia 7: 49. 1859), on the basis of the description, reduced *Camellia drupifera* Lour. to *Pyrenaria serrata* Bl. which is an erroneous disposition of it. Pierre in discussing *Thea drupifera* notes that Loureiro describes the ovary as 4-celled and the style as having 4 branches, but this does not invalidate the present identification of Loureiro's species, as Pierre also (Fl. Forest. Cochinch. 2: *pl. 116. f. B 1*) figures *Thea sasanqua* Nois. var. *oleosa* Pierre as having a 4-celled ovary and a 4-branched style, although the normal number of each is 3. The type of *Thea oleosa* Lour. is preserved in the herbarium of the Paris Museum and regarding it Seemann (Trans. Linn. Soc. Bot. 22: 344. 1859) states: " What is preserved in the Parisian Museum as the original specimen of *Thea oleosa*, Lour., is *Thea Chinensis*, var. *Bohea* "; but Doctor Gagnepain who recently examined it at my request, states that it is *Thea sasanqua* as figured by Pierre. The species is very common in Kwangtung Province where it is usually (always?) planted, the Cantonese name appearing on recent collections as *cha tsai* and *cha yao*. Rehder & Wilson (Pl. Wils. 2: 393. 1915) accept the binomial *Thea oleifera* (Abel) Rehd. & Wils. for this species, based on *Camellia oleifera* Abel (1818), but Loureiro's name is much older and in my opinion appertains to the same species which, like most cultivated plants, is distinctly variable.

---

[33] A Loureiro herbarium name here first published by Gomes.

Thea sinensis Linn. Sp. Pl. 515. 1753.

Thea cantoniensis Lour. Fl. Cochinch. 339. 1790, ed. Willd. 414. 1793, Anamese *chè taù*, Chinese *hŏ nâm châ yòng*.

Thea cochinchinensis Lour. Fl. Cochinch. 338. 1790, ed. Willd. 413. 1793, Anamese *chè ăn năm*.

Camellia thea Link Enum. Hort. Berol. 2: 73. 1822.

Thea chinensis Seem. in Trans. Linn. Soc. 22: 349. *pl. 61.* 1859; Pitard in Lecomte Fl. Gén. Indo-Chine 1: 341. 1910.

Thea chinensis Seem. var. *cantoniensis* Pierre Fl. Forest. Cochinch. 2: *pl. 113.* 1887; Pitard op. cit. 342 (based on *Thea cantoniensis* Lour.).

Camellia sinensis O. Ktz. in Act. Hort. Petrop. 10: 195. 1887.

Theaphylla cantoniensis Raf. Sylva Tellur. 139. 1838 (based on *Thea cantoniensis* Lour.).

Theaphylla annamensis Raf. l.c. (based on *Thea cochinchinensis* Lour.).

For *Thea cantoniensis* Loureiro states: " Habitat tam culta, quam inculta prope Cantonem Sinarum "; and for *T. cochinchinensis:* " Habitat culta, incultaque in provinciis Borealibus Cochinchinae." The type of the former is preserved in the herbarium of the Paris Museum, and definitely is *Thea sinensis* Linn., more commonly known as *Camellia thea* Link. I see no reason for considering that *T. cochinchinensis* Lour. represents other than a form of the common tea plant. For critical discussions of *Thea cantoniensis* Lour. and *T. cochinchinensis* Lour., see Cohen-Stuart (Bull. Jard. Bot. Buitenzorg III 1: 217, 251. 1919).

GUTTIFERAE

Hypericum Linnaeus

Hypericum japonicum Thunb. Fl. Jap. 295. *pl. 31.* 1784; Gagnepain in Lecomte Fl. Gén. Indo-Chine 1: 286. 1909.

Reseda chinensis Lour. Fl. Cochinch. 299. 1790, ed. Willd. 367. 1793, Chinese *thin kí hoâm*.

Reseda cochinchinensis Lour. Fl. Cochinch. 299. 1790, ed. Willd. 366. 1703, Anamese *hoa phân*.

For *Reseda chinensis* Loureiro states: " Habitat suburbia Cantoniensia in Sinis," and for *R. cochinchinensis:* " Habitat in agris Cochinchinae." Both descriptions apply to forms of the very common and widely distributed *Hypericum japonicum* Thunb., an abundant species both in the vicinity of Canton and in Indo-China. The chief differences in Loureiro's two descriptions are in the number of stamens, 18 and 30, and in the false character of laciniate petals for *R. cochinchinensis;* doubtless the " petala . . . laciniata " was added to make the description conform to the characters of the genus in which it was erroneously placed. Gagnepain does not account for Loureiro's binomial, but cites the significantly similar Anamese name *cây ban* under *Hypericum japonicum* Thunb.

Hypericum chinense Linn. Syst. ed. 10, 1184. 1759.

Hypericum aureum Lour. Fl. Cochinch. 472. 1790, ed. Willd. 578. 1793, Chinese *guéi thoung hõa*.

" Habitat incultum prope Cantonem Sinarum: colitur etiam ob florum nitorem, & pulchritudinem." Loureiro's description applies unmistakably to the Linnaean species as the latter is currently interpreted.

Cratoxylon Blume

Cratoxylon cochinchinense (Lour.) Blume Mus. Bot. Lugd.-Bat. **2**: 17. 1852; Merr. Enum. Philip. Fl. Pl. **3**: 77. 1923 (based on *Hypericum cochinchinense* Lour.).
> *Hypericum cochinchinense* Lour. Fl. Cochinch. 472. 1790, ed. Willd. 577. 1793, Anamese *le nganh do.*
> *Vismia cochinchinensis* Spreng. Syst. **3**: 350. 1826 (based on *Hypericum cochinchinense* Lour.).
> *Elodea formosa* Jack in Malay. Miscel. 2(7): 24. 1822.
> *Tridesmis formosa* Korth. Verh. Nat. Gesch. Nederl. Overz. Bezit. Bot. 179. *pl. 37.* 1839–42.
> *Cratoxylon formosum* Dyer in Hook. f. Fl. Brit. Ind. **1**: 258. 1874; Gagnep. in Lecomte Fl. Gén. Indo-Chine **1**: 288. 1909.

" Habitat in sylvis Cochinchinae." Loureiro's description is definite and I believe applies to the very common and widely distributed Indo-Malaysian species currently known as *Cratoxylon formosum* Dyer. It is represented by *Squires 303* from Hue, the classical locality.

Cratoxylon ligustrinum (Spach) Blume Mus. Bot. Lugd.-Bat. **2**: 16. 1852; Merr. Enum. Philip. Fl. Pl. **3**: 77. 1923.
> *Ancistrolobus ligustrinus* Spach Hist. Nat. Vég. Phan. **5**: 361. 1836, Ann. Sci. Nat. II Bot. **5**: 352. *pl. 6.* 1836.
> *Hypericum biflorum* Lam. Encycl. **4**: 170. 1796.
> *Cratoxylon biflorum* Turcz.[34] in Bull Soc. Nat. Mosc. **36**(1): 580. 1863.
> *Cratoxylon chinense* Merr. in Philip. Journ. Sci. Bot. **4**: 292. 1909.
> *Hypericum chinense* Retz. Obs. **5**: 27. 1789, non Linn.
> *Hypericum olympicum* (non Linn.) Lour. Fl. Cochinch. 471. 1790, ed. Willd. 577. 1793, Anamese *le nganh tláng,* Chinese *hôang xŏc.*
> *Hypericum petiolatum* (non Linn.) Lour. Fl. Cochinch. 472. 1790, ed. Willd. 577. 1793, Chinese *hoâng nièu thâu.*
> *Cratoxylon polyanthum* Korth. Verh. Nat. Gesch. Nederl. Overz. Bezit. Bot. 175. *pl. 36.* 1839–42; Gagnep. in Lecomte Fl. Gén. Indo-Chine **1**: 290. 1910.

For *Hypericum olympicum* Loureiro states: " Habitat incultum in Cochinchina, & Cantone Sinarum "; and for *H. petiolatum:* " Habitat incultum prope Cantonem Sinarum." Both descriptions I believe appertain to the widely distributed species currently known as *Cratoxylon polyanthum* Korth., the only species of *Cratoxylon* known from southeastern China, where it is common, and one that is equally common in Indo-China. This is supposedly *Hypericum chinense* Retz., non Linn., on which *Hypericum biflorum* Lam. (1797) was based. It may be that Lamarck's specific name should be accepted, but *Cratoxylon biflorum* Turcz. (1863), which represents the same species, was published independently and without reference to Lamarck's species. *Squires 333* and *Clemens 3155, 3844,* from

---

[34] This binomial was based wholly on *Wallich 4820* with no reference to *Hypericum biflorum* Lam. which, however, apparently represents the same species.

Hue and Tourane, represent *Hypericum olympicum* Lour., and very numerous collections from the vicinity of Canton represent it and also *H. petiolatum* Lour.

Calophyllum Linnaeus

Calophyllum inophyllum Linn. Sp. Pl. 513. 1753; Pitard in Lecomte Fl. Gén. Indo-Chine 1: 324. 1901.

Balsamaria inophyllum Lour. Fl. Cochinch. 470. 1790, ed. Willd. 574. 1793, Anamese *cây muu*.

"Habitat tam culta, quam agrestis in locis mediterraneis Cochinchinae. Vidi etiam frequentum in Cambodia, & in sylvis Malaiorum juxta fretum Malacense." The species for which Loureiro proposed the new generic name *Balsamaria* is identical with the very common and widely distributed littoral *Calophyllum inophyllum* Linn.

Garcinia Linnaeus

Garcinia cochinchinensis (Lour.) Choisy in DC. Prodr. 1: 561. 1824; Pierre Fl. Forest. Cochinch. 1: Enum. Garcinia XXVIII. 1883 (based on *Oxycarpus cochinchinensis* Lour.).

Oxycarpus cochinchinensis Lour. Fl. Cochinch. 648. 1790, ed. Willd. 796. 1793, Anamese *cay búa;* Moore in Journ. Bot. **63**: 289. 1925.

Garcinia loureiri Pierre Fl. Forest. Cochinch. 1: *pl. 66.* Enum. Garcinia XXVIII. 1883; Pitard in Lecomte Fl. Gén. Indo-Chine 1: 308. 1910.

"Habitat tam cultus, quam incultus in Cochinchina." Loureiro's type is preserved in the herbarium of the British Museum, and Moore, who examined it, states that it is too fragmentary to decide whether or not *Garcinia loureiri* Pierre is really distinct from *G. cochinchinensis* Choisy. Pierre described *G. loureiri* as new on the basis of material collected in Indo-China because Loureiro's description of the flowers of *Oxycarpus cochinchinensis* was not in full agreement with the characters of these of *G. loureiri;* yet both are cultivated and both have the same local name. Under the circumstances I can see no valid reason for considering that *Garcinia loureiri* Pierre is other than *G. cochinchinensis* Choisy. *Folium acidum majus* Rumph. (Herb. Amb. **3**: 58. *pl. 32*), cited by Loureiro, must be excluded as it represents *Garcinia amboinensis* Spreng.

Garcinia hanburyi Hook. f. in Journ. Linn. Soc. Bot. **14**: 485. 1875; Pierre Fl. Forest. Cochinch. 1: *pl. 74.* 1883; Pitard in Lecomte Fl. Gén. Indo-Chine 1: 312. 1910.

Cambogia gutta (non Linn.) Lour. Fl. Cochinch. 332. 1790, ed. Willd. 406. 1793, saltem pro parte, Anamese *cây vàng nhua, trân huỳnh*, Chinese *hôam lô*.

"Habitat non rara in sylvis Cochinchinae: abundantius vero in Siamo, & Cambodia." Loureiro's description applies to *Garcinia hanburyi* Hook. f., although he may have included in his conception of *Cambogia gutta* other than this species.

TAMARICACEAE

Tamarix Linnaeus

Tamarix chinensis Lour. Fl. Cochinch. 182. 1790, ed. Willd. 228. 1793, Chinese *cuǒn nham lâu*.

"Habitat in provincia Cantoniensi Sinarum." Loureiro undoubtedly had specimens from cultivated plants. Hemsley (Journ. Linn. Soc. Bot. 23: 346. 1888) states that there is no evidence that the tamarisk is anywhere wild in China and that Ehrenberg was probably correct in treating the Chinese form as a variety of the widely spread *Tamarix gallica* Linn. Loureiro's type is preserved in the herbarium of the Paris Museum.

WINTERANACEAE

Cinnamosma Baillon

Cinnamosma fragrans Baill. in Adansonia 7: 219. *pl. 5.* 1867.

? *Winterania canella* (non Linn.) Lour. Fl. Coch. 293. 1790, ed. Willd. 359. 1793.

"Habitat in altis montibus insulae Madgascariae, S. Laurentii dictae." This reduction was suggested to me by Doctor Humbert and Doctor Danguy. As they note, it is impossible to determine from Loureiro's imperfect description whether Baillon's species is represented, or the more recently described *Cinnamosma madagascariensis* Danguy (Not. Syst. 1: 236. *f. 12.* 1910). Loureiro's species seems safely to represent a *Cinnamosma*.

VIOLACEAE

Rinorea Aublet

Rinorea anguifera (Lour.) O. Ktz. Rev. Gen. Pl. 42. 1891 (based on *Medusa anguifera* Lour.).

Medusa anguifera Lour. Fl. Cochinch. 406. 1790, ed. Willd. 493. 1793, Anamese *cây chôm chôm dât.*

Jürgensia anguifera Spreng. Syst. 3: 50. 1826 (based on *Medusa anguifera* Lour.).

Alsodeia echinocarpa Korth. in Nederl. Kruidk. Arch. 1: 360. 1848; de Boissieu in Lecomte Fl. Gén. Indo-Chine 1: 214. 1909.

No locality is cited by Loureiro, but his specimens were manifestly from Cochinchina because of the local name cited. The species is clearly the same as *Alsodeia echinocarpa* Korth., which extends from Indo-China through the Malay Peninsula to Sumatra and Borneo. A specimen from Loureiro listed as being among the material received from him has not been located in the herbarium of the British Museum.

Rinorea sessilis (Lour.) O. Ktz. Rev. Gen. Pl. 42. 1891 (based on *Pentaloba sessilis* Lour.).

Pentaloba sessilis Lour. Fl. Cochinch. 154. 1790, ed. Willd. 192. 1793, Anamese *cây cuong tàu.*

Alsodeia sessilis Spreng. Syst. 1: 806. 1825 (based on *Pentaloba sessilis* Lour.).

"Habitat inculta in montibus Cochinchinae." In describing the new genus *Pentaloba*, Loureiro indicated the fruit as 5-lobed. His type is preserved in the herbarium of the British Museum and R. Brown, who has examined it (Tukey Congo 441. 1818, Misc. Bot. Works 1: 123. 1866), states that the flowering specimen is in all respects an *Alsodeia* [= *Rinorea*] even to the number of parietal placentae. De Boissieu does not mention Loureiro's species in his treatment of the Violaceae of Indo-China (Lecomte Fl. Gén. Indo-Chine 1: 212–217. 1909). In my original manuscript of 1919 I placed *Alsodeia membranacea* King as interpreted by de Boissieu, as a synonym of Loureiro's species. This is a manifest error, as Gagnepain (Not. Syst. 3: 251. 1916) notes that the specimens cited by de Boissieu, apparently all in fruit, do not represent an *Alsodeia* but rather *Casearia flexuosa* Craib. Loureiro's species is apparently represented by *Clemens 3357* from Tourane, near the classical locality Hue.

Viola (Tournefort) Linnaeus

Viola betonicifolia Sm. subsp. **nepalensis** (Ging.) W. Becker in Bot. Jahrb. **54**: Beibl. **120**: 166. 1917.

Viola patrinii Ging. var. *nepalensis* Ging. in DC. Prodr. **1**: 293. 1824.
Viola primulifolia (non Linn.) Lour. Fl. Cochinch. 513. 1790, ed. Willd. 628. 1793.
Viola chinensis G. Don Gen. Syst. **1**: 322. 1831 (based on *Viola primulifolia* Lour.);
Melch. in Notizbl. Bot. Gart. Berlin **11**: 376. 1932.
? Viola patrinii Ging. var. *chinensis* Ging. in DC. Prodr. **1**: 293. 1824.

" Habitat inculta prope Cantonem Sinarum." Melchior in his recent critical treatment of the species *Viola* of Kwangtung Province (Notizbl. Bot. Gart. Berlin **11**: 364–378. 1932), extensively discusses Loureiro's species and leaves *Viola chinensis* G. Don, which was based on Loureiro's description, among the *species incertae sedis*. While I have here followed Becker (Bot. Jahrb. **54**: Beibl. **120**: 166. 1917) in his interpretation of *Viola betonicifolia* Sm. I am by no means certain that the form Loureiro so inadequately described is really a synonym of Smith's species. De Gingen's variety *chinensis* of *Viola patrinii* was based on a specimen collected by Staunton, and to it he refers, with doubt, *Viola primulifolia* Lour. This must have been the specimen in the herbarium of the British Museum that Hemsley (Journ. Linn. Soc. Bot. **23**: 53. 1886) thought came from Loureiro, as Melchior quotes Exell to the effect that there is no Loureiro specimen of this species in the herbarium. Nakai (Bull. Soc. Bot. France **72**: 186. 1925) made the same error in apparently interpreting the Staunton specimen as a Loureiro one.

Viola alata Burgsd. subsp. **alata** W. Becker in Beih. Bot. Centralbl. **34**(2): 227. 1916.

? Viola odorata (non Linn.) Lour. Fl. Cochinch. 513. 1790, ed. Willd. 627. 1793, Anamese
kíet tuong hõa.

" Habitat Cantone Sinarum in hortis, & prope vias: inde in Cochinchinam delata, ubi raro culta." It is suspected that more than one species is included in Loureiro's description, a native one and perhaps an introduced and cultivated form.

FLACOURTIACEAE

Oncoba Forskål

Oncoba africana (Lour.) Planch. in Hook. Lond. Journ. Bot. **6**: 296. 1847 (based on *Heptaca africana* Lour.).

Heptaca africana Lour. Fl. Cochinch. 657. 1790, ed. Willd. 808. 1793.

" Habitat in sylvis Africae Orientalis." Planchon gives his reasons for reducing *Heptaca* Lour. to *Oncoba* Forskål and in proposing the section *Heptaca* of the genus *Oncoba*, he notes that Loureiro's description of the ovary and of the fruit is erroneous. There is no evidence, however, that Planchon saw material collected by Loureiro. Oliver (Fl. Trop. Afr. **1**: 114. 1868) places *Heptaca* as a synonym of *Oncoba*, but does not mention

Oncoba africana (Lour.) Planch. I cannot place it among the 14 species described by Oliver, but if it should be an *Oncoba* it should fall in the small group of four species with axillary or extra-axillary racemes, none of which is reported from East Africa.

Scolopia [35] Schreber

Scolopia chinensis (Lour.) Clos in Ann. Sci. Nat. IV Bot. **8**: 249. 1857; Gagnep. in Lecomte Fl. Gén. Indo-Chine 1: 229. 1909 (based on *Phoberos chinensis* Lour.).

Phoberos chinensis Lour. Fl. Cochinch. 318. 1790, ed. Willd. 389. 1793, Chinese cŏ tsû.

Phoberos cochinchinensis Lour. Fl. Cochinch. 318. 1790, ed. Willd. 389. 1793, Anamese *cây gai bôm.*

Scolopia cochinchinensis Clos in Ann. Sci. Nat. IV Bot. **8**: 253. 1857 (based on *Phoberos cochinchinensis* Lour.).

Scolopia germaini Briq. in Ann. Conserv. Jard. Bot. Genève 2: 42. 1898.

For *Phoberos chinensis* Loureiro states: " Habitat in China, ubi ex ea similiter [*P. cochinchinensis*] ordinantur sepes "; and for *P. cochinchinensis:* " Habitat in Cochinchina, ubi, connexis ejus ramis, fiunt sepes, cuilibet animali horribiles, & imperviae." I judge the latter to be the type of *Phoberos*, as it is the first species described and apparently was known to Loureiro many years before he observed the Chinese form. The type of *Phoberos chinensis* Lour. is preserved in the herbarium of the Paris Museum; the type of *P. cochinchinensis* is preserved in the herbarium of the British Museum. I can see no reason, however, for distinguishing the Indo-China form as described by Loureiro. He does not mention in the description of either species whether or not the leaves are glandular at the base, or whether or not the anther-connectives are bearded, characters used by various authors in distinguishing species, otherwise very similar to each other. *Scolopia germaini* Briq, type from Cochinchina, *Germain 29, 37*, was overlooked by Gagnepain, but seems to be referable to *S. chinensis* Clos as interpreted by Gagnepain. Whether or not the branches are spiny is an independable character, as the species is exceedingly variable in the presence or absence of spines at least on those branches ordinarily preserved as a part of herbarium specimens.

Homalium Jacquin

Homalium cochinchinense (Lour.) Druce in Rept. Bot. Exch. Club Brit. Isles 4: 628. 1917 (based on *Astranthus cochinchinensis* Lour.).

Astranthus cochinchinensis Lour. Fl. Cochinch. 222. 1790, ed. Willd. 274. 1793, Anamese *cây chây.*

Blackwellia fagifolia Lindl. in Trans. Hort. Soc. London 6: 269. 1826.

Blackwellia padiflora Lindl. Bot. Reg. **16**: *pl. 1308.* 1830.

Blackwellia padifolia Steud. Nomencl. ed. 2, 1: 208. 1840.

Blackwellia loureirii Benth. in Hook. Lond. Journ. Bot. **1**: 482. 1842 (based on *Astranthus cochinchinensis* Lour.).

Blackwellia cochinchinensis Blume Mus. Bot. Lugd.-Bat. **2**: 27. 1852 (based on *Astranthus cochinchinensis* Lour.).

Homalium fagifolium Benth. in Journ. Linn. Soc. Bot. **4**: 35. 1860; Gagnep. in Lecomte Fl. Gén. Indo-Chine 2: 1008. 1921.

[35] *Scolopia* Schreber (1789), conserved name, Brussels Code: an older one is *Aembilla* Adanson (1763).

" Habitat agrestis, non rara in Cochinchina." Bentham (Journ. Linn. Soc. Bot. **4**: 38. 1869) states regarding Loureiro's species: " Probably not distinct from *H. fagifolium.*" Gagnepain placed it definitely as a synonym of Bentham's species. The two are manifestly identical, the oldest specific name being here accepted. Loureiro's description of the fruits as 1-seeded is a manifest error of observation. His type is preserved in the herbarium of the British Museum.

Homalium sp.

> *Pythagorea cochinchinensis* Lour. Fl. Cochinch. 244. 1790, ed. Willd. 300. 1793, Anamese *xuong cá tiá nho lá*, non *Homalium cochinchinense* (Lour.) Druce.

" Habitat agrestis in Cochinchina." The description applies unmistakably to *Homalium* and to a species of the section *Blackwellia*. Gagnepain in his treatment of the Homaliaceae of Indo-China (Lecomte Fl. Gén. Indo-Chine **2**: 1005–1015. 1921) admits 11 species of *Homalium* from Indo-China, but does not mention *Pythagorea cochinchinensis* Lour., which is manifestly allied to *Homalium fagifolium* Benth. = *H. cochinchinense* (Lour.) Druce, and to *H. digynum* Gagnep. It is represented by *Clemens 3860* from Tourane. The specific name is invalidated by *Homalium cochinchinense* Druce (*Astranthus cochinchinensis* Lour.).

Xylosma [36] Forster f.

Xylosma congestum (Lour.) Merr. in Philip. Journ. Sci. **15**: 247. 1919 (based on *Croton congestum* Lour.).

> *Croton congestum* Lour. Fl. Cochinch. 582. 1790, ed. Willd. 714. 1793, Chinese *pă táu.*
> *Apactis japonica* Thunb. Nov. Gen. 66. 1783, Fl. Jap. 191. 1784.
> *Xylosma racemosum* Miq. Ann. Mus. Bot. Lugd. Bat. **2**: 155. 1866.
> *Hisingera racemosa* Sieb. & Zucc. Fl. Jap. **1**: 169, 189. *pl. 88, 100. f. III, 1–14.* 1826–1835.
> *Xylosma japonicum* A. Gray [37] in Mem. Am. Acad. **6**: 381. 1859.
> *Flacourtia japonica* Walp. Repert. **2**: 205. 1843.
> *Flacourtia chinensis* Clos in Ann. Sci. Nat. IV Bot. **8**: 219. 1857.
> *Casearia subrhombea* Hance in Journ. Bot. **23**: 323. 1885.
> *Xylosma apactis* Koidz. in Bot. Mag. Tokyo **39**: 316. 1925.

" Habitat agreste circa Cantonem Sinarum." Mueller-Arg. (DC. Prodr. **15**(2): 696. 1866) repeated Loureiro's short description, placing the species among those " excludendae, sed nondum recognitae, " noting that the description of the inflorescences and the apetalous flowers does not conform to *Croton*. As noted by me in transferring Loureiro's species to *Xylosma*, the description, except the statement " foeminei capsulae 3-coccae, pendulae," conforms entirely with *Xylosma racemosum* Miq., one not uncommon in the vicinity of Canton. Handel-Mazzetti (Symb. Sin. **7**: 383. 1931) does not accept Loureiro's specific name as valid for the reason that the characters of *Xylosma* (all of the description except the statement above quoted) are combined with those of a totally different species, in accordance with the rule that a species based on a mixture is invalid. If this rule be strictly followed, hundreds of accepted Linnaean binomials, as well as hundreds of others proposed by his

---

[36] *Xylosma* Forster f. (1786), conserved name, Vienna Code; an older one is *Myroxylon* Forster (1776).

[37] This binomial was based on *Hisingera japonica* Sieb. & Zucc., which in turn was based on *Hisingera racemosa* Sieb. & Zucc., not on *Apactis japonica* Thunb. These three binomials actually appertain to the same species.

contemporaries and successors, will automatically fall. Loureiro certainly added the statement quoted above to make his description conform more closely with the characters of the genus in which he erroneously placed it, without actually seeing fruiting specimens. I do not consider it to be a " mixture."

Flacourtia (Commerson) L'Héritier

Flacourtia indica (Burm. f.) Merr. Interpret. Herb. Amb. 377. 1917, Enum. Philip. Fl. Pl. 3: 112. 1923, Journ. Arnold Arb. 6: 137. 1925.

Gmelina indica Burm. f. Fl. Ind. 132. *pl. 39. f. 5.* 1768.

Stigmarota africana Lour. Fl. Cochinch. 634. 1790, ed. Willd. 779. 1793.

Flacourtia sepiaria Roxb. Pl. Coromand. **1**: 48. *pl. 68.* 1795; Gagnep. in Lecomte Fl. Gén. Indo-Chine **1**: 236. 1909.

Flacourtia ramontchi L'Hérit. Stirp. Nov. 59. *pl. 30, 30 B.* 1784–85; Oliv. in Fl. Trop. Afr. **1**: 120. 1868.

Flacourtia balansae Gagnep. in Bull. Soc. Bot. France **55**: 521. 1908, in Lecomte Fl. Gén. Indo-Chine **1**: 235. *f. 23.* 1909.

No locality is cited, but from his specific name it is clear that Loureiro had African material, probably from near Mozambique. Willdenow in 1793 states in a footnote: " Est *Flacourtia Ramontchi* l'Héritier." This is correct, as I understand the species, but the oldest specific name is that supplied by *Gmelina indica* Burm. f. For a note on the reduction of *Flacourtia balansae* Gagnep. see Merrill (Journ. Arnold Arb. **6**: 137. 1925). If it be desirable to retain Gagnepain's species as a distinct one, an older name is supplied by *Myroxylon decline* Blanco (1837) which is identical with *Flacourtia balansae* Gagnep.

Flacourtia jangomas (Lour.) Raeusch. Nomencl. ed. 3, 290. 1797; Steud. Nomencl. 343. 1821 (based on *Stigmarota jangomas* Lour.).

Stigmarota jangomas Lour. Fl. Cochinch. 634. 1790, ed. Willd. 779, 1793, Anamese *cây mu cŭon.*

Flacourtia cataphracta Roxb. ex Willd. Sp. Pl. **4**: 830. 1805; Gagnep. in Lecomte Fl. Gén. Indo-Chine **1**: 233. 1909.

Roumea jangomas Spreng. Syst. **2**: 632. 1825 (based on *Stigmarota jangomas* Lour.).

" Habitat culta in Cochinchina, puto, quod etiam agrestis." Willdenow in 1793 indicated that Loureiro's genus *Stigmarota* was referable to *Flacourtia* Commerson. The description conforms to the characters of *Flacourtia cataphracta* Roxb. *Spina spinarum* Rumph. (Herb. Amb. **7**: 36. *pl. 19. f. 1, 2*), cited by Loureiro as a synonym, belongs with *Flacourtia indica* (Burm. f.) Merr. (*F. sepiaria* Roxb.). The specific name *jangomas* was taken from *Jangomas* (Garc. Arom. lib. 2. cap. 5) also cited as a synonym. Loureiro's type is preserved in the herbarium of the British Museum.

PASSIFLORACEAE

Passiflora Linnaeus

Passiflora caerulea Linn. Sp. Pl. 959. 1753; Lour. Fl. Cochinch. 527. 1790, ed. Willd. 644. 1793, Chinese *ù sì hoā.*

Passiflora loureirii G. Don Gen. Syst. **3**: 54. 1834 (based on *Passiflora caerulea* Lour.).

" Habitat prope Cantonem Sinarum, agrestis." Loureiro's description conforms very closely to the characters of the Linnaean species, an introduced and naturalized one near Canton. The only possible objection that I can see to this disposition of the form Loureiro described is his description of the flowers as " luteo-viridis." The description applies to no other native or introduced species known to occur in China.

Passiflora cochinchinensis Spreng. Syst. **4**: Cur. Post. 346. 1827 (based on *Passiflora pallida* Lour.); Gagnep. in Lecomte Fl. Gén. Indo-Chine **2**: 1017. *f. 111. 4.* 1921.

Passiflora pallida (non Linn.) Lour. Fl. Cochinch. 527. 1790, ed. Willd. 644. 1793, Anamese *cây com lang.*

Passiflora ligulifolia Mast. in Trans. Linn. Soc. **27**: 632. 1871; Gagnep. in Lecomte Fl. Gén. Indo-Chine **2**: 1023. 1921.

Passiflora hainanensis Hance in Journ. Bot. **16**: 227. 1878.

" Habitat in dumetis Cochinchinae." Willdenow in a footnote indicated the discrepancy in the leaves being described as opposite, and de Candolle (Prodr. **3**: 331. 1828) stated: " Ab omnibus Passifloreis differt foliis oppositis! an forte Malpighiacea quaedam?" The description, except in the position of the leaves, is unmistakably that of a *Passiflora*, the opposite leaves being an error in observation or in recording the characters by Loureiro. The species is represented by *Clemens 3904*, and *Kuntze 3646*, from Tourane. Gagnepain admitted *Passiflora ligulifolia* Mast., of Hainan and Hongkong, as an insufficiently known species, but I can see no reason for distinguishing this from *P. cochinchinensis* Spreng. It is represented by *Ford s. n.*, *McClure 9251*, and *Tsang 17666* from Hainan, and by *Tsiang 2222, 2264, 2303, 2305, 2680* from Kwangtung Province. *Clemens 3904* from Tourane presents both the relatively broad elliptic leaves and the linear or linear-oblong ones on the same plant, as do some of the specimens from Hainan and Kwangtung.

CARICACEAE

Carica Linnaeus

Carica papaya Linn. Sp. Pl. 1036. 1753; Lour. Fl. Cochinch. 628. 1790, ed. Willd. 772. 1793, Anamese *cây du du*, Chinese *màn xèu cŏ.*

" Habitat culta, & a me observata in Cochinchina, in China, & in Africa." The Linnaean species was correctly interpreted by Loureiro.

ANCISTROCLADACEAE

Ancistrocladus [38] Wallich

Ancistrocladus tectorius (Lour.) Merr. in Lingnan Sci. Journ. **6**: 329. 1930 (based on *Bembix tectoria* Lour.).

Bembix tectoria Lour. Fl. Cochinch. 282. 1790, ed. Willd. 347. 1793, Anamese *lá trung cuŏn.*

Ancistrocladus extensus Wall. List no. 1052. 1829, *nomen nudum;* Planch. in Ann. Sci. Nat. III Bot. **13**: 318. 1849; Gagnep. in Lecomte Fl. Gén. Indo-Chine **1**: 395. 1910.

Ancistrocladus hainanensis Hayata Ic. Pl. Formos. **3**: 46. 1913.

" Habitat in sylvis Cochinchinae." The status of the genus *Bembix* was entirely unknown until 1927, when Moore located Loureiro's type in the herbarium of the British

[38] *Ancistrocladus* Wallich (1829), conserved name, Vienna Code; older ones are *Bembix* Loureiro (1790) and *Wormia* Vahl (1810).

Museum and published a critical note on it (Journ. Bot. **65**: 279–281. 1927). I refer to it *Clemens 3350* from near the classical locality and *Robinson 1300* from Nhatrang. I see no reason for considering that other than Wallich's species is represented.

CACTACEAE

Opuntia (Tournefort) Miller

Opuntia dillenii (Ker) Haw. Suppl. Pl. Succ. 79. 1819; Britt. & Rose Cact. **1**: 162. *pl. 28. f. 2. f. 201*. 1919.

> *Cactus dillenii* Ker in Bot. Reg. **3**: *pl. 255*. 1818.
> *Cactus ficus indica* (non Linn.) Lour. Fl. Cochinch. 306. 1790, ed. Willd. 373. 1793, Anamese *cây luoi roùng*.

"Habitat in Cochinchina. Vidi etiam in Bengala, & in aliis Indiae locis." Loureiro's description seems to apply to Haworth's species which is widely distributed in India, and is also naturalized in southern China. It cannot be the common *Nopalea cochenillifera* (Linn.) Salm-Dyck because of the floral characters indicated by Loureiro. Mr. I. H. Burkill (Records Bot. Surv. India **4**: 290. 1911) expressed the opinion that Loureiro's species might be *Opuntia monacantha* Haw. [= *O. vulgaris* Mill.], the most widely distributed of all the cacti in the Old World tropics. The characters given by Loureiro agree almost equally well with the descriptions and colored illustrations of both *O. dillenii* Haw. and *O. vulgaris* Mill. as given by Britton and Rose (Cactaceae **1**: 156, 162. *pl. 27, 28*. 1919); my sole reason for selecting *O. dillenii* Haw. is Loureiro's description of the spines as " confertis " which does not apply to *O. vulgaris* Mill. (*O. monacantha* Haw.), but does apply to *O. dillenii* Haw.

THYMELAEACEAE

Daphne (Tournefort) Linnaeus

Daphne odora Thunb. Fl. Jap. 159. 1784; Lour. Fl. Cochinch. 237. 1790, ed. Willd. 292. 1793, Chinese *nhuc môi, nún muêi*.

> *Daphne triflora* Lour. Fl. Cochinch. 236. 1790, ed. Willd. 291. 1793, Chinese *u si seng*.

For *Daphne odora* Loureiro states: " Colitur studiose Cantone Sinarum." The description applies unmistakably to Thunberg's species, which is widely distributed in eastern Asia. For *D. triflora* he states: " Habitat suburbia Cantoniensia Sinarum," but his specimens must have been from cultivated plants. I dismiss Lecomte's surmise (Not. Syst. **3**: 102. 1915) that a species of *Eriosolena* was represented: " ne peut évidemment être rapportée qu'au genre *Eriosolena*," because of the locality being so entirely out of range for that genus and further, in my judgment, because the description does not apply sufficiently well. The " calycibus 3-flores " is merely an attempt to interpret the bracts of *Daphne*. I concur with Rehder's opinion (Sargent Pl. Wils. **2**: 545. 1916) that *Daphne triflora* Lour. is that form of *D. odora* Thunb. that was briefly described by Carrière as *D. mazeli* Carr. (Rev. Hort. [Paris] 292. 1872) which is apparently a form of *D. odora* Thunb. Rehder (Sargent Pl. Wils. **2**: 546. 1916) placed *D. odora* Lour. as a synonym of *D. sinensis* Lam., expressing doubt as to whether or not Lamarck's species was distinct from *Daphne odora* Thunb.

Daphne sp.

Daphne indica (non Linn.) Lour. Fl. Cochinch. 237. 1790, ed. Willd. 292. 1793, Chinese *lu ha sin.*

" Habitat Cantone a Nankino oriunda, vulgoque audit frutex Nankinensis." The description does not apply to the Linnaean species, but does suggest *Wikstroemia nutans* Champ. except as to the number of stamens and the color of the flowers, " albus rubro conspersus "; Dunn and Tutcher report the flowers of Champion's species as yellow. The description may apply to some species of *Daphne.* The inference is that the plant was found only as a cultivated one in Canton.

Aquilaria [39] Lamarck

Aquilaria agallocha Roxb. Hort. Bengal 33. 1814, *nomen nudum*, Fl. Ind. ed. 2, 2: 422. 1832.

Aloexylum agallochum Lour. Fl. Cochinch. 267. 1790, ed. Willd. 327. 1793, Anamese *cây dêó bâù,* Chinese *chin hiàm, năn hiàm;* Moore in Journ. Bot. **63**: 281. 1925.

Cynometra agallocha Spreng. Syst. 2: 327. 1825 (based on *Aloexylum agallochum* Lour.).

" Habitat in altissimis montibus Cochinchinae prope magnum flumen Lavum, quod inter hoc regnum, & Laosios interfluit." A. Chevalier (Cat. Pl. Jard. Bot. Saigon 66. 1919) placed Loureiro's species as a synonym of *Aquilaria agallocha* Roxb., but in my preliminary manuscript of 1919 I thought the description applied to some leguminous genus, as it is clearly more nearly leguminous than thymeleaceous, according to the characters given by Loureiro. Baillon, judging by the characters given by Loureiro, surmised that it might belong in the Caesalpinoideae, while Sprengel definitely placed it in the genus *Cynometra.* Loureiro's type in the herbarium of the British Museum has been critically examined by Moore, who states that it is without flowers or fruits and that " it was identified by Dryander as *Aquilaria ovata* Cav., and it so closely resembles *Wallich 7250a* from Silhet (*A. agallocha* Roxb.) that both specimens might be supposed to come from the same tree." He notes, as did Baillon and myself, that Loureiro's description of the flowers applies to a leguminous plant. Here is a clear case of a mixture of material on which the description was based, but, as explained by Moore, essentially *Aquilaria* was the plant Loureiro intended to describe. It is to be noted that Roxburgh's use of the specific name *agallocha* under *Aquilaria* was independent of Loureiro's earlier binomial *Aloeoxylum agallochum;* both authors were considering the *calambac* or *agallochum* of the ancients.

Aquilaria sinensis (Lour.) Gilg in Bot. Jahrb. **28**: 145. 1900; Merr. in Philip. Journ. Sci. **15**: 248. 1919 (based on *Ophispermum sinense* Lour.).

Ophispermum sinense Lour. Fl. Cochinch. 281. 1790, ed. Willd. 344. 1793, Chinese *pă mŏu yong.*

Aquilaria ophispermum Poir. in Dict. Sci. Nat. **18**: 161. 1820 (based on *Ophispermum sinense* Lour.).

Aquilaria chinensis Spreng. Syst. 2: 356. 1825 (based on *Ophispermum sinense* Lour.).

Aquilaria grandiflora Benth. Fl. Hongk. 297. 1861.

Agallochum sinense O. Ktz. Rev. Gen. Pl. 583. 1891 (based on *Ophispermum sinense* Lour.).

Agallochum grandiflorum O. Ktz. l.c.

[39] *Aquilaria* Lamarck (1786), conserved name, Vienna Code; an older one is *Agallochum* Lamarck (1783).

No locality is given, but the specific name and Chinese name cited clearly indicate that the specimen came from China and probably from Canton, where the species is not uncommon. The description applies definitely to the form commonly known as *A. grandiflora* Benth. The "perianthium . . . 6-partitum" is an error, as the description of the stamens as 10 indicates clearly that a 5-merous flower was intended. Loureiro's type is preserved in the herbarium of the Paris Museum.

Wickstroemia [40] Endlicher

Wickstroemia indica (Linn.) C. A. Mey. in Bull. Acad. Sci. St. Pétersb. II **1**: 357. 1843; Ann. Sci. Nat. II Bot. **20**: 50. 1843.

Daphne indica Linn. Sp. Pl. 357. 1753.

Capura purpurata Linn. Mant. **2**: 225. 1771.

Wickstroemia purpurata Druce in Rept. Bot. Exch. Club Brit. Isles **4**: 652. 1917.

Daphne cannabina Lour. Fl. Cochinch. 236. 1790, ed. Willd. 291. 1793, Anamese *cây deó niet.*

Wickstroemia viridiflora Meisn. in Denkschr. Bot. Ges. Regensb. **3**: 286. 1841; Lecomte Fl. Gén. Indo-Chine **5**: 166. 1915.

"Habitat sylvas Cochinchinae." Meisner (DC. Prodr. **14**: 546. 1857) placed *Daphne cannabina* Lour. as a synonym of *Wickstroemia viridiflora* Meisn. = *W. indica* (Linn.) C. A. Mey. on the basis of Decaisne's definite reduction of Loureiro's species (Jacquem. Voy. **4**: 145. 1844) which, in turn, was based on an examination of Loureiro's type in the herbarium of the Paris Museum. The actual type of *Daphne indica* Linn. was collected near Canton. The type specimen of *Capura purpurata* Linn. in the Linnaean Herbarium is apparently a form of *Wickstroemia indica* (Linn.) C. A. Mey., the flowers erroneously described as purple.

ELAEAGNACEAE

Elaeagnus (Tournefort) Linnaeus

Elaeagnus glabra Thunb. Fl. Jap. 67. 1784.

Elaeagnus latifolia (non Linn.) Lour. Fl. Cochinch. 89. 1790, ed. Willd. 113. 1793, Chinese *pǎ poi tsù.*

"Habitat in China spontanea, prope Cantonem." I believe this to be represented by a series of specimens from the vicinity of Canton, including *Levine 243, 254,* and *1073.* Loureiro's species has previously been referred to *Elaeagnus loureiri* Champ. (Hook. Journ. Bot. Kew Gard. Miscel. **5**: 196. 1853), a species essentially based on Hongkong specimens and one characterized by having distinctly large flowers; *E. latifolia* Lour. is cited as a doubtful synonym. This large flowered form has not been found in the vicinity of Canton. Servattez in his monograph of the *Elaeagnaceae* (Beih. Bot. Centralbl. **25**(2): 69. 1909) thought that Loureiro's species might be near *E. ovata* Serv. or *E. oldhami* Maxim.

[40] *Wickstroemia* Endlicher (1833), conserved name (as *Wikstroemia*), Vienna Code; an older one is *Capura* Linnaeus (1771). Blake (Contr. Gray Herb. **53**: 36–41. 1918) replaced the theaceous genus *Laplacea* HBK. by the earlier *Wickstroemia* Schrader (1821), suggesting that *Capura* Linnaeus be accepted in place of *Wickstroemia* Endlicher for this thymeleaeceous genus, a proceeding that is inadmissible under the International Code, as shown by Rehder and by Sprague (Journ. Arnold Arb. **2**: 158. 1921; Sprague in Kew Bull. 175. 1921); see also Blake in Journ. Bot. **60**: 52. 1922, who, however, shows that Endlicher's original spelling was *Wickstroemia;* this spelling is accepted here.

Elaeagnus fruticosa (Lour.) A. Cheval. Cat. Pl. Jard. Bot. Saigon 66. 1919 (based on *Octarillum fruticosum* Lour.).

Octarillum fruticosum Lour. Fl. Cochinch. 90. 1790, ed. Willd. 113. 1793, Anamese *cây nhút;* Moore in Journ. Bot. **63**: 249. 1925.

Elaeagnus gaudichaudiana Schlecht. in DC. Prodr. **14**: 612. 1857.

"Habitat in sylvis Cochinchinae." Loureiro's type is preserved in the British Museum and Moore states that this specimen, being in leaf only, cannot be determined except that it is an *Elaeagnus.* Loureiro's description applies to *E. gaudichaudiana* Schlecht., the type of which was from Tourane, a short distance south of Hue, and which is the only species of *Elaeagnus* reported from Indo-China by Servattez. Servattez in his monograph of the Elaeagnaceae (Beih. Bot. Centralbl. **25**(2): 4. 1909) repeats Loureiro's description and excludes *Octarillum* from the *Elaeagnaceae* because Loureiro described the leaves as glabrous and the seeds as arillate. In describing the leaves Loureiro merely meant that they were not pubescent, which is true; he failed to mention the closely appressed scales so characteristic of *Elaeagnus.* Loureiro's description of the seeds reads: "Sem. 1, oblongum, arillatum membrana tenaci, oblonga," and he derives the generic name thus: "Octarillum ab arillo seminis octogono." The "aril" of Loureiro is merely the characteristic thin pericarp which surrounds the seed inside the thickened perianth-tube.

LYTHRACEAE

Lagerstroemia Linnaeus

Lagerstroemia indica Linn. Syst. ed. 10, 1076. 1759; Lour. Fl. Cochinch. 340. 1790, ed. Willd. 415. 1793, Anamese *cây tuòng vi,* Chinese *sát chú môi hõa.*

"Habitat tam culta, quam spontanea in Cochinchina, & China." Loureiro correctly interpreted the Linnaean species, as the specimen from him in the Paris Museum represents the common crêpe myrtle. *Tsjinkin* Rumph. (Herb. Amb. **7**: 61. *pl. 28. f. 1*), cited by Loureiro as a synonym, is correctly placed.

Lawsonia Linnaeus

Lawsonia inermis Linn. Sp. Pl. 349. 1753.

Lawsonia spinosa Linn. l.c.; Lour. Fl. Cochinch. 229. 1790, ed. Willd. 281. 1793, Anamese *cay maóng tay nhuom.*

"Habitat in hortis Cochinchinae." The Linnaean species was correctly interpreted by Loureiro; the well-known henna plant.

SONNERATIACEAE

Sonneratia [41] Linnaeus f.

Sonneratia caseolaris (Linn.) Engler in Engler & Prantl Nat. Pflanzenfam. Nachtr. 1: 261. 1897; Merr. Interpret. Herb. Amb. 383. 1917, Enum. Philip. Fl. Pl. **3**: 139. 1923.

[41] *Sonneratia* Linnaeus f. (1781), conserved name, Vienna Code; older ones are *Blatti* Adanson (1763) and *Pagapate* Sonnerat (1776).

Rhizophora caseolaris Linn. in Stickm. Herb. Amb. 13. 1754, Amoen. Acad. **4**: 123. 1759, Syst. ed. 10. 1043. 1759, Sp. Pl. ed. 2, 635. 1762; Lour. Fl. Cochinch. 296. 1790, ed. Willd. 363. 1793, Anamese *cày băn tlòn tlái.*

Sonneratia acida Linn. f. Suppl. 252. 1781; Gagnep. & Guillaumin in Lecomte Fl. Gén. Indo-Chine **2**: 979. 1921.

"Habitat prope ripas fluminum in Cochinchina." *Mangium caseolare* Rumph. (Herb. Amb. **3**: 111. *pl. 74*), cited by Loureiro as representing the species, is correctly placed. His description applies to the Linnaean species which is common and widely distributed along the seashore in the Indo-Malaysian region. He notes that he observed a smaller form of what he took to be the same species in East Africa, and this is doubtless the form listed and described from Mozambique by Hiern (Oliver Fl. Trop. Afr. **2**: 483. 1871) as *Sonneratia acida* Linn. f. (*S. mossambicensis* Klotzsch).

PUNICACEAE

Punica (Tournefort) Linnaeus

Punica granatum Linn. Sp. Pl. 472. 1753; Lour. Fl. Cochinch. 313. 1790, ed. Willd. 383. 1793, Anamese *cây thach luu*, Chinese *hăn xĕ liêu.*

Punica nana Linn. Sp. Pl. ed. 2, 676. 1762; Lour. Fl. Cochinch. 314. 1790, ed. Willd. 384. 1793, Anamese *luu chua thâp.*

For *Punica granatum* Loureiro states: "Habitat, & colitur in China, & Cochinchina." He correctly interpreted the Linnaean species, the common pomegranate. For *P. nana* he states: "Culta in hortis Cochinchinae." This is the dwarfed plant with small fruits not uncommonly cultivated for ornamental purposes, which apparently is a cultigen derived from *Punica granatum* Linn.

LECYTHIDACEAE

Barringtonia [42] J. R. & G. Forster

Barringtonia acutangula (Linn.) Gaertn. Fruct. **2**: 97. *pl. 101.* 1791; Gagnep. in Lecomte Fl. Gén. Indo-Chine **2**: 860. 1921.

Eugenia acutangula Linn. Sp. Pl. 471. 1753.

Meteorus coccineus Lour. Fl. Cochinch. 410. 1790, ed. Willd. 499. 1793, Anamese *rau bung.*

Stravadium coccineum DC. Prodr. **3**: 289. 1828 (based on *Meteorus coccineus* Lour.); Miers in Trans. Linn. Soc. Bot. **1**: 83. 1875.

Barringtonia coccinea Kostel. Allgem. Med.-Pharm. Fl. **4**: 1536. 1835 (based on *Meteorus coccineus* Lour.).

Careya coccinea A. Chev. Cat. Pl. Jard. Bot. Saigon 64. 1919 (based on *Meteorus coccineus* Lour.).

"Habitat spontaneus in sylvis planis Cochinchinae." Loureiro's type is preserved in the herbarium of the British Museum and Miers has given an ample description of the species based on this specimen. *Butonica terrestris* Rumph. (Herb. Amb. **3**: 181. *pl. 115*), cited by Loureiro as doubtfully representing his species, must be excluded as it represents

---

[42] *Barringtonia* Forster (1776), conserved name, Vienna Code; an older one is *Huttum* Adanson (1763).

Barringtonia racemosa (Linn.) Blume. Doctor Chevalier erred in transferring Loureiro's specific name to *Careya*, on the basis of the local name given by Loureiro; without reference to the extant type, Loureiro's description is most definitely that of a *Barringtonia*, not of a *Careya*. I follow Gagnepain in the broader interpretation of *Barringtonia acutangula* (Linn.) Gaertn.; for those who segregate on finer specific limits, Loureiro's name is available for the Indo-China form.

Barringtonia cochinchinensis (Miers) Merr. ex Gagnep. in Lecomte Fl. Gén. Indo-Chine
 2: 862. 1921 (based on *Doxomma cochinchinensis* Miers = *Eugenia acutangula* Lour.).
 Eugenia acutangula (non Linn.) Lour. Fl. Cochinch. 307. 1790, ed. Willd. 375. 1793,
  Anamese *cây tam lang.*
 Doxomma cochinchinensis Miers in Trans. Linn. Soc. Bot. 1: 101. 1875 (based on *Eugenia*
  acutangula Lour.).
 Barringtonia annamica Gagnep. in Not. Syst. 3: 383. 1918; Lecomte Fl. Gén. Indo-
  Chine 2: 858. *f. 94.* 1921.

" Habitat in sylvis Cochinchinae." Loureiro's type is preserved in the herbarium of the British Museum and Miers has given a detailed description based on this specimen. Gagnepain's description is based on those of Loureiro and Miers and he considers the species to be one of doubtful status near *B. longipes* Gagnep. I interpret *Clemens 4097, 4410* from Tourane and vicinity, " a common forest tree," to represent Loureiro's species. It may be noted that these specimens conform to Loureiro's description in having sessile fruits corresponding to Loureiro's description of the flowers as sessile, in which character the species differs markedly from *B. longipes* Gagnep. (Not. Syst. 3: 384. 1918). Gagnepain expressed the opinion that Loureiro's species was allied to *Barringtonia longipes* Gagnep., but I believe that *B. annamica* Gagnep. is the same as *B. acutangula* Lour., although in his original description of this, Gagnepain stated that it did not appear to be the same as Loureiro's species.

RHIZOPHORACEAE

Carallia [43] Roxburgh

Carallia brachiata (Lour.) Merr. in Philip. Journ. Sci. 15: 249. 1919 (based on *Diatoma*
 brachiata Lour.).
 Diatoma brachiata Lour. Fl. Cochinch. 296. 1790, ed. Willd. 362. 1793, Anamese *cây ma.*
 Carallia lucida Roxb. Hort. Beng. [92]. 1814, *nomen nudum*, Pl. Coromandel 3: 8.
  pl. 211. 1819; Guillaumin in Lecomte Fl. Gén. Indo-Chine 2: 732. 1920.
 Carallia integerrima DC. Prodr. 3: 33. 1828.
 Petalotoma brachiata DC. Prodr. 3: 295. 1828 (based on *Diatoma brachiata* Lour.).
 Karekandelia brachiata O. Ktz. Rev. Gen. Pl. 235. 1891 (based on *Diatoma brachiata*
  Lour.).

" Habitat in sylvis Cochinchinae." The species is common and of wide distribution in the Indo-Malaysian region. A specimen from Loureiro listed as being among those received from him has not been located in the herbarium of the British Museum.

[43] *Carallia* Roxburgh (1814), conserved name, Vienna Code; older ones are *Karekandel* Adanson (1763), *Diatoma* Loureiro (1790) and *Barraldeia* Thouin (1806).

Bruguiera Lamarck

Bruguiera conjugata (Linn.) Merr. in Philip. Journ. Sci. Bot. **9**: Bot. 118. 1914, Interpret. Herb. Amb. 388. 1917, Enum. Philip. Fl. Pl. **3**: 146. 1923.

Rhizophora conjugata Linn. Sp. Pl. 443. 1753.

Rhizophora gymnorhiza Linn. Sp. Pl. 443. 1753; Lour. Fl. Cochinch. 297. 1790, ed. Willd. 364. 1793, Anamese *cây deà*.

Bruguiera gymnorhiza Lam. Ill. **2**: *pl. 397*. 1797; Guillaumin in Lecomte Fl. Gén. Indo-Chine **2**: 728. 1920.

"Habitat ingentes tractus litorum ad Austrum Cochinchinae, & Cambodiae. Etiam frequenter occurit in freto Malacenci." Loureiro's description applies unmistakably to the species currently known as *Bruguiera gymnorhiza* Lam. (*Rhizophora gymnorhiza* Linn.) = *Rhizophora conjugata* Linn. = *Bruguiera conjugata* Merr. Most modern authors have misinterpreted *Rhizophora conjugata* Linn., treating it as a true *Rhizophora*.

Bruguiera sexangula (Lour.) Poir. in Lam. Encycl. Suppl. **4**: 262. 1816; Merr. Interpret. Herb. Amb. 389. 1917, Enum. Philip. Fl. Pl. **3**: 147. 1923 (based on *Rhizophora sexangula* Lour.).

Rhizophora sexangula Lour. Fl. Cochinch. 297. 1790, ed. Willd. 363. 1793, Anamese *cây băn deai tlái*.

Bruguiera eriopetala Wight & Arn. in Ann. Nat. Hist. **1**: 368. 1838; Guilaumin in Lecomte Fl. Gén. Indo-Chine **2**: 729. 1920.

"Habitat prope ripas fluminum in Cochinchina." Loureiro's description applies unmistakably to the common and widely distributed Indo-Malaysian species currently known as *Bruguiera eriopetala* Wight & Arn., the description of the "drupa" being based on a recently germinated seedling with the radicle but slightly produced. Loureiro's type is preserved in the herbarium of the British Museum.

COMBRETACEAE

Quisqualis Linnaeus

Quisqualis indica Linn. Sp. Pl. ed. 2, 556. 1762; Lour. Fl. Cochinch. 274. 1790, ed. Willd. 336. 1793, Anamese *cây tlun, su cuŏn tu*, Chinese *xi kiūn tsù*.

Quisqualis loureiri G. Don Gen. Syst. **2**: 667. 1832 (based on *Quisqualis indica* Lour.).

Quisqualis sinensis Lindl. Bot. Reg. **30**: *pl. 15*. 1844.

Quisqualis grandiflora Miq. in Journ. Bot. Néerl. **1**: 119. 1861.

Mekistus sinensis Lour.[44] ex Gomes in Mem. Acad. Sci. Lisb. Cl. Sci. Pol. Mor. Bel.-Let. n.s. **4**(1): 29. 1868.

"Habitat prope sepes, & ripas fluminum in Cochinchina, & Cantone Sinarum." Specimens from Loureiro are preserved in the herbaria of the British and the Paris Museums. They represent the Linnaean species as that is currently interpreted. The references to Rumphius and Burman, cited as synonyms by Loureiro, are correctly placed.

[44] A Loureiro herbarium name here first published by Gomes.

Terminalia Linnaeus [45]

Terminalia catappa Linn. Mant. **1**: 128. 1767, **2**: 519. 1771; Gagnep. in Lecomte Fl. Gén. Indo-Chine **2**: 750. 1920.

Juglans catappa Lour. Fl. Cochinch. 573. 1790, ed. Willd. 703. 1793, Anamese *cây mo cua.*

"Habitat in sylvis Cochinchinae montanis." Loureiro described *Juglans catappa* as a new species independently of the earlier *Terminalia catappa* Linn., which is the correct name for it. He derived his specific name from *Catappa domestica* Rumph. (Herb. Amb. **1**: 174. *pl. 68*) which is also the source of the Linnaean name.

MYRTACEAE

Rhodamnia Jack

Rhodamnia trinervia (Sm.) Blume Mus. Bot. Lugd.-Bat. **1**: 79. 1849; Gagnep. in Lecomte Fl. Gén. Indo-Chine **2**: 844. 1921.

Myrtus trinervia Sm. in Trans. Linn. Soc. **3**: 280. 1797.

Myrtus trinervia Lour. Fl. Cochinch. 312. 1790 (err. *triinervia*), ed. Willd. 381. 1793, Anamese *sim rúng lón.*

Myrtus dumetorum Poir. in Lam. Encycl. Suppl. **4**: 52. 1816 (based on *Myrtus trinervia* Lour.).

Nelitris trinervia Spreng. Syst. **2**: 488. 1825 (based on *Myrtus trinervia* Lour.).

Eugenia ? dumetorum DC. Prodr. **3**: 284. 1828 (based on *Myrtus trinervia* Lour.).

"Habitat dumeta Cochinchinae." Loureiro's description applies to the glabrous or nearly glabrous-leaved form of *Rhodamnia trinervia* (Sm.) Blume, which is abundant in Indo-China; it is represented by *Clemens 3329, 3689*, from near the classical locality. It is to be noted that Smith's binomial was proposed as new independently of Loureiro's earlier one. The binomial *Rhodamnia trinervia* Blume was based on *Myrtus trinervia* Smith, not on *M. trinervia* Lour., as one might infer from Gagnepain's synonymy. Gagnepain does not account for two of the synonyms, cited above, based on *Myrtus trinervia* Lour.

Psidium Linnaeus

Psidium guajava Linn. Sp. Pl. 470. 1753.

Psidium pomiferum Linn. Sp. Pl. ed. 2, 672. 1762; Lour. Fl. Cochinch. 310. 1790, ed. Willd. 379. 1793, Anamese *ôi rùng nho.*

Psidium pyriferum Linn. Sp. Pl. ed. 2, 672. 1762; Lour. Fl. Cochinch. 309. 1790, ed. Willd. 378. 1793, Anamese *cây ôi.*

For *P. pomiferum* Loureiro states: "Habitat in sylvis Cochinchinae, & Chinae," and for *P. pyriferum:* "Habitat in Cochinchina, & in China Australi." He correctly interpreted both of the Linnaean species which, however, are but forms of a single one, the older *Psidium guajava* Linn. This is the common guava.

Rhodomyrtus Reichenbach

Rhodomyrtus tomentosa (Ait.) Hassk. in Flora Beibl. **25**(2): 35. 1842; Wight Spicil. Neilgh. **1**: 60. *pl. 71.* 1846–51; Gagnep. in Lecomte Fl. Gén. Indo-Chine **2**: 794. f. *85. 1–8.* 1920.

---

[45] *Terminalia* Linnaeus (1767), conserved name, Brussels Code; older ones are *Adamaram* Adanson (1763) and *Panel* Adanson (1763).

Myrtus tomentosa Ait. Hort. Kew. **2**: 159. 1789.

Myrtus canescens Lour. Fl. Cochinch. 311. 1790, ed. Willd. 381. 1793, Anamese *cây sim nhà.*

"Habitat loca inculta ubique in Cochinchina." The description unmistakably applies to the very common and widely distributed *Rhodomyrtus tomentosa* Hassk. It may be noted that *Rhodomyrtus tomentosa* Hassk. is actually based on *Myrtus tomentosa* Blume Bijdr. 1051. 1826, which, however, Blume himself properly credited to Aiton.

Eugenia (Micheli) Linnaeus

Eugenia bullockii Hance in Journ. Bot. **16**: 227. 1878; Gagnep. in Lecomte Fl. Gén. Indo-Chine **2**: 817. 1920.

Myrtus androsaemoides (non Linn.) Lour. Fl. Cochinch. 312. 1790, ed. Willd. 382. 1793, Anamese *cây maóc hôt.*

"Habitat in dumetis Cochinchinae." Loureiro's species is safely represented by *Clemens 3716,* a shrub frequent in thickets at Tourane, which in turn is an excellent match for the type of *Eugenia bullockii* Hance in the herbarium of the British Museum. Further confirmation is found in the citation of the Anamese name *cay moc* by Gagnepain for Hance's species; he made no attempt to account for Loureiro's species. It is manifest that the statement "Bacca . . . polysperma" was added by Loureiro to make his description conform to the generic characters of *Myrtus.*

Eugenia corticosa Lour. Fl. Cochinch. 308. 1790, ed. Willd. 376. 1793, Anamese *cây tlâm bôi;* Moore in Journ. Bot. **63**: 283. 1925.

Myrtus corticosa Spreng. Syst. **2**: 488. 1825 (based on *Eugenia corticosa* Lour.).

"Habitat in sylvis Cochinchinae." The description apparently is of a species of the section *Syzygium.* Moore states that because of the unsatisfactory nature of Loureiro's specimen in the herbarium of the British Museum, it is impossible to tell to what species of *Eugenia* it should be referred. Gagnepain admits 55 species of *Eugenia* in his treatment of the Myrtaceae of Indo-China (Lecomte Fl. Gén. Indo-Chine **2**: 796–844. 1920–1921) and doubtless Loureiro's *E. corticosa* is the same as one of these, but this I am unable to determine by comparison of descriptions alone. Gagnepain does not attempt to determine the status of Loureiro's species.

Eugenia cumini (Linn.) Druce in Rept. Bot. Exch. Club Brit. Isles **3**: 418. 1914; Merr. Interpret. Herb. Amb. 394. 1917, Enum. Philip. Fl. Pl. **3**: 164. 1923.

Myrtus cumini Linn. Sp. Pl. 471. 1753.

Eugenia jambolana Lam. Encycl. **3**: 198. 1789.

Jambolifera pedunculata (non Linn.) Lour. Fl. Cochinch. 230. 1790, ed. Willd. 283. 1793.

Syzygium jambolanum DC. Prodr. **3**: 259. 1828.

Jambolifera chinensis Spreng. Syst. **2**: 216. 1825 (based on *J. pedunculata* Lour.).

"Habitat Macai in China, colitur in multisque Indiae locis." The description applies unmistakably to the very common and widely distributed species usually known as *Eugenia jambolana* Lam. Loureiro cites the local name *jamboloens* for it, as used by the Portuguese at Macao. For a discussion of *Jambolifera pedunculata* Linn. see p. 220.

Eugenia jambos Linn. Sp. Pl. 470. 1753; Lour. Fl. Cochinch. 307. 1790, ed. Willd. 375.
1793, Anamese *cây daò annam.*
 Eugenia malaccensis (non Linn.) Lour. Fl. Cochinch. 306. 1790, ed. Willd. 374. 1793,
  Anamese *daò huong taù.*
 For *E. jambos* Loureiro states: " Habitat in Cochinchina, & in multis Indiae locis."
The description applies to the Linnaean species which is commonly cultivated in the
Indo-Malaysian region. For *E. malaccensis* he states: " Habitat culta Malaccae, Macai
Sinarum, & in multis Indiae locis." His description, and especially that of the fruit,
" rosae odorem spirans, " applies to the rose-apple, *Eugenia jambos* Linn. not to *E. malac-*
censis Linn. *E. jambos* Linn. is commonly cultivated in Kwangtung Province and at
Macao, but *E. malaccensis* Linn. is not known from southeastern China.

Eugenia nervosa Lour. Fl. Cochinch. 308. 1790, ed. Willd. 376. 1793, Anamese *cây sòi.*
 Myrtus loureiri Spreng. Syst. **2**: 488. 1825 (based on *Eugenia nervosa* Lour.).
 Cleistocalyx nervosus Blume Mus. Bot. Lugd.-Bat. **1**: 85. 1849 (based on *Eugenia*
  nervosa Lour.).
 " Habitat in sylvis Cochinchinae." A species known only from Loureiro's description,
but clearly a representative of the section *Jambosa.* It is rather curious that Blume should
place this in his genus *Cleistocalyx*, which is characterized by having calyptrate calyces,
in view of Loureiro's statement that the large flowers of his species had 4-partite (lobed)
calyces, the lobes obtuse and concave. It is suspected that *E. nervosa* Lour. is redescribed
among the 55 species of *Eugenia* considered by Gagnepain, but I cannot place it satisfactorily by comparison of descriptions alone.

Eugenia sp.
 Caryophyllus aromaticus (non Linn.) Lour. Fl. Cochinch. 333. 1790, ed. Willd. 406.
  1793, Anamese *dinh huong rùng*, Chinese *xān tīm hiàm.*
 " Habitat in sylvis borealibus provinciae *Quang bình*, regni Cochinchinae." The
description is manifestly of some species of *Eugenia*, but is not that of the clove, *Eugenia*
caryophyllata Thunb., to which *Caryophyllum silvestre* Rumph. (Herb. Amb. **2**: 12. *pl. 3*)
appertains, Loureiro erroneously citing Rumphius' plate as illustrating his species. Gagnepain (Lecomte Fl. Gén. Indo-Chine **2**: 802. 1920) merely lists Loureiro's binomial as apparently representing some species of *Eugenia.*

Eugenia millettiana Hemsl. in Journ. Linn. Soc. Bot. **23**: 297. 1887.
 Opa odorata Lour. Fl. Cochinch. 309. 1790, ed. Willd. 377. 1793, Anamese *cây nô;*
  Moore in Journ. Bot. **63**: 283. 1925, non *Eugenia odorata* Berg.
 Syzygium odoratum DC. Prodr. **3**: 260. 1828; Hook. & Arn. Bot. Beechey's Voy. 187.
  1833 (based on *Opa odorata* Lour.).
 " Habitat in dumetis Cochinchinae." Loureiro's type, an excellent specimen of
which I have a photograph, is preserved in the herbarium of the British Museum and
Moore (Journ. Bot. **63**: 283. 1925) gives additional descriptive data based upon it, considering that it is not the same as *Eugenia millettiana* Hemsl., to which Hemsley reduced
it; he could not match it at the British Museum or at Kew and concluded that an endemic
species was represented by it. It is well matched by *Clemens 3778* from dune thickets
at Tourane and is the species described by Gagnepain (Lecomte Fl. Gén. Indo-Chine **2**:
804. 1920) as *Eugenia zeylanica* Wight (non Willd.) = *E. spicata* Lam. *Eugenia millettiana*

Hemsl. in Journ. Linn. Soc. Bot. **23**: 297. 1887 is a new name most casually published by Hemsley based on *Syzygium odoratum* Hook. & Arn., *S. odoratum* DC.? and *Opa odorata* Lour.; he gives no description, but cites five different collections *all with furfuraceous inflorescences*, none of which represents the species which Loureiro, Hooker and Arnott, and de Candolle describe. Essentially *Eugenia millettiana* Hemsl. must be interpreted from the actual descriptions on which the name was based, not on the specimens cited. Hooker and Arnott state that their plant represented a species very near to *Syzygium zeylanicum*, and that their description of the flowers and fruits was taken from specimens sent by Millett and Vachell. The two Millett sheets in the Kew herbarium represent the species near *Eugenia zeylanica* Wight interpreted here as true *Eugenia millettiana* Hemsl., the form with glabrous inflorescences and white fruits; the Vachell sheet has one specimen representing this same form and two specimens representing the one with furfuraceous inflorescences (*Eugenia millettiana* Dunn & Tutcher in Kew Bull. Add. Ser. **10**: 105. 1912, non Hemsl.). The latter, which is remote from *Eugenia zeylanica* Wight., *E. spicata* Lam., *Opa odorata* Lour., and *E. millettiana* Hemsl., is *Eugenia levinei* Merr. A specimen from the Beechey Voyage collection is in the Torrey Herbarium at the New York Botanical Garden, and is *E. millettiana* Hemsl. as here interpreted. Seemann (Journ. Bot. **1**: 280. 1863), who examined Loureiro's type, adds *S. lucidum* Gaertn. as a synonym, but Britten (Journ. Bot. **58**: 151. 1920) calls attention to the fact that Banks' specimen of *Eugenia lucida* Banks = *Syzygium lucidum* Gaertn. was from the Endeavour River, New South Wales, Australia, and that it is not the same species as *Opa odorata* Lour. Loureiro's specific name is invalidated in *Eugenia* by *E. odorata* Berg. This is the type of the genus *Opa* which, therefore, becomes a synonym of *Eugenia*. The other species described, *O. metrosideros* Lour., does not conform to the generic description and is the rosaceous *Raphiolepis indica* Lindl. *Opanea* Raf. (Sylva Tellur. 106. 1838) was apparently intended as a new generic name for *Opa* Lour. " Types *M*[*yrtus*] *trinervia* Sm. and *billardiana* K. . . . also the 2d sp. of *Opa* of Loureiro," but Rafinesque does not transfer Loureiro's specific name. I interpret *Myrtus trinervia* Sm. as the type of *Opanea* Raf., this generic name thus becoming a synonym of *Rhodamnia*.

Eugenia sp.

Psidium rubrum Lour. Fl. Cochinch. 311. 1790, ed. Willd. 380. 1793, Anamese *cây trâm*. " Habitat in sylvis Cochinchinae." The description, except that of the fruits, applies to *Eugenia*, section *Jambosa*, and the Anamese name in a measure confirms this disposition of Loureiro's species, as *trâm* appears as the name or as a part of the name of about nine species of *Eugenia* in Gagnepain's treatment of the genus (Lecomte Fl. Gén. Indo-Chine **2**: 796–844. 1920–21). Gagnepain lists *Psidium rubrum* Lour. under *Eugenia*, p. 802, as apparently a representative of this genus, but very obscure. The fruits described by Loureiro as " polysperma " cannot be *Eugenia* if Loureiro's description is correct, but this character was doubtless added by Loureiro to make his description conform with the generic characters of *Psidium*.

Melaleuca [46] Linnaeus

Melaleuca leucadendron Linn. Mant. **1**: 105. 1767 (*leucadendra*); Lour. Fl. Cochinch. 468. 1790, ed. Willd. 573. 1793 (*leucadendra*), Anamese *cây tlàm;* Gagnep. in Lecomte Fl. Gén. Indo-Chine **2**: 790. 1920 (var. *minor* Duthie).

[46] *Melaleuca* Linnaeus (1767), conserved name, Vienna Code; an older one is *Cajuputi* Adanson (1763).

"Habitat frequens in sylvis Cochinchinae." Loureiro correctly interpreted the Linnaean species, which is apparently common in Indo-China.

Baeckea Linnaeus

Baeckea frutescens Linn. Sp. Pl. 358. 1753.

> *Cedrela rosmarinus* Lour. Fl. Cochinch. 160. 1790, ed. Willd. 199. 1793, Anamese *rành rành chôi, dia phu tu*, Chinese *tí phủ pi.*
>
> *Itea rosmarinus* Schult. in Roem. & Schult. Syst. 5: 408. 1819 (based on *Cedrela rosmarinus* Lour.).
>
> *Drosodendron rosmarinus* M. Roem. Syn. 1: 138, 140. 1846 (based on *Cedrela rosmarinus* Lour.).

"Habitat frequenter in locis arenosis, ad boream sitis in Cochinchina. Vidi etiam sponte nascentem in insulis circa Macaum in China." The description applies unmistakably to *Baeckea frutescens* Linn., the type of which was from Kwangtung Province. The species is common in Kwangtung and occurs also in Indo-China (*Clemens 3854, Squires 149*, from Loureiro's classical locality), extending to the Malay Peninsula and Borneo.

MELASTOMATACEAE

Melastoma (Burman) Linnaeus

Melastoma dodecandrum Lour. Fl. Cochinch. 274. 1790, ed. Willd. 336. 1793, Anamese *cây mua thâp*, Chinese *pẽ giẽ hòng.*

> *Melastoma repens* Desr. in Lam. Encycl. 4: 54. 1796; Cogn. in DC. Monog. Phan. 7: 344. 1891; Guillaumin in Lecomte Fl. Gén. Indo-Chine 2: 885. 1921.

"Habitat inculta in Cochinchina, & Cantone Sinarum." Loureiro's description applies unmistakably to the species currently known as *Melastoma repens* Desr. which is common in the vicinity of Canton and which is widely distributed in Indo-China. Loureiro's specific name is the oldest valid one for this common, well-known and very characteristic species.

Melastoma septemnervium Lour. Fl. Cochinch. 273. 1790, ed. Willd. 335. 1793, Anamese *cây mua.*

> *Melastoma candidum* D. Don in Mem. Wern. Soc. 4: 288. 1823; Cogn. in DC. Monog. Phan. 7: 347. 1891; Guillaumin in Lecomte Fl. Gén. Indo-Chine 2: 880. 1921.

"Habitat in Cochinchina, ubique obvia per colles, & agros minus cultos." Loureiro's description is good and, in my judgment, applies unmistakably to the well-known *Melastoma candidum* D. Don, which is common in Indo-China. Guillaumin (Not. Syst. 2: 315. 1913) thought that *Melastoma decemfidum* Roxb. may have been the species intended by Loureiro, and in Lecomte (Fl. Gén. Indo-Chine 2: 883. 1921) cites *M. septemnervium* Lour. as a doubtful synonym of Roxburgh's species. However, Loureiro's description of the leaves as 7-nerved and as pilose on both surfaces points unmistakably to *M. candidum* D. Don and not to *M. decemfidum* Roxb. I suspect that Guillaumin was influenced by Loureiro's description of the stems and calyces as hispid, which does apply rather better to *M. decemfidum* Roxb. than to *M. candidum* D. Don, but which can scarcely be construed to eliminate the latter because it has conspicuous strigose paleae on the branches and calyces.

Osbeckia Linnaeus

Osbeckia chinensis Linn. Sp. Pl. 345. 1753; Lour. Fl. Cochinch. 228. 1790, ed. Willd. 281. 1793, Chinese *kām yòng lù*.

"Habitat inculta Cantone Sinarum." Loureiro correctly interpreted the Linnaean species, the type of which was from Canton. He cites Osbeck's illustration of it, which, in turn, was based on the type collection. The species is a common, well-known, and widely distributed one.

Blastus Loureiro

Blastus cochinchinensis Lour. Fl. Cochinch. 527. 1790, ed. Willd. 643. 1793, Anamese *cây mua rừng;* Guillaumin in Lecomte Fl. Gén. Indo-Chine 2: 896. 1921.

"Habitat in sylvis Cochinchinae." A well-known species, the type of the genus, extending from India to southern China, the Riu Kiu Islands and Formosa. Loureiro's type is preserved in the herbarium of the British Museum (Seemann Journ. Bot. 1: 281. 1863; Cogniaux in DC. Monog. Phan. 7: 476. 1891).

Memecylon Linnaeus

Memecylon scutellatum (Lour.) Naud. in Ann. Sci. Nat. III Bot. 18: 282. 1852; Cogn. in DC. Monog. Phan. 7: 1157. 1891 (based on *Scutula scutellata* Lour.).

Scutula scutellata Lour. Fl. Cochinch. 235. 1790, ed. Willd. 290. 1793, Anamese *cây ran;* Moore in Journ. Bot. 63: 255. 1925.

Memecylon lurerii Cogn. in DC. Monog. Phan. 7: 1158. 1891, non Naud.

Memecylon umbellatum Guillaumin in Lecomte Fl. Gén. Indo-Chine 2: 927. 1921, non Kostel, nec Burm. f., non *Scutula umbellata* Lour.

Memecylon edule Roxb. var. *scutellatum* C. B. Clarke in Hook. f. Fl. Brit. Ind. 2: 564. 1879; Guillaumin in Lecomte Fl. Gén. Indo-Chine 2: 935. 1921 (based on *Scutula scutellata* Lour.).

"Habitat agrestis in Cochinchina." Loureiro's type is preserved in the herbarium of the British Museum, and Britten prepared a note published by Moore (Journ. Bot. 63: 255. 1925) explaining the confusion with *Scutula umbellata* Lour. This has been reduced to *Memecylon edule* Roxb. (var. *scutellatum* C. B. Clarke), but Cogniaux retains it as a distinct species. It is represented by numerous collections from Indo-China, including *Pierre 968, 990, Thorel 190, Harmand 198, Clemens 3313, 3734, 4488,* the last three from Tourane and Hue. It may be noted that *Memecylon scutellatum* (Lour.) Naud. is a much older name than *M. edule* Roxb. Guillaumin's broad interpretation of *M. edule* Roxb. will scarcely be followed by most authors as he has included a very wide range of forms in this as a collective species. His short description of *Memecylon umbellatum* Kostel. (Lecomte Fl. Gén. Indo-Chine 2: 927. 1921) was based on Cogniaux's error in misreading *Scutula scutellata* on Loureiro's specimen in the herbarium of the British Museum as *Scutula umbellata.* Cogniaux states under *Memecylon lurerii* (DC. Monog. Phan. 7: 1158. 1891): "In Cochinchina (Loureiro in hb. Brit. Mus.)." His error is explained by Britten (ex Moore Journ. Bot. 63: 255. 1925).

Memecylon lurerii Naud. in Ann. Sci. Nat. III Bot. 18: 282. 1852 (based on *Scutula umbellata* Lour.).

Memecylon loureirii Triana in Trans. Linn. Soc. 28: 156. 1871 (based on *Scutula umbellata* Lour.).

Memecylon umbellatum Kostel. Allg. Med.-Pharm. Fl. **4**: 1517. 1835 (based on *Scutula umbellata* Lour.), non Burm. f.

Scutula umbellata Lour. Fl. Cochinch. 236. 1790, ed. Willd. 290. 1793, Anamese *cây maóe com.*

" Habitat in dumetis Cochinchinae." Naudin states that he did not find Loureiro's specimen in the herbarium of the British Museum, and Cogniaux's reference to it (DC. Monog. Phan. **7**: 1158. 1891) is explained by Britten (ex Moore Journ. Bot. **63**: 255. 1925) by Cogniaux having misread the specific name *scutellata* as *umbellata*. *Scutula scutellata* Lour. is represented in the herbarium of the British Museum by a specimen from Loureiro, but there is no evidence that *Scutula umbellata* Lour. was ever received from him.

ONAGRACEAE

Jussiaea [47] Linnaeus

Jussiaea repens Linn. Sp. Pl. 388. 1753; Gagnep. in Lecomte Fl. Gén. Indo-Chine **2**: 987. 1921.

Cubospermum palustre Lour. Fl. Cochinch. 275. 1790, ed. Willd. 337. 1793, Anamese *rau jùa.*

Ludwigia palustris A. Chevalier Cat. Pl. Jard. Bot. Saigon 65. 1919 (based on *Cubospermum palustre* Lour.).

" Habitat in locis aquosis Cochinchinae." Doctor Chevalier, writing in July, 1919, states that the Anamese name cited by Loureiro does not belong to *Jussiaea repens* Linn., but appertains to a *Ludwigia* common in marshy places and often cultivated by the Anamese. It may be noted that *Jussiaea* and *Ludwigia* are distinguished only by the number of stamens and that Gagnepain, apparently with good reason, reduces *Ludwigia* to *Jussiaea*. Loureiro's description of the stamens as twice the number of petals is a *Jussiaea* character, for in *Ludwigia* the floral parts are isomerous. Loureiro's type is preserved in the herbarium of the British Museum.

Jussiaea erecta Linn. Sp. Pl. 388. 1753; Ridl. in Journ. Bot. **59**: 258. 1921, Fawcett in Journ. Bot. **64**: 11. 1926; Lewin in Fedde Repert. **23**: 129. 1926.

Epilobium tetragonum Lour. Fl. Cochinch. 225. 1790, ed. Willd. 276. 1793, Anamese *cây muòng núoc*, Chinese *sòy hoâng teng.*

Jussiaea tetragona Spreng. Syst. **2**: 231. 1825 (based on *Epilobium tetragonum* Lour.).

" Habitat loca aquosa tam in Cochinchina, quam in China." See the discussion of *Jussiaea suffruticosa* Linn., below, appertaining to the literature and problems involved regarding this species and *J. suffruticosa* Linn. I have rather arbitrarily placed the glabrous form under *J. erecta* Linn., as this is one of the distinguishing characters given by Linnaeus in the original description, and Loureiro describes *Epilobium tetragonum* as glabrous.

Jussiaea suffruticosa Linn. Sp. Pl. 388. 1753; Fawcett in Journ. Bot. **64**: 12. 1926; Lewin in Fedde Repert. **23**: 128. 1926; Gagnep. in Lecomte Fl. Gén. Indo-Chine **2**: 986. 1921.

Epilobium fruticosum Lour. Fl. Cochinch. 226. 1790, ed. Willd. 277. 1793, Anamese *cây muòng dât.*

Oenothera fruticosa Lour. ex G. Don Gen. Syst. **2**: 695. 1832 in syn. (error for *Epilobium fruticosum* Lour.).

[47] Often spelled *Jussieua*, but the original form is here retained; see Sprague, Kew Bull. 355. 1928.

Jussiaea villosa Lam. Encycl. **3**: 331. 1789; Ridl. Journ. Bot. **59**: 259. 1921.

Jussiaea fruticosa DC. Prodr. **3**: 57. 1828 (based on *Epilobium fruticosum* Lour.).

"Habitat in arvis incultis Cochinchinae." The synonymy and specific limits in this group are peculiarly complicated. I have rather arbitrarily referred the pubescent form to *J. suffruticosa* described by Linnaeus as "villosa" and by Loureiro (*Epilobium fruticosum*) as "hirsuta." It seems highly probable that *J. suffruticosa* Linn. and the preceding one, *J. erecta* Linn., which is glabrous, are merely forms of a single species. For a discussion of the problems involved and varying opinions expressed, see Ridley, H. N., The Indo-Malayan species of *Jussiaea* (Journ. Bot. **59**: 257–260. 1921); Fawcett, W., Linnaeus's species of *Jussieua* (Journ. Bot. **64**: 10–13. 1926); and Lewin, K., Die indomaliischen Jussieua-Arten (Fedde Repert. **23**: 128–130. 1926).

HYDROCARYACEAE

Trapa Linnaeus

Trapa bicornis Osbeck Dagbok Ostind. Resa 191. 1757; Linn. f. Suppl. 128. 1781.

> *Trapa cochinchinensis* Lour. Fl. Cochinch. 86. 1790, ed. Willd. 108. 1793, Anamese *cây aú*.
>
> *Trapa chinensis* Lour. Fl. Cochinch. 86. 1790, ed. Willd. 109. 1793, Anamese *linh that*, Chinese *kī xi, leng cŏ*.

For *Trapa cochinchinensis* Loureiro states: "Habitat & fluctuat in paludibus Cochinchinae," and for *T. chinensis:* "Habitat fluctuans prope Cantonem," stating under the latter "Fortasse est varietas Trapae Cochinchinensis, sed ab Europaea specie differens." Essentially *Trapa chinensis* Lour. is but a new name for *T. bicornis*, as Loureiro cites "Linn. jun. suppl. plant. p. 128. Trapa Bicornis." There seems to be little agreement as to the exact status of this southeastern Asiatic form, the latest consideration of it by Flerov [48] being as follows: "*Trapa bicornis* mihi (*T. bispinosa* Roxb. ex par.)—*T. bicornis* L. fil.?" He overlooked *Trapa bicornis* Osbeck.

HALORAGIDACEAE

Haloragis [49] J. R. & G. Forster

Haloragis chinensis (Lour.) comb. nov.

> *Gaura chinensis* Lour. Fl. Cochinch. 225. 1790, ed. Willd. 276. 1793, Chinese *sán sí tsào*.
>
> *Goniocarpus scaber* Koenig in Koenig & Sims Ann. Bot. **1**: 547. *pl. 12. f. 6.* 1805.
>
> *Ludwigia octandra* Banks ex Koenig l.c. in syn.
>
> *Haloragis scabra* Benth. Fl. Hongk. 139. 1861.
>
> *Haloragis scabra* Benth. var. *elongata* Schindl. in Pflanzenreich **23**(IV–225): 29. 1905.

"Habitat inculta Cantone Sinarum." I have found no previous suggestions as to what *Gaura chinensis* Lour. might represent. The description is a most excellent one for *Haloragis scabra* Benth., and the species is a common one in the vicinity of Canton.

[48] Flerov, A. De genere Trapa L. Bull. Jard. Bot. Republ. Russe **24**: 13–45. 1925.

[49] *Halorrhagis*, used by some authors, is a philologically better form, but the original spelling is here retained; see Sprague Kew Bull. 354. 1928.

ARALIACEAE

Schefflera J. R. & G. Forster

Schefflera octophylla (Lour.) Harms in Engler & Prantl Nat. Pflanzenfam. **3**(8): 38. 1894 (based on *Aralia octophylla* Lour.); Viguier in Lecomte Fl. Gén. Indo-Chine **2**: 1178. 1923.

Aralia octophylla Lour. Fl. Cochinch. 187. 1790, ed. Willd. 233. 1793, Anamese *chim chim nhà*.

Heptapleurum octophyllum Benth. ex Hance in Journ. Linn. Soc. Bot. **13**: 105. 1873 (based on *Aralia octophylla* Lour.).

Agalma octophyllum Seem. in Journ. Bot. **2**: 298. 1864 (based on *Aralia octophylla* Lour.).

Paratropia cantoniensis H. & A. Bot. Beechey's Voy. 189. 1833.

" Habitat in Cochinchina, tam culta, quam spontanea." This species is abundant in Indo-China, and I can detect no essential differences between the Indo-China form and that of Kwangtung. Loureiro definitely based his description on the Indo-China plant and notes regarding this Kwangtung form: " Nascitur prope Cantonem in Siniis hujus plantae varietas, ni velis species, caule, foliis, et habitu florum omnino similis: differt autem staminibus 10, et stigmate 10-fido."

Acanthopanax Miquel

Acanthopanax trifoliatus (Linn.) Merr. in Philip. Journ. Sci. **1**: Suppl. 217. 1906; Schneider Ill. Handb. Laubholzk. **2**: 427. 1909; Nakai in Journ. Arnold Arb. **5**: 1. 1924.

Zanthoxylum trifoliatum Linn. Sp. Pl. 270. 1753.

Panax aculeatum Ait. Hort. Kew. **3**: 448. 1789.

Plectronia chinensis Lour. Fl. Cochinch. 162. 1790, ed. Willd. 201. 1793, Chinese *pă lac hōa.*

Panax loureirianum DC. Prodr. **4**: 252. 1830 (based on *Plectronia chinensis* Lour.).

Acanthopanax aculeatus Seem. in Journ. Bot. **5**: 238. 1867; Viguier in Lecomte Fl. Gén. Indo-Chine **2**: 1166. 1923.

" Habitat agrestis in provincia Cantoniensi Sinarum." This species is very common in thickets near Canton, the Linnaean type being from the same general locality. Loureiro's description applies unmistakably to *Acanthopanax trifoliatus* (Linn.) Merr. His type is preserved in the herbarium of the Paris Museum.

Nothopanax Miquel

Nothopanax fruticosus (Linn.) Miq. Fl. Ind. Bat. **1**(1): 765. 1857.

Panax fruticosum Linn. Sp. Pl. ed. 2, 1513. 1763; Lour. Fl. Cochinch. 656. 1790, ed. Willd. 806. 1793, Anamese *cây ca la va, dinh lang.*

Tieghemopanax fruticosus Viguier in Ann. Sci. Nat. IX Bot. **4**: 61. 1906, Lecomte Fl. Gén. Indo-Chine **2**: 1163. 1923.

Polyscias fruticosa Harms in Engl. & Prantl Nat. Pflanzenfam. **3**(8): 45. 1894.

" Habitat cultum in hortis Cochinchinae, & Chinae." Loureiro manifestly was correct in his interpretation of the widely distributed and commonly planted Linnaean species. *Scutellaria tertia* Rumph. (Herb. Amb. **4**: 78. *pl. 33*), cited by Loureiro as a synonym, is correctly placed. A specimen from Loureiro is preserved in the herbarium of the British Museum, as noted by Seemann (Fl. Vit. 114. 1865, *in nota*). Viguier used the name *Tieghemopanax fruticosus* in a key published in Ann. Sci. Nat. IX Bot. **4**: 61. 1906, but there failed to indicate the binomial on which it was based; the publication was validated by Lecomte in 1923. *Tieghemopanax* was based on a series of 26 species, mostly from New Caledonia and Australia, which I do not consider to be congeneric with *Nothopanax fruticosus* (L.) Miq. If, however, Viguier is correct in transferring *Panax fruticosum* Linn. to *Tieghemopanax*, then, at least in part, *Tieghemopanax* is the same as *Nothopanax*. I have interpreted this Linnaean species as typifying *Nothopanax* Miquel, but Sprague & Green (Kew Bull. 154–155. 1933) interpret the type as *N. cochleatus* (Lam.) Miq. In my judgment the two species are congeneric.

Kalopanax Miquel

Kalopanax septemlobus (Thunb.) Koidz. in Bot. Mag. Tokyo **39**: 306. 1925; Hand.-Maz. Symb. Sin. **7**: 699. 1933.

Acer septemlobum Thunb. Fl. Jap. 162. 1784.

Aralia palmata (non Linn.) Lour. Fl. Cochinch. 187. 1790, ed. Willd. 233. 1793, Anamese *ngu già bi*, Chinese *ù kīa pī*.

Aralia scandens Poir. in Lam. Encycl. Suppl. **1**: 419. 1811 (based on *Aralia palmata* Lour.).

Hedera scandens DC. Prodr. **4**: 264. 1830 (based on *Aralia palmata* Lour.).

Kalopanax ricinifolius Miq. Ann. Mus. Bot. Lugd.-Bat. **1**: 16. 1863.

"Habitat agrestis in China." Hemsley notes (Journ. Linn. Soc. Bot. **23**: 338. 1888) that *Aralia palmata* Lour. was an obscure species, doubtfully (and apparently erroneously) referred by Seemann to the Indian *Brassaiopsis hainla* (Ham.) Seem., a species unknown from China. Thunberg's species is not scandent, nor are the inflorescences lateral or of simple umbels, as Loureiro states. While Loureiro merely credits the species to China, he also cites an Anamese name, perhaps indicating an herbalist as the source of his material, especially in view of the fact that none of the Indo-China species of Araliaceae admitted by Viguier (Lecomte Fl. Gén. Indo-Chine **2**: 1158–1182. 1923) conforms to the characters indicated by Loureiro.

Aralia (Tournefort) Linnaeus

Aralia armata (Wall.) Seem. in Journ. Bot. **6**: 134. 1868; Viguier in Lecomte Fl. Gén. Indo-Chine **2**: 1162. *f. 137, 1–4.* 1923.

Panax armatum Wall. List. no. 4933. 1832, *nomen nudum*.

Aralia chinensis (non Linn.) Lour. Fl. Cochinch. 187. 1790, ed. Willd. 234. 1793, Anamese *cây quòng*.

"Habitat per sepes scandens in Cochinchina: valde infensa viatoribus, quorum vestes vel carnes mordicus tenet, & lacerat aculeis plurimis, aduncis." Loureiro's description applies unmistakably to Wallich's species. His misinterpreted *Aralia chinensis* is not accounted for in Viguier's treatment of the Araliaceae of Indo-China (Lecomte Fl. Gén. Indo-Chine **2**: 1158–1182. 1923). The species is represented by *Clemens 4145* from Tourane, near the classical locality Hue. *Frutex aquosus mas* Rumph. (Herb. Amb. **4**: 102. *pl. 45*), cited by Loureiro as a synonym, must be excluded as it represents the vitaceous *Leea* *aculeata* Blume. A specimen listed as being among the Loureiro plants in the herbarium of the British Museum has not been located.

UMBELLIFERAE

Hydrocotyle (Tournefort) Linnaeus

Hydrocotyle sibthorpioides Lam. Encycl. **3**: 153. 1789; Merr. Enum. Philip. Fl. Pl. **3**: 237. 1923.

Hydrocotyle rotundifolia Roxb. Hort. Beng. 21. 1814, *nomen nudum*, Fl. Ind. ed. 2, **2**: 88. 1832; DC. Prodr. **4**: 64. 1830; Chermez. in Lecomte Fl. Gén. Indo-Chine **2**: 1137. 1923.

Hydrocotyle umbellata (non Linn.) Lour. Fl. Cochinch. 177. 1790, ed. Willd. 220. 1793, Anamese *rau má mo.*

" Habitat in Cochinchina, frequens in locis humidis, hyberno tempore." In my original manuscript of 1919 I placed this as a probable synonym of *Centella asiatica* Urban, but Chermezon is unquestionably correct in identifying Loureiro's species as *Hydrocotyle rotundifolia* Roxb., which I consider to be a synonym of the older *H. sibthorpioides* Lam.

Centella Linnaeus

Centella asiatica (Linn.) Urban in Mart. Fl. Bras. **11**(1): 287. *pl. 78. f. 1.* 1879; Chermez. in Lecomte Fl. Gén. Indo-Chine **2**: 1134. 1923.

Hydrocotyle asiatica Linn. Sp. Pl. 234. 1753.

Trisanthus cochinchinensis Lour. Fl. Cochinch. 176. 1790, ed. Willd. 219. 1793, Anamese *rau má.*

" Habitat frequenter ad sepes in Cochinchina, non cultus, quamvis esculentus. Si bene memini, etiam in China, & in aliis Indiae locis." Specimens from Loureiro are preserved in the herbaria of the British and the Paris Museums, and these represent the Linnaean species. *Pes equinus* Rumph. (Herb. Amb. **5**: 455. *pl. 169. f. 1*) is correctly placed as a synonym.

Foeniculum (Tournefort) Linnaeus

Foeniculum vulgare Gaertn. Fruct. **1**: 105. *pl. 23. f. 5.* 1788.

Anethum foeniculum Linn. Sp. Pl. 263. 1753; Lour. Fl. Cochinch. 181. 1790, ed. Willd. 226. 1793, Anamese *tieo hôi*, Chinese *hòei hiàm.*

" Habitat abundanter in China: colitur etiam in Cochinchina." The description manifestly applies to the widely distributed Linnaean species.

Torilis Adanson

Torilis anthriscus (Linn.) Gmel. Fl. Bad. **1**: 615. 1805; Chermez. in Lecomte Fl. Gén. Indo-Chine **2**: 1157. 1923.

Tordylium anthriscus Linn. Sp. Pl. 240. 1753.

Caucalis anthriscus Huds. Fl. Angl. 99. 1762.

Caucalis orientalis (non Linn.) Lour. Fl. Cochinch. 177. 1790, ed. Willd. 221. 1793, Anamese *cây hôt gai.*

" Habitat inculta Cantone, & alibi." Loureiro's description applies unmistakably to *Torilis anthriscus* Gmel., which is common near Canton and which is very widely distributed in China, occurring also in Indo-China.

Coriandrum (Tournefort) Linnaeus

Coriandrum sativum Linn. Sp. Pl. 256. 1753; Lour. Fl. Cochinch. 180. 1790, ed. Willd. 225. 1793, Anamese *rau ngò tàu*, Chinese *xĕ hû yû;* Chermez. in Lecomte Fl. Gén. Indo-Chine **2**: 1156. 1923.

Coriandrum testiculatum (non Linn.) Lour. Fl. Cochinch. 180. 1790, ed. Willd. 225. 1793, Anamese *rau ngò, hô noi.*

Bifora loureirii Kostel. Allgem. Med.-Pharm. Fl. **4**: 1183. 1835 (based on *Coriandrum testiculatum* Lour.).

For *Coriandrum sativum* Loureiro states: " Habitat, & colitur in China, raro in Cochinchina." The description apparently applies to the common coriander which is not uncommon in cultivation in China and is also cultivated in Indo-China. For *C. testiculatum* he states: " Habitat, & colitur abundanter per totam Cochinchinam," and notes that the fruit in odor and flavor is superior to those characters in the other species he described. I can see little reason for considering this other than a form of the common coriander. G. Don (Gen. Syst. **3**: 382. 1834) discusses both of Loureiro's species under *Coriandrum sativum* Linn., but does not admit either as synonyms of it. Chermezon does not account for *Bifora loureirii* Kostel.

Carum (Ruppius) Linnaeus

Carum involucratum (Roxb.) Baill. ex Jacks. Ind. Kew. **1**: 445. 1895.

Apium involucratum Roxb. ex Flem. in Asiat. Research **11**: 157. 1810; Fl. Ind. ed. 2, **2**: 97. 1832 (without specific name).

Carum roxburghianum Benth. ex C. B. Clarke in Hook. f. Fl. Brit. Ind. **2**: 682. 1879; Chermez. in Lecomte Fl. Gén. Indo-Chine **2**: 1144. 1923.

Ptychotis roxburghiana DC. Prodr. **4**: 109. 1830.

Bubon macedonicum (non Linn.) Lour. Fl. Cochinch. 179. 1790, ed. Willd. 224. 1793, Anamese *hoa khóm.*

" Habitat cultum, nec raro in hortis metropolis Huaeae Cochinchinae." The form Loureiro erroneously referred to *Bubon macedonicum* Linn. seems clearly to be referable to *Carum involucratum* Baill. *Levisticum indicum* Rumph. (Herb. Amb. **5**: 269. *pl. 93. f. 3*), which Loureiro cites as representing his species, probably represents the same species; in my consideration of Rumphius' species (Interpret. Herb. Amb. 411. 1917) I left it as *Anthriscus* sp.? Chermezon gives no local name for *Carum roxburghianum* Benth., but states that the species is found only in cultivation in Indo-China. The fruit characters indicated by Loureiro are apparently those of this species. There is a possibility that *Carum aromaticum* (Linn.) Druce (*Bunium aromaticum* Linn., *C. copticum* Benth. & Hook. f., *Ptychotis coptica* DC.) may be the form that Loureiro had.

Oenanthe (Tournefort) Linnaeus

Oenanthe stolonifera (Roxb.) Wall. List no. 585. 1829, *nomen nudum;* DC. Prodr. **4**: 138. 1830; Chermez. in Lecomte Fl. Gén. Indo-Chine **2**: 1149. 1923.

Phellandrium stoloniferum Roxb. Hort. Beng. 21. 1814, *nomen nudum*, Fl. Ind. ed. 2, **2**: 93. 1832.

Sium sisarum (non Linn.) Lour. Fl. Cochinch. 179. 1790, ed. Willd. 223. 1793, Anamese *rau kân nuòc*, Chinese *xuèi kín.*

Sium graecum (non Linn.) Lour. Fl. Cochinch. 179. 1790, ed. Willd. 223. 1793, Anamese *rau kân hoang*.

Ligusticum scoticum (non Linn.) Lour. Fl. Cochinch. 180. 1790, ed. Willd. 224. 1793,
Anamese *rau kân nhà*.

For *Sium sisarum* Loureiro states: " Habitat loca aquosa in China, & Cochinchina,
ubi colitur." *Sium sisarum* Linn. as represented in the Linnaean herbarium, which may
or may not be the type, although the specimen was in the herbarium in 1753, is the species
as interpreted by Hegi (Ill. Fl. Mittel-Eur. **5**: 1222. 1926). Linnaeus (Sp. Pl. 251. 1753)
based the species on five pre-Linnaean citations, and gives no additional description. He
states: " Habitat in China?," which is probably an error, as Hemsley (Journ. Linn. Soc.
Bot. **23**: 329. 1888) notes that this species has not been observed in China by recent botanists and quotes Bretschneider to the effect that in Japan the Chinese name given by
Loureiro is applied to *Oenanthe stolonifera* Wall.; *Sium sisarum* Lour., however, is safely a
form of *Oenanthe stolonifera* Wall. For *Sium graecum* Loureiro states: " Habitat ubique
incultum in hortis Cochinchinae: puto, quod etiam in China, quamvis non viderim." While
Hemsley (Journ. Linn. Soc. Bot. **23**: 329. 1888) states that *Sium graecum* Lour. is altogether an obscure plant, I am of the opinion that what Loureiro had is the common *Oenanthe
stolonifera* Wall. and do not hesitate to make this reduction. For *Ligusticum scoticum*
Loureiro states: " Habitat cultum in Cochinchina." The reduction to *Oenanthe stolonifera*
Wall. was made by Chermezon, and I see no reason for doubting the correctness of this
disposition of it.

Selinum Linnaeus

Selinum monnieri Linn. Cent. Pl. **1**: 9. 1755; Chermez. in Lecomte Fl. Gén. Indo-Chine **2**:
1151. 1923.

Athamanta chinensis (non Linn.) Fl. Cochinch. 178. 1790, ed. Willd. 222. 1793, Anamese *xà sàng*, Chinese *xê chōan*.

" Habitat in Cochinchina, & China culta, spontaneaque." Loureiro's description applies sufficiently well to *Selinum monnieri* Linn. which is widely distributed in China, to
warrant this reduction, a disposition of it that I made in 1919, and independently arrived
at by Chermezon in 1923. Linnaeus states that he received the seeds of *Athamanta chinensis* from Bartram [of Philadelphia]: " Habitat . . . Chinensem dixit Barthram, qui semina
misit ex Virginia." A. Gray (Journ. Bot. **19**: 325–326. 1881) has conclusively shown that
the Linnaean species is identical with *Selinum canadense* Michx.; Linnaeus' error was
probably due to his mistaking the locality Genesee, as Bartram visited the Genesee region
in New York. The species does not occur in China.

Daucus (Tournefort) Linnaeus

Daucus carota Linn. Sp. Pl. 242. 1753; Lour. Fl. Cochinch. 178. 1790, ed. Willd. 222. 1793,
Anamese *hô la bac*, Chinese *hû lû pā*.

" Habitat Cantone Sinarum: indeque in Cochinchinam translata." The description
applies unmistakably to the Linnaean species, the common carrot, which is widely cultivated in China.

ALANGIACEAE

Alangium [50] Lamarck

Alangium chinense (Lour.) Rehder in Sargent Pl. Wils. **2**: 552. 1916 (based on *Stylidium chinense* Lour.); Evrard in Lecomte Fl. Gén. Indo-Chine **2**: 1187. 1923.

Stylidium chinense Lour. Fl. Cochinch. 221. 1790, ed. Willd. 273. 1793, Chinese *pàu tsàu*.

Stylis chinensis Poir. in Lam. Encycl. Suppl. **5**: 260. 1817 (based on *Stylidium chinense* Lour.).

Karangolum chinense O. Ktz. Rev. Gen. Pl. 273. 1891 (based on *Stylidium chinense* Lour.).

Stylidium bauthas Lour.[51] ex Gomes in Mem. Acad. Sci. Lisb. Cl. Sci. Pol. Mor. Bel.-Let. n.s. **4**(1): 28. 1868.

Marlea begonifolia Roxb. Hort. Beng. 28. 1914, *nomen nudum*, Pl. Coromand. **3**: 80. *pl. 283*. 1819.

Alangium begoniifolium Baill. ex Harms in Engler & Prantl Nat. Pflanzenfam. 3(8): 261. *f. 80 A–G.* 1898; Wang. in Pflanzenreich 41(IV–220b): 20. *f. 5*. 1910.

Marlea chinensis Druce in Rept. Bot. Exch. Club Brit. Isles **4**: 634. 1917 (based on *Stylidium chinense* Lour.).

" Habitat incultum suburbia Cantonis Sinarum." Loureiro's type is preserved in the herbarium of the Paris Museum. The species is more generally known under Roxburgh's binomial, but Loureiro's specific name is much older. The species is one of very wide geographic distribution extending from East Africa to China and Malaysia.

ERICACEAE

Rhododendron Linnaeus

Rhododendron simsii Pl. in Fl. Serr. **9**: 78. 1854; Wilson in Wilson & Rehd. Monogr. Azalea. 45. 1921, cum syn.

Enkianthus biflorus Lour. Fl. Cochinch. 277. 1790, ed. Willd. 340. 1793, pro majore parte.

Loureiro's species is considered more extensively under *Enkianthus quinqueflorus* below. The description is in part manifestly of a *Rhododendron*, and undoubtedly a form of *R. simsii* Pl. which is commonly cultivated in China for ornamental purposes. Wilson and Rehder give a very full list of synonyms of Planchon's species, it being the one that appears in most Chinese botanical literature as *R. indicum* (non Linn.).

Enkianthus Loureiro

Enkianthus quinqueflorus Lour. Fl. Cochinch. 277. 1790, ed. Willd. 339. 1793, Chinese *tsiàu tsung hōa*.

Enkianthus biflorus Lour. Fl. Cochinch. 277. 1790, ed. Willd. 340. 1793, pro minore parte, Chinese *sān liéo hōa*.

[50] *Alangium* Lamarck (1783), conserved name, Vienna Code; older ones are *Angolam* Adanson (1763), *Karo-Angolam* Adanson (1763) (= *Karangolum* O. Kuntze 1892) and *Angolamia* Scopoli (1777).

[51] A Loureiro herbarium name here first published by Gomes.

For *E. quinqueflorus* Loureiro states: " Colitur Cantone Sinarum." This is the type of the genus *Enkianthus* Lour. It is a well understood species occurring in Kwangtung Province, both native and cultivated. For *E. biflorus* Loureiro states: " Habitat Cantone Sinarum." Hemsley (Journ. Linn. Soc. Bot. **26**: 18. 1889) retains this as *Enkianthus biflorus* Lour. with the statement: " Perhaps a reduced state of *E. quinqueflorus*, or it may be *Rhododendron indicum*." The actual *description* in my judgment applies largely to *Rhododendron simsii* Pl. (*R. indicum* auct. non Linn.). A very fragmentary specimen from Loureiro in the herbarium of the Paris Museum labelled *san lieo hoa*, consisting of two leafless branchlets and a few detached bracts, apparently represents *Enkianthus quinqueflorus* Lour. Sprengel (Syst. **2**: 287. 1825) spells the generic name *Encyanthus*, and de Candolle (Prodr. **7**: 732. 1839) uses the form *Enkyanthus*.

Vaccinium Linnaeus

Vaccinium bracteatum Thunb. Fl. Jap. 156. 1784; Dop in Lecomte Fl. Gén. Indo-Chine **3**: 710. 1930.

Acosta spicata Lour. Fl. Cochinch. 276. 1790, ed. Willd. 338. 1793, Anamese *cây men.*

Vaccinium acosta Raeusch. Nomencl. ed. 3, 109. 1797 (based on *Acosta spicata* Lour.).

Vaccinium orientale Sw. in Weber & Mohr Beitr. Naturk. **1**: 6. *pl. 2.* 1805 (based on *Acosta spicata* Lour.).

Vaccinium spicatum Poir. in Lam. Encycl. Suppl. **1**: 280. 1810 (based on *Acosta spicata* Lour.); Moore in Journ. Bot. **63**: 282. 1925.

Agapetes (*?*) *acosta* Dunal in DC. Prodr. **7**: 556. 1839 (based on *Acosta spicata* Lour.).

Epigynum acosta Klotzsch in Linnaea **24**: 53. 1851 (based on *Acosta spicata* Lour.).

Agapetes (*?*) *spicata* Moore in Journ. Bot. **63**: 282. 1925, spahlm.!, *A. acosta* intended (based on *Acosta spicata* Lour.).

" Habitat in sylvis planis Cochinchinae." Bentham (Benth. & Hook. f. Gen. Pl. **2**: 572. 1876) mentions *Acosta* under *Agapetes* as a genus of somewhat doubtful status. O. Kuntze saw Loureiro's specimen in the herbarium of the British Museum and reduced it without question to *Vaccinium bracteatum* Thunb. (Rev. Gen. Pl. 384. 1891). Moore published an amplified description of the species in 1925, thus removing all doubt as to the generic status of *Acosta*. He accepted *Vaccinium spicatum* (Lour.) Poir. as a valid species, stating that his endeavors to match the type met with no success. I personally examined Loureiro's type in 1930 and find that it is well matched by *Clemens 3965, 4458* from the general vicinity of Hue. I do not think that Loureiro's species can be distinguished from *Vaccinium bracteatum* Thunb. by any constant characters. Dop (Lecomte Fl. Gén. Indo-Chine **3**: 710. 1930) admits *Vaccinium bracteatum* Thunb., giving a full description, citing numerous synonyms and collections. Curiously he overlooked *Acosta spicata* Lour. entirely and the five additional synonyms based on it, although Loureiro's type is extant and the species was based on material from Anam (vicinity of Hue). The long standing doubt regarding the exact status of Loureiro's genus and species, and Dop's oversight of it, is all the more curious, because Swartz [52] as early as 1805 published an amplified description of it with a good illustration based on a Loureiro specimen. This specimen is not in

[52] Swartz, O. *Acosta spicata* Loureiro, eine neue Art von *Vaccinium.* Weber & Mohr Beitr. Naturk. **1**: 4–7. *pl. 2.* 1805.

Swartz's herbarium and Doctor Samuelsson suspects that his data were based on the specimen in the Banksian herbarium, now in the British Museum.

MYRSINACEAE

Maesa Forskål

Maesa perlarius (Lour.) comb. nov.

Dartus perlarius Lour. Fl. Cochinch. 124. 1790, ed. Willd. 153. 1793, Anamese *cây don*.

Maesa sinensis A. DC. in Ann. Sci. Nat. II Bot. **16**: 80. 1841; Prodr. **8**: 82. 1844; Mez in Pflanzenreich 9(IV–236): 34. 1902; Pitard in Lecomte Fl. Gén. Indo-Chine **3**: 774. 1930.

" Habitat agrestis ad fluminum ripas, & loca umbrosa in Cochinchina." Loureiro's genus has remained as one of uncertain status, generally placed in the Solanaceae where it cannot possibly belong. The description applies unmistakably to *Maesa*, and the species is clearly identical with *Maesa sinensis* A. DC. If further verification of the correctness of the generic reduction be needed, this is supplied by the local names, *cây don, cây don nem, cay don an goi*, cited by Pitard for two species of *Maesa*. *Perlarius alter* Rumph. (Herb. Amb. **4**: 122. *pl. 57*), cited by Loureiro, and whence he took his specific name, is *Maesa tetrandra* (Roxb.) A. DC. Loureiro's species is represented from the classical locality, Hue and vicinity, by *Squires 108*, erroneously identified by Pitard as *Maesa indica* Wall.; it is *M. sinensis* A. DC. A specimen from Loureiro listed as being among the plants received from him has not been located in the herbarium of the British Museum.

Ardisia [53] Swartz

Ardisia humilis Vahl Symb. **3**: 40. 1794 (non auct. plur.); Merr. in Lingnan Sci. Journ. **11**: 50. 1932.

Pyrgus racemosa Lour. Fl. Cochinch. 121. 1790, ed. Willd. 149. 1793, Anamese *cây luc chôt*.

Tinus racemosa O. Ktz. Rev. Gen. Pl. 974. 1891 (based on *Pyrgus racemosa* Lour.).

Ardisia racemosa Mez in Pflanzenreich 9(IV–236): 138. 1902 (based on *Pyrgus racemosa* Lour.); Pitard in Lecomte Fl. Gén. Indo-Chine **3**: 834. 1930, non Spreng.

Ardisia pyrgus Roem. & Schult. Syst. **4**: 518. 1819 (based on *Pyrgus racemosa* Lour.).

Ardisia pyrgina St. Lag. in Ann. Soc. Bot. Lyon **7**: 119. 1880 (based on *A. pyrgus* R. & S., *Pyrgus racemosa* Lour.).

Ardisia hainanensis Mez in Pflanzenerich 9(IV–236): 138. 1902.

" Habitat in dumetis Cochinchinae." The synonymy is simple, yet has been curiously confused. *Ardisia humilis* Vahl was totally misinterpreted by Mez, who apparently followed current usage in spite of the fact that he examined Vahl's type. This specimen in the Copenhagen Herbarium, examined by me in 1930, is identical with *Ardisia hainanensis* Mez, and furthermore represents the same species as *Pyrgus racemosa* Lour., the type of which I also examined in the herbarium of the British Museum in 1930. Pitard (Lecomte Fl. Gén. Indo-Chine **3**: 848. 1930) follows Mez in the current misinterpretation of *Ardisia humilis* Vahl, but not one of the synonyms cited by Mez or by Pitard belongs with

[53] *Ardisia* Swartz (1788), conserved name, Vienna Code; older ones are *Kathoutheka* Adanson (1763), ? *Vedela* Adanson (1763), *Icacorea* Aublet (1775), and *Bladhia* Thunberg (1781).

this binomial. The widely distributed collective species described by Mez as *Ardisia humilis* (non Vahl) Pflanzenreich (9(IV–236): 127. 1902) should be known as *Ardisia elliptica* Thunb. (*A. littoralis* Andr.), type from Ceylon.

Ardisia loureiriana (G. Don) comb. nov.

    Rhododendron loureirianum G. Don Gen. Syst. 3: 846. 1834 (based on *Azalea punctata* Lour.); Dop in Lecomte Fl. Gén. Indo-Chine 3: 746. 1930.

    Azalea punctata Lour. Fl. Cochinch. 113. 1790, ed. Willd. 139. 1793, Anamese *cây maóc neò*, non *Rhododendron punctatum* Andr.

    Ardisia expansa Pitard in Lecomte Fl. Gén. Indo-Chine 3: 844. 1930.

" Habitat in sylvis Cochinchinae." The description applies in all respects to *Ardisia* except in the indicated fruit character " caps. 5-locularis "; this was certainly added by Loureiro to make the description conform to the characters of the genus in which *Azalea punctata* was erroneously placed. The reddish-punctate sepals, petals, anthers, and ovary point unmistakably to *Ardisia*. Dop merely compiled a description from Loureiro's original, without comment, leaving the species under *Rhododendron*. The species is represented by *Clemens 3234, 3706*, from Hue and Tourane, both identified by Pitard as *Ardisia expansa* Pitard.

<h3 style="text-align:center">Embelia [54] Burman f.</h3>

Embelia scandens (Lour.) Mez in Pflanzenreich 9(IV–236): 317. 1902 (based on *Calispermum scandens* Lour.); Pitard in Lecomte Fl. Gén. Indo-Chine 3: 792. *f. 87. 3. 89, 1.* 1930.

    Calispermum scandens Lour. Fl. Cochinch. 156. 1790, ed. Willd. 194. 1793, Anamese *bo gie deei.*

    Embelia nervosa A. DC. in Ann. Sci. Nat. II Bot. **16**: 81. 1841.

    Ribesoides nervosum O. Ktz. Rev. Gen. Pl. 403. 1891.

" Habitat in sylvis Cochinchinae." A sufficiently well-known species represented by numerous Indo-China collections; it also occurs in Hainan. Loureiro's type is preserved in the herbarium of the British Museum.

<h3 style="text-align:center">Rapanea Aublet</h3>

Rapanea linearis (Lour.) Moore in Journ. Bot. **63**: 249. 1925 (based on *Athruphyllum lineare* Lour.).

    Athruphyllum lineare Lour. Fl. Cochinch. 120. 1790, ed. Willd. 148. 1793, Anamese *cây ma ca.*

    Myrsine linearis Poir. in Lam. Encycl. Suppl. **3**: 709. 1814 (based on *Athruphyllum lineare* Lour.).

    Myrsine athyrophyllum (error for *athruphyllum*) R. Br. ex Roem. & Schult. Syst. **4**: 509. 1819 (based on *Athruphyllum lineare* Lour.).

    Athrurophyllum lineare G. Don Gen. Syst. **4**: 10. 1838, in syn. (error for *Athruphyllum lineare* Lour.).

    Myrsine athrurophyllum G. Don l.c. in syn., sphalm.

    Myrsine playfairii Hemsl. in Journ. Linn. Soc. Bot. **26**: 61. 1889.

    Rapanea playfairii Mez in Pflanzenreich 9(IV–236): 361. 1902.

[54] *Embelia* Burman f. (1768), conserved name, Vienna Code; older ones are *Ghesaembilla* Adanson (1763) and *Pattara* Adanson (1763).

" Habitat in sylvis montanis Cochinchinae." Mez (Pflanzenreich 9(IV–236): 361. 1902) discusses Loureiro's species as a probable synonym of *Rapanea neriifolia* (S. & Z.) Mez which is not known to occur in Indo-China. Moore, after examining Loureiro's authentic specimen in the herbarium of the British Museum, concluded that it represents a distinct species. I examined the type in 1930 and found it was most excellently matched by the type of *Rapanea playfairii* (Hemsl.) Mez from southern Kwangtung and I do not hesitate in reducing the latter to *R. linearis* (Lour.) Moore. Pitard, in his treatment of the Myrsinaceae of Indo-China (Lecomte Fl. Gén. Indo-Chine 3: 765–786. 1930) overlooked Moore's note and failed to account for Loureiro's species; the form, represented by *Clemens 3551, 4253, 4415*, from Tourane, near the classical locality Hue, was placed by him under *Rapanea capitellata* (Wall.) Mez and this is his identification of the Clemens specimens cited above. I cannot agree with him in this interpretation of Wallich's species. The species occurs in Kwangtung Province, Hainan, and Indo-China. De Pirey's specimens of *ma ca, ma ca chup*, and *ma ca tia* from Quang Tri, Anam, *Chevalier 41256, 41257, 41193*, all represent *Rapanea capitellata* (Wall.) Mez.

PRIMULACEAE

Lysimachia (Tournefort) Linnaeus

Lysimachia decurrens Forst. f. Prodr. 12. 1786; Pax & Knuth in Pflanzenreich 22(IV–237): 296. 1905; Bonati in Lecomte Fl. Gén. Indo-Chine 3: 762. 1930.

Cerium spicatum Lour. Fl. Cochinch. 136. 1790, ed. Willd. 168. 1793, Anamese *cây chăt chăt;* Moore in Journ. Bot. 63: 250. 1925.

" Habitat absque cultura in hortis Cochinchinae." Loureiro's type is preserved in the herbarium of the British Museum and although the genus, because of Loureiro's erroneous description, remained in the limbo of *genera incertae sedis* from 1790 to 1925, Moore (Journ. Bot. 63: 250. 1925) states the solution of the riddle was ridiculously easy with the specimens before one. *Cerium spicatum* Lour. is identical with the older *Lysimachia decurrens* Forst. f. Thus another enigma is solved and *Cerium*, which no botanist had hitherto been able to refer to its proper family, falls as a synonym of *Lysimachia*. Bonati overlooked Moore's note on Loureiro's type and failed to account for *Cerium spicatum* Lour. in his treatment of the Primulaceae of Indo-China (Lecomte Fl. Gén. Indo-Chine 3: 753–764. 1930) although admitting *Lysimachia decurrens* Forst. f.

Androsace (Tournefort) Linnaeus

Androsace umbellata (Lour.) Merr. in Philip. Journ. Sci. 15: 237. 1919.

Drosera umbellata Lour. Fl. Cochinch. 186. 1790, ed. Willd. 232. 1793, Anamese *koŭc tinh thao,* Chinese *ku tsĭm tsào.*

Androsace saxifragifolia Bunge in Mém. Sav. Étr. Acad. St. Pétersb. 2: 127. 1833; Pax & Knuth in Pflanzenreich 22(IV–237): 179. 1905.

" Habitat in China." Planchon (Ann. Sci. Nat. III Bot. 9: 304. 1848) indicated that *Drosera umbellata* Lour. represented some species of *Androsace*. This is clearly the case and, further, *Drosera umbellata* Lour. is unquestionably identical with *Androsace saxifragifolia* Bunge. The species is common and widely distributed, occurring at low altitudes in Kwangtung Province and extending from India to Japan, and southward to Indo-China and Luzon.

PLUMBAGINACEAE

Plumbago (Tournefort) Linnaeus

Plumbago indica Linn. in Stickm. Herb. Amb. 24. 1754, Amoen, Acad. **4**: 133. 1759; Merr. Interpret. Herb. Amb. 414. 1917, Enum. Philip. Fl. Pl. **3**: 275. 1923.

Plumbago rosea Linn. Sp. Pl. ed. 2, 215. 1762; Pellegr. in Lecomte Fl. Gén. Indo-Chine **3**: 752. 1930.

Plumbago coccinea Salisb. Prodr. 122. 1796 (based on *P. rosea* Linn.).

Thela coccinea Lour. Fl. Cochinch. 119. 1790, ed. Willd. 147. 1793, Anamese *xích hoa xà*, Chinese *chè hŏa tân*.

This species was observed by Loureiro in Indo-China, and China. The description clearly applies to the Linnaean species, a common and widely distributed one in the Old World tropics. Loureiro's type is preserved in the herbarium of the British Museum.

Plumbago zeylanica Linn. Sp. Pl. 151. 1753; Pellegr. in Lecomte Fl. Gén. Indo-Chine **3**: 753. 1930.

Thela alba Lour. Fl. Cochinch. 119. 1790, ed. Willd. 147. 1793, Anamese *bach hoa xà*, Chinese *pă hŏa tân*.

"Habitat ambae species inter sepes arundinis Indicae, per quas scandunt tam in Cochinchina, quam in China." The description applies unmistakably to the common Linnaean species. Loureiro's type is preserved in the herbarium of the Paris Museum.

EBENACEAE

Maba J. R. & G. Forster

Maba buxifolia (Rottb.) Pers. Syn. **2**: 606. 1807; Lecomte Fl. Gén. Indo-Chine **3**: 973. 1930.

Pisonia buxifolia Rottb. in Nye Samml. Danske Vidensk. Selsk. Skr. **2**: 536. *pl. 4. f. 2.* 1783.

Ehretia ferrea Willd. Phyt. 4. *pl. 2. f. 2.* 1794.

Ebenoxylum verum Lour. Fl. Cochinch. 613. 1790, ed. Willd. 752. 1793, Anamese *cây mun, ô mouc*, Chinese *ū múen mŏ*.

Ferreola buxifolia Roxb. Pl. Coromandel **1**: 35. *pl. 45.* 1795.

Maba ebenoxylon G. Don Gen. Syst. **4**: 43. 1838 (based on *Ebenoxylon verum* Lour.).

Diospyros vera A. Chev. Cat. Pl. Jard. Bot. Saigon 65. 1919 (based on *Ebenoxylon verum* Lour.).

Diospyros ferrea Bakh. Gard. Bull. Straits Settl. **7**: 162. 1933.

"Habitat vastas sylvas Cochinchinae, maxime prope confinia Cambodiae." The generic description, in its 3 sepals and 3 petals, conforms to *Maba*, not to *Diospyros*. The specific description conforms reasonably well with the widely distributed *Maba buxifolia* Pers. as this species is currently interpreted; this is also apparently the correct disposition of *Ebenus* Rumph. (Herb. Amb. **3**: 1. *pl. 1*) which Loureiro cites as representing his species. I do not consider that Hiern was correct (Trans. Cambr. Philos. Soc. **12**: 122. 1873) in referring *Ebenoxylum verum* Lour. to *Maba elliptica* Forst., a species unknown from Indo-China. Lecomte (Fl. Gén. Indo-Chine **3**: 972. 1930) at the end of *Diospyros* mentions

Ebenoxylum verum Lour. (in a footnote) thus: ". . . cette espèce paraît appartenir au genre *Maba* [*M. elliptica* Forst.] d'après Hiern, *Ebénacées*, p. 122; mais la description étant très incomplète, ce rapprochement doit être accepté avec doute." Loureiro definitely states that he had incomplete material: " sed tunc minus cogitans de Botanica flores non collegi." His description of the leaves and fruits is very definitely *Maba buxifolia* Pers. as currently interpreted; his description of the inflorescences may be ignored as he must have added these data from hearsay or from some source not indicated.

Diospyros Linnaeus

Diospyros decandra Lour. Fl. Cochinch. 227. 1790, ed. Willd. 279. 1793, Anamese *cây thi*,
Chinese *hoâm sí;* Hiern in Trans. Cambr. Philos. Soc. **12**: 160. 1873; Lecomte Fl.
Gén. Indo-Chine **3**: 961. *f. 109.* 1930.

" Habitat frequens in provinciis Borealibus Cochinchinae: ubi fructus exponuntur venales in foro." Loureiro's type is preserved in the herbarium of the British Museum. Hiern gave an amplified description of it in 1873 and very recently Lecomte published a still more extensive one and an excellent illustration. The species is known only from Indo-China and is allied to *D. melanoxylon* Roxb. *Hebenaster* Rumph. (Herb. Amb. **3**: 13. *pl. 6*), cited by Loureiro as representing his species, must be excluded as it apparently represents *Diospyros ebenum* Koen.

Diospyros dodecandra Lour. Fl. Cochinch. 228. 1790, ed. Willd. 280. 1793, Anamese *cày thi
trâm;* Hiern in Trans. Cambr. Philos. Soc. **12**: 264. 1873.
Embryopteris loureiriana G. Don Gen. Syst. **4**: 41. 1838 (based on *Diospyros dodecandra* Lour.).

" Habitat spontanea, cultaque in Cochinchina." In spite of Loureiro's rather detailed description, I am unable to place this species satisfactorily among the 64 species recognized by Lecomte (Fl. Gén. Indo-Chine **3**: 915–972. 1930), yet I suspect that it is there redescribed under some other name. Loureiro notes that it was used for supporting pepper vines, so it would seem that further field work might yield data and material which would render the positive identification of this species possible. Lecomte does not mention Loureiro's species or Don's synonym based upon it.

Diospyros kaki Linn. f. Suppl. 439. 1781; Lour. Fl. Cochinch. 226. 1790, ed. Willd. 278.
1793, Anamese *cây houng*, Chinese *sú xú.*

" Colitur frequenter in China, & Cochinchina, aliisque locis Indiae extra Gangem." Loureiro was correct in his interpretation of this very commonly cultivated species.

Diospyros lobata Lour. Fl. Cochinch. 227. 1790, ed. Willd. 279. 1793, Anamese *cây cây.*
Irvingia lobata A. Cheval. Cat. Pl. Jard. Bot. Saigon 64. 1919 (based on *Diospyros
lobata* Lour.).
Diospyros odoratissima Lecomte in Not. Syst. **4**: 109. *f. 4.* 1928, Fl. Gén. Indo-Chine
3: 938. *f. 105.* 1930.

" Habitat culta, incultaque in Cochinchina." Hiern (Trans. Cambr. Philos. Soc. **12**: 228. 1873) suggested that *Diospyros lobata* Lour. might be a synonym of *D. lotus* Linn. but Loureiro's definite statement that his species had eight stamens would seem to preclude the correctness of this reduction, for *D. lotus* Linn. has 16 stamens. Among the 64 species of the genus admitted by Lecomte (Fl. Gén. Indo-Chine **3**: 915–972. 1930) the characters given by Loureiro best conform to those of *Diospyros odoratissima* Lecomte. It may be noted that Loureiro describes the flowers as axillary and solitary, which is not strictly the case at least with the ♀ flowers of Lecomte's species. The 8-celled ovary, 8 staminodes, conspicuous calyx lobes (whence Loureiro apparently derived his specific name) and edible fruits are significant characters in Lecomte's species, conforming with Loureiro's description. Lecomte does not mention Loureiro's species.

Diospyros loureiriana G. Don Gen. Syst. **4**: 39. 1838; Hiern in Trans. Cambr. Philos. Soc. **12**: 194. 1873, Oliver Fl. Trop. Afr. **3**: 522. 1877 (based on *Diospyros lotus* Lour.).
    Diospyros lotus (non Linn.) Lour. Fl. Cochinch. 226. 1790, ed. Willd. 278. 1793.
    " Habitat agrestis ad litora orae Orientalis Africae." A well-known African species, well represented in various herbaria by ample collections. Hiern has supplied amplified descriptions of it.

Diospyros ? pilosa (Lour.) A. DC. ex Hiern in Trans. Cambr. Philos. Soc. **12**: 265. 1873 (based on *Euclea pilosa* Lour.).
    Euclea pilosa Lour. Fl. Cochinch. 629. 1790, ed. Willd. 773. 1793, Anamese *cây nhaoc.*
    " Habitat in altis sylvis Cochinchinae." A species known only from Loureiro's description, suggested by A. de Candolle (Prodr. **8**: 219. 1844), to belong in *Diospyros* with doubt; de Candolle, however, does not there publish the binomial as credited to him by Hiern. In all essentials the description agrees with the characters of *Diospyros* and I believe the species to be correctly placed as to the genus. It may eventually prove to be the same as *Diospyros pilosella* H. Lecomte (Not. Syst. **4**: 104. 1928, Fl. Gén. Indo-Chine **3**: 932. 1930), but unfortunately the two descriptions are not directly comparable as Lecomte had no flowers and Loureiro did not describe the fruit. Lecomte does not account for *Diospyros pilosa* A. DC. in his treatment of the Ebenaceae (Fl. Gén. Indo-Chine **3**: 915–978. 1930). A specimen listed as being among those received from Loureiro has not been located in the herbarium of the British Museum.

STYRACEAE

Styrax (Tournefort) Linnaeus

Styrax agrestis (Lour.) G. Don Gen. Syst. **4**: 5. 1838 (based on *Cyrta agrestis* Lour.); Perk. in Pflanzenreich **30**(IV–241): 27. 1907; Guillaumin in Lecomte Fl. Gén. Indo-Chine **3**: 980. *f. 113. 1–3.* 1933.
    Cyrta agrestis Lour. Fl. Cochinch. 278. 1790, ed. Willd. 341. 1793, Anamese *cây cau cau.*
    " Habitat in dumetis Cochinchinae." Loureiro's type in the herbarium of the British Museum was examined by Doctor J. Perkins, who gives an amplified description of the species based on this specimen.

SYMPLOCACEAE

Symplocos Jacquin

Symplocos chinensis (Lour.) Druce in Rept. Bot. Exch. Club Brit. Isles **4**: 650. 1917; Merr. in Philip. Journ. Sci. **15**: 252. 1919; Guillaumin in Bull. Soc. Bot. France **71**: 276. 1924, Lecomte Fl. Gén. Indo-Chine **3**: 1030. 1933; Rehd. & Wils. in Journ. Arnold Arb. **8**: 188. 1927 (based on *Myrtus chinensis* Lour.).

Myrtus chinensis Lour. Fl. Cochinch. 313. 1790, ed. Willd. 382. 1793, Chinese *tàn quắt xiong.*

Myrtus sinensis Lour.[55] ex Gomes in Mem. Acad. Sci. Lisb. Cl. Sci. Mor. Pol. Bel.-Let. n.s. 4(1): 29. 1868.

Symplocos sinica Ker in Bot. Reg. 9: *pl. 710.* 1823; Brand in Pflanzenreich 6(IV–242): 34. 1901.

" Habitat inculta prope Cantonem Sinarum." This species is apparently common in the White Cloud hills, practically in the suburbs of Canton. Loureiro's type is preserved in the herbarium of the Paris Museum. *Symplocos sinica* Ker = *S. chinensis* (Lour.) Druce is hardly more than a pubescent form or variety of the variable and widely distributed *S. paniculata* (Thunb.) Miq. (*S. crataegoides* Ham.). In fact Guillaumin (Bull. Soc. Bot. France 71: 276. 1924) reduces *Symplocos crataegoides* Ham. and *S. paniculata* Wall. to *S. chinensis* (Lour.) Druce. With this wide interpretation of specific limits the proper binomial would be *Symplocos paniculata* (Thunb.) Miq. (Ann. Mus. Bot. Lugd.-Bat. 3: 102. 1867) as explained by Rehder & Wilson (Journ. Arnold Arb. 8: 188. 1927). *S. paniculata* Wall. which was *not* based on *Prunus paniculata* Thunb., was not effectively published by Wallich, and has no standing. Loureiro's type is preserved in the herbarium of the Paris Museum.

Symplocos cochinchinensis (Lour.) Moore in Journ. Bot. 52: 148. 1914 (based on *Dicalyx cochinchinensis* Lour.); Guillaumin in Bull. Soc. Bot. France 71: 277. 1924, Lecomte Fl. Gén. Indo-Chine 3: 998. 1933.

Dicalyx cochinchinensis Lour. Fl. Cochinch. 663. 1790, ed. Willd. 816. 1793, Anamese *cây deung bôp;* Brand in Pflanzenreich 6(IV–242): 90. 1901 (sp. dub.).

Symplocos ferruginea Roxb. Hort. Beng. 40. 1814, *nomen nudum,* Fl. Ind. ed. 2, 2: 542. 1832; Brand in Pflanzenreich 6(IV–242): 40. 1901.

Dicalyx javanicus Blume Bijdr. 1117. 1826.

Symplocos javanica Kurz in Journ. As. Soc. Bengal 40(2): 64. 1871.

" Habitat in sylvis montanis Cochinchinae." Moore (Journ. Bot. 52: 148. 1914) gives an amplified description of Loureiro's type which is preserved in the herbarium of the British Museum. He considered it to be a distinct species allied to *Symplocos floridissima* Brand. I personally examined the type in 1930 and in my opinion Guillaumin (Bull. Soc. Bot. France 71: 277. 1924) is correct in accepting Loureiro's binomial as representing the widely distributed and variable Asiatic form commonly known as *Symplocos ferruginea* Roxb., or as *S. javanica* Kurz. He notes (page 274) that Chevalier in 1919 also transferred *Dicalyx cochinchinensis* Lour. to *Symplocos*, overlooking Moore's transfer of the same name five years earlier, but that Chevalier was wrong in reducing *Symplocos dung* Eberh. & Duby to *S. cochinchinensis* (Lour.) Moore. *Arbor rediviva* Rumph. (Herb. Amb. 3: 165. *pl. 104),* cited by Loureiro as representing his species, must be excluded as it represents a species of *Elaeocarpus*, perhaps *E. rumphii* Merr. On this Rumphian figure, but without reference to Loureiro, Rafinesque (Sylva Tellur. 154. 1838) based the genus *Ayparia*, with a single species *A. crenata* Raf. *Ayparia*, currently placed as a synonym of *Symplocos*, is thus really a synonym of *Elaeocarpus*.

Symplocos dung Eberh. & Duby in Agron. Colon. 1: 79. 1913; Guillaumin in Bull. Soc. Bot. France 71: 278. 1924, Lecomte Fl. Gén. Indo-Chine 3: 1009. 1933.

[55] A Loureiro herbarium name here first published by Gomes.

Decadia aluminosa Lour. Fl. Cochinch. 315. 1790, ed. Willd. 385. 1793, Anamese *cây deung se;* Moore in Journ. Bot. **52**: 146. 1914, **63**: 284. 1925, non *Symplocos aluminosa* Brand.

? Myrtus zeylanica (non Linn.) Lour. Fl. Cochinch. 312. 1790, ed. Willd. 382. 1793, Anamese *sim rûng nho.*

For *Decadia aluminosa* Loureiro states: " Habitat in sylvis Cochinchinae." His type is preserved in the herbarium of the British Museum, and Moore has published a long note on it, concluding when he first examined it in 1914 (Journ. Bot. **52**: 146–148. 1914) that it represented a species near *Symplocos spicata* Roxb., perhaps *S. syringoides* Brand; but in 1925 (op. cit. **63**: 284) on re-examination of it, he was convinced that his earlier views were erroneous and that the relationships of *Decadia aluminosa* remained doubtful. Guillaumin (Bull. Soc. Bot. France **71**: 281. 1924), influenced by Moore's earlier opinion, erroneously reduced *Decadia aluminosa* Lour. to *Symplocos laurina* (Retz.) Wall. I examined Loureiro's type in 1930; it is beautifully matched by *Clemens 3686, 3825, 3951,* from Hue and Tourane, a small tree, common in thickets. These specimens represent neither *Symplocos spicata* Roxb. nor *S. laurina* Wall., but they are safely referable to *Symplocos dung* Eberh. & Duby, which is common and widely distributed in Indo-China. *Arbor aluminosa* Rumph. (Herb. Amb. **3**: 160. *pl. 100*), cited by Loureiro as representing his species and whence he took his specific name, is either *Symplocos javanica* (Blume) Kurz = *S. cochinchinensis* (Lour.) Moore, or *S. syringoides* Brand. The Rumphian description and illustration do not typify Loureiro's species, which was based on an actual specimen. Loureiro's specific name would seem to be invalidated in *Symplocos* by *S. aluminosa* (Blume) Brand (Pflanzenreich 6(IV–242): 35. 1901) which was specifically based on *Dicalix aluminosus* Blume (1826) and *Dicalyx aluminosus* Hassk. (1844), excluding Blume's reference to Loureiro, in spite of the fact that *Decadia aluminosa* Lour. was the name-bringing synonym of Blume's species. Some taxonomists doubtless would consider Brand to be in error in his treatment of *Symplocos aluminosa*, and would adopt this binomial for the species Loureiro actually described. For *Myrtus zeylanica* Loureiro states: " Habitat in dumetis Cochinchinae." I suspect this to be the same as *Decadia aluminosa* Lour. with erroneous fruit characters added, these perhaps taken from some published description of the Linnaean species, which is a *Eugenia*. The alternate serrate leaves eliminate any of the Asiatic Myrtaceae; the 4-merous flowers, if Loureiro be correct in his observation of this character, eliminate what I would consider to be the only other possibility, *Photinia* in the Rosaceae. I do not think that Guillaumin's suggestion (Bull. Soc. Bot. France **71**: 287. 1924) that perhaps some lecythidaceous species near *Careya* was represented, is well taken.

Symplocos laurina (Retz.) Wall.[56] List no. 4416. 1830; Rehd. & Wils. in Sargent Pl. Wils. 2: 594. 1916; Guillaumin in Bull. Soc. Bot. France **71**: 281. 1924, et in Lecomte Fl. Gén. Indo-Chine **3**: 1005. 1933.

[56] Wallich's binomial is not definitely based on *Myrtus laurinus* Retz. He cites as synonyms "Myrtus laurina Hb. Madr. e Colombo et ex itinere Travancoreaei," "Myrtus laurina Hb. Wight" and "Eugenia laurina Rottl. Hb. Heyn." These doubtless represent the same species as *Myrtus laurina* Retz., which was based on a Ceylon specimen from Koenig and on *Laurus serrata, floribus spicatis* Burm. (Thesaurus 139. *pl. 62.* 1737). If Rehder & Wilson's interpretation of *Symplocos laurina* Wall., as based on *Myrtus laurina* Retz. be not accepted, *Symplocos loba* Ham. (1825) would be the oldest valid name, as it antedates the actual valid publication of *Symplocos spicata* Roxb. by 7 years. Loureiro's specific name of 1790 is invalidated by *S. cochinchinensis* Moore.

Myrtus laurina Retz. Obs. **4**: 26. 1786.

Symplocos spicata Roxb. Hort. Beng. **40**. 1814, *nomen nudum*, Fl. Ind. ed. 2, **2**: 541. 1832; Brand in Pflanzenreich 6(IV–242): 39. 1901.

Drupatris cochinchinensis Lour. Fl. Cochinch. 314. 1790, ed. Willd. 385. 1793, Anamese *cây deung*.

"Habitat in altis sylvis Cochinchinae." Gürke (Engler & Prantl Nat. Pflanzenfam. 4(1): 168. 1890) placed *Drupatris* as a synonym of *Symplocos*, although Brand in his monographic treatment of the Symplocaceae (Pflanzenreich 6(IV–242): 1–100. 1901) does not mention it. I believe this to be the correct disposition of Loureiro's genus, the very slight discrepancies between his description and the characters of *Symplocos* being due to faulty observation on Loureiro's part. It may be noted that for *Decadia* Lour. = *Symplocos* Jacq., the local name *cây deung se* is given; while that for *Drupatris* is *cây deung*, this in a measure confirming this reduction of the latter. Guillaumin (Bull. Soc. Bot. France **71**: 281. 1924) reduced *Decadia aluminosa* Lour. to *Symplocos laurina* Wall., but I do not agree in this reduction, considering *Decadia aluminosa* Lour. to be a synonym of *Symplocos dung* Eberhardt & Duby. Loureiro's description of *Drupatris cochinchinensis* conforms sufficiently well with the characters of the common, widely distributed, and the somewhat variable *Symplocos laurina* Wall. to warrant this reduction of the former to the latter; Rehder & Wilson (Sargent Pl. Wils. **2**: 594. 1916) cite no less than 23 synonyms for *Symplocos laurina* Wall. A specimen listed as being among those received from Loureiro has not been located in the herbarium of the British Museum.

OLEACEAE

Osmanthus Loureiro

Osmanthus fragrans Lour. Fl. Cochinch. 29. 1790, ed. Willd. 35. 1793, Anamese *hoa mouc tây*, Chinese *mŏ sĩ hõa*, *guéi hõa;* Gagnep. in Lecomte Fl. Gén. Indo-Chine **3**: 1062. 1933.

Olea fragrans Thunb. Fl. Jap. 18. *pl. 2.* 1784.

"Habitat in hortis Cochinchinae: frequentius in China, puto, quod etiam agrestis." This is the type of the genus. *Olea fragrans* Thunb. (Fl. Jap. 18. *pl. 2.* 1784), which is identical with Loureiro's species as noted by Willdenow, was published independently. Loureiro's type is preserved in the herbarium of the British Museum.

Linociera [57] Swartz

Linociera sp.

Cylindria rubra Lour. Fl. Cochinch. 69. 1790, ed. Willd. 87. 1793, Anamese *cây lo nge;* Moore in Journ. Bot. **63**: 247. 1925.

"Habitat in sylvis Cochinchinae." *Cylindria* remained a genus of entirely doubtful status until 1925, when Moore published a critical note on it based on an examination of Loureiro's type preserved in the herbarium of the British Museum; he notes Dryander's detailed description preserved in the Solander manuscript (1: 317), Dryander concluding

[57] *Linociera* Swartz (1791), conserved name, Vienna Code; older ones are *Mayepea* Aublet, 1775 (= *Majepea* O. Kuntze 1892), *Thouinia* Linnaeus f. (1781), *Freyeria* Scopoli (1777), *Ceranthus* Schreber (1789), and *Cylindria* Loureiro (1790).

that *Olea* was represented. Moore, who gives additional descriptive data, states that the species, represented by somewhat fragmentary material, cannot be matched among the other collections at the British Museum. I am unable to place it among the nine species admitted by Gagnepain (Lecomte Fl. Gén. Indo-Chine **3**: 1066–1074. 1933). *Blimbingum sylvestre* Rumph. (Herb. Amb. **4**: 138. *pl. 73*), cited by Loureiro as representing his species, must be excluded as it represents *Elaeocarpus oppositifolius* Miq. = *Aceratium oppositifolium* DC.

Olea (Tournefort) Linnaeus

Olea brachiata (Lour.) Merr. in Lingnan Sci. Journ. **5**: 147. 1927 (based on *Tetrapilus brachiatus* Linn.).

Tetrapilus brachiatus Lour. Fl. Cochinch. 611. 1790, ed. Willd. 750. 1793, Anamese *cây câu;* DC. Prodr. **8**: 299. 1844.

Pachyderma javanicum Blume Bijdr. 682. 1826.

Tetrapilus axillaris Raf. Sylva Tellur. 138. 1838 (based on *Tetrapilus brachiatus* Lour.).

Olea maritima Wall. List. no. 2813. 1831, *nomen nudum;* G. Don Gen. Syst. **4**: 49. 1838; C. B. Clarke in Hook. f. Fl. Brit. Ind. **3**: 612. 1882; Gagnep. in Lecomte Fl. Gén. Indo-Chine **3**: 1078. 1933.

" Habitat in dumetis Cochinchinae." Loureiro's species is represented by *McClure 7398* and *Clemens 3932*, from the classical locality and I believe *Olea maritima* Wall. to represent the same species. This interpretation is based on Loureiro's description, excluding the fruit characters indicated by him; the latter characters " bacca . . . polysperma " must have been based on material from some other species or an erroneous observation. In the next sentence he mentions the seeds as few, inconsistent with his description of the fruit. *Clemens 3932* from Hue, the classical locality, has both entire and sparingly toothed leaves.

Ligustrum (Tournefort) Linnaeus

Ligustrum indicum (Lour.) comb. nov.

Phillyrea indica Lour. Fl. Cochinch. 19. 1790, ed. Willd. 23. 1793, Anamese *cây râm*.

Olea microcarpa Vahl Enum. **1**: 43. 1804; DC. Prodr. **8**: 287. 1844 (based on *Phillyrea indica* Lour.).

Ligustrum nepalense Wall. Pl. As. Rar. **3**: 44. *pl. 270*. 1832; Gagnep. in Lecomte Fl. Gén. Indo-Chine **3**: 1080. *f. 123. 1-4*. 1933.

" Habitat in Cochinchina tam culta, quam agrestis." Some authors have considered that Loureiro's description appertains to *Olea maritima* Wall. = *Olea brachiata* (Lour.) Merr., but *Ligustrum* is clearly indicated by the terminal inflorescences; C. B. Clarke (Hook. f. Fl. Brit. Ind. **3**: 613. 1882) expressed the opinion that *Ligustrum* was represented rather than *Olea*. The species is represented by *Clemens 3705*, and *Squires 103, 346*, from near the classical locality, Hue. The " bacca-parva, rotunda " indicates Wallich's species, as interpreted by Gagnepain, rather than *L. robustum* Blume. Neither Mansfeld (Vorarbeiten zu einer Monographie der Gattung Ligustrum. Bot. Jahrb. **59**: Beibl. **132**: 19–75. 1924) nor Gagnepain accounts for Loureiro's species or Vahl's synonym based upon it.

Ligustrum sinense Lour. Fl. Cochinch. 19. 1790, ed. Willd. 23. 1793.

Faulia sinensis Raf. Fl. Tellur. **2**: 84. 1837 (based on *Ligustrum sinense* Lour.).

" Habitat agreste prope Cantonem Sinarum." Loureiro's type is preserved in the herbarium of the Paris Museum. This is a well-known species common in thickets near Canton. Local names on recently collected specimens are *shan tsz kah* and *shan lap shii.*

Jasminum (Tournefort) Linnaeus

Jasminum nervosum Lour. Fl. Cochinch. 20. 1790, ed. Willd. 25. 1793, Anamese *cây vang.*
Jasminum anastomosans Wall. List no. 2863. 1831, *nomen nudum;* de Candolle Prodr.
8: 305. 1844; Kobuski in Journ. Arnold Arb. **13**: 169. 1932; Gagnep. in Lecomte Fl.
Gén. Indo-Chine **3**: 1049. 1933.

"Habitat inter sepes in Cochinchinae." Loureiro misinterpreted the branchlets and distichous simple leaves as imparipinnate leaves, which probably is the chief reason why his otherwise well described species has not been well understood. It is represented by rather numerous collections, including *McClure 7308, 7357, 7297* and *Clemens 3136* from near Hue, the classical locality, and *McClure 9193, 9530, 9792* from Hainan. De Pirey's specimen of *day vang hau,* Chevalier *40210,* is the species as here interpreted; his specimen of *day vang trau* is *Jasminum amplexicaule* Ham. (*J. undulatum* Ker); Gagnepain does not account for Loureiro's species. Kobuski [58] briefly discusses Loureiro's species under *Jasminum pentaneurum* Hand.-Maz., and more extensively under *J. anastomosans* Wall. (p. 169), placing it as a doubtful synonym of the latter. He seemed to be influenced chiefly by Loureiro's erroneous description of the leaves as pinnate in his failure to understand Loureiro's really good description. To me Loureiro's error is readily understandable, and I should more seriously question the statement "pedunculis polyfloris" which is not truly descriptive of *J. anastomosans* Wall. The several collections from near Hue that I have examined are so uniform that I unhesitatingly accept Loureiro's name for the form that Kobuski placed under Wallich's species. He questions my identification of *Nyctanthes sambac* Blanco, the basis of *Jasminum blancoi* Hassk., with *Jasminum sambac* Ait. because Blanco (Fl. Filip. 9. 1837) described the leaves as compound: "Hojas opuestas aladas sin impar en numero de tres pares." Here Blanco made exactly the same error that Loureiro did, but in the second edition of the Flora de Filipinas (6. 1845) he eliminated the above erroneous statement and described the leaves of *Nyctanthes sambac* as simple. *Nyctanthes sambac* Linn. was correctly interpreted by Blanco; it is *Jasminum sambac* Ait., and *Jasminum blancoi* Hassk. is a synonym of it. Both Loureiro and Blanco not infrequently described distichously arranged simple leaves as pinnate leaves, and occasionally distichously arranged leaves and branchlets as bipinnate leaves!

Jasminum officinale Linn. Sp. Pl. 7. 1753; Lour. Fl. Cochinch. 20. 1790, ed. Willd. 24.
1793, Chinese *sú hán hōa.*

"Habitat agreste prope Cantonem Sinarum." It is impossible to determine from Loureiro's description whether he had specimens of *Jasminum officinale* Linn. or of the allied *J. grandiflorum* Linn., both of which occur in China as cultivated plants. It is improbable that Loureiro's specimens were from wild plants.

Jasminum sambac (Linn.) Ait. Hort. Kew. **1**: 8. 1789; Gagnep. in Lecomte Fl. Gén. Indo-
Chine **3**: 1054. 1933.
Nyctanthes sambac Linn. Sp. Pl. 6. 1753; Lour. Fl. Cochinch. 21. 1790, ed. Willd. 25.
1793, Anamese *hoa lài,* Chinese *mŏ lî hōa.*

[58] Kobuski, E. Synopsis of the Chinese species of Jasminum. Journ. Arnold Arb. **13**: 145–179. 1932.

Nyctanthes grandiflora Lour. Fl. Cochinch. 21. 1790, ed. Willd. 26. 1793, Anamese *hoa lài taù*, Chinese *tá mo li*.

For *Nyctanthes sambac* Loureiro states: " Habitat in hortis Cochinchinae, & Chinae." The description applies to this very commonly cultivated species. *Flos manorae* Rumph. (Herb. Amb. **5**: 52. *pl. 30*), cited by Loureiro, after Linnaeus, as a synonym, is correctly placed. For *Nyctanthes grandiflora* Loureiro states: " Habitat culta tam in Cochinchina, quam in China." Roxburgh placed this as a synonym of *Jasminum arborescens* Roxb. (Fl. Ind. ed. 2, **1**: 95. 1832), but I have no record of his species as a Chinese one. G. Don (Gen. Syst. **4**: 59. 1838) correctly placed *Nyctanthes grandiflora* Lour. as a synonym of *Jasminum sambac* Ait. var. *trifoliatum* (Vahl) G. Don, but Roemer & Schultes, Syst. **1**: 78. 1817, had already made this combination 21 years earlier. It is the form with some of the leaves in whorls of three's, described by Loureiro as: " Folia stellata, terna, oppositaque."

LOGANIACEAE

Strychnos Linnaeus

Strychnos ignatii Berg. Mat. Med. **1**: 146. 1778.

Ignatia amara Linn. f. Suppl. 149. 1781, pro parte.

Ignatiana philippinica Lour. Fl. Cochinch. 126. 1790, ed. Willd. 155. 1793, Anamese *hôt dâut giô; yasug* (Philippine).

" Habitat agrestis in insulis Philippinis: inde in Cochinchinam, & aliorsum translata." *Ignatiana philippinica* Lour. is essentially a new binomial for *Ignatia amara* Linn. f., the generic name *Ignatiana* being apparently a misprint of *Ignatia* Linn. f. The species is known only from the central and southern Philippines, its most common Bisayan name being *igasud*, given by Loureiro as *yasug*. The seeds have had a place in oriental commerce for many centuries and it is improbable that Loureiro saw more than the seeds; his description otherwise was probably taken from Linnaeus f. or from Bergius.

Strychnos nux vomica Linn. Sp. Pl. 189. 1753; Lour. Fl. Cochinch. 125. 1790, ed. Willd. 154. 1793, Anamese *cây cu chi*; A. W. Hill in Kew Bull. 183. 1917; Dop in Lecomte Fl. Gén. Indo-Chine **4**: 168. 1914.

" Habitat in sylvis provinciae *Binh Khang*, olim ad regnum Champavam, nunc ad Cochinchinam pertinentis." Loureiro in all probability had specimens representing the Linnaean species, which occurs wild in Indo-China; see J.M.H. (Kew Bull. 238. 1919) who records the true *S. nux vomica* Linn. from several localities in Indo-China. It is possible that Loureiro may have had material representing the allied *S. nux blanda* A. W. Hill.

Strychnos umbellata (Lour.) Merr. in Philip. Journ. Sci. **15**: 252. 1919.

Cissus umbellata Lour. Fl. Cochinch. 84. 1790, ed. Willd. 106. 1793, Chinese *yòng cŏ loi*.

Strychnos paniculata Champ. in Hook. Journ. Bot. Kew Gard. Miscel. **5**: 56. 1853.

" Habitat Cantone Sinarum spontanea." The reduction to *Strychnos* follows Planchon's suggestion (DC. Monog. Phan. **5**: 626. 1887). Loureiro's description agrees entirely with the characters of *Strychnos paniculata* Champ., which occurs near Canton and which is characterized by its 4-merous flowers.

Buddleia Houstoun

Buddleia asiatica Lour. Fl. Cochinch. 72. 1790, ed. Willd. 90. 1793, Anamese *sâu dâu chuot;* Britten ex Moore in Journ. Bot. **63**: 248. 1925.

"Habitat agrestis in Cochinchina." This is a widely distributed and well-known species. A specimen from Loureiro is preserved in the herbarium of the British Museum.

GENTIANACEAE

Gentiana (Tournefort) Linnaeus

Gentiana loureiri (G. Don) Griseb. in DC. Prodr. **9**: 108. 1845 (based on *Gentiana aquatica* Lour.); Dop & Gagnep. in Lecomte Fl. Gén. Indo-Chine **4**: 185. 1914.

Ericala loureiri G. Don Gen. Syst. **4**: 192. 1838 (based on *Gentiana aquatica* Lour.).

Gentiana indica Steud. Nomencl. ed. 2, 1: 674. 1840 (based on *Gentiana aquatica* Lour.).

Gentiana aquatica (non Linn.) Lour. Fl. Cochinch. 172. 1790, ed. Willd. 214. 1793, Chinese *xí kām xiong.*

"Habitat in locis humidis prope Cantonem Sinarum." This species is not uncommon in the vicinity of Canton and is also known from Indo-China. It is allied to *Gentiana squarrosa* Ledeb.

Limnanthemum S. G. Gmelin

Limnanthemum hydrophyllum (Lour.) Griseb. Gen. Sp. Gent. 348. 1838 (based on *Menyanthes hydrophyllum* Lour.); Dop & Gagnep. in Lecomte Fl. Gén. Indo-Chine **4**: 195. 1914.

Menyanthes hydrophyllum Lour. Fl. Cochinch. 105. 1790, ed. Willd. 129. 1793, Anamese *cây raong tlòn lá.*

Villarsia hydrophylla Roem. & Schult. Syst. **4**: 181. 1819 (based on *Menyanthes hydrophyllum* Lour.).

Nymphodes hydrophyllum O. Ktz. Rev. Gen. Pl. 429. 1891 (based on *Menyanthes hydrophyllum* Lour.).

"Habitat spontanea loca aquosa Cochinchinae." The species is definitely known only from Indo-China and Dop and Gagnepain have given a detailed description of it, based on numerous collections. *Robinson 1170,* cited by them under both *L. hydrophyllum* and *L. indicum,* was reported by Ostenfeld to represent a variety of *Limnanthemum indicum* Griseb., and from Loureiro's description it seems scarcely possible to separate his species from *L. indicum;* both species, as interpreted by Dop & Gagnepain, l.c., occur in Indo-China.

APOCYNACEAE

Melodinus J. R. & G. Forster

Melodinus cochinchinensis (Lour.) comb. nov.

Oncinus cochinchinensis Lour. Fl. Cochinch. 123. 1790, ed. Willd. 152.. 1793, Anamese *cây deom.*

Melodinus sylvaticus Pitard in Lecomte Fl. Gén. Indo-Chine **3**: 1108. 1933.

Theophrasta cochinchinensis Spreng. Syst. **1**: 671. 1825 (based on *Oncinus cochinchinensis* Lour.).

"Habitat in sylvis Cochinchinae." In Index Kewensis this is reduced to *Melodinus monogynus* Roxb., but I consider it to be represented by *Clemens 3427* from Hue, Loureiro's classical locality; this is the Clemens plant cited by Pitard in the original description of *Melodinus sylvaticus.* Pitard does not account for Loureiro's species or Sprengel's synonym based upon it.

Carissa [59] Linnaeus

Carissa africana A. DC. Prodr. **8**: 332. 1844; Stapf in Dyer Fl. Trop. Afr. **4**(1): 92. 1902 (based on *C. carandas* Lour.).

Carissa carandas (non Linn.) Lour. Fl. Cochinch. 124. 1790, ed. Willd. 153. 1793.

"Planta a me descripta nascitur agrestis in litore Africae Orientalis." Loureiro's type is preserved in the herbarium of the Paris Museum, where it was examined by A. de Candolle, who states that the species is more closely allied to *C. xylopron* Thouars than to *C. carandas* Linn. The species is as yet known only from the original collection and description.

Plumeria [60] (Tournefort) Linnaeus

Plumeria acuminata Ait. Hort. Kew. ed. 2, **2**: 70. 1811.

Plumeria acutifolia Poir. in Lam. Encycl. Suppl. **2**: 667. 1812; Pitard in Lecomte Fl. Gén. Indo-Chine **3**: 1160. 1933.

Plumeria obtusa (non Linn.) Lour. Fl. Cochinch. 117. 1790, ed. Willd. 144. 1793, Anamese *hoa sú tláng.*

"Habitat in Cochinchina, & China, cultura facilis, odore, & colore floris grata." The description is ample and applies unmistakably to the commonly cultivated temple flower. *Flos convolutus* Rumph. (Herb. Amb. **4**: 85. *pl. 38*) is correctly placed as a synonym. The original spelling of the generic name is here retained, although some authors use the forms *Plumiera* and *Plumieria*, as the genus was named in honor of Charles Plumier.

Catharanthus G. Don

Catharanthus roseus (Linn.) G. Don Gen. Syst. **4**: 95. 1838; Hubb. & Rehd. in Bot. Mus. Leafl. Harvard Univ. **1**: 4. 1932.

Vinca rosea Linn. Syst. ed. 10, 944. 1759; Lour. Fl. Cochinch. 118. 1790, ed. Willd. 146. 1793, Anamese *hoa hai dàng;* Pitard in Lecomte Fl. Gén. Indo-Chine **3**: 1140. 1933.

Lochnera rosea Reichb. Conspect. 134. 1828.

Hottonia litoralis Lour. Fl. Cochinch. 105. 1790, ed. Willd. 128. 1793 (*littoralis*), Anamese *cay bô bô bien.*

Erythraea cochinchinensis Spreng. **1**: 580. 1825 (based on *Hottonia litoralis* Lour.).

Centaurium cochinchinense Druce in Rept. Bot. Exch. Club Brit. Isles **4**: 613. 1917 (based on *Erythraea cochinchinensis* Spreng, *Hottonia litoralis* Lour.).

Ammocallis rosea Small Fl. Southeast. U. S. 936. 1903.

For *Vinca rosea* Loureiro states: "Habitat in Cochinchina, & China, tam agretis, quam culta." Loureiro correctly interpreted the very common and widely distributed

[59] *Carissa* Linnaeus (1767), conserved name, Vienna Code; older ones are *Arduina* Miller (1760), Linnaeus (1767), and *Carandas* Adanson (1763).

[60] Often spelled *Plumiera*, but the original form is here retained; see Sprague, Kew Bull. 49. 1929.

Linnaean species; a specimen from him is preserved in the herbarium of the British Museum. For *Hottonia litoralis* he states: " Habitat in litore arenoso portus Eo, prope Huaeam metropolim Cochinchinae." In spite of minor discrepancies in the description this can scarcely be other than the somewhat dwarfed form of *Catharanthus roseus* G. Don that frequently occurs in abundance in sandy soil near the sea. *Lochnera* Reichb. (1828) is invalidated by *Lochneria* Scop. (1771).

Tabernaemontana (Plumier) Linnaeus

Tabernaemontana bovina Lour. Fl. Cochinch. 118. 1790, ed. Willd. 145. 1793, Anamese *cây sùng bò;* Dub. & Eberh. in Ann. Sci. Agron. IV **2**: 135. *fig.* 1913; Pitard in Lecomte Fl. Gén. Indo-Chine **3**: 1159. 1933.

Tabernaemontana tonkinensis Pitard in Lecomte Fl. Gén. Indo-Chine **3**: 1154. 1933.

" Habitat inculta in locis planis Cochinchinae." The description is clearly that of *Tabernaemontana* and I refer to this species *Clemens 3353, 4055, 4056,* from Tourane and vicinity, although Pitard identified all three as representing *Tabernaemontana tonkinensis* Pitard. Loureiro's description of the fruits as " folliculi 2, horizontales, divaricatis, recurvi, breves, ventricosi, acuminati, nec torulosi " conforms to the fruit characters of the specimens cited above. De Pirey's specimen of *sung trau, Chevalier 41172,* in flower, conforms to the characters of this species rather than with those of *T. bufalina* Lour. which is described as having geminate 1-flowered peduncles.

Tabernaemontana bufalina Lour. Fl. Cochinch. 117. 1790, ed. Willd. 145. 1793, Anamese *cây sùng tlâu;* Pitard in Lecomte Fl. Gén. Indo-Chine **3**: 1160. 1933.

" Habitat in dumetis Cochinchinae." The description is definite and applies to *Tabernaemontana* as that genus is currently interpreted. The species is distinguished by Loureiro from his *T. bovina* by its having 1-flowered peduncles and somewhat elongated, torulose follicles. I have seen no material from Indo-China that conforms entirely with the description. *Capsicum sylvestre* Rumph. (Herb. Amb. **4**: 133. *pl. 67*), cited by Loureiro as representing his species, must be excluded as it represents the Moluccan *Tabernaemontana capsicoides* Merr., although it is possible that Loureiro took his fruit characters from this illustration. Pitard gives only a brief description compiled from Loureiro.

Rauwolfia (Plumier) Linnaeus

Rauwolfia verticillata (Lour.) Baill. in Bull. Soc. Linn. Paris **1**: 768. 1888 (based on *Dissolena verticillata* Lour.).

Dissolena verticillata Lour. Fl. Cochinch. 138. 1790, ed. Willd. 171. 1793, Chinese *mat sa;* Baillon Adansonia **4**: 378. *pl. 12.* 1864.

Cerbera chinensis Spreng. Syst. **1**: 642. 1825 (based on *Dissolena verticillata* Lour.).

Ophioxylon chinense Hance in Journ. Bot. **3**: 380. 1865.

Rauwolfia chinensis Hemsl. in Journ. Linn. Soc. Bot. **26**: 95. 1889.

" Habitat in provincia Cantoniensi Sinarum." Loureiro's type is preserved in the herbarium of the Paris Museum. Baillon (Bull. Soc. Linn. Paris **1**: 768. 1888) examined this specimen, and at an earlier date gave an excellent description and illustration of the species (Adansonia **4**: 378. *pl. 12.* 1864). A. de Candolle (Prodr. **8**: 318. 1844) quotes from the manuscript of his father, who had seen Loureiro's type in Paris, to the effect that *Dissolena* belonged in the Verbenaceae, and Bentham (Benth. & Hook. f. Gen. Pl. **2**: 690) followed him in this erroneous disposition of it. Loureiro's species is manifestly the same as *Rauwolfia chinensis* Hemsl., as Baillon notes that *Ophioxylon chinense* Hance, the name-bringing synonym of *R. chinensis* Hemsl., is the same as *Dissolena verticillata* Lour. = *R. verticillata* Baill. The reference given above is an earlier publication of the latter binomial than that indicated in Index Kewensis Suppl. 1: 359. 1906.

Cerbera Linnaeus

Cerbera manghas Linn. Sp. Pl. 208. 1753.

Cerbera odollam Gaertn. Fruct. **2**: 193. *pl. 124*. 1791; Pitard in Lecomte Fl. Gén. Indo-Chine **3**: 1137. 1933.

Cerbera salutaris Lour. Fl. Cochinch. 136. 1790, ed. Willd. 168. 1793, pro majore parte, Anamese *cây múop sác*.

"Habitat prope maris litora in Cochinchina." Loureiro's description, particularly that of the fruit, applies in part to the very common littoral *Cerbera manghas* Linn., more commonly known as *C. odollam* Gaertn.; it may have been modified somewhat from Rumphius' illustration. The characters are in small part those of *Scaevola frutescens* (Mill.) Krause, and Loureiro's specimen in the herbarium of the British Museum represents the latter species; the description essentially is of *Cerbera*.

Ichnocarpus [61] R. Brown

Ichnocarpus volubilis (Lour.) Merr. in Philip. Journ. Sci. **21**: 506. 1922 (based on *Gardenia volubilis* Lour.); Enum. Philip. Fl. Pl. **3**: 336. 1923.

Gardenia volubilis Lour. Fl. Cochinch. 148. 1790, ed. Willd. 184. 1793, Chinese *xang lan thân*.

Ichnocarpus ovatifolius A. DC. Prodr. **8**: 435. 1844; Pitard in Lecomte Fl. Gén. Indo-Chine **3**: 1254. 1933.

Gardenia sinensis Lour.[62] ex Gomes in Mem. Acad. Sci. Lisb. Cl. Sci. Pol. Mor. Bel.-Let. n.s. **4**(1): 28. 1868.

"Habitat inculta apud Sinas, extra suburbia Cantoniensis." A specimen from Loureiro is preserved in the herbarium of the Paris Museum and was identified by Pierre as *Ichnocarpus* sp. and by Gagnepain in August, 1920, as *Ichnocarpus ovatifolius* A. DC. This is the only species of the genus known from Kwangtung Province and it may not prove to be distinct from the older *I. frutescens* (Linn.) R. Br. Here is a very definite case where Loureiro described a species from a flowering specimen and added the fruit characters: "bacca infera, parva, subrotunda, 2-locularis, polysperma; seminibus paucis, rotundis," to make his description conform to the characters of the genus in which he erroneously placed it.

Nerium Linnaeus

Nerium indicum Mill. Gard. Dict. ed. 8, no. 2. 1768.

Nerium oleander (non Linn.) Lour. Fl. Cochinch. 115. 1790, ed. Willd. 141. 1793, Anamese *cây daò lê*.

Nerium odorum Soland. in Ait. Hort. Kew. **1**: 297. 1789; Pitard in Lecomte Fl. Gén. Indo-Chine **3**: 1195. 1933.

[61] *Ichnocarpus* R. Brown (1809), conserved name, Vienna Code; an older one is *Quirivelia* Poir. (1804).

[62] A Loureiro herbarium name here first published by Gomes.

" Habitat in Cochinchina, & China, multisque aliis Asiae locis." *Oleander sinicus* Rumph. (Herb. Amb. **7**: 15. *pl. 9. f. 1*) and *Belluta areli* Rheed. (Hort. Malabar. **9**: 1. *pl. 1-2*), cited by Loureiro as illustrating the species, are correctly placed as synonyms. Although *Nerium odorum* Soland. = *N. indicum* Mill. is usually maintained as a species distinct from *N. oleander* Linn., it is not entirely clear that more than one species is represented. A specimen from Loureiro is preserved in the herbarium of the British Museum.

Strophanthus De Candolle

Strophanthus divaricatus (Lour.) Hook. & Arn. Bot. Beechey's Voy. 199. 1836 (based on *Pergularia divaricata* Lour.).

Pergularia divaricata Lour. Fl. Cochinch. 169. 1790, ed. Willd. 210. 1793, Chinese *hú muôn*.

Nerium chinense Hunter in Roxb. Hort. Beng. 84. 1814, *nomen nudum*, Fl. Ind. ed. 2, **2**: 9. 1832.

Periploca divaricata Spreng. Syst. **1**: 836. 1825 (based on *Pergularia divaricata* Lour.).

Emericia divaricata Roem. & Schult. Syst. **4**: 401. 1819 (based on *Pergularia divaricata* Lour.).

Vallaris divaricata G. Don Gen. Syst. **4**: 79. 1838 (based on *Pergularia divaricata* Lour.).

Strophanthus divergens Grah. in Edinb. New Philos. Journ. 177. 1827.

Strophanthus chinensis G. Don Gen. Syst. **4**: 85. 1838.

Streptocaulon divaricatum G. Don op. cit. 162 (based on *Pergularia divaricata* Lour.).

Loureiro states: " Habitat inculta apud Sinas." His description applies unmistakably to the single species of *Strophanthus* known from Kwangtung Province and one that is distinctly common there; the statement that the seeds are naked, i.e., without a coma, is an error, probably due to Loureiro's having observed seeds from which the coma had fallen. It may be noted that although the binomial *Strophanthus divaricatus* Wall. (List no. 1642. 1829) is older than *S. divaricatus* Hook. & Arn., yet it has no standing, being merely a *nomen nudum*, and no description of it appeared until two years after Hooker and Arnott's transfer of Loureiro's name to *Strophanthus*, when G. Don (Gen. Syst. **4**: 85. 1838) supplied the description.

Strophanthus caudatus (Burm. f.) Kurz in Journ. As. Soc. Beng. **46**(2): 257. 1877; Pitard [63] in Lecomte Fl. Gén. Indo-Chine **3**: 1199. 1933.

Echites caudata Burm. f. Fl. Ind. 68. *pl. 26*. 1768.

Nerium scandens (non Linn.) Lour. Fl. Cochinch. 116. 1790, ed. Willd. 143. 1793, Anamese *cây bòi boi*.

Strophanthus scandens Roem. & Schult. Syst. **4**: 412. 1819 (based on *Nerium scandens* Lour.).

Strophanthus dichotomus DC. var. *cochinchinensis* Ker in Bot. Reg. **6**: 469. 1820 (based on Loureiro's specimen in herb. Banks).

Strophanthus dichotomus DC. in Bull. Soc. Philom. **3**: 123. 1802.

[63] Pitard here makes the Indo-China form *Strophanthus caudatus* Kurz var. *macrophyllus* Franch. in Nouv. Arch. Mus. Hist. Nat. [Paris] III **5**: 265. 1783. If a varietal name is needed there are two earlier ones, var. *cochinchinensis* Ker (1819) and var. *loureiri* A. DC. (1844).

Faskia divaricata Lour.[64] ex Gomes in Mem. Acad. Sci. Lisb. Cl. Sci. Pol. Mor. Bel.-
Let. n.s. **4**(1): 28. 1868.

Strophanthus dichotomus DC. var. *loureiri* A. DC. Prodr. **8**: 417. 1844 (based on *Nerium scandens* Lour.).

" Habitat in dumetis Cochinchinae." Loureiro's description is ample and definite.
I can see no reason for considering his species other than a form of *Strophanthus caudatus*
(Burm. f.) Kurz, of which *S. dichotomus* DC. is a synonym, a species extending from Tenasserim and Indo-China, through the Malay Peninsula, to Java. Specimens from Loureiro
are preserved in the herbaria of the British and the Paris Museums.

Wrightia R. Brown

Wrightia annamensis Eberh. & Duby in Agron. Colon. **1**: 38. 1913; Pitard in Lecomte Fl.
Gén. Indo-Chine **3**: 1191. 1933.

Nerium antidysentericum (non Linn.) Lour. Fl. Cochinch. 116. 1790, ed. Willd. 142.
1793, Anamese *cây móc hoa tláng.*

" Tam culta quam inculta in Cochinchina." The Linnaean species is *Wrightia anti-
dysenterica* (Linn.) R. Br. (*W. zeylanica* R. Br.), one that does not extend to Indo-China.
Loureiro's description applies to *Wrightia* and rather definitely to *W. annamensis* Eberh.
& Duby. Loureiro describes the flowers as " albo-viridis," while Mrs. Clemens notes on
her *no. 3367* from Tourane that the flowers are red; in this character the specimen agrees
better with *Nerium divaricatum* as described by Loureiro, but the corolla appendages " laciniis oblongis, alternatim 3-fidis et linearibus " are those of *Wrightia annamensis;* for *Nerium divaricatum* Loureiro describes them as " laciniis subulatis, brevibus." The species is
represented by *Clemens 3367* in flower and *4124* in fruit, from the vicinity of Tourane.
Nerium indicum Burm. (Thes. Zeyl. 167. *pl. 77*) and *Codaga pala* Rheede (Hort. Malabar.
1: 85. *pl. 47*), cited by Loureiro as illustrating his species, are to be excluded, as they do not
represent the form he described.

Wrightia sp.

Nerium divaricatum (non Linn.) Lour. Fl. Cochinch. 115. 1790, ed. Willd. 142. 1793,
Anamese *cây móc hoa do.*

Tabernaemontana ? divaricata G. Don Gen. Syst. **4**: 91. 1838 (based on *Nerium divaricatum* Lour.).

" Tam culta quam inculta in Cochinchina " (sub *Nerium antidysentericum*). The
Linnaean species is *Tabernaemontana divaricata* (Linn.) R. Br., but Loureiro's description
is that of a *Wrightia.* It may be *W. annamensis* Eberh. & Duby is the species represented
by Loureiro's description; see above.

ASCLEPIADACEAE

Streptocaulon Wight & Arnott

Streptocaulon juventas (Lour.) comb. nov.

Apocynum juventas Lour. Fl. Cochinch. 167. 1790, ed. Willd. 208. 1793, Anamese *hà
thu ô nam;* Moore in Journ. Bot. **63**: 252. 1925.

[64] A Loureiro herbarium name here first published by Gomes.

Streptocaulon tomentosum Wight Contrib. Bot. Ind. 64. 1834; Costantin in Lecomte
Fl. Gén. Indo-Chine **4**: 31. 1912.

Tylophora juventas Woodson in Ann. Missouri Bot. Gard. **17**: 146. 1930 (based on
Apocynum juventas Lour.).

"Habitat in locis agrestibus Cochinchinae, saepius procumbens." Loureiro's specimen, which agrees with his description, is in the herbarium of the British Museum, and
Moore (Journ. Bot. **63**: 252. 1925) has supplied a critical note regarding it, concluding that
there is no reasonable doubt as to the identity of *Apocynum juventas* Lour. with *Streptocaulon tomentosum* Wight. His note should be consulted because of the confusion that has
existed between that species and *Tylophora indica* (Burm. f.) Merr. (*Apocynum reticulatum*
Lour.), apparently due to a mixture of specimens and labels, or by accrediting the medicinal
qualities to the wrong plant; see also *Tylophora indica* (Burm. f.) Merr. (p. 319). Costantin (Lecomte Fl. Gén. Indo-Chine **4**: 113. 1912) placed Loureiro's species as a synonym of
Tylophora ovata Hook. (*Diplolepis ovata* Lindl.) following Hooker & Arnott's and Decaisne's
erroneous disposition of it, admitting that species as an Indo-China one solely on the basis of
this erroneous reduction of *Apocynum juventas* Lour. *Streptocaulon juventas* (Lour.) Merr.
is represented by *Clemens 4016* from the classical locality Hue. Handel-Mazzetti thinks
that this number represents *S. griffithii* Hook. f., a species very closely allied to *S. tomentosum* Wight; both are recorded from Indo-China.

Cryptolepis R. Brown

Cryptolepis sinensis (Lour.) Merr. in Philip. Journ. Sci. **15**: 254. 1919 (based on *Pergularia sinensis* Lour.).

Pergularia sinensis Lour. Fl. Cochinch. 169. 1790, ed. Willd. 211. 1793, Chinese *fí sí
thân*.

Emericia sinensis Roem. & Schult. Syst. **4**: 402. 1819 (based on *Pergularia sinensis*
Lour.).

Periploca chinensis Spreng. Syst. **1**: 836. 1825 (based on *Pergularia sinensis* Lour.).

Vallaris sinensis G. Don Gen. Syst. **4**: 79. 1838 (based on *Pergularia sinensis* Lour.).

Streptocaulon ? chinensis G. Don Gen. Syst. **4**: 162. 1838 (based on *Pergularia sinensis* Lour.).

Cryptolepis elegans Wall. List no. 1639. 1829, *nomen nudum;* G. Don Gen. Syst. **4**: 82.
1838.

Periploca sinensis Steud. Nomencl. ed. 2, **1**: 552. 1840, in syn. (based on *Emericia sinensis* Roem. & Schult. = *Pergularia sinensis* Lour.).

Aganosma edithae Hance in Ann. Sci. Nat. V Bot. **5**: 227. 1866.

Cryptolepis edithae Benth. & Hook. f. ex Maxim in Bull. Acad. St. Pétersb. **23**: 352.
1877, Mel. Biol. **9**: 774. 1877.

"Habitat apud Sinas inculta." Loureiro's description, in all but his brief statement
regarding the fruits and the naked seeds, applies unmistakably to the species currently
known as *Cryptolepis elegans* Wall. which is common in thickets near Canton.

Cryptolepis sp.

Apocynum africanum Lour. Fl. Cochinch. 168. 1790, ed. Willd. 209. 1793.

Ichnocarpus loureirii Spreng. Syst. **1**: 635. 1825 (based on *Apocynum africanum* Lour.).

Ichnocarpus africanus Woodson in Ann. Missouri Bot. Gard. **17**: 144. 1930 (based
on *Apocynum africanum* Lour.).

" Habitat in planitie arenosa orae Zanguebariae in Africa." Loureiro's species is not
mentioned by Stapf or by N. E. Brown in their treatment of the Apocynaceae and Asclepiadaceae (Thiselton-Dyer Fl. Trop. Afr. **4**(1): 24–503. 1902–03). It can scarcely be an
Ichnocarpus where it was placed by Sprengel and by Woodson, as no species of this genus
is known from tropical Africa. The imperfect description suggests *Cryptolepis*, but some
genus near *Strophanthus* of the Apocynaceae may possibly be represented.

Calotropis R. Brown

Calotropis gigantea (Linn.) Dryand. in Ait. Hort. Kew. ed. 2, **2**: 78. 1811; Costantin in
Lecomte Fl. Gén. Indo-Çhine **4**: 70. *f. 11. 1–4.* 1912.

Asclepias gigantea Linn. Sp. Pl. 214. 1753.

Periploca cochinchinensis Lour. Fl. Cochinch. 167. 1790, ed. Willd. 207. 1793, Anamese *cây bup bup.*

Streptocaulon ? cochinchinensis G. Don, Gen. Syst. **4**: 162. 1838 (based on *Periploca
cochinchinensis* Lour.).

" Habitat ad portum Turanum, indigenis dictum *Hàn* in Cochinchina, ad Austrum
metropolis Huaeae. Vidi etiam, sed minori forma, in Benghala." This reduction of
Loureiro's species was made by Dr. A. Chevalier (Cat. Pl. Jard. Bot. Saigon 65. 1919) and
is certainly the correct disposition of it. It is represented by *Clemens 3166* from Tourane.

Ceropegia Linnaeus

Ceropegia loureiri G. Don Gen. Hist. **4**: 112. 1838 (based on *Ceropegia candelabrum* Lour.).

Ceropegia candelabrum (non Linn.) Lour. Fl. Cochinch. 114. 1790, ed. Willd. 140. 1793,
Anamese *cây saong kêm;* Costantin in Lecomte Fl. Gén. Indo-Chine **4**: 152. 1912.

" Habitat in dumetis Cochinchinae." Following the description of the next species
(*C. obtusa* Lour.) Loureiro states that this occurred in the suburbs of Hue. The Linnaean
species is Indian and there is little or no reason for considering that Loureiro's plant represents the same species. The reference to Rheede (Hort. Malabar. **9**: 27. *pl. 16*), cited,
after Linnaeus, typifies the Linnaean species of which Hooker f. (Fl. Brit. Ind. **4**: 70. 1883)
states he had seen no specimen. It is then rare or local in India, or represents a species
that is perhaps a synonym of some later described one; yet Costantin admits it in his treatment of the Asclepiadaceae (Lecomte Fl. Gén. Indo-Chine **4**: 152. 1912) as a species of
doubtful status. He does not account for Don's binomial based on Loureiro's description.
He gives an amplified description, but cites only Loureiro, although he gives an additional
local name *lay saong.* Loureiro's type is not extant and I do not know on what material
Costantin's additional data were based. A little reasonably intensive field work in the
immediate vicinity of Hue should yield material and data that would solve the identity
of this and the next two doubtful species of *Ceropegia*.

Ceropegia obtusa Lour. Fl. Cochinch. 114. 1790, ed. Willd. 140. 1793, Anamese *rau kem.*

" Habitat . . . in suburbiis metropolis Cochinchinae Huaeae." This is noted by
Loureiro as differing more in its leaves than in its flowers from the specimen he referred
to *Ceropegia candelabrum.* Costantin admits it in his treatment of the Asclepiadaceae

(Lecomte Fl. Gén. Indo-Chine **4**: 153. 1912) as a species of doubtful status. It is known only from Loureiro's short and imperfect description.

Ceropegia cordata Lour. Fl. Cochinch. 114. 1790, ed. Willd. 141. 1793, Anamese *deei di mo.*
   " Habitat scandens per sepes in Cochinchina." Like the other species of the genus described by Loureiro, this is known only from his description; Costantin (Lecomte Fl. Gén. Indo-Chine **4**: 153. 1912) admits it as an imperfectly known species. From the description it is suspected that no *Ceropegia* is represented; the statement " umbellis magnis, hemisphaericis, axillaribus " distinctly does not suggest *Ceropegia* characters. Willdenow queries: " An *Cynanchum cordifolium* Retz. ?," but it agrees with none of the species of this genus admitted by Costantin. In some respects the description suggests the common *Dregea volubilis* (Linn. f.) Benth., but the indicated floral characters, if correctly given, eliminate this species as a possibility.

Gymnema R. Brown

Gymnema alterniflorum (Lour.) comb. nov.
   Apocynum alterniflorum Lour. Fl. Cochinch. 168. 1790, ed. Willd. 209. 1793, Chinese *fû muôn thân.*
   Asclepias curassavica (non Linn.) Lour. Fl. Cochinch. 170. 1790, ed. Willd. 211. 1793, Chinese *yong cŏ láy.*
   Strophanthus alterniflorus Spreng. Syst. 1: 638. 1825 (based on *Apocynum alterniflorum* Lour.).
   Gymnema affine Decne. in DC. Prodr. **8**: 622. 1844.
   Gymnema parviflorum Wall. List no. 8184b. 1847, *nomen nudum.*
   For *Apocynum alterniflorum* Loureiro states: " Habitat in insulis prope Cantonem Sinarum." Sprengel was misled by Loureiro's description of the coronal scales as 10, and assuming that Loureiro had an apocynaceous plant, scarcely any other reference than to *Strophanthus* was possible. In making current identifications of Kwangtung plants, I accepted Sprengel's binomial for the only *Strophanthus* known from southern China and in my original manuscript of 1919 adopted *Strophanthus alterniflorus* Spreng. for the species. A more attentive study of Loureiro's description clearly indicates that no *Strophanthus* is represented, but rather *Gymnema affine* Decne., which is abundant in thickets near Canton. The chief objection to this interpretation is Loureiro's statement: " Nectarium lobis decem "; the number of coronal scales in *Gymnema* is 5. The number given by Loureiro was probably based on an erroneous observation on his part. Woodson (Ann. Missouri Bot. Gard. **17**: 144. 1930) followed Hooker & Arnott (Bot. Beechey's Voy. 200. 1836) and referred Loureiro's species to *Gymnema sylvestris* (Willd.) R. Br. (Mem. Wern. Soc. 1: 33. 1809), a species greatly resembling *G. affine* Decne., and manifestly allied to it, but one unknown from China. The presence of small umbels in alternate leaf-axils is characteristic of many collections representing Decaisne's species, but the character is far from being a constant one, as in other specimens inflorescences occur in opposite axils. The actual type of *Gymnema affine* Decne., *Wallich 8184b*, was from a plant cultivated in Calcutta, originating in China. For *Asclepias curassavica* Loureiro states: " Habitat prope Cantonem Sinarum." The description conforms to the characters of *Gymnema affine* Decne. = *G. alterniflorum* (Lour.) Merr., but not at all to *Asclepias curassavica* Linn. The statement

" Folliculi . . . ventricoso-conici," together with the other data given, unmistakably place the species as here reduced.

Gymnema inodorum (Lour.) Decne. in DC. Prodr. **8**: 551. 1844 (based on *Cynanchum inodorum* Lour.); Costantin in Lecomte Fl. Gén. Indo-Chine **4**: 87. 1912.

   Bidaria inodora Decne. in DC. Prodr. **8**: 624. 1844 (based on *Cynanchum inodorum* Lour.).

   Ganosma inodora Lour. ex Decne. l.c. in syn. (Loureiro's herbarium name).

   Cynanchum inodorum Lour. Fl. Cochinch. 166. 1790, ed. Willd. 207. 1793, Anamese *ti yong thân.*

No locality is cited, but the Anamese name (erroneously indicated by Loureiro as Chinese) clearly indicates an Indo-China specimen. Decaisne examined Loureiro's type in the herbarium of the Paris Museum, which bears his identification as *Gymnema inodorum.* In his monograph of the Asclepiadaceae (DC. Prodr. **8**: 551), in excluding the species from *Cynanchum*, he published the new binomial *Gymnema inodorum*, apparently not realizing that on page 624 of the same work he had based *Bidaria inodora* on Loureiro's species. In the description of *Bidaria inodora* he also cites an Indian specimen collected by Perrottet and thus this description may have been in part based on that specimen. Hooker f. (Fl. Brit. Ind. **4**: 33. 1883) enumerated *Bidaria inodora* Decne. as a species of doubtful status with the statement that he had no idea as to what it represented. Costantin apparently did not know of Loureiro's extant type and admits *Gymnema inodorum* Decne. solely on the basis of Loureiro's published record. He quotes *Cynanchum reticulatum* Retz. as a synonym which, if correctly placed, would supply the oldest valid name for the species; however, *Cynanchum reticulatum* Retz. is the name bringing synonym of the very different *Leptadenia reticulata* (Retz.) Wight & Arn. and apparently has nothing to do with Loureiro's species; Willdenow (Fl. Cochinch. ed. 2, 207. 1793) originally suggested this erroneous reduction. There is also a fragmentary specimen of Loureiro's species in the herbarium of the British Museum.

Tylophora R. Brown

Tylophora indica (Burm. f.) Merr. in Philip. Journ. Sci. **19**: 373. 1921.

   Cynanchum indicum Burm. f. Fl. Ind. 70. 1768.

   Asclepias asthmatica Linn. f. Suppl. 171. 1781.

   Tylophora asthmatica Wight & Arn. in Wight Contrib. Bot. Ind. 51. 1834; Costantin in Lecomte Fl. Gén. Indo-Chine **4**: 109. 1912.

   Apocynum reticulatum (non Linn.) Lour. Fl. Cochinch. 167. 1790, ed. Willd. 208. 1793, Anamese *deei xanh;* Moore in Journ. Bot. **63**: 253. 1925, sub *Apocynum juventas* Lour.

" Habitat in dumetis Cochinchinae." Moore was doubtless correct in his assumption that the second specimen of *Apocynum juventas* Lour. in the herbarium of the British Museum, which does not agree with Loureiro's description of that species, really represents *Apocynum reticulatum* Lour. This specimen is, according to him (Journ. Bot. **63**: 253. 1925), a small-leaved form of *Tylophora asthmatica* Wight & Arn. As explained by him, Loureiro, apparently by error, affixed the wrong names to the specimens representing *Apocynum juventas* and *A. reticulatum;* but these names stand for the species actually described by Loureiro, one a *Tylophora*, the other a *Streptocaulon*. Hooker & Arnott (Bot. Beechey's Voy. 197. 1836) placed Loureiro's species as a doubtful synonym of *Parsonsia ? helicandra* Hook. & Arn. (*Helicandra sinensis* Hook. & Arn. op. cit. 198), a species that Hemsley (Journ. Linn. Soc. Bot. 26: 96. 1889) considered to be of doubtful status. A fragmentary specimen of this exists in the Glasgow herbarium which I have examined. It is a true *Parsonsia*, the same as *Henry 306* from Formosa, which is referred to *P. spiralis* Wall. (Merrill in Brittonia 1: 233–237. 1933). In any case, even if *Heligia javanica* Blume (1826) proves to represent the same species, this specific name is invalid in *Parsonsia*, and Hooker & Arnott's name will stand, for *Parsonsia spiralis* Wall. was not effectively published until 1838 (G. Don Gen. Syst. 4: 80); *Apocynum reticulatum* Linn. = *Parsonsia reticulata* Woodson (Ann. Missouri Bot. Gard. 17: 147. 1930) is wholly unknown except for Royen's very short descriptive phrase on which the Linnaean species is based: " *Apocynum* caule volubili perenni, foliis ovatis venosis " (Royen Fl. Leyd. Prodr. 412. 1740); Linnaeus had no specimen. There is no reason for considering that *Apocynum reticulatum* Linn. has anything to do with *Parsonsia helicandra* Hook. & Arn. *Olus crudum minus* Rumph. (Herb. Amb. 5: 75. *pl. 40. f. 2*), cited by Loureiro as a synonym, probably represents *Gymnema syringaefolium* (Decne.) Boerl., although Moore thinks a *Tylophora* may be represented. The species as described by Loureiro is represented by *Clemens 3362, 4256*, from Tourane, which on preliminary examination I referred to *T. glabra* Cost., but which probably do not represent Costantin's species; they might pass for a small-leaved form of *Tylophera asthmatica* Wight & Arn. = *T. indica* (Burm. f.) Merr. Moore's extensive note should be consulted; he gives some additional descriptive data based on Loureiro's plant. Mr. J. E. Dandy and Mr. G. Taylor have kindly compared *Clemens 3362* with Loureiro's specimen in the herbarium of the British Museum and both consider it to represent the same species, also closely matching a specimen of *Tylophora asthmatica* Wight & Arn. collected by Roxburgh in India. From material available to me, named as representing this species, it is suspected that a collective species is involved and that on a critical revision several of more or less closely allied ones will be found to be represented. The oldest specific name for the species as currently interpreted is apparently that of Burman f.

Tylophora sp.

> *? Cynanchum fuscum* Schult. in Roem. & Schult. Syst. 6: 111. 1820 (based on *Asclepias fusca* Lour.).
>
> *? Asclepias fusca* Lour. Fl. Cochinch. 170. 1790, ed. Willd. 211. 1793, Anamese *cây so dua*.

" Habitat in antiquis muris, & ruderibus Cochinchinae." Judging from the description this may be a *Tylophora*, and intensive collecting at Hue should readily solve the problem of its status. Costantin in his treatment of the Asclepiadaceae of Indo-China (Lecomte Fl. Gén. Indo-Chine 4: 66. 1912) cites *Asclepias fusca* Lour. as a synonym of *Vincetoxicum medium* Decne., which is placed as a doubtful species of *Cynanchum*. His synonymy is badly confused, his accepted name *Vincetoxicum medium* Decne., and the synonyms *V. luteum* Hoffm. & Link, *Asclepias fuscata* Willd., and *Cynanchum fuscatum* Link, all appertaining to a Portuguese species, having nothing whatever to do with *Asclepias fusca* Lour. His description was compiled from Loureiro.

Hoya R. Brown

Hoya chinensis (Lour.) Traill in Trans. Hort. Soc. London **7**: 27. 1827 (based on *Stapelia chinensis* Lour.).

Stapelia chinensis Lour. Fl. Cochinch. 165. 1790, ed. Willd. 205. 1793, Chinese *yong sau khau*.

" Habitat prope Cantone Sinarum." R. Brown (Mem. Wern. Soc. 1: 27. 1809) mentions a specimen from Loureiro in the Banksian herbarium, but Traill and Britten have shown that the extant specimen is labeled *Stapelia cochinchinensis* Lour., is not *S. chinensis* Lour.; see below. Except for the description of the leaves as lanceolate, I should be inclined to refer *Hoya chinensis* (Lour.) Traill to *H. carnosa* (Linn. f.) R. Br. (*Asclepias carnosa* Linn. f. Suppl. 170. 1781), the type of which was from China.

Hoya cochinchinensis (Lour.) Schult. in Roem. & Schult. Syst. **6**: 52. 1820; Costantin in Lecomte Fl. Gén. Indo-Chine **4**: 141. 1912 (based on *Stapelia cochinchinensis* Lour.).

Triplosperma cochinchinensis G. Don Gen. Syst. **4**: 134. 1838 (based on *Stapelia cochinchinensis* Lour.).

Stapelia cochinchinensis Lour. Fl. Cochinch. 165. 1790, ed. Willd. 206. 1793, Anamese *deei luoi tlâu*.

" Habitat in montibus Cochinchinae." Decaisne (DC. Prodr. **8**: 640. 1844) suggested *Ceropegia* as the proper generic disposition of Loureiro's species. Most serious objections to this are the description of the leaves as fleshy, the thick peduncles, the large umbels, and the white flowers. The description of the corolla as " 1-petala, infundibuliformis: tubo gracili, longo " is erroneous for *Hoya*. A specimen from Loureiro labeled *Stapelia cochinchinensis* is preserved in the herbarium of the British Museum. Regarding this Britten (Journ. Bot. **36**: 414. 1898) notes: " To *H. carnosa* Brown also (Mem. Wern. Soc. i. 27) referred a plant of Loureiro's which he cites as ' *Stapelia Chinensis* Lour. Cochin. i.p. 205, fide specim. ab auctore missi in Herb. Banks.' The only specimen of ' *Stapelia* ' from Loureiro in Herb. Banks is named *S. cochinchinensis*, so it seems clear that this was the name Brown had intended to cite. Having arrived at this conclusion, I found I had been anticipated by Traill (l.c. 20), who gives a very careful note, based on an examination by Brown, showing that Loureiro's specimen cannot be identified, save in part, with either of his descriptions and that both remain obscure." Traill's more extensive note (Trans. Hort. Soc. London **7**: 20–21. 1827) should be consulted. He notes that the specimen consists of a leafy branch with a few detached flowers, the latter being certainly those of a *Hoya*. Mr. A. H. G. Alston had kindly re-examined the specimen for me and reports that there is no reason to suppose that the flowers and leaves do not belong to the same plant. He further states that by Costantin's arrangement of the Indo-China species (Lecomte Fl. Gén. Indo-Chine **4**: 125–141. 1912) *H. cochinchinensis* (Lour.) Schult. falls under *Hoya macrophylla* Blume, although Loureiro's specimen agrees better with the illustration of *Hoya pottsii* Traill (Curtis's Bot. Mag. **62**: *pl. 3425.* 1835) than with Blume's illustration of *Hoya macrophylla* Blume (Rumphia **4**: 32. *pl. 185.* 1848). R. Brown identified Loureiro's specimen as representing *Hoya carnosa* R. Br., but it has glabrous petals and 3-nerved leaves. It is possible that Costantin misinterpreted *Hoya macrophylla* Blume, and that Loureiro's specific name should be accepted for the Indo-China form he referred to that species.

Telosma Coville

(*Prageluria* N. E. Brown)

Telosma cordata (Burm. f.) Merr. in Philip. Journ. Sci. **19**: 372. 1921.

Asclepias cordata Burm. f. Fl. Ind. 72. *pl. 27. f. 2.* 1768.

Cynanchum odoratissimum Lour. Fl. Cochinch. 166. 1790, ed. Willd. 206. 1793, Anamese *hoa lí.*

Pergularia odoratissima Sm. Ic. Pict. *pl. 16.* 1793 (based on *Cynanchum odoratissimum* Lour.); Andrews Bot. Repos. **3**: *pl. 185.* 1801.

Telosma odoratissima Coville in Contr. U. S. Nat. Herb. **9**: 384. 1905 (based on *Cynanchum odoratissimum* Lour.).

" Habitat, & colitur in viridariis Cochinchinae. . . . Colitur etiam Cantone Sinarum."
Although I adopted Burman's specific name for this plant in 1921, I later doubted its correctness because of certain manifest discrepancies in Burman's illustration, particularly the inflorescence and the stamens, and his description of the flowers as " purpurea." To settle the matter I asked Doctor Hochreutiner to examine Burman's extant type at Geneva, and he reports that, in spite of the poor illustration, in which the flowers are erroneously drawn as having exserted stamens, my interpretation was correct. Mr. A. W. Exell kindly examined Loureiro's type in the herbarium of the British Museum and reports that the specimen agrees with Smith's illustration cited above. It is to be noted that Loureiro described the flowers as " luteus," and because of this I thought that possibly *Pergularia minor* Andr., as described by Costantin (Lecomte Fl. Gén. Indo-Chine **4**: 102. *f. 15.* 1912), might be the species Loureiro described. Mr. Exell states that he finds it difficult to separate the British Museum specimens into two different species, but if such a separation be possible, Loureiro's type would certainly be distinguished from *Pergularia minor* Andr., the plate (Bot. Repos. **3**: *pl. 184.* 1801) showing much shorter corolla lobes than those of *Pergularia odoratissima* Sm. = *Telosma cordata* (Burm. f.) Merr. Coville in 1905 and N. E. Brown in 1907 independently determined that the generic name *Pergularia* of Linnaeus appertained strictly to the South African *Daemia* (*Doemia*) R. Brown, *Prageluria* being proposed by N. E. Brown (Kew Bull. 323. 1907) as a new generic name for the group here considered; no new binomial was proposed.

CONVOLVULACEAE

Cuscuta (Tournefort) Linnaeus

Cuscuta chinensis Lam. Encycl. **2**: 229. 1786; Choisy in Mém. Soc. Phys. Hist. Nat. Genève **9**: 279. 1841, DC. Prodr. **9**: 457. 1845; Engelm. in Trans. Act. Sci. St. Louis **1**: 479. 1859; Yuncker in Mem. Torr. Bot. Club **18**: 209. *f. 80. A–G.* 1932.

Cuscuta carinata R. Br. Prodr. 491. 1810.

Cuscuta chinensis Lam. var. *carinata* Engelm. in Trans. Acad. Sci. St. Louis **1**: 480. 1859.

Grammica aphylla Lour. Fl. Cochinch. 171. 1790, ed. Willd. 212. 1793, Anamese *to haong tláng.*

" Habitat inculta in hortis Cochinchinae, & inter plantas humiles volutatur, duratione brevi paucorum mensium, tempore verno." Gagnepain & Courchet (Lecomte Fl. Gén.

Indo-Chine **4**: 311. 1915) admit for Indo-China only *Cuscuta hygrophilae* Pearson, under which they erroneously cite *Cuscuta chinensis* Lam. as a doubtful synonym, and *C. japonica* Choisy; Pearson's species is *Cuscuta australis* R. Br. They do not account for *Grammica aphylla* Lour. *Clemens 4390*, from Tourane, det. Yuncker, and apparently *Squires 404* from Hue, represent *Cuscuta australis* R. Br. (*C. hygrophilae* Pearson), and I accordingly suspected that *Grammica aphylla* Lour. might represent *Cuscuta australis* R. Br. rather than *C. chinensis* Lam. to which all authors have reduced it. Mr. Ramsbottom kindly sent me a flower from Loureiro's type in the herbarium of the British Museum, which on dissection proves to be identical with *Cuscuta chinensis* Lam. as interpreted and illustrated by Yuncker. Apparently no collector since Loureiro has discovered *Cuscuta chinensis* Lam. in Indo-China. R. Brown (Prodr. 491. 1810) identified Loureiro's type as representing *Cuscuta carinata* R. Br. = *C. chinensis* Lam. var. *carinata* Engelm. = *C. chinensis* Lam.

Merremia Dennstaedt

Merremia gemella (Burm. f.) Hall. f. in Bot. Jahrb. **16**: 552. 1892; Merr. Enum. Philip. Fl. Pl. **3**: 360. 1923.

> *Convolvulus gemellus* Burm. f. Fl. Ind. 46. *pl. 21. f. 1*. 1768.
>
> *Ipomoea chryseides* Ker in Bot. Reg. **4**: *pl. 270*. 1818; Gagnep. & Courchet in Lecomte Fl. Gén. Indo-Chine **4**: 254. 1915.
>
> *Ipomoea campanulata* (non Linn.) Lour. Fl. Cochinch. 112. 1790, ed. Willd. 138. 1793, Anamese *rau mòi*.

No locality is given, but the Anamese name cited indicates an Indo-China specimen. The description agrees in all essentials with the common and widely distributed *Merremia gemella* Hall. f., the only possible exception to this reduction being Loureiro's description of the leaves as "integerrima." *Adamboe* Rheede (Hort. Malabar. **11**: 115. *pl. 56*), cited by Loureiro as a synonym, after Linnaeus, must be excluded as it represents the true *Ipomoea campanulata* Linn. = *Stictocardia campanulata* (Linn.) House, not at all the species that Loureiro described.

Merremia hastata (Desr.) Hall. f. in Bot. Jahrb. **16**: 552. 1892.

> *Convolvulus hastatus* Desr. in Lam. Encycl. **3**: 542. 1791.
>
> *Ipomoea angustifolia* C. B. Clarke in Hook. f. Fl. Brit. Ind. **4**: 205. 1883, non Jacq.
>
> *Convolvulus medium* (non Linn.) Lour. Fl. Cochinch. 106. 1790, ed. Willd. 130. 1793.
>
> *Convolvulus arvensis* (non Linn.) Lour. Fl. Cochinch. 106. 1790, ed. Willd. 130. 1793, Anamese *bìm bìm dât*.

For both species Loureiro states: "Habitat inter sepes Cochinchinae." The descriptions apply to the very common and widely distributed *Merremia hastata* Hall. f. This is by some botanists reduced to *Ipomoea angustifolia* Jacq. (Coll. **2**: 367. 1788, Ic. Pl. Rar. **2**: 10. *pl. 317*. 1793) and Gagnepain & Courchet (Lecomte Fl. Gén. Indo-Chine **4**: 265. 1915) so dispose of it. Jacquin's illustration does not appear to me to represent the species described by Desrousseaux; his species was based on material from Guinea.

Merremia umbellata (Linn.) Hall. f. in Bot. Jahrb. **16**: 552. 1892, **18**: 114. 1893, Versl. Lands Plant. Buitenz. 1895. 127. 1896.

> *Convolvulus umbellatus* Linn. Sp. Pl. 155. 1753.
>
> *Ipomoea umbellata* G. F. W. Mey. Prim. Fl. Esseq. 99. 1818, non Linn. (1759).

Ipomoea polyanthes Roem. & Schult. Syst. **4**: 234. 1819.

Ipomoea cymosa Roem. & Schult. Syst. **4**: 241. 1819; Gagnep. & Courchet in Lecomte
Fl. Gén. Indo-Chine **4**: 251. 1915, non G.F.W.Mey. (1818).

Convolvulus loureiri G. Don Gen. Syst. **4**: 290. 1838 (based on *Convolvulus scammonia*
Lour.).

Convolvulus scammonia (non Linn.) Lour. Fl. Cochinch. 106. 1790, ed. Willd. 130.
1793, Anamese *khoai cà hoa vàng*.

"Habitat in sylvis Cochinchinae." Loureiro's description unmistakably applies to
the very common pantropic species, usually known as *Ipomoea cymosa* Roem. & Schult.,
or as *I. polyanthes* Roem. & Schult.; it is the form with yellow flowers, *Merremia umbellata*
Hall. f., the more common one in the Old World tropics being the one with white flowers,
Merremia umbellata var. *orientalis* Hall. f. (Versl. Lands Plant. Buitenz. 1895. 132. 1896).
In proposing the binomial *Merremia umbellata*, Hallier based it on *Ipomoea umbellata* Mey.
and *I. cymosa* Roem. & Schult., but Meyer's binomial was based on *Convolvulus umbellatus*
Willd. (Sp. Pl. **1**: 857. 1797) which in turn is *Convolvulus umbellatus* Linn. (Sp. Pl. 155.
1753). For those who do not recognize *Merremia* as generically distinct from *Ipomoea*,
the oldest valid specific name in *Ipomoea* is apparently *I. polyanthes* Roem. & Schult.
Ipomoea umbellata Linn. (Syst. ed. 10, 924. 1759) is a totally different species.

Operculina S. Manso

Operculina turpethum (Linn.) S. Manso Enum. Subst. Braz. 16. 1836.

Convolvulus turpethum Linn. Sp. Pl. 155. 1753.

Ipomoea turpethum R. Brown Prodr. 485. 1810; Gagnep. & Courchet in Lecomte Fl.
Gén. Indo-Chine **4**: 263. 1915.

? *Convolvulus panduratus* (non Linn.) Lour. Fl. Cochinch. 107. 1790, ed. Willd. 131.
1793, Anamese *bìm bìm vàng*.

? *Ipomoea panduraeformis* Choisy in Mém. Soc. Phys. Hist. Nat. Genève **6**: 476. 1833
(based on *Convolvulus panduratus* Lour.).

"Habitat inter sepes Cochinchinae." I refer Loureiro's species to the very common
and widely distributed *Operculina turpethum* S. Manso, with doubt, as being the most
probable reduction of it, in spite of Loureiro's description of the flowers as "nitide luteus";
in *O. turpethum* S. Manso the flowers are white. *Convolvulus megalorrizos flore amplo*
Dill. (Hort. Elth. 101. *pl. 85. f. 99*), cited by Loureiro as representing the species, is a form
of *Ipomoea batatas* Poir.

Lepistemon Blume

Lepistemon binectariferum (Wall.) O. Ktz. Rev. Gen. Pl. 446. 1891.

Convolvulus binectariferus Wall. in Roxb. Fl. Ind. **2**: 47. 1824.

Lepistemon flavescens Blume Bijdr. 722. 1826; Gagnep. in Lecomte Fl. Gén. Indo-
Chine **4**: 287. 1915.

Convolvulus subglobosus G. Don Gen. Syst. **4**: 293. 1838 (based on *Convolvulus canariensis* Lour.).

Convolvulus canariensis (non Linn.) Lour. Fl. Cochinch. 108. 1790, ed. Willd. 133.
1793. Anamese *bìm bìm tláng*.

"Habitat inter sepes Cochinchinae." Loureiro's description applies unmistakably to *Lepistemon* and with very little doubt to the sole representative of this genus known from Indo-China.

Ipomoea Linnaeus

Ipomoea batatas (Linn.) Poir. in Lam. Encycl. **6**: 14. 1804; Gagnep. & Courchet in Lecomte Fl. Gén. Indo-Chine **4**: 240. 1915.

> *Convolvulus batatas* Linn. Sp. Pl. 154. 1753; Lour. Fl. Cochinch. 107. 1790, ed. Willd. 131. 1793, Anamese *khoai lang*, Chinese *hoân xý*.
> *Convolvulus mammosus* Lour. Fl. Cochinch. 108. 1790, ed. Willd. 132. 1793, Anamese *khoai tù*.
> *Ipomoea mammosa* Choisy in Mém. Soc. Phys. Hist. Nat. Genève **6**: 475. 1833 (based on *Convolvulus mammosus* Lour.).
> *Merremia mammosa* Hall. f. ex Prain in Journ. Asiat. Soc. Bengal **74**: 307. 1906, in obs. (based on *Convolvulus mammosus* Lour.).

For *Convolvulus batatas* Loureiro states: "Habitat in omnibus fere locis utruisque Indiae, extra; & intra Gangem." This is the common sweet-potato, the Linnaean species having been correctly interpreted by Loureiro. For *Convolvulus mammosus* Loureiro states: "Habitat frequenter cultus in agris Cochinchinae." The description is manifestly based on a mixture of material, the vegetative and flower characters of *Ipomoea batatas* Poir., and the tuber characters and local name based on *Dioscorea esculenta* (Lour.) Burkill. Gagnepain & Courchet (Lecomte Fl. Gén. Indo-Chine **4**: 254. 1915) place *Convolvulus mammosus* Lour., and the two binomials based upon it, as doubtful synonyms of *Ipomoea gomezii* C. B. Clarke. The original description does not apply to Clarke's species. It is unreasonable to consider that Loureiro's species, which he says was *frequently cultivated* in Indo-China, can possibly be referred to *I. gomezii* Clarke, which is never cultivated, does not have tubers, and which has been collected but once in Indo-China, and then on the Island of Pulu Condor off the extreme south end of Indo-China. *Batatta mammosa* Rumph. (Herb. Amb. **5**: 370. *pl. 131*), cited by Loureiro as representing his species, and from whence he derived his specific name, is based in part on the tubers of a *Dioscorea*, perhaps *D. esculenta* (Lour.) Burkill, and on the vegetative characters of a convolvulaceous plant, perhaps *Operculina* or some species of *Ipomoea;* see Merrill, Interpret. Herb. Amb. 442. 1917.

Ipomoea aquatica Forsk. Fl. Aeg.-Arab. 44. 1775; Gagnep. & Courchet in Lecomte Fl. Gén. Indo-Chine **4**: 248. 1915.

> *Ipomoea reptans* Poir. in Lam. Encycl. Suppl. **3**: 460. 1813, non *Convolvulus reptans* Linn.
> *Convolvulus reptans* (non Linn.) Lour. Fl. Cochinch. 109. 1790, ed. Willd. 133. 1793, Anamese *rau muóng*.

"Habitat in Cochinchina, & China, passim obvius in locis aquosis." This is the very common tropical Asiatic species, the young shoots of which are commonly used as a pot-herb. *Olus vagum* Rumph. (Herb. Amb. **5**: 419. *pl. 155. f. 1*), cited by Loureiro as representing his species, is *Ipomoea aquatica* Forsk. It should be noted that the actual type in the Linnaean herbarium, the specimen on which the original description is definitely based, is *Merremia caespitosa* Hallier f. = *M. hirta* (Linn.) Merr. Linnaeus erred in referring *Ballel* Rheede (Hort. Malabar. **11**: 107. *pl. 52*) to his species. Rheede's illustration represents *Ipomoea aquatica* Forsk., from which most authors have erroneously interpreted the Linnaean species. The Rheedian reference cannot be interpreted as the type of *Convolvulus reptans* Linn. in view of the fact that the Linnaean description was based on an actual specimen which is still extant and which represents a form of the totally different *Merremia hirta* (Linn.) Merr.

Ipomoea digitata Linn. Syst. ed. 10, 924. 1759; Gagnep. & Courchet in Lecomte Fl. Gén. Indo-Chine **4**: 239. 1915.

> *Convolvulus paniculatus* Linn. Sp. Pl. 156. 1753, non *Ipomoea paniculata* Burm. f.
> *Ipomoea paniculata* R. Br. Prodr. 486. 1810.
> *Batatas loureirii* G. Don Gen. Syst. **4**: 262. 1838 (based on *Ipomoea tuberosa* Lour.).
> *Ipomoea tuberosa* (non Linn.) Lour. Fl. Cochinch. 112. 1790, ed. Willd. 138. 1793, Anamese *khoai xiem*.

"Habitat in Cochinchina a Siamo oriunda." The description is definite and applies unmistakably to the widely distributed species sometimes known as *Ipomoea paniculata* R. Br., the valid name for which is *Ipomoea digitata* Linn.

Ipomoea hederacea Jacq. Ic. Pl. Rar. **1**: 4. *pl. 36*. 1786, Coll. **1**: 124. 1786; Gagnep. & Courchet in Lecomte Fl. Gén. Indo-Chine **4**: 241. 1915.

> *Convolvulus hederaceus* Linn. Sp. Pl. 154. 1753.
> *Ipomoea hepaticifolia* (non Linn.) Lour. Fl. Cochinch. 112. 1790, ed. Willd. 138. 1793, Anamese *bìm bìm biéc*.
> ? *Convolvulus tomentosus* (non Linn.) Lour. Fl. Cochinch. 108. 1790, ed. Willd. 133. 1793, Anamese *bìm bìm loung*, Chinese *khiēn nieu*.

For *Ipomoea hepaticifolia* Loureiro states: "Habitat inculta in Cochinchina." The description clearly applies to *Ipomoea hederacea* Jacq., which is common and widely distributed in tropical Asia, not to *Ipomoea pes-tigridis* Linn., which is the species represented by Burman's illustration (Fl. Ind. 50. *pl. 20. f. 2*) which Loureiro erroneously cites as illustrating the species he described. *Ipomoea hepaticifolia* Linn. is the same as *I. pes-tigridis* Linn. It may be noted that *Ipomoea hederacea* Jacq., although representing the species described by Linnaeus as *Convolvulus hederaceus*, was independently published by Jacquin, as in neither the "Collectanea" nor in the "Icones" does he cite any synonym. It may be that *Ipomoea nil* (Linn.) Roth should be accepted as the proper binomial for this common species. For *Convolvulus tomentosus* Loureiro states: "Habitat incultus in Cochinchina, & China." Hemsley (Journ. Linn. Soc. Bot. **26**: 163. 1890) suggested that Loureiro's species might be *Ipomoea hederacea* Jacq. or *I. congesta* R. Br. The description is short, and is unsatisfactory for either, yet because of Loureiro's long note regarding its medicinal properties, it is suspected that he had imperfect material secured from an herbalist, which would account for his description of the flowers as "pallidis."

Ipomoea obscura (Linn.) Ker in Bot. Reg. **3**: *pl. 239*. 1817; Gagnep. & Courchet in Lecomte Fl. Gén. Indo-Chine **4**: 246. 1915.

> *Convolvulus obscurus* Linn. Sp. Pl. ed. 2, 220. 1762.
> *Convolvulus malabaricus* (non Linn.) Lour. Fl. Cochinch. 108. 1790, ed. Willd. 132. 1793, Anamese *bìm bìm doùng tièn*.

"Habitat agrestis in Cochinchina." I believe this to represent the species that Loureiro so inadequately described. *Kattu Kelengu* Rheed. (Hort. Malabar. **11**: 105. *pl. 51*), cited by Loureiro as representing his species, is the totally different *Argyreia malabarica* (Linn.) Choisy.

Ipomoea pes-caprae (Linn.) Sweet Hort. Suburb. Lond. 35. 1818; Roth Nov. Sp. Pl. 109. 1821.

> *Convolvulus pes-caprae* Linn. Sp. Pl. 159. 1753; Lour. Fl. Cochinch. 109. 1790, ed. Willd. 134. 1793, Anamese *cây muong bien.*
>
> *Ipomoea biloba* Forsk. Fl. Aeg.-Arab. 44. 1775; Gagnep. & Courchet in Lecomte Fl. Gén. Indo-Chine **4**: 259. 1915.

"Habitat ad maris litora arenosa in Cochinchina, China, & in ora Africae Orientali." Loureiro correctly interpreted the Linnaean species; it is a very common pantropic strand plant.

Ipomoea pes-tigridis Linn. Sp. Pl. 162. 1753; Gagnep. & Courchet in Lecomte Fl. Gén. Indo-Chine **4**: 268. 1915.

> *Convolvulus aggregatus* Lour. Fl. Cochinch. 109. 1790, ed. Willd. 135. 1793, Anamese *bìm bìm lá dua.*
>
> *Ipomoea aggregata* Poir. in Lam. Encycl. Suppl. **3**: 471. 1814 (based on *Convolvulus aggregatus* Lour.).
>
> *Cleisostoma villosa* Raf. Fl. Tellur. **4**: 80. 1838 (based on *Convolvulus aggregatus* Lour.).

"Habitat in Cochinchina." Loureiro's description is definite and applies unmistakably to the common and widely distributed Linnaean species. The genus *Cleisostoma* Raf. is typified by this species.

Ipomoea sp.

> *Convolvulus obscurus* (non Linn.) Lour. Fl. Cochinch. 107. 1790, ed. Willd. 131. 1793, Anamese *bìm bìm rùng,* Chinese *că phân xý.*

"Habitat in Cochinchina, & China, in locis sylvestribus." The description is short and rather indefinite. I should be inclined to refer it to *Ipomoea obscura* Lindl., but for Loureiro's description of the peduncles as many-flowered and the leaves as pubescent on both surfaces. In some respects the description suggests *Hewittia sublobata* (Linn.) O. Ktz., but in *Hewittia* the seeds are glabrous, not hirsute. The illustration cited by Loureiro (Dill. Hort. Elth. *pl. 83. f. 95*) is *Ipomoea obscura* Lindl.

Ipomoea (vel *Merremia?*) **sp.**

> *Convolvulus bufalinus* Lour. Fl. Cochinch. 109. 1790, ed. Willd. 134. 1793, Anamese *bìm bìm tlâu.*
>
> *Ipomoea bufalina* Choisy in Mém. Soc. Phys. Hist. Nat. Genève **6**: 452. 1833 (based on *Convolvulus bufalinus* Lour.).
>
> *Nemanthera bufalina* Raf. Fl. Tellur. **4**: 80. 1836 (based on *Convolvulus bufalinus* Lour.).
>
> *Merremia bufalina* Merr. & Rolfe in Philip. Journ. Sci. Bot. **3**: 122. 1908 (based on *Convolvulus bufalinus* Lour.).
>
> *Operculina bufalina* Hall. f. in Meded. Rijks Herb. Leiden **1**: 26. 1910 (based on *Convolvulus bufalinus* Lour.).

"Habitat in sylvis Cochinchinae: nec alibi vidi." Hallier f. in adopting Loureiro's specific name under *Operculina*, placed *Ipomoea riedeliana* Oliv. (Hook. Ic. **15**: *pl. 1424.* 1883) as a synonym, which I followed (Enum. Philip. Fl. Pl. **3**: 363. 1923). I now consider this disposition of Oliver's species to be erroneous, first, because *Ipomoea riedeliana* Oliv. is not known from Indo-China and, second, because its anthers are not spirally contorted. This anther character first described as "antheris cochleatis" in the diagnosis and later in the description as "antheris 5, filiformibus, spiraliter contortis" suggests *Operculina turpethum* S. Manso, but the "Flos luteus . . . pedunculis polyfloris" are not characters of *Operculina turpethum*. *Ipomoea staphylina* Roem. & Schult. which, as interpreted by Gagnepain and Courchet, occurs near Hue, and is represented by *Clemens 3238* from Mount Bana, seems to be eliminated as a possibility as its leaves are broadly ovate, and, while cordate, are not at all "cordato-sagittata," nor are its flowers large, while its anthers are straight. The flowers of this species in Indo-China and in Hainan are yellow, although Clarke (Hooker f. Fl. Brit. Ind. **4**: 210. 1883) describes them as purplish-white or purplish in the tube, the mouth being nearly white. Gagnepain & Courchet in their treatment of the Convolvulaceae of Indo-China (Lecomte Fl. Gén. Indo-Chine **4**: 228–313. 1915) do not account for Loureiro's species or any of the synonyms based upon it, nor can I definitely refer it to any of the 51 species of *Ipomoea* admitted by them.

Quamoclit Moench

Quamoclit pennata (Desr.) Bojer Hort. Maurit. 224. 1837.

  Convolvulus pennatus Desr. in Lam. Encycl. **3**: 567. 1791.

  Ipomoea quamoclit Linn. Sp. Pl. 159. 1753; Lour. Fl. Cochinch. 111. 1790, ed. Willd. 137. 1793, Anamese *cây deuong leo*, Chinese *kam peng fung;* Gagnep. in Lecomte Fl. Gén. Indo-Chine **4**: 235. 1915.

  Quamoclit vulgaris Choisy in Mém. Soc. Phys. Hist. Nat. Genève **6**: 434. 1833.

"Habitat culta in Cochinchina, & in China." The description unmistakably applies to the very common Linnaean species, one of American origin, but now widely naturalized in the Old World tropics. *Flos cardinalis* Rumph. (Herb. Amb. **5**: 421. *pl. 155. f. 2*) and *Tsiuri-acranti* Rheede (Hort. Malabar. **11**: 123. *pl. 60*), cited by Loureiro as synonyms, are correctly placed.

Argyreia Loureiro

Argyreia acuta Lour. Fl. Cochinch. 135. 1790, ed. Willd. 167. 1793, Chinese *pă hŏ thân;* Gagnep. in Bull. Soc. Bot. France **62**: 4. 1915; Gagnep. & Courchet in Lecomte Fl. Gén. Indo-Chine **4**: 279. 1915.

  Argyreia festiva Wall. Pl. As. Rar. **1**: 68. *pl. 76.* 1830.

"Habitat in Sinis, a praecedenti [*A. obtusifolia*] non admodum differens." Choisy (DC. Prodr. **9**: 333. 1845) examined Loureiro's type and reduced *A. festiva* Wall. to *A. acuta* Lour.; Gagnepain (Bull. Soc. Bot. France **62**: 4. 1915) confirms this. Loureiro's type is preserved in the herbarium of the Paris Museum.

Argyreia obtusifolia Lour. Fl. Cochinch. 134. 1790, ed. Willd. 166. 1793, Anamese *lá bac thau;* Gagnep. in Bull. Soc. France **62**: 5. 1915; Gagnep. & Courchet in Lecomte Fl. Gén. Indo-Chine **4**: 283. 1915.

  ? Convolvulus obtectus Wall. List no. 1416. 1829, *nomen nudum.*

? *Rivea* ? *obtecta* Choisy in Mém. Soc. Phys. Hist. Nat. Genève **6**: 410. 1833.

? *Argyreia obtecta* C. B. Clarke in Hook. f. Fl. Brit. Ind. **4**: 186. 1883.

"Habitat in Cochinchina per sepes scandens." A very fragmentary specimen from Loureiro is preserved in the herbarium of the British Museum, consisting of branchlets, fruits and a fragment of one leaf. This was examined by Choisy (DC. Prodr. **9**: 333. 1845). Clarke referred Loureiro's species to *Argyreia obtecta* Clarke var. *obtusifolia* (Lour.) Clarke (Hook. f. Fl. Brit. Ind. **4**: 186. 1883), but in my list of Loureiro's British Museum plants the statement appears: "This C. B. Clarke did not see; seems different from *obtecta*." Prain (Journ. As. Soc. Beng. **74**: 320. 1906) affirms Clarke's disposition of it. For discussion of the problems involved see Gagnepain (Bull. Soc. Bot. France **62**: 5. 1915). Although Gagnepain & Courchet (Lecomte Fl. Gén. Indo-Chine **4**: 275. 1915) place *Argyreia obtusifolia* Lour., the type of the genus, as a synonym of the much more recent *A. obtecta* (Wall.) C. B. Clarke, they at the same time (p. 283) admit Loureiro's species as an independent one, with a short description compiled from his original. This procedure is reversed in the case of *A. acuta* Lour. where they properly accept Loureiro's binomial and place the much better described *A. festiva* Wall. as a synonym of it.

HYDROPHYLLACEAE

Hydrolea [65] Linnaeus

Hydrolea zeylanica (Linn.) Vahl Symb. **2**: 46. 1791; Brand in Pflanzenreich **59**(IV–251): 174. 1913.

Nama zeylanica Linn. Sp. Pl. 226. 1753.

Hydrolea inermis Lour. Fl. Cochinch. 172. 1790, ed. Willd. 214. 1793, Chinese *xiong fung*.

"Habitat Cantone, in locis humidis." Loureiro's description conforms to the characters of the widely distributed Indo-Malaysian Linnaean species. It is abundant in low wet places, rice paddies, etc., in the vicinity of Canton. Loureiro's type is preserved in the Paris Museum.

BORAGINACEAE

Cordia Linnaeus

Cordia dichotoma Forst. f. Prodr. 18. 1786; Merr. Enum. Philip. Fl. Pl. **3**: 373. 1923.

Cordia obliqua Willd. Phytogr. 4. *pl. 4*. 1794; Hutch. in Kew Bull. 221. 1918.

Varronia sinensis Lour. Fl. Cochinch. 138. 1790, ed. Willd. 171. 1793, Anamese *son châu duu*, Chinese *xān chū yû*, non *Cordia sinensis* Lam.

Cordia loureiri Roem. & Schult. Syst. **4**: 466. 1819 (based on *Varronia sinensis* Lour.).

Argyreia arborea Lour. Fl. Cochinch. 135. 1790, ed. Willd. 167. 1793, Anamese *truong xuĕn hoā*.

Cordia myxa auct. plur., non Linn.

For *Varronia sinensis* Loureiro states: "Habitat in variis locis imperii Sinensis," and for *Argyreia arborea*: "Habitat in sylvis, & colitur in hortis tam in Cochinchina, quam in China." Hutchinson (Kew Bull. 219. 1918) has shown that *Cordia myxa* Linn. has

---

[65] *Hydrolea* Linnaeus (1763), conserved name, Vienna Code; an older one is *Nama* Linnaeus (1753).

been misinterpreted by all modern authors, and that the Linnaean species is one confined to Arabia, Egypt, and tropical Africa. For the very common and widely distributed Indo-Malaysian species, currently but erroneously identified as *Cordia myxa* Linn., he adopted the binomial *Cordia obliqua* Willd. (1794); I adopt the still older *Cordia dichotoma* Forst. f. (1786), as I believe it to represent the same species. *Argyreia arborea* Lour. supplies a specific name which would be valid in *Cordia*, and which is older than *Cordia obliqua* Willd. in case it be found that *C. dichotoma* Forst. f. does not apply to this species. Hemsley (Journ. Linn. Soc. Bot. **26**: 143. 1890) merely mentions *Varronia sinensis* Lour. as an obscure plant, that is, one not definitely placed either as to its proper genus or species, but at that time he had no Chinese specimens of *Cordia; C. dichotoma* Forst. f. is now known from rather numerous collections in Kwangtung and is not uncommon in the immediate vicinity of Canton. Loureiro's description of *Varronia sinensis* applies unmistakably to the species here interpreted as *Cordia dichotoma* Forst. f. It may be noted here that *Cordia sinensis* Lam. was published independently of *Varronia sinensis* Lour. and Doctor Gagnepain, who kindly looked up Lamarck's type at my request, informs me that it is not a representative of the Boraginaceae. *Argyreia arborea* Lour. was reduced by Hallier f. (Bull. Herb. Boiss. **6**: 716–719. 1898, Meded. Rijks Herb. Leiden **36**: 3. 1918) to *Cordia myxa* as he understood the latter species (= *C. dichotoma* Forst. f.) and I see no valid reason for not accepting this reduction.

Rotula Loureiro

Rotula aquatica Lour. Fl. Cochinch. 121. 1790, ed. Willd. 150. 1793, Anamese *rì rì cát;* Bunting in Journ. Bot. **47**: 270. 1909.

> *Rhabdia lycioides* Mart. Nov. Gen. Sp. **2**: 137. *pl. 195.* 1826; Gagnep. & Courchet in Lecomte Fl. Gén. Indo-Chine **4**: 214. *f. 25, 12–18.* 1914.

"Habitat in paludibus, & prope ripas fluminum in Cochinchina," Bunting's critical note on Loureiro's type in the herbarium of the British Museum definitely settles the status of Loureiro's genus and species, he finding that *Rotula aquatica* Lour. is identical with *Rhabdia lycioides* Mart. Loureiro's generic name antedates that of Martius by 36 years.

Tournefortia Linnaeus

Tournefortia montana Lour. Fl. Cochinch. 122. 1790, ed. Willd. 150. 1793, Anamese *cây thuóc moi;* Gagnep. & Courchet in Lecomte Fl. Gén. Indo-Chine **4**: 219. 1914 (sp. dub.); Johnst. in Journ. Arnold Arb. **16**: 153. 1935.

> *Messerschmidtia montana* Roem. & Schult. Syst. **4**: 544. 1819 (based on *Tournefortia montana* Lour.).
>
> *Tournefortia gaudichaudii* Gagnep. in Lecomte Fl. Gén. Indo-Chine **4**: 217. *f. 26, 1–4.* 1914.

"Habitat in sylvis montanis, ubi illa pro pharmacho utuntur incolae agrestis, qui vocantur Moii, Cochinchinae tributarii." Although Gagnepain and Courchet treat *Tournefortia montana* Lour. as a species of doubtful status, I interpret it as representing the one Gagnepain described as *Tournefortia gaudichaudii*, which is apparently the most common and most widely distributed species of the genus in Indo-China. Loureiro's description agrees in all essentials with the characters of Gagnepain's species. The only other possibilities, *T. sarmentosa* Lam. and *T. boniana* Gagnep., are eliminated by their more or less pubescent leaves. Loureiro definitely describes *T. montana* as having glabrous leaves, a character also of *T. gaudichaudii* Gagnep.

Heliotropium (Tournefort) Linnaeus

Heliotropium indicum Linn. Sp. Pl. 130. 1753; Lour. Fl. Cochinch. 103. 1790, ed. Willd. 126. 1793, Anamese *cây bòi boi.*

Under *H. tetrandrum* Loureiro states: "Habitat inter herbas luxuriantes in hortis Cochinchinae tam ista, quam prima [*H. indicum*] species." His description applies to the very common and widely distributed Linnaean species.

Cynoglossum (Tournefort) Linnaeus

Cynoglossum lanceolatum Forsk. Fl. Aeg.-Arab. 41. 1775; Brand in Pflanzenreich 78(IV-252): 137. *f. 18.* 1921.

Cynoglossum micranthum Desf. Tabl. 220. 1804.

Echium vulgare (non Linn.) Lour. Fl. Cochinch. 102. 1790, ed. Willd. 125. 1793, Anamese *cây lo buòn.*

No locality is given, but from the local name cited, Loureiro's specimens were from Indo-China. The short description agrees with the characters of the widely distributed *Cynoglossum lanceolatum* Forsk.

Lithospermum (Tournefort) Linnaeus

Lithospermum officinale Linn. Sp. Pl. 132. 1753.

Anchusa officinalis (non Linn.) Lour. Fl. Cochinch. 103. 1790, ed. Willd. 127. 1793, Anamese *tu thao,* Chinese *tsù tsào.*

"Habitat in variis locis imperii Sinensis." Hemsley (Journ. Linn. Soc. Bot. **26**: 155. 1890) cites Henry and Bretschneider to the effect that the Chinese name given by Loureiro belongs to *Lithospermum officinale* Linn., a species that occurs in China and which is extensively used by the Chinese as a drug plant. There is little doubt as to the correctness of this reduction of *Anchusa officinalis* Lour.

VERBENACEAE

Verbena Linnaeus

Verbena officinalis Linn. Sp. Pl. 20. 1753; Lour. Fl. Cochinch. 27. 1790, ed. Willd. 33. 1793, Anamese *co roi ngua, ma tien thao,* Chinese *mà pién tsào.*

"Habitat agrestis, cultaque tam in Cochinchina, quam in China." Loureiro correctly interpreted the Linnaean species, which is a very common weed in southern China.

Lippia Houstoun

Lippia nodiflora (Linn.) L. C. Rich. in Michx. Fl. Bor. Amer. **2**: 15. 1803.

Verbena nodiflora Linn. Sp. Pl. 20. 1753.

Phyla chinensis Lour. Fl. Cochinch. 66. 1790, ed. Willd. 83. 1793, Chinese *lién fuēn.*

Piarimula chinensis Raf. Fl. Tellur. **2**: 102. 1837 (based on *Phyla chinensis* Lour.).

No locality is given, but the Chinese name cited indicates specimens from Canton, where the species is very common. There are some curious discrepancies in the description, as Loureiro interpreted the bracts as calyx-segments and apparently the bracteoles as the " perianth," indicating the inflorescence as many-flowered, yet in the description of the species referring to the peduncles as 1-flowered. In spite of these discrepancies, or errors in interpretation of floral parts, Loureiro's description applies only to *Lippia nodiflora* L. C. Rich. among all the known species of southern China. His type is preserved in the herbarium of the Paris Museum.

Callicarpa Linnaeus

Callicarpa dichotoma (Lour.) K. Koch [66] Dendr. 2: 336. 1872 (based on *Porphyra dichotoma* Lour.).

Porphyra dichotoma Lour. Fl. Cochinch. 70. 1790, ed. Willd. 87. 1793, Chinese *tsù hōa uôn*.

Callicarpa purpurea Juss. in Ann. Mus. Hist. Nat. (Paris) 7: 67. 1806 (based on *Porphyra dichotoma* Lour.).

" Habitat in collibus provinciae Cantoniensis Sinarum." The species is not uncommon in thickets near Canton and is frequently known as *Callicarpa purpurea* Juss. Willdenow's suggestion: " Genus vix a *Callicarpa* diversum," led to Raeuschel's abortive attempt to transfer the specific name to *Callicarpa*. Loureiro's type is preserved in the herbarium of the Paris Museum.

Callicarpa loureiri Hook. & Arn. Bot. Beechey's Voy. 205. 1836, in nota (based on *Callicarpa americana* Lour.).

Callicarpa americana (non Linn.) Lour. Fl. Cochinch. 70. 1790, ed. Willd. 88. 1793, Anamese *cây nang nang*.

Callicarpa tomentosa Hook. & Arn. Bot. Beechey's Voy. 205. 1836, non Murr.

Callicarpa longiloba Merr. in Philip. Journ. Sci. 13: Bot. 156. 1918.

Callicarpa roxburghii H. Lam in Bull. Jard. Bot. Buitenzorg III 3: 22. 1921, non Wall.

" Habitat ad sepes in Cochinchina." The description in general and particularly that of the 4-fid calyx applies unmistakably to *Callicarpa loureiri* Hook. & Arn., a species which is very common in southeastern China. I have, however, seen no material from Indo-China representing it. The synonymy has been much confused, but I believe the oldest valid name to be *C. loureiri* Hook. & Arn. This was incidentally published in a note following Hooker & Arnott's consideration of *Callicarpa tomentosa* Willd. thus: " If our species prove distinct it may be called *C. loureiri*, for it is undoubtedly the plant of Loureiro [*C. americana* Lour.]." This binomial was overlooked by the compilers of Index Kewensis, and by myself when I proposed the new name *Callicarpa longiloba* in 1919. Doctor Lam has, erroneously I believe, adopted the binomial *C. roxburghii* Wall. for the Chinese species with 4-fid calyces. The history of this binomial is as follows: Wallich, List no. 1833. 1829, states: " 1833 *Callicarpa Roxburghii* Wall. *C. incana* Roxb. non *C. cana* L. HBC. [Hort. Bot. Calcutt.]." Roxburgh's description of *C. incana* (Fl. Ind. 1:

---

[66] Raeuschel, Nomenclator ed. 3, 37. 1797, lists *Callicarpa dichotoma*, but here it is a *nomen nudum* with no indication whatever on which it was based. The binomial appears again in synonymy under *Callicarpa purpurea* Juss. in Steud. Nomencl. 137. 1821, which is not a valid publication under the rule that a name appearing in synonymy is not validly published. K. Koch seems to have been the first author to make the valid transfer of Loureiro's specific name to *Callicarpa*.

407. 1820) was based on material from Bengal. His description of " *C. cana* Linn. Sp. Pl. ed. Willd. I. 620 " on the preceding page was based on material grown in the Botanic Garden at Calcutta, introduced from the Moluccas in 1798, and to this he refers *C. americana* Lour. as a synonym. In describing the calyx he merely states that it is woolly and 4-toothed. The description does not apply to the 4-fid calyces of Loureiro's species. I suspect that this Moluccan plant will prove to be the same as *Callicarpa macrophylla* Vahl as interpreted by Lam. Walpers (Repert. **4**: 128. 1844) in describing *C. roxburghii* Wall. does not mention the calyx lobes or teeth and queries: " An re vera a *C. macrophylla* diversa? (v.s. sp.)." I enlisted the services of Mr. T. A. Sprague at Kew in reference to the problem involved and below is quoted the report prepared by him and Mr. C. E. C. Fischer, August 18, 1931.

" (1) The name *Callicarpa Roxburghii* was published by Wallich, Cat. n. 1833 (1828–29) as a new name for *C. incana* Roxb., non *C. cana* L. It was effectively published since it is associable with the description of *C. incana* Roxb., but it is an illegitimate name because it was superfluous. (2) Walpers, Rep. iv. 127 (1844–48) published a description of *C. Roxburghii* apparently based on Wall. Cat. n. 1833, *specimen*. A much better description of *Callicarpa Roxburghii* Wall. Cat. n. 1833, *specimen*, was published by Schauer in DC. Prodr. xi. 640 (1847). This mentions the setaceous calyx-lobes. (3) C. B. Clarke (F.B.I. iv. 568) and Lam (Bull. Jard. Bot. Buitenz. ser. 3, iii. 23) reduce *C. incana* Roxb. to *C. macrophylla* Vahl, apparently correctly. (4) *Callicarpa Roxburghii* Wall. (1828–29) is accordingly a taxonomic synonym of *C. macrophylla* Vahl. (5) *The specimen of Callicarpa Roxburghii* Wall. Cat. n. 1833 described by Walpers (?) and Schauer belongs, however, to a different species, namely to the South Chinese *Callicarpa* included in Index Fl. Sin. ii. 255 (1890) as *C. tomentosa* Willd. It has the characteristic calyx-lobes of this South Chinese plant. (6) The South Chinese plant concerned should therefore, as suggested by Dr. Merrill, bear the name *C. Loureiri* Hook. et Arn. with ' *C. Roxburghii* Wall. ex Schauer (1847), excl. syn.' as a synonym. Even if Lam (Bull. Jard. Bot. Buitenz. ser. 3, iii. **22**: 1921) is correct in identifying it with *C. tomentosa* Willd. (1809), that name cannot be used since it is a later homonym of *C. tomentosa* (L.) Murr. 1774)." Doctor Diels informed me in May 1930 that Willdenow's type of *Callicarpa tomentosa* does not exist in the Berlin herbarium.

Tectona [67] Linnaeus f.

Tectona grandis Linn. f. Suppl. 151. 1781.

> *Tectona theka* Lour. Fl. Cochinch. 137. 1790, ed. Willd. 169. 1793 (*T. theca*), Anamese *cây sao.*

" Habitat in vastis sylvis provinciae *Doùng nai,* ad confinia Cochinchinae, & Cambodiae." The short description applies to the common teak and the pre-Linnaean synonyms cited appertain to it.

Vitex (Tournefort) Linnaeus

Vitex negundo Linn. Sp. Pl. 638. 1753; Lour. Fl. Cochinch. 390. 1790, ed. Willd. 474. 1793, Anamese *thuóc ôn rùng,* Chinese *mûen kĭm.*

[67] *Tectona* Linnaeus f. (1781), conserved name, Vienna Code; an older one is *Theka* Adanson (1763).

Vitex spicata Lour. Fl. Cochinch. 390. 1790, ed. Willd. 475. 1793, Anamese *cây ngu trao*, Chinese *ù chū kīm*.

For *Vitex spicata* Loureiro states: " Habitat inculta, cultaque in Cochinchina, & China." His excellent description applies unmistakably to the common, widely distributed and somewhat variable *Vitex negundo* Linn. For *Vitex negundo* he states: " Habitat in locis mediterraneis sylvestribus Cochinchinae, & Chinae," and while the description is not good, it can scarcely refer to other than the Linnaean species.

Vitex payos (Lour.) comb. nov.
>*Allasia payos* Lour. Fl. Cochinch. 85. 1790, ed. Willd. 107. 1793.
>*Vitex hildebrandtii* Vatke in Linnaea 43: 534. 1882; Baker in Thistelton-Dyer Fl. Trop. Afr. 5: 326. 1900.

" Habitat in ora Africae Orientale." Loureiro gives the local name as *muringuiringue*. His type is preserved in the herbarium of the Paris Museum and Planchon (Ann. Sci. Nat. IV Bot. 2: 262. 1854) states that the specimen is a *Vitex* probably very near *V. lanigera* Schauer of Madagascar. He further states that the description of the fruits is incorrect. In my judgment, from comparisons of descriptions, *Vitex hildebrandtii* Vatke is the same as *Allasia payos* Lour., and I have accordingly accepted Loureiro's specific name, reducing Vatke's species as a synonym.

Vitex quinata (Lour.) F. N. Will. in Bull. Herb. Boiss. II 5: 431. 1905; Druce in Rept. Bot. Exch. Club Brit. Isles 4: 652. 1917 (based on *Cornutia quinata* Lour.).
>*Cornutia quinata* Lour. Fl. Cochinch. 387. 1790, ed. Willd. 470. 1793, Chinese *ù sī kām*.
>*Vitex loureiri* Hook. & Arn. Bot. Beechey's Voy. 206. *pl. 48.* 1836 (based on *Cornutia quinata* Lour.).
>*Vitex heterophylla* Roxb. Hort. Beng. 46. 1814, *nomen nudum*, Fl. Ind. ed. 2, 3: 75. 1832.

" Habitat in sylvis prope Cantonem Sinarum." This species is still common in the immediate vicinity of Canton, in and near villages, and is one of the largest native trees still growing in the region. It is represented by numerous recent collections. I am not sure that the Indian form actually described by Roxburgh is identical with *Vitex quinata* F. N. Will., but the Chinese form so-named is certainly Loureiro's species. Loureiro's specimen in the herbarium of the Paris Museum was identified by Desvaux as representing *Vitex leucoxylon* Linn., which is apparently erroneous, as the Linnaean species is known only from British India. H. Lam (Bull. Jard. Bot. Buitenzorg III 3: 55. 1921) retains *Vitex heterophylla* Roxb., citing *Vitex quinata* F. N. Will. as a synonym, and gives the species, including two varieties, an extended range from India to southeastern China, to Sumatra, Java, Philippines, Celebes, and the Moluccas. Loureiro's name is much older than Roxburgh's.

Vitex trifolia Linn. Sp. Pl. 638. 1753 (*trifoliis*); Lour. Fl. Cochinch. 390. 1790, ed. Willd. 474. 1793, Anamese *thuóc òn, quan âm bien, man kinh*.

" Habitat frequentissima ad maris litora, per arenam repens in Cochinchina, sicut etiam in China." Loureiro's description and other data, including *Lagondium vulgare* Rumph. (Herb. Amb. 4: 48. *pl. 18*, cited as a synonym), appertain unmistakably to the very common, widely distributed, littoral Linnaean species.

Vitex tripinnata (Lour.) comb. nov.

Tripinna tripinnata Lour. Fl. Cochinch. 391. 1790, ed. Willd. 476. 1793, Anamese
cây den; Moore in Journ. Bot. **63**: 286. 1925.

Tanaecium tripinna Raeusch.[68] Nomencl. ed. 3, 178. 1797, *nomen nudum.*

Tripinnaria cochinchinensis Pers. Syn. **2**: 173. 1806 (based on *Tripinna tripinnata*
Lour.).

Tripinnaria asiatica Spreng. Syst. **2**: 842. 1825 (based on *Tripinna tripinnata* Lour.).

Tripinnaria tripinnata Steud. Nomencl. ed. 2, **2**: 712. 1841 (based on *Tripinna tripinnata* Lour.).

Colea tripinnata Seem. in Bonplandia **4**: 128. 1856 (based on *Tripinna tripinnata*
Lour.).

Vitex annamensis Dop in Bull. Soc. Hist. Nat. Toulouse **57**: 203. 1928.

" Habitat in sylvis montanis Cochinchinae." Baillon (Bull. Soc. Linn. Paris **1**: 714.
1888) states that Loureiro's specimen in the herbarium of the British Museum is a *Vitex*,
very near, if not identical with, numerous specimens collected by Pierre in Indo-China.
Moore confirms this and states that Loureiro's type may be conspecific with *Pierre 1864.*
Loureiro erred in describing the fruits as many-seeded and the leaves as *tripinnate;* it is
suspected that *trifoliolate* was intended, as his type specimen has 3-foliolate leaves. These
two erroneous characters led to Seemann's attempt to place *Tripinna* in the Bignoniaceae.
Here we have a rather long list of synonyms based on what was, to all authors proposing
them, an unknown species or at least one known only from Loureiro's incomplete and
inaccurate description. Dop[69] does not mention Loureiro's genus and species, but I
believe *Vitex annamensis* Dop to represent the species Loureiro originally described.

Gmelina Linnaeus

Gmelina philippensis Cham. in Linnaea **7**: 109. 1832; Merr. Enum. Philip. Fl. Pl. **3**: 399.
1923.

Gmelina hystrix Kurz in Journ. As. Soc. Bengal **39**(2): 81. 1870.

Gmelina asiatica Linn. var. *philippinensis* Bakh. in Bull. Jard. Bot. Buitenz. III **3**:
70. 1921.

Gmelina asiatica (non Linn.) Lour. Fl. Cochinch. 376. 1790, ed. Willd. 456. 1793,
Anamese *cây gang tu hú.*

" Habitat in sylvis Cochinchinae." Loureiro referred his material to *Gmelina asiatica*
Linn. with expressed doubt. His description conforms better with the characters of
Gmelina philippensis Cham. than with *G. asiatica* Linn., and the former is well represented
by *Clemens 3152* from near Loureiro's classical locality. Loureiro describes the bracts as
red, but in the Philippine form they are always yellow. Doctor Lam (Verb. Malay.
Archipel. 221. 1919) reduced *G. philippensis* to *G. asiatica* Linn., but two years later
Bakhuizen assigned the former to varietal rank under *G. asiatica* Linn. *Radix deiparae
spuria* Rumph. (Herb. Amb. **2**: 125. *pl. 40),* cited by Loureiro as a synonym, must be
excluded as it represents *Gmelina elliptica* Sm. (*G. villosa* Roxb.).

[68] Raeuschel does not cite the binomial on which his new one is based. He followed Willdenow's
suggestion, (Fl. Cochinch. ed. Willd. 476. 1793) footnote, and intended *Tanaecium tripinna* to represent
Tripinna tripinnata Lour.

[69] Dop, P. Les Vitex de l'Indo-Chine. Bull. Soc. Hist. Nat. Toulouse **57**: 197–211. *pl. 2–4.* 1928.
Reprinted Trav. Lab. Forest. Toulouse **1**(1): 1–15. *pl. 2–4.* 1928.

Gmelina racemosa (Lour.) comb. nov.

Lantana racemosa Lour. Fl. Cochinch. 376. 1790, ed. Willd. 457. 1793, Anamese *cây tlai*.
Gmelina hainanensis Oliv. in Hook. Ic. **19**: sub *pl. 1874*. 1889, *nomen subnudum*.
Gmelina balansae Dop in Bull. Soc. Bot. France **61**: 322. 1915.

"Habitat in sylvis Cochinchinae." Loureiro's description definitely applies to *Gmelina* and to a species in the group with *G. chinensis* Benth., *G. hainanensis* Oliv., *G. balansae* Dop and *G. lecomtei* Dop. Among these it agrees best with the characters of *G. hainanensis* Oliv., from which I do not think that *G. balansae* Dop can be distinguished. *Gmelina hainanensis* Oliv. was most casually and inadequately described by Oliver in the discussion following a description of *G. chinensis* Benth. In recent Hainan collections the flowers are described as "white and yellow" and "white or pinkish outside, lavender and orange inside." Oliver describes them as "white-edged with purple centre," while Loureiro describes them as white. *Clemens 3980* from Mount Bana, near Tourane, "flowers yellow with purple," probably represents Loureiro's species.

Clerodendrum [70] (Burman) Linnaeus

Clerodendrum fortunatum Linn. ex Torner Cent. Pl. **2**: 23. 1756 (*Clerodendrum fortunata*), Amoen. Acad. **4**: 320. 1759 (*Clerodendrum fortunata*); Osbeck Dagbok Ostind Resa 228. *pl. 11*. 1757 (*Clerodendron fortunata*), Voy. China **1**: 369. *pl. 11*. 1771; Merr. in Sunyatsenia **1**: 30. 1930.

Volkameria pumila Lour. Fl. Cochinch. 388. 1790, ed. Willd. 472. 1793, Chinese *sān tâng lûng*.
Clerodendron lividum Lindl. Bot. Reg. **11**: *pl. 945*. 1826.
Clerodendron pumilum Spreng. Syst. **2**: 759. 1825 (based on *Volkameria pumila* Lour.).
Clerodendron castaneifolium Hook. & Arn. Bot. Beechey's Voy. 205. 1836.
Clerodendron pentagonum Hance in Walp. Ann. **3**: 238. 1852–53.
Clerodendron oxysepalum Miq. in Journ. Bot. Néerl. **1**: 114. 1861.

"Habitat inculta prope Cantonem Sinarum." Hemsley (Journ. Linn. Soc. Bot. **26**: 262. 1890) merely lists Loureiro's species as an obscure plant. The description is good and it applies in essentials to the Linnaean species, the type of which was a specimen from the vicinity of Canton; it may be noted however, that Seemann (Fl. Vit. 188. 1866) states that Loureiro's species is not the same as *Clerodendrum fortunatum* Linn. This, however, may possibly be due to a misinterpretation of the Linnaean species itself. Mr. J. E. Dandy informs me that he could find no specimen of *Volkameria pumila* from Loureiro in the herbarium of the British Museum, nor is the species checked in the Museum copy of Loureiro's Flora Cochinchinensis as having been received from him. Loureiro describes the leaves as large and tomentose; they are relatively small, as compared with those of many species of *Clerodendrum*, and are nearly glabrous in all specimens of the Linnaean species that I have seen. In the original description of *Clerodendrum fortunatum* Linnaeus states: "Habitat in India;" the specimen on which the species was based was collected by Osbeck near Canton, China. For a species that presents comparatively little variation, and one that is limited in distribution, it has accumulated a rather long list of synonyms, all based on material from southeastern China where it is common.

[70] Under the provisions of the International Code of Botanical Nomenclature *Clerodendrum* must be retained although most authors spell it *Clerodendron*. *Clerodendrum* was the form originated by Burman, was consistently used by Linnaeus, and is philologically correct.

Clerodendrum inerme (Linn.) Gaertn. Fruct. **1**: 271. *pl. 57.* 1788.
  Volkameria inermis Linn. Sp. Pl. 637. 1753; Lour. Fl. Cochinch. 388. 1790, ed. Willd.
    471. 1793, Chinese *sān fú mûn.*
  Volkameria commersonii Poir. in Lam. Encycl. **8**: 688. 1808.
  Clerodendron commersoni Spreng. Syst. **2**: 758. 1825.
  Volkameria nereifolia Roxb. Fl. Ind. ed. 2, **3**: 64. 1832.
  Clerodendron neriifolium Wall. List. No. 1789. 1829.
  " Habitat inculta prope Cantonem Sinarum." Loureiro's description applies to the
Linnaean species, which is common in suitable habitats near Canton. A specimen from
him is preserved in the herbarium of the Paris Museum.

Clerodendrum kaempferi (Jacq.) Sieb. in Verh. Bat. Genoots. **12**:31. 1830; Fisch. ex Steud.
    Nomencl. ed. 2, **1**: 383. 1840.
  Volkameria kaempferi Jacq. Collect. Bot. **3**: 207. 1789, Ic. Pl. Rar. **3**: 7. 1793.
  Volkameria kaempferiana Jacq. Ic. Pl. Rar. **3**: sub. *pl. 500.* 1786-93 (text printed in
    1793, plates issued at intervals between 1786 and 1793).
  Clerodendrum squamatum Vahl Symb. **2**: 74. 1791; H. Lam in Bull. Jard. Bot. Buitenzorg
    III **3**: 93. 1921.
  Clerodendrum infortunatum (non Linn.) Lour. Fl. Cochinch. 387. 1790, ed. Willd. 471.
    1793, Chinese *fung m̄i chū.*
  " Habitat Cantone Sinarum." Loureiro's description applies unmistakably to the
widely distributed species currently known as *Clerodendrum squamatum* Vahl, for which
H. Lam cites about twenty synonyms. *Clerodendron japonicum* (Thunb.) Sweet Hort.
Brit. 322. 1826, Makino in Bot. Mag. Tokyo **17**: 91. 1903 is the oldest binomial, if Doctor
Lam be followed in treating this as a collective species, as it was based on *Volkameria japonica*
Thunb. which dates from 1784. Doctor Carl G. Alm kindly supplied me with excellent
photographs of Thunberg's type with critical notes. Thunberg's statement: " Arbor
vasta, excelsa " is an error; the species is a small shrub. The plant is not " tota glabra,"
the branches of the inflorescences being densely hairy and with numerous intermixed
glandular hairs but the pilosity is not visible to the naked eye. The leaves are glabrous.
This form differs from *C. squamatum* Vahl, among other characters, by its much larger
calyces. The form with smaller calyces, which is not uncommon near Canton, is *C.
kaempferi* (Jacq.) Sieb. (*C. squamatum* Vahl), and this I believe to be specifically distinct
from *C. japonicum* (Thunb.) Sweet.

Clerodendrum paniculatum Linn. Mant. **1**: 90. 1767.
  Volkameria angulata Lour. Fl. Cochinch. 389. 1790, ed. Willd. 473. 1793, Anamese
    nu trinh thao, cây bây.
  Cleianthus coccineus Lour. ex Gomes in Mem. Acad. Sci. Lisb. Cl. Sci. Mor. Pol.
    Bel.-Let. n.s. **4**(1): 28. 1868.[71]
  " Habitat ubique in collibus, & in hortis minus cultis in Cochinchina." Loureiro's
description is an excellent one and it conforms to the characters of the Linnaean species.
His type is preserved in the herbarium of the Paris Museum and it has been identified by
Desvaux as *Clerodendrum paniculatum* Linn.

[71] A Loureiro herbarium name here first published by Gomes.

Clerodendrum petasites (Lour.) Moore in Journ. Bot. **63**: 285. 1925 (based on *Volkameria petasites* Lour.).

Volkameria petasites Lour. Fl. Cochinch. 388. 1790, ed. Willd. 472. 1793, Anamese *cay boung rôi*.

Clerodendron subpandurifolium O. Ktz. Rev. Gen. Pl. 506. 1891.

" Habitat in dumetis Cochinchinae." Loureiro took his specific name from *Petasites agrestis* Rumph. (Herb. Amb. **4**: 108. *pl. 49*) which he cites as illustrating his species, but which, however, represents a species very different from *Clerodendrum petasites* Moore. Schauer, perhaps interpreting the species from the Rumphian illustration, erroneously reduced *V. petasites* Lour. to *C. infortunatum* Gaertn. Loureiro's type is preserved in the herbarium of the British Museum, which on examination Moore found to be identical with *Clerodendrum subpandurifolium* O. Ktz., a species based on specimens collected by Kuntze at Tourane, Anam; Kuntze's actual type is preserved in the herbarium of the New York Botanical Garden; the species is also represented by *Squires 329*, from the classical locality Hue, and by *Robinson 1290* from Nha Trang. *Petasites agrestis* Rumph. which I (Interpret. Herb. Amb. 455. 1917) referred to *Clerodendrum speciosissimum* Van Geert is placed by H. Lam (Bull. Jard. Bot. Buitenzorg III **3**: 91. 1921) as a synonym of *Clerodendrum buchanani* (Roxb.) Walp., this apparently being the correct disposition of it.

Caryopteris Bunge

Caryopteris incana (Thunb.) Miq. Ann. Mus. Bot. Lugd.-Bat. **2**: 97. 1865.

Nepeta incana Thunb. Fl. Jap. 244. 1784.

Barbula sinensis Lour. Fl. Cochinch. 367. 1790, ed. Willd. 445. 1793, Chinese *sat song kim*.

Mastacanthus sinensis Endl. in Walp. Rep. **4**: 3. 1844 (based on *Barbula sinensis* Lour.).

Caryopteris mastacanthus Schauer in DC. Prodr. **11**: 625. 1847.

Caryopteris ovata Miq. in Journ. Bot. Néerl. **1**: 114. 1861.

Caryopteris sinensis Dippel Handb. Laubholzk. **1**: 59. 1889 (based on *Barbula sinensis* Lour.).

" Habitat Cantone Sinarum inculta." This species is locally abundant in open dry places in the vicinity of Canton. *Barbula* Lour. is older than *Caryopteris* Bunge, but is invalidated by the still older *Barbula* Hedw., a genus of mosses. Loureiro's type is preserved in the herbarium of the Paris Museum.

LABIATAE

Ajuga Linnaeus

Ajuga decumbens Thunb. Fl. Jap. 243. 1784.

Ajuga reptans (non Linn.) Lour. Fl. Cochinch. 363. 1790, ed. Willd. 441. 1793.

" Habitat inculta circa Cantonem Sinarum." Loureiro cites *Ajuga decumbens* Thunb. as a synonym of his misinterpreted *Ajuga reptans* Linn., and hence Loureiro's description might be typified by Thunberg's species. His description applies closely to the Kwangtung form that is currently referred to *Ajuga decumbens* Thunb. and I believe this to represent the species that he described.

Rosmarinus (Tournefort) Linnaeus

Rosmarinus officinalis Linn. Sp. Pl. 23. 1753; Lour. Fl. Cochinch. 28. 1790, ed. Willd. 34. 1793, Anamese *tây duong chôi*, Chinese *yong tsao*.

"Haec planta in China, & Cochinchina studiose colitur, sed rara est: crediturque aliunde oriundam industria Lusitanorum huc fuisse delatam." The description applies to the European rosemary. Doubtless Loureiro was correct in his statement that the plant was introduced by the Portuguese. At an early date the Spaniards introduced it into the Philippines where it is still cultivated, although of very limited distribution there.

Leonotis R. Brown

Leonotis nepetaefolia (Linn.) R. Br. in Ait. Hort. Kew. ed. 2, **3**: 409. 1811; Benth. in DC. Prodr. **12**: 535. 1848; Baker in Thistelton-Dyer Fl. Trop. Afr. **5**: 491. 1900.

Phlomis nepetaefolia Linn. Sp. Pl. 586. 1753.

Leonurus marrubiastrum (non Linn.) Lour. Fl. Cochinch. 360. 1790, ed. Willd. 436. 1793.

"Habitat prope litora Africae orientalis in Zanguebaria." Loureiro's description appertains to *Leonotis nepetaefolia* R. Br., a species which is apparently common in East Africa and adjacent islands.

Leucas (Burman) Linnaeus

Leucas mollissima Wall. Pl. As. Rar. **1**: 62. 1830.

Leucas benthamiana Hook. & Arn. Bot. Beechey's Voy. 204. 1836.

Leucas stachyoides Spreng.[72] Syst. **2**: 743. 1825 (quoad syn. Loureiro).

Ballota pilosa Lour. Fl. Cochinch. 364. 1790, ed. Willd. 442. 1793, Anamese *rau nhaong*.

"Habitat inculta in Cochinchina." Loureiro's description applies closely to *Leucas mollissima* Wall. His specific name is invalidated in *Leucas* by *L. pilosa* (Roxb.) Benth.

Leucas zeylanica (Linn.) R. Br. ex Spreng. Syst. **2**: 742. 1825 (*ceylanica*).

Phlomis zeylanica Linn. Sp. Pl. 586. 1753.

Nepeta hirsuta (non Linn.) Lour. Fl. Cochinch. 366. 1790, ed. Willd. 444. 1793, Anamese *cây mè dât*.

"Habitat spontanea in agris Cochinchinae." The reduction has been made from the description alone. The statement "pilis multis intermixtis" in reference to the flowers applies to the characteristically ciliate sepals. R. Brown (Prodr. 504. 1810) does not actually publish the binomial as is usually accredited to him. The species is represented by *Squires 89* from Hue and by *Kuntze 3609* from Tourane.

Leonurus Linnaeus

Leonurus sibiricus Linn. Sp. Pl. 584. 1753.

Stachys artemisia Lour. Fl. Cochinch. 365. 1790, ed. Willd. 443. 1793, Anamese *cây ich mâu*, Chinese *kĕ hoéi, sung úy*.

"Habitat culta, incultaque in Cochinchina, & China." Loureiro's description clearly applies to the common and very widely distributed Linnaean species. The Cantonese name appears on recently collected material as *hung fa i* and *hung fa ngai*. Loureiro's type is preserved in the herbarium of the Paris Museum.

[72] This binomial is based on "*Stachys decemdentata* Forst., *Phlomis* [*decemdentata*] Willd., *Ballota pilosa* Lour."; they do not appear to be conspecific.

Anisomeles R. Brown

Anisomeles indica (Linn.) O. Ktz. Rev. Gen. Pl. 512. 1891.

Nepeta indica Linn. Sp. Pl. 571. 1753.

Anisomeles ovata R. Br. in Ait. Hort. Kew ed. 2, **3**: 364. 1811.

Lamium garganicum (non Linn.) Lour. Fl. Cochinch. 365. 1790, ed. Willd. 442. 1793, Anamese *co cút lon*, Chinese *hĭ kiem tsào*.

" Habitat incultum in Cochinchina, & China." Loureiro's description seems clearly to apply to the very common and widely distributed *Anisomeles indica* (Linn.) O. Ktz.

Origanum (Tournefort) Linnaeus

Origanum vulgare Linn. Sp. Pl. 590. 1753.

Origanum heracleoticum (non Linn.) Lour. Fl. Cochinch. 373. 1790, ed. Willd. 453. 1793, Anamese *kinh giái taù*.

Origanum creticum (non Linn.) Lour. Fl. Cochinch. 373. 1790, ed. Willd. 453. 1793, Chinese *quām tūm kĭm kiái*.

Origanum loureiri Kostel. Allgem. Med.-Pharm. Fl. **3**: 768. 1834 (based on *O. heracleoticum* Lour.).

For *Origanum heracleoticum* Loureiro states: " Habitat in Cochinchina, & China," and for *O. creticum:* " Habitat incultum prope Cantonem Sinarum." Although I follow Hemsley (Journ. Linn. Soc. Bot. **26**: 282. 1890) in these reductions, it should be noted that Indo-China and Kwangtung are out of range for the Linnaean species. It is suspected that Loureiro may have secured his material from herbalists.

Thymus (Tournefort) Linnaeus

Thymus serpyllum Linn. Sp. Pl. 590. 1753.

? Origanum majorana (non Linn.) Lour. Fl. Cochinch. 374. 1790, ed. Willd. 454. 1793.

For *Origanum majorana* Loureiro states: " Habitat in Cochinchina, & China, rara tamen, forsan aliunde advecta." *Thymus serpyllum* Linn. is suggested as a possible reduction of the species Loureiro described; this species is of course out of range for Indo-China. It is suspected that Loureiro had either material from cultivated plants or that his specimens were secured from an herbalist.

Mentha (Tournefort) Linnaeus

Mentha arvensis Linn. Sp. Pl. 577. 1753.

Mentha crispa (non Linn.) Lour. Fl. Cochinch. 360. 1790, ed. Willd. 437. 1793, Anamese *rau húng, rau thom taù*.

Mentha pulegium (non Linn.) Lour. Fl. Cochinch. 361. 1790, ed. Willd. 437. 1793, Anamese *cây bac hà*, Chinese *pŏ hó*.

Mentha hirsuta (non Linn.) Lour. Fl. Cochinch. 361. 1790, ed. Willd. 437. 1793, Anamese *rau thom nam*, Chinese *hiám tsái*.

In sequence of Loureiro's description, the habitats are as follows: " Culta in Cochinchina, & China ubique obvia "; " agrestis, cultaque in China, & Cochinchina "; " in Cochinchina, & China similiter culta." *M. crispa* as described by Loureiro is apparently only a form of *M. arvensis* Linn. and I believe the other two species should also be reduced to this one. Prain (Journ. As. Soc. Bengal **74**: 710. 1908) distinguishes the form that occurs in southern Asia from *Mentha arvensis* Linn., retaining it as *M. javanica* Blume.

Perilla Linnaeus

Perilla frutescens (Linn.) Britton in Mem. Torr. Bot. Club **5**: 277. 1894.

Ocimum frutescens Linn. Sp. Pl. 597. 1753, ed. 2, 832. 1763.

Perilla ocymoides Linn. Gen. Pl. ed. 6, 578. 1764.

Dentidia nankinensis Lour. Fl. Cochinch. 369. 1790, ed. Willd. 448. 1793, Chinese *kiām năn tsù sū.*

Plectranthus nankinensis Spreng. Syst. **2**: 691. 1825 (based on *Dentidia nankinensis* Lour.).

Perilla nankinensis Decne. in Rev. Hort. IV. **1**: 61. 1852 (based on *Dentidia nankinensis* Lour.).

Perilla frutescens Britton var. *nankinensis* (Lour.) Britton in Mem. Torr. Bot. Club **5**: 277. 1894; L. H. Bailey Stand. Cyclop. Hort. 2553. 1916 (based on *Dentidia nankinensis* Lour.).

Melissa cretica (non Linn.) Lour. Fl. Cochinch. 368. 1790, ed. Willd. 446. 1793, Anamese *tu tó*, Chinese *tsù sū.*

Melissa rugosa Lour. Fl. Cochinch. 368. 1790, ed. Willd. 447. 1793, Anamese *tiá tô nham lá.* '

For *Dentidia nankinensis* Loureiro states: " Habitat Nankini in Sinis: inde Cantonem delata ob pulchritudinem studiose colitur "; for *Melissa cretica:* " Habitat culta in China & Cochinchina "; and *Melissa rugosa:* " Habitat culta, & spontanea in Cochinchina." I believe that all three of Loureiro's descriptions appertain to *Perilla frutescens* (Linn.) Britt., a species very extensively cultivated in China and which, like most commonly cultivated species, has numerous more or less distinct horticultural forms, these frequently with definite local names. In opposition to this view most authors, more or less automatically, I am afraid, retain *Perilla nankinensis* Decne. as a valid and distinct species. Recently collected specimens of the common form cultivated about Canton bear the Cantonese names *tsie so* and *chi so.*

Pogostemon Desfontaines

Pogostemon cablin (Blanco) Benth. in DC. Prodr. **12**: 156. 1848.

Mentha cablin Blanco Fl. Filip. 473. 1837.

Betonica officinalis (non Linn.) Lour. Fl. Cochinch. 364. 1790, ed. Willd. 441. 1793, Anamese *hoac huong*, Chinese *hŏ hiàm.*

" Habitat non raro culta in Cochinchina, & in China." This is the true patchouli plant, the reduction of Loureiro's species having been made by A. Chevalier (Cat. Pl. Jard. Bot. Saigon 66. 1919). Loureiro states that he never saw flowers, although the plant was cultivated in his own garden for many years. This statement corroborates the correctness of this reduction as patchouli in cultivation characteristically is rarely found in flower. Bretschneider states that the Chinese name cited by Loureiro is the same as that for *Lophanthus rugosus* Fisch. & Mey., but this is a northern species and Loureiro's description scarcely applies to it. For a critical note on patchouli see Prain, D. Journ. Asiat. Soc. Bengal **66**(2): 519. 1897.

Dysophylla Blume

Dysophylla benthamiana Hance in Ann. Sci. Nat. V Bot. **5**: 234. 1866.

Dysophylla verticillata Benth. in Wall. List. no. 1544. 1829, *nomen nudum*, Wall. Pl. As. Rar. **1**: 30. 1830.

Mentha verticillata Roxb. Hort. Beng. 44. 1814, *nomen nudum*, Fl. Ind. ed. 2, **3**: 5. 1832, non Linn.

Dysophylla ramosissima Benth. in Wall. List no. 1543. 1829, *nomen nudum*, DC. Prodr. **12**: 157. 1848, in syn.

Mentha stellata Lour. Fl. Cochinch. 361. 1790, ed. Willd. 438. 1793, Anamese *rau ngu hoang;* Moore in Journ. Bot. **63**: 285. 1925, non *Dysophylla stellata* Benth.

" Habitat loca humida inculta in Cochinchina." Loureiro's type is preserved in the herbarium of the British Museum and was examined by Bentham (DC. Prodr. **12**: 158. 1848). The description conforms closely with the characters of the common and widely distributed species currently known as *Dysophylla verticillata* Benth. Britten, quoted by Moore (Journ. Bot. **63**: 285. 1925), gives the reasons why *Dysophylla ramosissima* Benth. cannot be maintained. *Dysophylla verticillata* Benth. was based on an invalid binomial and I accordingly adopt Hance's specific name.

Dysophylla auricularia (Linn.) Blume Bijdr. 826. 1826; Benth. in DC. Prodr. **12**: 156. 1848.

Mentha auricularia Linn. Mant. **1**: 81. 1767.

Heliotropium tetrandrum Lour. Fl. Cochinch. 103. 1790, ed. Willd. 126. 1793, Anamese *cây cò cò.*

" Habitat inter herbas luxuriantes in hortis Cochinchinae tam ista, quam prima species." A. de Candolle (Prodr. **9**: 549. 1845) admits this as a doubtful species of *Heliotropium* with the statement: " Videtur a genere, forte ex ordine removenda. An Labiata? an Verbena? " I believe it to be the common *Dysophylla auricularia* Blume, which Loureiro otherwise does not describe, but if so, Loureiro errs in describing the leaves as glabrous. The Linnaean species is represented by *Clemens 3037* from Tourane.

Plectranthus [73] L'Héritier

Plectranthus sp.

Clinopodium asiaticum Lour. Fl. Cochinch. 374. 1790, ed. Willd. 454. 1793, Anamese *cây cò.*

Melissa asiatica G. Don Gen. Syst. **4**: 784. 1838 (based on *Clinopodium asiaticum* Lour.).

" Habitat incultum in Cochinchina." The description in some respects suggests *Plectranthus ternifolius* D. Don, but Loureiro describes the leaves as opposite, while in Don's species they are usually whorled.

Coleus Loureiro

Coleus amboinicus Lour. Fl. Cochinch. 372. 1790, ed. Willd. 452. 1793, Anamese *rau thom loung, tia tô taù.*

[73] *Plectranthus* L'Héritier (1785 vel 1788?) conserved name, Vienna Code; an older one is *Germanea* Lamarck (1786 vel 1787? = *Germainia* O. Kuntze 1892).

Plectranthus amboinensis Spreng. Syst. **2**: 690. 1825 (based on *Coleus amboinicus* Lour.).

Coleus aromaticus Benth. in Wall. Pl. As. Rar. **2**: 15. 1831.

Majana amboinica O. Ktz. Rev. Gen. Pl. 524. 1891 (based on *Coleus amboinicus* Lour.).

" Habitat in hortis Cochinchinae, & in variis Indiae locis, praesertim humidis." The type of the genus *Coleus*. The species, one with fleshy, very aromatic leaves, is widely cultivated in the Indo-Malaysian region. *Marrubium album amboinicum* Rumph. (Herb. Amb. **5**: 294. *pl. 102. f. 2*), cited by Loureiro as a synonym, and whence he took his specific name, is correctly placed. In habit and general appearance this species is remarkably distinct from the other species of *Coleus*.

Nosema Prain

Nosema cochinchinensis (Lour.) comb. nov.

Dracocephalum cochinchinense Lour. Fl. Cochinch. 371. 1790, ed. Willd. 450. 1793, Anamese *cây co cò.*

Geniostoma holocheilum Hance in Journ. Bot. **17**: 13. 1879, ex descr.

Mesona prunelloides Hemsl. in Journ. Linn. Soc. Bot. **26**: 267. 1890.

Nosema prunelloides Prain in Journ. As. Soc. Beng. **73**(2): 21. 1904.

" Habitat agreste in Cochinchina." Loureiro's description in my judgment applies unmistakably to the species described from Hainan by Hemsley in 1890 as *Mesona prunelloides*. The species is represented by *Squires 156, 383*, from Hue, Loureiro's classical locality. From the description I cannot distinguish *Geniostoma holocheilum* Hance.

Ocimum Linnaeus

Ocimum basilicum Linn. Sp. Pl. 597. 1753; Lour. Fl. Cochinch. 370. 1790, ed. Willd. 449. 1793, Anamese *rau é tiá, rau que.*

" Habitat cultum in hortis Cochinchinae." Loureiro was apparently correct in his interpretation of the Linnaean species. *Basilicum indicum* Rumph. (Herb. Amb. **5**: 263. *pl. 92. f. 1*) is correctly placed as a synonym.

Ocimum sanctum Linn. Mant. **1**: 85. 1767.

Ocimum gratissimum (non Linn.) Lour. Fl. Cochinch. 369. 1790, ed. Willd. 448. 1793 (*Ocymum*), Anamese *rau é lón lá.*

" Habitat in hortis Cochinchinae." The description seems to apply to *Ocimum sanctum* Linn. rather than to *O. gratissimum* Linn.

Ocimum africanum Lour. Fl. Cochinch. 370. 1790, ed. Willd. 449. 1793 (*Ocymum*).

Ocimum canum Sims in Curtis's Bot. Mag. **51**: *pl. 2452.* 1823; Baker in Thiselton-Dyer Fl. Trop. Afr. **5**: 337. 1900.

" Habitat incultum in arenariis Africae Orientalis, ubi illud virens ad examen revocavi." Loureiro's description conforms to the characters of Sims' species as interpreted by Baker, and I believe that his specific name should replace that of Sims. *Ocimum africanum* Lour. is not accounted for by Baker (Thiselton-Dyer Fl. Trop. Afr. **5**: 332–502. 1900).

Ocimum minimum (non Linn.?) Lour. Fl. Cochinch. 370. 1790, ed. Willd. 449. 1793 (*Ocymum*), Anamese *rau é nho lá.*

" Habitat in hortis Cochinchinae raro." This is probably a form of *Ocimum sanctum* Linn., or *O. basilicum* Linn., the description being very short and rather indefinite. *Ocimum minimum* Linn. is generally considered to be a synonym of *O. basilicum* Linn.

Orthosiphon Bentham

Orthosiphon spiralis (Lour.) Merr. in Lingnaam Agr. Rev. 2: 137. 1925 (based on *Trichostema spiralis* Lour.).

Trichostema spiralis Lour. Fl. Cochinch. 371. 1790, ed. Willd. 451. 1793, Anamese *cây râu meò*.

Ocimum aristatum Blume Bijdr. 833. 1826.

Orthosiphon stamineus Benth. in Wall. Pl. As. Rar. 2: 15. 1831, DC. Prodr. 12: 52. 1848.

Orthosiphon aristatus Miq. Fl. Ind. Bat. 2: 943. 1858; Merr. Enum. Philip. Fl. Pl. 3: 422. 1923.

" Habitat inculta in Cochinchina." Bentham (DC. Prodr. 12: 574. 1848) queries regarding Loureiro's species: " *Orthosiphon stamineus?* vel Clerodendri species? " Some may object to my interpretation because of Loureiro's description of the leaves as entire; they are exceedingly variable, ranging from prominently toothed to entire. If Loureiro's specific name be abandoned the next older one is *O. aristatus* (Blume) Miq.

LABIATAE OF UNCERTAIN STATUS

Ajuga orientalis (non Linn.) Lour. Fl. Cochinch. 363. 1790, ed. Willd. 440. 1793, Anamese *thích thao*.

" Habitat inculta in Cochinchina." Doubtless some well-known species of the Labiatae is represented by the description, but I am unable to suggest a reduction; it is not an *Ajuga*.

Origanum dictamnus (non Linn.) Lour. Fl. Cochinch. 373. 1790, ed. Willd. 452. 1793, Chinese *quăm tŭm fâm fŭm*.

" Habitat incultum prope Cantonem Sinarum." This probably represents some well-known species in some genus other than *Origanum*. I am unable to place it from the data at present available.

Origanum syriacum (non Linn.) Lour. Fl. Cochinch. 374. 1790, ed. Willd. 454. 1793, Anamese *kính giái năm*.

" Habitat agreste in Cochinchina." Apparently no *Origanum* is represented by the description. I can make no suggestion as to its probable identity.

Scutellaria albida (non Linn.) Lour. Fl. Cochinch. 367. 1790, ed. Willd. 445. 1793, Anamese *thiét côt taù*.

" Habitat spontanea in Cochinchina." I do not recognize this from the description; it is apparently not a *Scutellaria*.

Scutellaria alpina (non Linn.) Lour. Fl. Cochinch. 367. 1790, ed. Willd. 446. 1793, Anamese *hán săn thao*.

" Habitat in Cochinchina, agrestis." I do not recognize the species from the description. It is apparently not a *Scutellaria*.

Scutellaria altissima (non Linn.) Lour. Fl. Cochinch. 367. 1790, ed. Willd. 445. 1793.
Anamese *cây thiét côt, tiá tô rùng.*
" Habitat agrestis in Cochinchina." This is not recognized from the description, but apparently no *Scutellaria* is represented.

Teucrium massiliense (non Linn.) Lour. Fl. Cochinch. 362. 1790, ed. Willd. 439. 1793,
Anamese *tiá tô dât.*
" Habitat incultum in Cochinchina." The description applies to some representative of the Labiatae, but I cannot suggest a reduction at the present time; it is not a *Teucrium.*

Teucrium polium (non Linn.) Lour. Fl. Cochinch. 362. 1790, ed. Willd. 439. 1793, Anamese
dia tam thao, Chinese *tí tam tsào.*
" Habitat incultum prope Cantonem Sinarum." I have not been able to reduce this to any of the known Kwangtung Labiatae. It is not a *Teucrium.*

Teucrium thea Lour. Fl. Cochinch. 363. 1790, ed. Willd. 440. 1793, Anamese *cày chè baong.*
" Habitat in dumetis Cochinchinae." Apparently some species of Labiatae is represented, but certainly no *Teucrium.* The local name and indicated uses should eventually supply a clue to its identity.

Teucrium undulatum Lour. Fl. Cochinch. 362. 1790, ed. Willd. 439. 1793, Anamese *cây*
dôm dôm.
" Habitat ad sepes Cochinchinae." I can suggest no reduction. Some species of the Labiatae is represented, but no *Teucrium.*

SOLANACEAE

Lycium Linnaeus

Lycium chinense Mill. Gard. Dict. ed. 8, no. 5. 1768; Dunal in DC. Prodr. **13**(1): 510. 1852.
Lycium barbarum (non Linn.) Lour. Fl. Cochinch. 133. 1790, ed. Willd. 165. 1793,
Anamese *câu khi,* Chinese *kèu kì.*
" Habitat Cantone Sinarum." Loureiro's description applies unmistakably to Miller's species, which is widely distributed in China. The Cantonese name on recently collected material appears as *kau ki* and *kau kee tsz.*

Withania [74] Pauquy

Withania somnifera (Linn.) Dunal in DC. Prodr. **13**(1): 453. 1852; C. H. Wright in
Thiselton-Dyer Fl. Trop. Afr. **4**(2): 249. 1906.
Physalis somnifera Linn. Sp. Pl. 182. 1753; Lour. Fl. Cochinch. 132. 1790, ed. Willd.
163. 1793.
" Habitat agrestis in ora Africae Zanguebar." Loureiro's description clearly applies to the Linnaean species, which is widely distributed in tropical Africa.

Physalis Linnaeus

Physalis alkekengi Linn. Sp. Pl. 183. 1753; Lour. Fl. Cochinch. 133. 1790, ed. Willd. 164.
1793, Anamese *toan tuong,* Chinese *soàn tsiām.*
" Habitat in China, & Cochinchina." It is suspected from the medicinal qualities

[74] *Withania* Pauquy (1824), conserved name, Vïenna Code; an older one is *Physaloides* Moench (1794).

mentioned, and the reversed positions of the Chinese and Anamese names, that Loureiro's material was secured from an herbalist and was of Chinese origin. The Linnaean species occurs in northern China, but not in Kwangtung nor in Indo-China. The description is short and rather indefinite, and may or may not have been based on material representing *Physalis alkekengi* Linn. *Halicacabus indicus* Rumph. (Herb. Amb. **6**: 60. *pl. 26. f. 1*), cited by Loureiro as representing the species, is *Physalis minima* Linn.

Physalis angulata Linn. Sp. Pl. 183. 1753; Lour. Fl. Cochinch. 132. 1790, ed. Willd. 164.
1793, Anamese *lu lu cái;* Bonati in Lecomte Fl. Gén. Indo-Chine **4**: 335. 1915.
"Habitat spontanea in Cochinchina." The description applies to the Linnaean species, which is widely distributed in the warmer parts of the Old World and which is apparently common in Indo-China.

Physalis minima Linn. Sp. Pl. 183. 1753.
Physalis pubescens (non Linn.) Lour. Fl. Cochinch. 133. 1790, ed. Willd. 164. 1793,
Anamese *lu lu loung.*
"Habitat inculta in Cochinchina." *Physalis minima* Linn., as currently interpreted, is a common and widely distributed species in the warmer parts of the Old World. Bonati (Lecomte Fl. Gén. Indo-Chine **4**: 334-335. 1915) admits only *Physalis peruviana* Linn. and *P. angulata* Linn. as occurring in Indo-China, reducing *Physalis minima* Linn. to *Physalis angulata* Linn. var. *villosa* Bonati. Whatever name be used, this form with pubescent leaves is the one to which Loureiro's description applies.

Capsicum (Tournefort) Linnaeus

Capsicum annuum Linn. Sp. Pl. 188. 1753; Lour. Fl. Cochinch. 127. 1790, ed. Willd. 157.
1793, Anamese *ót tàu;* Irish in Rept. Missouri Bot. Gard. **9**: 65. *pl. 8-12.* 1898.
This was indicated by Loureiro as cultivated in China and in Cochinchina, a form with large oblong fruits. Loureiro's description applies to this protean species.

Capsicum annuum Linn. var. **cerasiforme** (Mill.) Irish Rept. Missouri Bot. Gard. **9**: 92.
1898.
Capsicum cerasiforme Mill. Gard. Dict. ed. 8, no. 5. 1768.
Capsicum baccatum (non Linn.) Lour. Fl. Cochinch. 127. 1790, ed. Willd. 157. 1793,
Anamese *ót tlòn tlái.*
This was reported by Loureiro as cultivated in both China and Cochinchina. His description applies to the cherry pepper rather than to *Capsicum frutescens* Linn. var. *baccatum* (Linn.) Irish (Rept. Missouri Bot. Gard. **9**: 99. 1898).

Capsicum frutescens Linn. Sp. Pl. 189. 1753; Lour. Fl. Cochinch. 128. 1790, ed. Willd.
158. 1793, Anamese *cây ót,* Chinese *lat tsiāo.*
This, like the preceding species, was observed by Loureiro in cultivation in China and Cochinchina. He states that the fruits were conical, curved, and an inch long, perhaps indicating a form of *Capsicum annuum* Linn., rather than a form of *C. frutescens* Linn.; the latter is widely naturalized in the orient and normally has somewhat smaller fruits than Loureiro describes. In any case it is highly probable that the protean *Capsicum annuum* Linn. is derived from *C. frutescens* Linn.

Solanum (Tournefort) Linnaeus

Solanum album Lour. Fl. Cochinch. 129. 1790, ed. Willd. 159. 1793, Anamese *cà co;* Bonati
in Lecomte Fl. Gén. Indo-Chine **4**: 328. 1915.

" Habitat spontaneum in Cochinchina." Bonati has given an amplified description
of this species, citing various collections from Anam (including Hue) and Tonkin.

Solanum biflorum Lour. Fl. Cochinch. 129. 1790, ed. Willd. 159. 1793, Chinese *thien pháo;*
Bonati in Lecomte Fl. Gén. Indo-Chine **4**: 320. 1915.

Lycianthes biflora Bitter in Abh. Nat. Ver. Bremen **24**: 461. 1919 (based on *Solanum
biflorum* Lour.).

Solanum osbeckii Dunal in DC. Prodr. **13**(1): 179. 1852 et var. *stauntoni* Dunal l.c.

Solanum calleryanum Dunal in DC. Prodr. **13**(1): 178. 1852.

" Habitat in Cochinchina, & China." This species is of wide geographic distribution
in southeastern Asia and Malaysia. The form common in the vicinity of Canton, China,
agrees with the original description and should be accepted as typical.

Solanum ferox Linn. Sp. Pl. ed. 2, 267. 1762.

Solanum mammosum (non Linn.) Lour. Fl. Cochinch. 131. 1790, ed. Willd. 162. 1793,
Anamese *cà ung.*

" Habitat agreste in Cochinchina." The description applies unmistakably to the
common and widely distributed *Solanum ferox* Linn.

Solanum indicum Linn. Sp. Pl. 187. 1753; Lour. Fl. Cochinch. 131. 1790, ed. Willd. 162.
1793, Anamese *cà hoang gai;* Bonati in Lecomte Fl. Gén. Indo-Chine **4**: 326. 1915.

" Habitat agreste in Cochinchina." Loureiro's description conforms closely to the
characters of the common and widely distributed Linnaean species.

Solanum lyratum Thunb. Fl. Jap. 92. 1784.

Solanum dulcamara Linn. var. *lyratum* Bonati in Lecomte Fl. Gén. Indo-Chine **4**:
317. 1915.

Solanum dichotomum Lour. Fl. Cochinch. 129. 1790, ed. Willd. 160. 1793, Chinese
kam ngi van.

" Habitat Cantone Sinarum, spontaneum." Loureiro's description is short and
rather unsatisfactory and it does not conform at all with the characters of any other species
of *Solanum* known from southern China. Forms currently referred to *Solanum lyratum*
Thunb. often have entire leaves, this being a character indicated by Loureiro. This
common Chinese species is frequently referred to the collective species of *Solanum dulcamara*
Linn., but most authors now recognize Thunberg's species as a distinct one. ·

Solanum melongena Linn. Sp. Pl. 186. 1753; Lour. Fl. Cochinch. 130. 1790, ed. Willd.
161. 1793, Anamese *cà ăn,* Chinese *kie tsù;* Bonati in Lecomte Fl. Gén. Indo-Chine
4: 325. 1915.

Solanum aethiopicum (non Linn.) Lour. Fl. Cochinch. 130. 1790, ed. Willd. 160. 1793,
Anamese *cà tién.*

For *S. melongena* Loureiro states: " Habitat in China, & Cochinchina: imo in toto
fere orbe non nimis frigido." It is the common brinjal or egg-plant. Loureiro describes
five different forms of the fruit. *Trongum hortense* Rumph. (Herb. Amb. **5**: 238. *pl. 85*)
is correctly placed as a synonym. For *S. aethiopicum* Loureiro states: " Habitat in Co-

chinchina, & China, in hortis cultum." I believe this to be one of numerous forms of *S. melongena* Linn. The fruit is described as globose, 6-lobed, large, glabrous, edible and white or purple. Dunal (DC. Prodr. **13**(1): 351. 1852) leaves it as *Solanum aethiopicum* Linn. var. *violaceum* Dunal and queries: " An spec. diversa? ". Bonati does not mention Loureiro's species.

Solanum nigrum Linn. Sp. Pl. 186. 1753; Lour. Fl. Cochinch. 129. 1790, ed. Willd. 160.
  1793, Anamese *cây lu lu duc;* Bonati in Lecomte Fl. Gén. Indo-Chine **4**: 317. 1915.
  " Habitat incultum in Cochinchina." The description applies to the ubiquitous Linnaean species.

Solanum procumbens Lour. Fl. Cochinch. 132. 1790, ed. Willd. 163. 1793, Anamese *cà*
  quănh; Nees in Trans. Linn. Soc. **17**: 58. 1834; Dunal in DC. Prodr. **13**(1): 281.
  1852.
  Solanum hainanense Hance in Journ. Bot. **6**: 331. 1868; Bonati in Lecomte Fl. Gén.
  Indo-Chine **4**: 329. 1915.
  " Habitat in Cochinchina inter sepes, & in locis agrestibus." The amplified descriptions of Nees and Dunal were based on a specimen collected by Finlayson at Hue, the classical locality, cited by Nees as " E Hué, in Herb. Finlays. (Wall. l.c.) " and " Wall. Catal., Suppl. n. 214." The species is also represented by *Squires 27, Clemens 3453, 4004,* and *Kuntze 3766,* all from Hue. I cannot see any reason for distinguishing *Solanum hainanense* Hance. Bonati curiously overlooked *Solanum procumbens* Lour. in his treatment of the Solanaceae of Indo-China (Lecomte Fl. Gén. Indo-Chine **4**: 313–341. 1915–27).

Solanum verbascifolium Linn. Sp. Pl. 184. 1753; Lour. Fl. Cochinch. 128. 1790, ed. Willd.
  159. 1793, Anamese *cây chià bòi;* Bonati in Lecomte Fl. Gén. Indo-Chine **4**: 319. 1915.
  " Habitat incultum in Cochinchina." The description applies unmistakably to the common and widely distributed Linnaean species.

Lycopersicon (Tournefort) Miller

Lycopersicon esculentum Mill. Gard. Dict. ed. 8, no. 2. 1768.
  Solanum lycopersicum Linn. Sp. Pl. 185. 1753; Lour. Fl. Cochinch. 130. 1790, ed.
  Willd. 161. 1793, Anamese *cà tàu tláng.*
  Solanum peruvianum (non Linn.) Lour. Fl. Cochinch. 131. 1790, ed. Willd. 162. 1793,
  Anamese *cà tàu vàng.*
  For the first Loureiro states: " Habitat incultum in hortis, & agris Cochinchinae."
This is a garden form of the common tomato; *Pomum amoris* Rumph. (Herb. Amb. **5**: 416. *pl. 154. f. 1*), cited as a synonym, is correctly placed. For the secondhe states: " Habitat incultum in hortis Cochinchinae," this being the normal wild or semi-wild form of the common tomato with small fruits usually about 1 cm. in diameter.

Datura Linnaeus

Datura metel Linn. Sp. Pl. 179. 1753; Lour. Fl. Cochinch. 110. 1790, ed. Willd. 135. 1793,
  Anamese *cà duoc,* Chinese *nǎo hiēn hōa.*
  Datura fastuosa Linn. Syst. ed. 10, 932. 1759; Bonati in Lecomte Fl. Gén. Indo-Chine
  4: 339. 1927.
  " Habitat inculta per vias, & hortos in Cochinchina, in China, & in Africa, ubi a me saepe examinata." Safford (Journ. Washington Acad. Sci. **11**: 178. *f. 2.* 1921) has definitely

shown that the species currently known as *Datura fastuosa* Linn. is *D. metel* Linn., and that the latter name is the proper one for this widely distributed species. *Stramonia indica* Rumph. (Herb. Amb. **5**: 242. *pl. 87. f. 1–2*), cited by Loureiro as a synonym, is correctly placed.

Nicotiana Linnaeus

Nicotiana tabacum Linn. Sp. Pl. 180. 1753; Bonati in Lecomte Fl. Gén. Indo-Chine **4**: 341. 1927.

Nicotiana fruticosa (non Linn.) Lour. Fl. Cochinch. 111. 1790, ed. Willd. 136. 1793, Anamese *cây thŭóc ăn*, Chinese *yēn yĕ.*

"Habitat ubique culta in Cochinchina, & China." The description clearly applies to the common tobacco plant, *Nicotiana tabacum* Linn.

SCROPHULARIACEAE

Linaria (Tournefort) Miller

Linaria vulgaris Mill. Gard. Dict. ed. 8, no. 1. 1768.

Antirrhinum linaria Linn. Sp. Pl. 616. 1753; Lour. Fl. Cochinch. 383. 1790, ed. Willd. 465. 1793, Chinese *sòy kúe hoā.*

"Habitat Cantone Sinarum, cultum ob speciem floris." The description applies to the widely distributed Linnaean species, which is common in northern China. Loureiro notes that he saw only cultivated specimens at Canton.

Linaria spuria (Linn.) Mill. Gard. Dict. ed. 8, no. 15. 1768.

Antirrhinum spurium Linn. Sp. Pl. 613. 1753; ? Lour. Fl. Cochinch. 383. 1790, ed. Willd. 465. 1793.

"Habitat incultum Cantone Sinarum." The description, although short, agrees very well with the characters of the Linnaean species. An objection to this reference of Loureiro's species is the fact that the Linnaean species is not recorded from China, although it may well have occurred near Canton in Loureiro's time as an introduced weed. The description does not apply to any known Kwangtung species in this or in any other family of plants.

Mazus Loureiro

Mazus japonicus (Thunb.) O. Ktz. Rev. Gen. Pl. 462. 1891.

Lindernia japonica Thunb. Fl. Jap. 253. 1784.

Mazus rugosus Lour. Fl. Cochinch. 385. 1790, ed. Willd. 468. 1793, Anamese *rau dáng lóng lá;* Bonati in Lecomte Fl. Gén. Indo-Chine **4**: 355. 1927.

"Habitat in agris Cochinchinae." This is the type of the genus. The species is a well-known one widely distributed in Asia and Malaysia.

Limnophila [75] R. Brown

Limnophila aromatica (Lam.) Merr. Interpret. Herb. Amb. 466. 1917.

Ambulia aromatica Lam. Encycl. **1**: 128. 1783.

Limnophila punctata Blume Bijdr. 750. 1826.

[75] *Limnophila* R. Brown (1810), conserved name, Vienna Code; older ones are *Ambulia* Lamarck (1783), *Diceros* Loureiro (1790) and *Hydropityon* Gaertner f. (1805).

Limnophila gratissima Blume op. cit. 749; Bonati in Lecomte Fl. Gén. Indo-Chine **4**: 379. 1927.

Antirrhinum aquaticum Lour. Fl. Cochinch. 384. 1790, ed. Willd. 466. 1793, Anamese *rau chiéo núoc.*

" Habitat in paludibus Cochinchinae." The habitat is that of *Limnophila* and the description conforms with the characters of the common and widely distributed Indo-Malaysian species currently known as *Limnophila gratissima* Blume = *L. aromatica* (Lam.) Merr. Bonati does not account for Loureiro's species in his treatment of the Scrophulariaceae of Indo-China (Lecomte Fl. Gén. Indo-Chine **4**: 341–461. 1927).

Limnophila chinensis (Osbeck) Merr. in Am. Journ. Bot. **3**: 581. 1916.

Columnea chinensis Osbeck Dagbok Ostind. Resa 230. 1757.

Diceros cochinchinensis Lour. Fl. Cochinch. 381. 1790, ed. Willd. 463. 1793, Anamese *rau ngu.*

Stemodia hirsuta Heyne in Wall. List no. 3930. 1831, *nomen nudum;* Benth. Scroph. Ind. 24. 1835.

Limnophila hirsuta Benth. in DC. Prodr. **10**: 388. 1846; Bonati in Lecomte Fl. Gén. Indo-Chine **4**: 378. 1927.

? *Aponoa repens* Raf. Sylva Tellur. 84. 1838 (based on *Columnea stellata* Lour.).

? *Columnea stellata* Lour. Fl. Cochinch. 384. 1790, ed. Willd. 467. 1793, Anamese *hõa kách.*

For the *Diceros* Loureiro states: " Habitat loca humida in Cochinchina." His description is definite and unmistakably applies to the widely distributed Indo-Malaysian species currently known as *Limnophila hirsuta* Benth., for which the oldest valid specific name is that supplied by *Columnea chinensis* Osbeck. For *Columnea stellata* Loureiro states: " Habitat in multis locis Cochinchinae: colitur in vasis fictilibus et ligneis aqua plenis, substrato luto." The description is apparently that of a *Limnophila*, and possibly a form of *L. chinensis* (Osbeck) Merr. The point should be readily solved by a little field work with reference to the local name and indicated method of cultivation.

Adenosma R. Brown

Adenosma glutinosum (Linn.) Druce in Rept. Bot. Exch. Club Brit. Isles **3**: 413. 1914; Merr. in Philip. Journ. Sci. Bot. **12**: 109. 1917.

Gerardia glutinosa Linn. Sp. Pl. 611. 1753; Osbeck Dagbok Ostind. Resa 229. *pl. 9.* 1757.

Pterostigma grandiflorum Benth. Scroph. Ind. 21. 1835; Hook & Arn. Bot. Beechey's Voy. 204. *pl. 45.* 1836.

Adenosma grandiflorum Benth. ex Hance in Journ. Linn. Soc. Bot. **13**: 114. 1873.

Digitalis sinensis Lour. Fl. Cochinch. 378. 1790, ed. Willd. 459. 1793, Chinese *tsú hõa yong.*

No locality is mentioned, but from the Chinese local name cited it is evident that Loureiro's specimens came from China, and probably from the vicinity of Canton, where the species is common. The type of *Gerardia glutinosa* Linn. was a specimen collected by Osbeck near Canton. Loureiro's description applies to the Linnaean species. Bentham (DC. Prodr. **10**: 380. 1846) cites *Digitalis sinensis* Lour. as a doubtful synonym of *Pterostigma grandiflorum* Benth., but the question mark should be removed.

Adenosma indianum (Lour.) comb. nov.

Manulea indiana Lour. Fl. Cochinch. 386. 1790, Anamese *non trân nam.*

Manulea indica Willd. in Lour. Fl. Cochinch. ed. 2, 469. 1793 (based on *Manulea indiana* Lour.).

Erinus bilabiatus Roxb. Hort. Beng. 47. 1814, *nomen nudum,* Fl. Ind. ed. 2, 3 : 92. 1832.

Stemodia capitata Benth. in Wall. List no. 3926. 1831, *nomen nudum,* Bot. Reg. **17**: sub. *pl. 1470.* 1831.

Pterostigma capitatum Benth. Scroph. Ind. 21. 1835, DC. Prodr. **10**: 380. 1846.

Adenosma capitatum Benth. in Hook. f. Fl. Brit. Ind. **4**: 264. 1884; Bonati in Lecomte Fl. Gén. Indo-Chine **4**: 365. 1927.

Adenosma bilabiatum Merr. Enum. Philip. Fl. Pl. **3** : 434. 1923.

" Habitat prope maris litora arenosa in Cochinchina." Loureiro's description is definite and manifestly applies to the widely distributed Indo-Malaysian species commonly known as *Adenosma capitatum* Benth.

Dopatrium Buchanan-Hamilton

Dopatrium junceum (Roxb.) Ham. ex Benth. Scroph. Ind. 31. 1835.

Gratiola juncea Roxb. Pl. Coromandel **2**: 16. *pl. 129.* 1798.

? Antirrhinum porcinum Lour. Fl. Cochinch. 384. 1790, ed. Willd. 466. 1793, Anamese *rau chiéo heo.*

" Habitat loca humida Cochinchinae: ubi colligitur pro nutrimento suum, ex quo vernaculum nomen Porcinum." The description is not good for *Dopatrium,* yet I know of no other scrophulariaceous plant to which it may apply. From Loureiro's note as to its economic uses, the plant should be a relatively abundant one in Indo-China. In *Dopatrium* the leaves are not subserrate, neither are the calyx-lobes pilose; the fresh leaves are fleshy and more or less mucilaginous, which might explain the term " viscosa " applied to them by Loureiro. In some respects the description suggests *Limnophila,* but probably no *Limnophila* would be used for the purpose indicated by Loureiro.

Bramia [76] Lamarck

Bramia monnieri (Linn.) Pennell in Proc. Acad. Nat. Sci. Phila. **71**: 243. 1919.

Lysimachia monnieri Linn. Cent. Pl. **2**: 9. 1756.

Gratiola monniera Linn. Amoen. Acad. **4**: 306. 1759.

Herpestis monnieria H.B.K. Nov. Gen. Sp. Pl. **2**: 366. 1817; Bonati in Lecomte Fl. Gén. Indo-Chine **4**: 356. 1927.

Bramia monniera Drake Fl. Polyn. Franç. 142. 1893.

Septas repens Lour. Fl. Cochinch. 392. 1790, ed. Willd. 477. 1793, Chinese *pă tsī hién.*

Lepidagathis repens Spreng. Syst. **2**: 827. 1825 (based on *Septas repens* Lour.).

" Habitat in suburbiis Cantoniensibus apud Sinas." Loureiro's type is preserved in the herbarium of the Paris Museum. The description clearly applies to the widely distributed species currently known as *Herpestis monniera* H.B.K. *Septilia* Raf. (Fl. Tellur. **4**: 68. 1838) is a new generic name for *Septas* Lour.; Rafinesque, however, published no binomial under it.

[76] I follow Pennell in treating this as generically distinct from *Bacopa* Aublet. For the more comprehensive group Aublet's generic name is conserved under the Vienna Code, against *Moniera (Monniera)* P. Browne (1756, Adanson 1763) and *Brami* Adanson (1763).

Torenia Linnaeus

Torenia peduncularis Benth. in Wall. List no. 3956. 1831, *nomen nudum;* Hook. f. Fl. Brit. Ind. **4**: 276. 1884; Bonati in Lecomte Fl. Gén. Indo-Chine **4**: 394. 1927.

Ruellia ciliaris (non Linn.) Lour. Fl. Cochinch. 381. 1790, ed. Willd. 462. 1793, Anamese *rau thom dât.*

" Habitat in agris Cochinchinae spontanea." The description conforms very closely with the characters of the common *Torenia peduncularis* Benth., except in the " Capsula . . . 2 locularis . . . apice elastice dissiliens." Its shape and its being described as many-seeded conform to *Torenia peduncularis* Benth. Doubtless the above quoted words were added by Loureiro to make his description conform to the characters of the genus in which he erroneously placed his plant. The doubtful reference to *Ruellia persica* Burm. f. (Fl. Ind. 135. *pl. 42. f. 1*. 1768) must be excluded.

Picria Loureiro
(*Curanga* A. L. de Jussieu)

Picria fel-terrae Lour. Fl. Cochinch. 393. 1790, ed. Willd. 478. 1793, Anamese *cây mât dât;* Baill. in Bull. Soc. Linn. Paris **1**: 699. 1887.

Caranga amara Vahl Enum. **1**: 100. 1804.

Curanga amara Juss. in Ann. Mus. Hist. Nat. [Paris] **9**: 320. 1807; Bonati in Lecomte Fl. Gén. Indo-Chine **4**: 426. 1927.

Curania amara Roem. & Schult. Syst. **1**: 138. 1817.

Pikria fel-terrae G. Don Gen. Syst. **4**: 617. 1836 (based on *Picria fel-terrae* Lour.).

Curanga fel-terrae Merr. Interpret. Herb. Amb. 467. 1917, Enum. Philip. Fl. Pl. **3**: 439. 1923 (based on *Picria fel-terrae* Lour.).

Curanga melissaefolia Juss. in Ann. Mus. Hist. Nat. [Paris] **9**: 320. 1807, in syn.

" Habitat culta in hortis Cochinchinae, & Chinae." The description, and the bitter properties of the plant (whence Loureiro's generic name), clearly apply to the species currently known as *Curanga amara* Juss., one of wide distribution in the Indo-Malaysian region. Loureiro's type is preserved in the herbarium of the British Museum. Baillon (Bull. Soc. Linn. Paris **1**: 699. 1887) quotes Bentham as follows: " *Picria* . . ., ex auctoris descriptione et specimine valde manco in Herb. Mus. Brit. servato, est verisimiliter *Curangae* species, a *C. amara* Juss. parum diversa," but calls attention to the fact that Loureiro's type specimen is an ample and excellent one, expressing the opinion that his binomial should be maintained. Loureiro's generic name is older than *Caranga* or *Curanga*, and I have accordingly accepted it as the valid one for the genus.

Scoparia Linnaeus

Scoparia dulcis Linn. Sp. Pl. 116. 1753; Lour. Fl. Cochinch. 71. 1790, ed. Willd. 89. 1793, Anamese *cam thao dât.*

" Habitat agrestis in Cochinchina, nec alibi vidi." The description applies unmistakably to the Linnaean species; a very common pantropic weed of American origin.

Ilysanthes Rafinesque

Ilysanthes antipoda (Linn.) Merr. Interpret. Herb. Amb. 467. 1917, Enum. Philip. Fl. Pl. **3**: 439. 1923.

Ruellia antipoda Linn. Sp. Pl. 635. 1753; Lour. Fl. 380. 1790, ed. Willd. 462. 1793, Anamese *rau chon vit.*

Ruellia anagallis Burm. f. Fl. Ind. 135. 1768.

Gratiola veronicifolia Retz. Obs. **4**: 8. 1786.

Bonnaya veronicaefolia Spreng. Syst. 1: 41. 1825; Bonati in Lecomte Fl. Gén. Indo-Chine **4**: 436. 1927.

" Habitat spontanae in pratis, & hortis Cochinchinae." Loureiro correctly interpreted the Linnaean species which is very common and widely distributed in the Indo-Malaysian region. *Crusta ollae major* Rumph. (Herb. Amb. **5**: 460. *pl. 170. f. 2*), cited by Loureiro as representing the species, is correctly placed.

Centranthera R. Brown

Centranthera cochinchinensis (Lour.) comb. nov.

Digitalis cochinchinensis Lour. Fl. Cochinch. 378. 1790, ed. Willd. 459. 1793, Anamese *cây bo loung.*

Centranthera hispida R. Br. Prodr. 438. 1810; Bonati in Lecomte Fl. Gén. Indo-Chine **4**: 448. 1927.

" Habitat in agris, & hortis Cochinchinae inculta." Loureiro's description definitely applies to the southern Asiatic form currently identified as *Centranthera hispida* R. Br. and I do not hesitate in adopting the older specific name.

Striga Loureiro

Striga asiatica (Linn.) O. Ktz. Rev. Gen. Pl. 466. 1891.

Buchnera asiatica Linn. Sp. Pl. 630. 1753, pro parte.

Striga lutea Lour. Fl. Cochinch. 22. 1790, ed. Willd. 27. 1793, Chinese *thŏc chiŏ kam;* Bonati in Lecomte Fl. Gén. Indo-Chine **4**: 457. 1927.

Striga hirsuta Benth. in DC. Prod. **10**: 502. 1846.

" Habitat inculta in suburbiis Cantoniensibus." This is the type of the genus *Striga,* and the species is common in open grassy places near Canton. Bentham (DC. Prodr. **10**: 502. 1846) reduced *Striga lutea* Lour. to *S. hirsuta* Benth. *Buchnera asiatica* Linn. was based on actual specimens from China and Ceylon, Bentham (*l.c.*) stating that the material in the Linnaean herbarium is in part *Striga hirsuta* Benth. and in part *S. densiflora* Benth. Linnaeus' description applies to the former, *i.e.,* the form described by Loureiro as *Striga lutea.* Trimen (Fl. Ceyl. **3**: 256. 1895) adopts the binomial *Striga lutea* Lour., citing as synonyms *Striga hirsuta* Benth. and *Buchnera asiatica* Linn. (in part). Loureiro's type is preserved in the herbarium of the Paris Museum. Under the " rule " that a specific name based on a mixture of two or more species is invalid, the Linnaean binomial would be dropped. But this " rule," if strictly followed, would invalidate many hundreds of species proposed by Linnaeus and his successors that are now accepted as valid.

BIGNONIACEAE

Oroxylum Ventenat

Oroxylum indicum (Linn.) Vent. Dec. Gen. Nov. 8, 1808; Dop in Lecomte Fl. Gén. Indo-Chine **4**: 570. 1930.

Bignonia indica Linn. Sp. Pl. 625. 1753.

Bignonia pentandra Lour. Fl. Cochinch. 379. 1790, ed. Willd. 460. 1793, Anamese *cây nguc ngnac.*

Hippoxylon indica Raf. Sylva Tellur. 78. 1838 (based on *Bignonia indica* Linn. and *Bignonia pentandra* Lour.).

"Habitat prope flumina in Cochinchina." Loureiro's description unmistakably applies to the common, widely distributed, and well-known *Oroxylum indicum* Vent. Loureiro's description in part typifies the genus *Hippoxylon* Raf.

Campsis Loureiro

Campsis grandiflora (Thunb.) K. Schum. in Engler & Prantl Nat. Pflanzenfam. 4(3b): 230. 1895.

Bignonia grandiflora Thunb. Fl. Jap. 253. 1784.

Bignonia chinensis Lam. Encycl. 1: 423. 1785.

Campsis adrepens Lour. Fl. Cochinch. ·378. 1790, ed. Willd. 458. 1793, Chinese *lién siéu.*

Tecoma grandiflora Loisel. Herb. Gén. Amat. 4: *pl. 286.* 1820.

Incarvillea chinensis Spreng. Syst. 2: 836. 1825 quoad syn. Loureiro (essentially based on *Campsis adrepens* Lour.).

Campsis chinensis Voss in Vilmorin Blumengärt. ed. 3, 1: 801. 1895; Rehder in Sargent Pl. Wils. 1: 303. 1912.

Campsis longiflora Dop in Lecomte Fl. Gén. Indo-Chine 4: 574. 1930 (sphalm., *C. grandiflora* K. Schum. intended).

"Habitat in sylvis prope Cantonem Sinarum." Loureiro's type is preserved in the herbarium of the Paris Museum. The species is a well-known one. The Cantonese name on recently collected material appears as *ling shiu fa.* Thunberg's binominal is one year older than Lamarck's, pages 369 to 752 of Lamarck's Encyclopédie not having been published until 1785.

Markhamia Seemann

Markhamia stipulata (Roxb.) Seem. ex C. B. Clarke in Hook. f. Fl. Brit. Ind. 4: 379. 1884; Dop in Lecomte Fl. Gén. Indo-Chine 4: 603. *f. 66, 3-5.* 1930.

Bignonia stipulata Roxb. Hort.'Beng. 47. 1814, *nomen nudum,* Fl. Ind. ed. 2, 3: 108. 1832.

Spathodea stipulata Wall. List. no. 6518. 1832, *nomen nudum,* Pl. As. Rar. 3: 20. *pl. 238.* 1832.

Dolichandrone stipulata Benth. ex C. B. Clarke in Hook. f. Fl. Brit. Ind. 4: 379. 1884.

Bignonia indica (non Linn.) Lour. Fl. Cochinch. 379. 1790, ed. Willd. 460. 1793, Anamese *cây dôi muong, cây do do.*

"Habitat agrestis Cochinchina, imprimis prope flumina." In spite of Loureiro's description of this as having bipinnate leaves, I believe Dop to be correct in his reduction of *Bignonia indica* Lour. (non Linn.) to *Markhamia stipulata* (Roxb.) Seem. It is to be noted that Seemann does not publish this binomial (Journ. Bot. 1: 226. 1863) as indicated in Index Kewensis, and by Dop. The first publication of the binomial that I have found is in the synonymy of *Dolichandrone stipulata* C. B. Clarke (Hook f. Fl. Brit. Ind. 4: 379. 1884).

Dolichandrone Fenzl

Dolichandrone spathacea (Linn. f.) K. Schum. Fl. Kaiser Wilhelmsland 123. 1889.

Bignonia spathacea Linn. f. Suppl. 283. 1781.

Bignonia longissima Lour. Fl. Cochinch. 380. 1790, ed. Willd. 461. 1793, Anamese *cây quao*.

Spathodea loureiriana DC. Prodr. **9**: 209. 1845 (based on *Bignonia longissima* Lour.).

Dolichandrone rheedii Seem. in Journ. Bot. **8**: 380. 1870; Dop in Lecomte Fl. Gén. Indo-Chine **4**: 601. 1930.

Dolichandrone longissima K. Schum. in Engler & Prantl Nat. Pflanzenfam. 4(3b): 240. 1894 (based on *Bignonia longissima* Lour.).

" Habitat ad ripas fluminum in Cochinchina." This common, characteristic, and well-known species occurs along tidal streams throughout the Indo-Malaysian region. *Lignum equinum* Rumph. (Herb. Amb. **3**: 73. *pl. 46*), cited by Loureiro as representing his species, is *Dolichandrone spathacea* K. Schum.

PEDALIACEAE

Sesamum Linnaeus

Sesamum orientale Linn. Sp. Pl. 634. 1753; Lour. Fl. Cochinch. 382. 1790, ed. Willd. 464. 1793, Anamese *cây mè*, Chinese *mâ chī*.

Sesamum indicum Linn. *l.c.*

" Habitat affatim cultum in Cochinchina, & China." The Linnaean species was correctly interpreted by Loureiro; it is the common and well-known sesame.

Dicerocaryum Bojer

(*Pretrea* Gay)

Dicerocaryum zanguebarium (Lour.) comb. nov.

Martynia zanguebaria Lour. Fl. Cochinch. 386. 1790, ed. Willd. 469. 1793.

Pretrea zanguebarica Gay ex Schauer in DC. Prodr. **9**: 256. 1845; Stapf in Thiselton-Dyer Fl. Trop. Afr. **4**(2): 565. 1906 (based on *Martynia zanguebaria* Lour.).

Dicerocaryum sinuatum Bojer in Ann. Sci. Nat. II Bot. **4**: 269. *pl. 10*. 1835.

" Habitat in litoribus Zanguebar Africae Orientalis." Loureiro's type is preserved in the herbarium of the Paris Museum. He gives the local name in Zanzibar as *biri viri.* While this is currently maintained as *Pretrea zanguebarica* Gay (Ann. Sci. Nat. **1**: 457. 1824) attention is called to the fact that neither the genus nor the binomial is there published, nor do I find an actual publication before 1845. In the meantime Bojer had described and illustrated the species as *Dicerocaryum sinuatum* in 1836 and under all rules of nomenclature his generic name should be retained. All that is said in 1824 regarding *Pretrea* [77] is as follows: " Dans un mémoire qui sera incessament publié, M. Gay donne la monographie des plantes que R. Brown avait détachées des *Bignoniacées* sous les noms des *Sesamées* et de *Pedalinées.* Deux nouveaux genres sonte partie de ce travail; le *Pretrea* qui est intermédiare entre le *Sesamum* et le *Josephinia*, et dont le seule espèce a été décrite par Loureiro sous le nom de *Martynia zanguebarica.*" Gay's memoir was never published.

[77] Sur le Pretrea et le Rogeria, deux nouveaux genres de plantes. Ann. Sci. Nat. **1**: 457. 1824.

LENTIBULARIACEAE

Utricularia Linnaeus

Utricularia aurea Lour. Fl. Cochinch. 26. 1790, ed. Willd. 32. 1793, Anamese *cây raong*.
> *Utricularia flexuosa* Vahl Enum. 1: 198. 1804.

" Habitat in fluviis lentioris cursus in Cochinchina." Loureiro's description is definite and unmistakably applies to the common Asiatic species currently known as *Utricularia flexuosa* Vahl. He interpreted the finely divided submerged leaves as roots, describing them thus: " Folia nulla. Radix plurima, capillaris, viridis, ramosa, utriculata."

Utricularia bifida Linn. Sp. Pl. 18. 1753.
> *Utricularia recurva* Lour. Fl. Cochinch. 26. 1790, ed. Willd. 32. 1793.
> *Askofake recurva* Raf. Fl. Tellur. 4: 108. 1838 (based on *Utricularia recurva* Lour.).

" Habitat in rivo *Hòn mô*, non procul ab urbe regia Cochinchinae." The description apparently applies to the common and widely distributed *Utricularia bifida* Linn.

ACANTHACEAE

Hygrophila R. Brown

Hygrophila phlomoides Nees in Wall. Pl. Asiat. Rar. 3: 80. 1832; DC. Prodr. 11: 90. 1847.
> *Antirrhinum molle* (non Linn.) Lour. Fl. Cochinch. 383. 1790, ed. Willd. 466. 1793, Anamese *rau chiéo loung*.

" Habitat spontaneum in pratis Cochinchinae." The description applies closely to the common and widely distributed *Hygrophila phlomoides* Nees. Blanco (Fl. Filip. 503. 1837) described Nees' species under the same Linnaean binomial, probably making his identification from Loureiro's description; see Merrill, E. D., Species Blancoanae 352. 1918.

Hemigraphis Nees

Hemigraphis procumbens (Lour.) Merr. in Philip. Journ. Sci. 15: 256. 1919 (based on *Barleria procumbens* Lour.).
> *Barleria procumbens* Lour. Fl. Cochinch. 377. 1790, ed. Willd. 458. 1793, Chinese *kām qūa tsù*.
> *Ruellia chinensis* Nees in DC. Prodr. 11: 147. 1847.
> *Strobilanthes scaber* Hance in Journ. Bot. 16: 231. 1878, non Nees.
> *Hemigraphis chinensis* T. Anders. ex Hemsl. in Journ. Linn. Soc. Bot. 26: 238. 1891.

" Habitat inculta prope Cantonem Sinarum." The description applies unmistakably to the species currently known as *Hemigraphis chinensis* T. Anders., which is rather conspicuous and not uncommon in the vicinity of Canton, growing in dry thickets on sterile slopes.

Barleria Linnaeus

Barleria acanthophora (Roem. & Schult.) Nees in DC. Prodr. 11: 726. 1847 (based on *Eranthemum spinosum* Lour.); C. B. Clarke in Thiselton-Dyer Fl. Trop. Afr. 5: 169. 1900.
> *Eranthemum acanthophorum* Roem. & Schult. Syst. 1: Mant. 154. 1822 (based on *Eranthemum spinosum* Lour.).

Eranthemum spinosum Lour. Fl. Cochinch. 19. 1790, ed. Willd. 24. 1793, non *Barleria spinosa* Hook.

" Habitat agreste in suburbiis Mozambicci in Africa." This species is known only from Loureiro's description and was placed by Nees among the species of *Barleria* of uncertain status. If correctly placed, as to the genus, then Loureiro's description of the calyx as " 2-phyllus " was probably due to his interpretation of the bracts as sepals, or to his observation of the outer two sepals only. In Index Kewensis Loureiro's species is reduced to *Haplanthus verticillaris* Nees, but I cannot find any record that this genus occurs in Africa. Clarke places it among the imperfectly known species of *Barleria*.

Acanthus (Tournefort) Linnaeus

Acanthus ebracteatus Vahl Symb. **2**: 75. *pl. 40.* 1791.

> *Acanthus ilicifolius* (non Linn.) Lour. Fl. Cochinch. 375. 1790, ed. Willd. 455. 1793, Anamese *cây ô rô.*

" Habitat ad ripas fluminum in Cochinchina, & China." Loureiro's interpretation of *Acanthus ilicifolius* was placed by Nees as a doubtful synonym of *A. ebracteatus* Vahl. From Loureiro's description of the flowers as white and from the appended note that the Canton form differed by the presence of bracteoles, it seems clear that the Cochinchina form Loureiro had is *Acanthus ebracteatus* Vahl. *Aquifolium indicum* Rumph. (Herb. Amb. **6**: 163. *pl. 71. f. 1*), cited by Loureiro as a synonym, is *Acanthus ilicifolius* Linn. The Canton form, for which Loureiro cites the Chinese name *lâo chú lăc* is unquestionably the Linnaean species, this being the only species of the genus known from Kwangtung Province, and is the form with pale blue flowers. Loureiro definitely states that his description was based on the Cochinchina form. A specimen from Loureiro, preserved in the herbarium of the Paris Museum, is identified as *Acanthus ilicifolius* Linn.; one in the herbarium of the British Museum is identified as *A. ebracteatus* Vahl; and this was sent by Loureiro from Cochinchina. Most of the Loureiro specimens in the Paris Museum are from the vicinity of Canton (see page 21).

Graptophyllum Nees

Graptophyllum pictum (Linn.) Griff. Notul. **4**: 139. 1854.

> *Justicia picta* Linn. Sp. Pl. ed. 2, 21. 1762; Lour. Fl. Cochinch. 24. 1790, ed. Willd. 29. 1753, Anamese *ngaoc diep.*
>
> *Graptophyllum hortense* Nees in Wall. Pl. As. Rar. **3**: 102. 1832.

" Habitat in hortis Cochinchinae." The description applies to the commonly culti-vated form with variegated leaves illustrated by Rumphius as *Folium bracteatum* Rumph. (Herb. Amb. **4**: 73. *pl. 30*) which Loureiro correctly cites as a synonym.

Pseuderanthemum Radlkofer

Pseuderanthemum sp.

> *Gratiola stricta* Lour. Fl. Cochinch. 23. 1790, ed. Willd. 28. 1793, Anamese *cây tu hít.*

" Habitat agrestis in Cochinchina." It is suspected that Loureiro's species is repre-sented by *Clemens 3178* from Tourane, " flowers white, tinged with pink-purple," Lou-reiro's description being " flos albus, rubro punctatus." The description of the leaves as " subserrate " however does not apply, while the " calyx nullus praeter bracteam triplicem flori suppositam " does not well apply to *Pseuderanthemum*. The fruit characters are manifestly those of an acanthaceous plant.

Andrographis Wallich

Andrographis sp.

? Justicia fastuosa (non Linn.) Lour. Fl. Cochinch. 24. 1790, ed. Willd. 30. 1793.

" Habitat loca agrestis in Cochinchina." If Loureiro's description be correct as to the many-seeded capsule, this cannot be a *Justicia*. The Linnaean species is *Hypoestes fastuosa* (Linn.) Soland. and Loureiro's description does not at all apply to it as noted by Willdenow. *Gratiola affinis Maderaspatana, digitalis aemula* Pluk. (Phyt. *pl. 193. f. 3*) cited by Loureiro (after Linnaeus) is according to Vahl, as noted by Willdenow, *Ruellia patula* Jacq. = *Dipteracanthus patulus* Nees. This figure does not remotely resemble *Andrographis paniculata* (Burm. f.) Nees, which I thought might be the species represented by Loureiro's description.

Peristrophe montana (Wall.) Nees in Wall. Pl. As. Rar. **3**: 113. 1832.

Ruellia montana Wall. List no. 2391. 1830, *nomen nudum*.

Justicia tinctoria Lour. Fl. Cochinch. 25. 1790, ed. Willd. 31. 1793, Anamese *kim loung nhuom*.

" Habitat inculta in Cochinchina." The description in general seems to apply to *Peristrophe*, and to Tourane specimens, *Kuntze 3789, Clemens 3193*, identified as *P. montana* (Wall.) Nees; the chief objection to this reduction of Loureiro's species is his description of it as procumbent. *Peristrophe tinctoria* Nees was based on *Justicia tinctoria* Roxb., not on Loureiro's binomial; I interpret it to be a synonym of *P. bivalvis* (Linn.) Merr. Interpret. Herb. Amb. 476. 1917.

Dicliptera Jussieu

Dicliptera chinensis (Linn.) Nees in DC. Prodr. **11**: 477. 1847.

Justicia chinensis Linn. Sp. Pl. 16. 1753; Lour. Fl. Cochinch. 25. 1790, ed. Willd. 30. 1793, Anamese *kim loung bánh*.

" Habitat inculta in hortis, & sepibus Cochinchinae." Loureiro's description agrees in essentials with the characters of the Linnaean species, the type of which was from southeastern China, where the species is common. *Dicliptera burmanni* Nees, to which Nees (DC. Prodr. **11**: 483. 1847) referred Loureiro's species, with doubt, as well as *Justicia chinensis* as described by Linnaeus and by Burman f., does not appear to be distinct from *Dicliptera chinensis;* at any rate *Dicliptera chinensis* Nees was based on *Justicia chinensis* Vahl (Symb. **2**: 13. 1791, Enum. **1**: 110. 1804) which in turn was the Linnaean species as Vahl understood it, as he cites both Linnaeus and Burman f.

Hypoestes Solander

Hypoestes purpurea (Linn.) Soland. ex Roem. & Schult. Syst. **1**: 140. 1817.

Justicia purpurea Linn. Sp. Pl. 16. 1753; Lour. Fl. Cochinch. 25. 1790, ed. Willd. 31. 1793, Chinese *chi chăp hoa*.

" Recentem observavi Cantone Sinarum, sicut etiam fecerat Osb. it. p. 230." Loureiro's description conforms to the characters of the Linnaean species, the type of which was from the vicinity of Canton; it is abundant near Canton. Nees reduced Loureiro's species to *Peristrophe tinctoria* Nees in which he has been followed by all subsequent authors, but this is certainly not the correct disposition of the plant Loureiro actually described. This reduction was doubtless made from *Folium tinctorum* Rumph. (Herb. Amb. **6**: 51. *pl. 22. f. 1*) which Loureiro mentions not as representing his species but as differing from it in the number and disposition of its flowers; this is *Peristrophe bivalvis* (Linn.) Merr. (*P. tinctoria* Nees). R. Brown (Prodr. 474. 1810) does not actually make the combination *Hypoestes purpurea* as is indicated in current literature.

Ecbolium Kurz

Ecbolium viride (Forsk.) comb. nov.

Justicia viridis Forsk. Fl. Aeg.-Arab. 5. 1775.

Justicia ligustrina Vahl Enum. **1**: 118. 1804.

Justicia ecbolium Linn. Sp. Pl. 15. 1753; Lour. Fl. Cochinch. 23. 1790, ed. Willd. 29. 1793, Anamese *cây dao laong*.

Ecbolium linneanum Kurz in Journ. As. Soc. Bengal **40**(2): 75. 1871.

"Habitat loca minus culta in Cochinchina." Loureiro's description applies in all essentials to the widely distributed species commonly known as *Justicia ecbolium* Linn. and although I have not seen any Indo-China material representing this species, it seems probable that Loureiro's interpretation of the Linnaean species is correct.

Clinacanthus Nees

Clinacanthus sp.?

Ziziphora siliquosa Lour. Fl. Cochinch. 27. 1790, ed. Willd. 33. 1793, Anamese *cây tút mât*.

Justicia ? obscura Vahl Enum. **1**: 170. 1805 (based on *Ziziphora siliquosa* Lour.).

"Habitat agrestis in Cochinchina." Bentham (DC. Prodr. **12**: 367. 1848) in excluding Loureiro's species from *Ziziphora*, referred it to *Justicia obscura* Vahl, which was based on Loureiro's description. In Index Kewensis it is reduced to *Justicia diffusa* Willd., but Loureiro's description does not apply to Willdenow's species. The description, however, does agree closely with material from Hue, *Squires 323*, which may or may not be referable to *Clinacanthus*. In any case *Ziziphora siliquosa* Lour. does not appear to be the same as *Clinacanthus nutans* (Burm. f.) Lundau, where I placed it in my original manuscript of 1919.

Rhinacanthus Nees

Rhinacanthus nasuta (Linn.) Kurz in Journ. Asiat. Soc. Bengal **39**(2): 79. 1870.

Justicia nasuta Linn. Sp. Pl. 16. 1753.

Rhinacanthus communis Nees in Wall. Pl. Asiat. Rar. **3**: 109. 1832; DC. Prodr. **11**: 442. 1847.

Dianthera paniculata Lour. Fl. Cochinch. 26. 1790, ed. Willd. 32. 1793, Anamese *thuóc lác nho lá*.

"Habitat sylvestris in Cochinchina." This reduction of Loureiro's species was made by Nees and is certainly correct, as Loureiro's description applies very closely to the widely distributed and well-known species.

Justicia Houstoun

Justicia championi T. Anders. in Benth. Fl. Hongk. 264. 1861.

Adhatoda chinensis Benth. Hook. Journ. Bot. Kew Gard. Miscel. **5**: 134. 1853, non *Justicia chinensis* Linn., nec Vahl.

Justicia chinensis Druce in Rept. Bot. Exch. Club Brit. Isles **4**: 629. 1917 (based on *Justicia chinensis* Benth., non Linn.).

Gratiola hyssopioides (non Linn.) Lour. Fl. Cochinch. 22. 1790, ed. Willd. 27. 1793, Chinese *pí pă tsāo.*

" Habitat inculta Cantonem Sinarum." Loureiro's description conforms in all essentials to *Justicia championi* T. Anders., a rather common and widely distributed species in Kwangtung Province.

Justicia gendarussa Burm. f. Fl. Ind. 10. 1768; Linn. f. Suppl. 85. 1781.

Justicia nigricans Lour. Fl. Cochinch. 24. 1790, ed. Willd. 30. 1793, Anamese *truong sinh cây.*

" Habitat in locis agrestis Cochinchinae." The reduction has been made from Loureiro's description alone and may not prove to be correct. Loureiro's statement " rami et folia lineis nigris notata " does not apply to Burman's species, but otherwise the description agrees reasonably well. *Justicia gendarussa* Burm. f. is represented by *Squires 202* from Hue.

Justicia sp.

? Gratiola rugosa Lour. Fl. Cochinch. 23. 1790, ed. Willd. 28. 1793, Anamese *cây lâu bac.*

" Habitat agrestis in Cochinchina." The description apparently applies to some acanthaceous plant. *Justicia* is suspected, but more data from the classical locality are needed to solve the status of Loureiro's species.

ACANTHACEAE OF UNCERTAIN GENERIC STATUS

Chelone obliqua (non Linn.) Lour. Fl. Cochinch. 382. 1790, ed. Willd. 464. 1793.

" Habitat agrestis in Cochinchina." No *Chelone* is represented by the description, but rather some acanthaceous genus and species as indicated by the capsule characters. I am unable to suggest what genus may be represented.

PLANTAGINACEAE

Plantago (Tournefort) Linnaeus

Plantago major Linn. Sp. Pl. 112. 1753; Lour. Fl. Cochinch. 71. 1790, ed. Willd. 90. 1793, Anamese *ma dê, xa tien,* Chinese *chē tsiên tsào.*

Plantago loureiri Roem. & Schult. Syst. **3**: 112. 1818 (based on *Plantago major* Lour.).

" Habitat ad vias, & in hortis, in Cochinchina, & China passim obvia." The Linnaean species, as currently interpreted, includes the form not uncommon in southeastern Asia. Willdenow notes that the European form differs from the one Loureiro had in that the leaves are 7-nerved rather than 5-nerved, the capsules are 6-seeded, not 4-seeded, and that the seeds are oblong, not globose. Loureiro's description of the plant as an annual must have been based on an erroneous observation.

RUBIACEAE

Dentella J. R. & G. Forster

Dentella repens (Linn.) Forst. Char. Gen. 26. *pl. 13.* 1776; Pitard in Lecomte Fl. Gén. Indo-Chine **3**: 76. 1922.

> *Oldenlandia repens* Linn. Mant. **1**: 40. 1767; Lour. Fl. Cochinch. 78. 1790, ed. Willd. 98. 1793, Chinese *há kīm tsào.*
>
> *Campanula repens* Lour. Fl. Cochinch. 139. 1790, ed. Willd. 173. 1793.
>
> *Hedyotis repens* G. Don Syst **3**: 526. 1834.

For *Oldenlandia repens* Loureiro states: " Habitat prope Cantonem Sinarum." The description conforms to the characters of the common and widely distributed Linnaean species. *Crusta ollae minima* Rumph. (Herb. Amb. **5**: 460. *pl. 170. f. 4*) and [*Oldenlandia repens*] Burm. f. (Fl. Ind. 38. *pl. 15. f. 2*) cited as synonyms, are correctly placed. For *Campanula repens* Loureiro states: " Habitat inculta in agris Cochinchinae." I am convinced that this new species is none other than the common *Dentella repens* (Linn.) Forst. with its small fruit erroneously described as 3-celled and 1-seeded, and the styles as cleft.

Oldenlandia Linnaeus

Oldenlandia biflora Linn. Sp. Pl. 119. 1753.

> *Oldenlandia paniculata* Linn. Sp. Pl. ed. 2, 1667. 1763; Lour. Fl. Cochinch. 78. 1790, ed. Willd. 99. 1793, Anamese *co naoc;* Pitard in Lecomte Fl. Gén. Indo-Chine **3**. 153. 1923.

The Anamese name indicates an Indo-Chinese specimen; no locality is cited. The description applies to the common and widely distributed *Oldenlandia paniculata* Linn., but *Oldenlandia biflora* Linn. is an older binomial for the same species (Trimen Fl. Ceyl. **2**: 317. 1894). [*Oldenlandia paniculata*] Burm. f. (Fl. Ind. 38. *pl. 15. f. 1.* 1768), cited by Loureiro as a synonym, is the Linnaean species, but *Mollugo zeylanica sylvestris* Burm. (Thes. Zeyl. 161. *pl. 71. f. 2.* 1737), also mentioned by Loureiro, represents the aizoaceous *Mollugo disticha* (Linn.) Seringe; regarding the latter Loureiro states: " dubitoque Oldenlandiam esse."

Oldenlandia corymbosa Linn. Sp. Pl. 119. 1753; Pitard in Lecomte Fl. Gén. Indo-Chine **3**: 146. 1923.

> *Pharnaceum incanum* (non Linn.) Lour. Fl. Cochinch. 185. 1790, ed. Willd. 231. 1793, Anamese *co mè.*

" Habitat in hortis, & agris Cochinchinae." The description in all essentials, except in the statement that the capsule was 3-celled, applies to the very common and widely distributed *Oldenlandia corymbosa* Linn. The correctness of this reduction is in a measure verified by Pitard, who cites *coc man* as one of the local names of *Oldenlandia corymbosa* Linn., apparently a cognate form of *co mè.* Loureiro expressed doubt in his interpretation of *Pharnaceum incanum* Linn. It is suspected that he took those parts of the description that do not conform to *Oldenlandia corymbosa* from some description of *Pharnaceum incanum* Linn.

Oldenlandia diffusa (Willd.) Roxb. Hort. Beng. 11. 1814, Fl. Ind. **1**: 444. 1820; Pitard in Lecomte Fl. Gén. Indo-Chine **3**: 145. 1923.

Hedyotis diffusa Willd. Sp. Pl. **1**: 566. 1797.

Hedyotis herbacea (non Linn.) Lour. Fl. Cochinch. 77. 1790, ed. Willd. 98. 1793, Anamese *co luoi răn.*

" Habitat spontanea in Cochinchina." Loureiro's description conforms closely to the characters of Willdenow's species, which is very common and widely distributed in Indo-China and throughout the Indo-Malaysian region.

Oldenlandia zanguebariae Lour. Fl. Cochinch. 78. 1790, ed. Willd. 99. 1793.

Hedyotis zanguebariae Roem. & Schult. Syst. **3**: 192. 1818 (based on *Oldenlandia zanguebariae* Lour.).

Oldenlandia obtusiloba Hiern in Oliv. Fl. Trop. Afr. **3**: 56. 1877.

" Habitat in ora Zanguebaria Africae Orientalis." I believe that Loureiro's species is the same as the one much later described by Hiern as *Oldenlandia obtusiloba*, based on specimens from Zanzibar and Mozambique. Hiern queries if this may not be the case. The two descriptions agree in all essentials.

Hedyotis Linnaeus

Hedyotis simplicissima (Lour.) comb. nov.

Petesia simplicissima Lour. Fl. Cochinch. 77. 1790, ed. Willd. 97, 1793, Anamese *cây bô bô.*

Hedyotis subdivaricata Drake ex Pitard in Lecomte Fl. Gén. Indo-Chine **3**: 124. 1922, in syn.

Oldenlandia subdivaricata Drake *l.c.*

Under the second species, *Petesia trifida*, Loureiro states: " Habitat utraque species in Cochinchina: illa [*P. simplicissima*] in agris, haec [*P. trifida*] in montibus *Son coung.*" The description appertains to *Hedyotis* and I do not hesitate in replacing Drake's specific name by Loureiro's much older one in spite of the latter's description of the fruiting calyces as 5-fid, they being 4-fid in Drake's species. Pitard does not mention Loureiro's species in his treatment of the Rubiaceae of Indo-China.

Hedyotis sp.

? Petesia trifida Lour. Fl. Cochinch. 77. 1790, ed. Willd. 97. 1793, Anamese *cây hang the.*

" Habitat . . . in Cochinchina . . . in montibus *Son coung.*" If Loureiro's description of the leaves as tricuspidate be correct no rubiaceous plant is represented by his description. Otherwise the data apply closely to *Hedyotis*. Pitard does not mention Loureiro's species.

Uncaria [78] Schreber

Uncaria cordata (Lour.) Merr. Interpret. Herb. Amb. 479. 1917.

Restiaria cordata Lour. Fl. Cochinch. 639. 1790, ed. Willd. 785. 1793, Anamese *cham bia tlon;* Moore in Journ. Bot. **63**: 289. 1925.

Uncaria pedicellata Roxb. Hort. Beng. 86. 1814, *nomen nudum*, Fl. Ind. **2**: 128. 1824; Pitard in Lecomte Fl. Gén. Indo-Chine **3**: 45. 1922.

Nauclea lanosa Poir. in Lam. Encycl. Suppl. **4**: 64. 1816.

[78] *Uncaria* Schreber (1789), conserved name, Vienna Code; an older one is *Ourouparia* Aublet (1775) = *Uruparia* O. Kuntze 1892.

" Habitat in sylvis Cochinchinae." Loureiro's type is preserved in the herbarium of the British Museum where it was examined by Haviland (Journ. Linn. Soc. Bot. **33**: 77. 1897) who placed the species as a synonym of *Uncaria pedicellata* Roxb. This was verified by Moore (Journ. Bot. **63**: 289. 1925) who explains certain discrepancies in Loureiro's description. *Restiaria nigra* Rumph. (Herb. Amb. **3**: 188), cited by Loureiro as doubtfully representing his species, and whence he derived his generic name, must be excluded as it may represent the tiliaceous *Colona scabra* Burret (*Microcos scabra* Sm., *Grewia scabra* DC., *Columbia subobovata* Hochr.). *Restiaria alba* Rumph. is *Commersonia bartramia* (Linn.) Merr. Pitard fails to account for Loureiro's genus and species.

Nauclea Linnaeus

(*Sarcocephalus* Afzelius)

Nauclea orientalis Linn. Sp. Pl. ed. 2, 243. 1762; Lour. Fl. Cochinch. 141. 1790, ed. Willd. 174. 1793, Anamese *cây gáo.*

Cephalanthus orientalis Linn. Sp. Pl. 95. 1753.

Sarcocephalus cordatus Miq. Fl. Ind. Bat. **2**: 133. 1856; Pitard in Lecomte Fl. Gén. Indo-Chine **3**: 27. 1922.

" Habitat in sylvis planis, & prope rivos in Cochinchina." The description conforms to the characters of the Linnaean species. *Bancalus latifolia* (i.e., *Arbor noctis*) Rumph. (Herb. Amb. **3**: 82. *pl. 54*), cited by Loureiro as representing the species, is the very closely allied *Naulcea undulata* Roxb. (*Sarcocephalus undulatus* Miq.). For a critical consideration of the application of the generic name *Nauclea*, see Merrill, E. D., Journ. Washington Acad. Sci. **5**: 534. 1915.

Cephalanthus Linnaeus

Cephalanthus angustifolius Lour. Fl. Cochinch. 67. 1790, ed. Willd. 83. 1793, Anamese *rì rì cây;* Havil. in Journ. Linn. Soc. Bot. **33**: 39. 1897 (amplified description based on Loureiro's type).

Acrodryon angustifolium Spreng. Syst. **1**: 386. 1825 (based on *Cephalanthus angustifolius* Lour.).

Axolus angustifolius Raf. Sylva Tellur. 61. 1838 (based on *Cephalanthus angustifolius* Lour.).

Cephalanthus stellatus Lour. Fl. Cochinch. 68. 1790, ed. Willd. 85. 1793, Anamese *rì rì boung gáo;* Pitard in Lecomte Fl. Gén. Indo-Chine **3**: 30. *f. 3, 6–13.* 1922.

Nauclea stellata Wall. List no. 6102. 1832 (based on *Cephalanthus stellatus* Lour.).

Eresimus stellatus Raf. Sylv. Tellur. 61. 1838 (based on *Cephalanthus stellatus* Lour.).

Cephalanthus mekongensis Pierre ex Pitard in Lecomte Fl. Gén. Indo-Chine **3**: 30. 1922, in syn.

Mimosa stellata Lour. Fl. Cochinch. 651. 1790, ed. Willd. 800. 1793, Anamese *cây rí rí.*

Mimosa ternata Pers. Syn. **2**: 261. 1806 (based on *Mimosa stellata* Lour.).

Acacia taxifolia Willd. Sp. Pl. **4**: 1050. 1805 (based on *Mimosa stellata* Lour.).

The types of both of Loureiro's species of *Cephalanthus* were from Indo-China, as indicated by him under *Cephalanthus stellatus.* *C. angustifolius* was indicated by him as having opposite leaves, while *C. stellatus* was so named because its leaves were in whorls of three's; yet Haviland definitely describes Loureiro's type of *C. angustifolius* in the herbarium of the British Museum as having *ternate* leaves. His appended note is somewhat ambiguous: " In Loureiro's description *C. angustifolius* is distinguished from *C. stellatus* by the leaves being opposite and not whorled, but this is not in the British Museum specimen." The descriptions manifestly appertain to a single species. *Cephalanthus angustifolius* Lour. typifies the genus *Axolus* Raf., *C. stellatus* typifies the genus *Eresimus* Raf., and the former is the second species cited under *Acrodryon* Spreng. Pitard overlooked Haviland's full description of the type of *Cephalanthus angustifolius* Lour.; he failed to account for this species and all of the synonyms cited above based on Loureiro's two descriptions, and yet adds an unnecessary new synonym by publishing *Cephalanthus mekongensis* Pierre in syn. In this case, under the rules of the International Code, either of Loureiro's specific names may be used, but I follow Haviland and not Pitard, first, because of the former's full description of *C. angustifolius* Lour. in 1897 which was based on Loureiro's extant type, thus establishing " usage," because of place priority and, finally, because *angustifolius* is a more descriptive specific name for the normal plant than is *stellatus*. For *Mimosa stellata* Loureiro states: " Habitat in montibus Cochinchinae." The description does not apply to any leguminous species, but conforms with the characters of the rubiaceous *Cephalanthus angustifolius* Lour. as far as it goes. The local name *cây rí rí* confirms this reduction.

Cephalanthus occidentalis Linn. Sp. Pl. 95. 1753; Lour. Fl. Cochinch. 67. 1790, ed. Willd. 83. 1793, Chinese *sòy yòng mâi*.

>   *Cephalanthus orientalis* (non Linn.) Lour. *l.c.* in nota: " Si cum Ceph. Americano (mihi non obvio) non conveniat, vocetur Ceph. Orientalis "; Roem. & Schult. Syst. **3**: 105. 1818.
>
>   *Acrodryon orientale* Spreng. Syst. **1**: 386. 1825 (based on *Cephalanthus orientalis* Lour. = *C. occidentalis* Lour.).
>
>   *Cephalanthus montanus* Lour. Fl. Cochinch. 67. 1790, ed. Willd. 84. 1793, Chinese *sān yòng mâi*.
>
>   *Gilipus montanus* Raf. Sylva Tellur. 61. 1838 (based on *Cephalanthus montanus* Lour.).
>
>   *Cephalanthus monas* Lour. ex Gomes in Mem. Acad. Sci. Lisb. Cl. Sci. Pol. Mor. Bel-Let. n.s. 4(1): 28. 1868.[79]

Loureiro's specimens of both *Cephalanthus occidentalis* and *C. montanus* were from China, as noted by him under *C. stellatus*. His description of *C. occidentalis* (*C. orientalis*) typifies the genus *Acrodryon* Sprengel. This form is not uncommon in Kwangtung Province and it is represented by numerous collections. In spite of the discontinuous distribution the Chinese form presents no appreciable differences when critically compared with the American *C. occidentalis* Linn. *Cephalanthus montanus* Lour. is described as having *alternate, crenate* leaves and *dioecious, apetalous* flowers, which are non-rubiaceous characters; these statements were apparently due to faulty observation on Loureiro's part, as his type in the Paris Museum is, according to Gagnepain who re-examined it for me, a true *Cephalanthus*, and is identical with other specimens in the Paris herbarium identified by Haviland as *C. occidentalis* Linn. The genus *Gilipus* Raf., typified by *Cephalanthus montanus* Lour. and since its publication in 1838 one of entirely doubtful status, thus falls as a synonym of *Cephalanthus*.

[79] A Loureiro herbarium name here first published by Gomes.

Mussaenda (Burman) Linnaeus

Mussaenda cambodiana Pierre ex Pitard in Lecomte Fl. Gén. Indo-Chine **3**: 188. 1923,
var. **annamensis** Pitard *op. cit.* 189.

Mussaenda frondosa (non Linn.) Lour. Fl. Cochinch. 151. 1790, ed. Willd. 188. 1793,
Anamese *cây buóm bac.*

" Habitat frequens, per dumeta scandens in Cochinchina." Loureiro's description
manifestly applies to the variety of Pierre's species described by Pitard. If further confirmation is needed it may be noted that one of Pitard's recorded Anamese names is exactly
the same as the one cited by Loureiro, while de Pirey's *buom bac, Chevalier 41254,* also
represents the same species; it is also represented by *Clemens 4466* from thickets at Tourane,
near Hue. The pre-Linnaean synonyms cited by Loureiro must be excluded. *Folium
principissae* Rumph. (Herb. Amb. **4**: 111. *pl. 51*) represents *Mussaenda reinwardtiana*
Miq. and *Mussaenda zeylanica flore rubro* Burm. (Thes. Zeyl. 165. *pl. 76*) represents the
true *Mussaenda frondosa* Linn. A specimen from Loureiro in the herbarium of the British
Museum is identified as representing *Mussaenda pubescens* Dryander, a species not admitted
by Pitard as occurring in Indo-China. I have not seen this, but I suspect that a form of
Mussaenda cambodiana Pierre is represented by it.

Randia Houstoun

Randia cochinchinensis (Lour.) comb. nov.

Aidia cochinchinensis Lour. Fl. Cochinch. 143. 1790, ed. Willd. 177. 1793, Anamese
cây tlai; Moore in Journ. Bot. **63**: 250. 1925.

Stylocoryna racemosa Cav. Ic. **4**: 46. *pl. 368.* 1797.

Randia racemosa F.-Vill. Novis. App. Fl. Filip. 108. 1880.

Randia densiflora Benth. Fl. Hongk. 155. 1861; Pitard in Lecomte Fl. Gén. Indo-
Chine **3**: 241. 1923.

Webera densiflora Wall. in Roxb. Fl. Ind. **2**: 536. 1824.

Webera oppositifolia Roxb. Fl. Ind. **2**: 535. 1824.

Stylocoryna densiflora Wall. List no. 8404. 1847–49.

Cupia densiflora DC. Prodr. **4**: 394. 1830.

Cupia oppositifolia DC. Prodr. **4**: 394. 1830.

Fagraea cochinchinensis A. Chev. Cat. Pl. Jard. Bot. Saigon 65. 1919 (based on *Aidia
cochinchinensis* Lour.); Merr. Enum. Philip. Pl. **3**: 314. 1923, quoad syn. Loureiro.

No locality is cited by Loureiro, but the Anamese name indicates an Indo-China
specimen. *Aidia* Lour. remained a genus of entirely doubtful status until 1925, when
Moore (Journ. Bot. **63**: 250. 1895) examined the extant type which definitely placed it as
a synonym of *Randia.* In my original manuscript of 1919 I followed Chevalier (Cat. Pl.
Jard. Bot. Saigon 65. 1919) in his reduction of *Aidia cochinchinensis* to *Fagraea fragrans*
Roxb., but manifestly Loureiro credited to *Aidia cochinchinensis* an erroneous local name,
that of *Fagraea fragrans* Roxb., and apparently accredited to his *Aidia* the wood characters
of the *Fagraea.* Loureiro's type in the herbarium of the British Museum is a species of
Randia, and Loureiro's description essentially agrees with it. Moore concluded that it
represented a species allied to or identical with *Randia eucodon* K. Schum., one known
only from Siam, and placed by Pitard in the section *Gardenioides.* I examined Loureiro's

type in 1930 and my conclusion is that it represents a species of the section *Gynopachys* and that it cannot be distinguished from the forms currently referred to *Randia densiflora* Benth. and which to me is identical with the older *Randia racemosa* (Cav.) F.-Vill.; the type of the latter was a Philippine specimen. Loureiro's type is well matched by *Clemens 3582, 3589, 3908* from Tourane, near the classical locality Hue. The species is one of rather wide geographic distribution and with numerous synonyms; Loureiro's specific name, being the oldest, is here adopted for it. Additional synonyms given by Pitard, after King and Gamble, are *Stylocoryna dimorphophylla* Teysm. & Binn., *Gynopachys axilliflora* Miq., *G. oblongata* Miq., *Urophyllum coriaceum* Miq., and *Ixora thozetia* F. Muell.

Randia esculenta (Lour.) comb. nov.

> *Genipa esculenta* Lour. Fl. Cochinch. 149. 1790, ed. Willd. 185. 1793, Anamese *cây gang com*.
> *Gardenia esculenta* Spreng. Syst. 1: 762. 1825 (based on *Genipa esculenta* Lour.).
> *Randia edulis* Kostel. Allgem. Med.-Pharm. Fl. 2: 579. 1831 (based on *Genipa esculenta* Lour.).
> *Posoqueria fasciculata* Roxb. Fl. Ind. 2: 568. 1824.
> *Randia fasciculata* DC. Prodr. 4: 386. 1830; Pitard in Lecomte Fl. Gén. Indo-Chine 3: 226. 1923 (var. *multiflora* Pitard).

"Habitat inculta in Cochinchina." Loureiro's description agrees closely with the characters of the widely distributed species currently known as *Randia fasciculata* DC. which extends from India to Indo-China, the Malay Peninsula, and Sumatra, and which is common in Indo-China. From Loureiro's statement " Flos . . . polynatus " I believe that the exact form that he described is *R. fasciculata* DC. var. *multiflora* Pitard. Pitard gives eight synonyms for *Randia fasciculata* DC., but does not include the first three cited above.

Randia horrida (Lour.) Schult. in Roem. & Schult. Syst. 5: 248. 1819 (based on *Oxyceros horrida* Lour.).

> *Oxyceros horrida* Lour. Fl. Cochinch. 151. 1790, ed. Willd. 187. 1793, Anamese *cây uŭt o, cây mo tló*.
> *Randia longiflora* Lam. Encycl. 3: 26. 1789, var. *horrida* Pierre ex Pitard in Lecomte Fl. Gén. Indo-Chine 3: 234. 1923 (based on *Oxyceros horrida* Lour.).
> *Gardenia horrida* Spreng. Syst. 1: 762. 1825 (based on *Oxyceros horrida* Lour.).

"Habitat in sylvis Cochinchinae." There is every reason to believe that this interpretation of Loureiro's species is correct. Loureiro's type is preserved in the herbarium of the British Museum. The same species is represented by *Squires 763*, and a specimen collected by de Pirey, *Chevalier 41209*. In both of these the corolla tubes are less than 5 mm. long. As Lamarck's type from Java has flowers with their corolla tubes " au moins un pouce de longueur," as indicated in the original description, I do not accept Pitard's interpretation of the species (Lecomte Fl. Gén. Indo-Chine 3: 234. 1923).

Randia sinensis (Lour.) Schult. in Roem. & Schult. Syst. 5: 248. 1819.

> *Oxyceros sinensis* Lour. Fl. Cochinch. 151. 1790, ed. Willd. 187. 1793, Chinese *cai tsoi lăc*.
> *Gardenia chinensis* Spreng. Syst. 1: 762. 1825 (based on *Oxyceros sinensis* Lour.).

" Habitat agrestis circa Cantonem Sinarum." This species is not uncommon in thickets near Canton and its status is well understood. Recently collected material bears the Cantonese name *lak tsoi sui*.

Randia spinosa (Thunb.) Poir. in Lam. Encycl. Suppl. 2: 829. 1812.

Gardenia spinosa Thunb. Diss. Gardenia 16. *pl. 2. f. 4.* 1780.

Gardenia dumetorum Retz. Obs. 2: 14. 1781.

Gardenia spinosa Linn. f. Suppl. 164. 1781.

Randia dumetorum Poir. in Lam. Encycl. Suppl. 2: 829. 1812, Ill. 2: 227. *pl. 156. f. 4.* 1819; Pitard in Lecomte Fl. Gén. Indo-Chine 3: 231. 1923.

Genipa ? flava Lour. Fl. Cochinch. 149. 1790, ed. Willd. 185. 1793, Chinese *uăt thau cay*.

Genipa buffalina Lour. Fl. Cochinch. 149. 1790, ed. Willd. 184. 1793, Anamese *cây gang tlâu*.

Gardenia buffalina Spreng. Syst. 1: 763. 1825 (based on *Genipa buffalina* Lour.).

For *Genipa ? flava* Loureiro states: " Habitat Cantone Sinarum." His description applies to the form, abundant in thickets near Canton, currently known as *Randia dumetorum* Poir. Poiret retained *Randia spinosa* and *R. dumetorum* as distinct species, the former with pubescent, the latter with glabrous calyx-tubes. Two species may be represented, but the Canton form is the one with the pubescent calyx-tubes and is apparently typical *Randia spinosa* (Thunb.) Poir. For *Genipa buffalina* Loureiro states: " Habitat agrestis in Cochinchina," and his description of this conforms to the characters of *Randia dumetorum* as interpreted by Pitard. Willdenow suggested that *Genipa buffalina* Lour. might be the same as *Gardenia spinosa* Thunb. (= *Randia spinosa* Poir.). If further proof of the correctness of this reduction is needed, it is supplied by the local name, one of those cited by Pitard being *cây cang trau*, a cognate form of *cây gang tlâu* as given by Loureiro for *Genipa buffalina*. Pitard lists 22 synonyms for *Randia dumetorum* Poir., but nowhere accounts for Loureiro's two binomials, although *Genipa buffalina* Lour. was based on an Indo-China specimen.

Gardenia [80] Ellis

Gardenia jasminoides Ellis in Philos. Trans. 51(2): 935. *pl. 23.* 1761; Sprague in Kew Bull. 15. 1929; Hubb. & Rehd. in Bot. Mus. Leafl. Harvard Univ. 1: 5. 1932.

Varneria augusta Linn. Amoen. Acad. 4: 136. 1759, *nomen*.

Warneria augusta Linn. op. cit. 138, *nomen*.

Gardenia florida Linn. Sp. Pl. ed. 2, 305. 1762; Lour. Fl. Cochinch. 147. 1790, ed. Willd. 183. 1793, Anamese *cây deanh tàu*, Chinese *chȳ tsú*; Pitard in Lecomte Fl. Gén. Indo-Chine 3: 249. 1923.

Gardenia grandiflora Lour. Fl. Cochinch. 147. 1790, ed. Willd. 182. 1793, Anamese *cây deạnh nam*.

Gardenia augusta Merr. Interpret. Herb. Amb. 485. 1917.

Mussaenda chinensis Lour. Fl. Cochinch. 152. 1790, ed. Willd. 189. 1793, Chinese *xān xïe lâu*.

[80] *Gardenia* Ellis (1761), conserved name, Cambridge Code. *Varneria* Linnaeus and *Warneria* Linnaeus (1759) are both older, but were not validly published.

For *Gardenia florida* Loureiro states: " Habitat tam in China, quam in Cochinchina." This is the double-flowered form commonly found in cultivation, exactly *Catsjopiri* Rumph. (Herb. Amb. **7**: 26. *pl. 14. f. 2*) cited by Loureiro as representing the species. For *Gardenia grandiflora* Loureiro states: " Habitat in Cochinchina prope fluvios: indeque in hortos transfertur. . . ." This is manifestly the normal single-flowered form of the preceding, which is widely distributed as a native plant in southern China and Indo-China. For *Mussaenda chinensis* Loureiro states: " Habitat inculta in suburbiis Cantoniensibus," and the description, as far as it goes, conforms to the not uncommon form of *Gardenia jasminoides* Ellis with small crowded leaves. For a discussion of the problems involved as to the proper specific name for this species see Sprague T. A. Gardenia or Warneria (Kew Bull. 12–16. 1929). He accepts *Gardenia jasminoides* Ellis and I now agree that I erred in accepting *Varneria augusta* Linn. as validly published, because no genus description of *Varneria* or *Warneria* was ever published by Linnaeus. *Varneria augusta* Linn. and *Warneria augusta* Linn.[81] appear as follows: " 14 Cathjopiri Varneria augusta " and " Warneria augusta 7–14." The figures represent the Rumphian illustrations, and these definitely place the two Linnaean binomials, for as published they were based solely on *Catsjopiri* Rumph. (Herb. Amb. **7**: 26. *pl. 14. f. 2*). I do not think that Sprague's further argument to the effect that *Catsjopiri* Rumph. represents a monstrosity (a double-flowered form) should be given much weight in spite of the rule invalidating species based on monstrosities. Before accepting a double flower of horticultural origin as a " monstrosity " I would prefer to wait until some one has compiled data on how many scores of currently accepted binomials would be invalidated by the general acceptance of this rule under the interpretation that a double flower of horticultural origin is a monstrosity. Incidentally *Gardenia jasminoides* Ellis was in part based on the double-flowered form as shown by his plate, as the habit sketch obviously represents a double flower, although the description and the dissected flower on the plate appertain to the normal single-flowered form.

Canthium Lamarck

Canthium dicoccum (Gaertn.) Merr. in Philip. Journ. Sci. **35**: 8. 1928.

> *Psydrax dicoccos* Gaertn. Fruct. **1**: 125. *pl. 26. f. 2.* 1788.
> *Canthium didymum* Gaertn. f. Fruct. **3**: 94. 1805; Gagnep. in Lecomte Fl. Gén. Indo-Chine **3**: 293. 1924.
> *Polyozus bipinnata* Lour. Fl. Cochinch. 74. 1790 (*bippinata*), ed. Willd. 94. 1793, Anamese *cây trâm ná;* Moore in Journ. Bot. **63**: 248. 1925.
> *Plectronia didyma* Kurz in Journ. As. Soc. Bengal **46**(2): 153. 1877; Elm. Leafl. Philip. Bot. **1**: 28. 1906.
> *Plectronia dicocca* Burck in Verslag Lands Plant. Buitenzorg 1897 37. 1898; Merr. Enum. Philip. Fl. Pl. **3**: 536. 1923.

" Habitat in sylvis Cochinchinae." In my original manuscript of 1919 I stated that a species of *Plectronia* [*Canthium*] was probably represented in spite of Loureiro's description of the leaves as bipinnate, he probably having misinterpreted the distichously arranged branchlets and leaves as representing a bipinnate leaf. Moore (Journ. Bot. **63**: 248, 1925), examining Loureiro's type preserved in the Herbarium of the British Museum, confirms this generic reduction of *Polyozus*, and further states that *P. bipinnata* is almost

[81] Amoen. Acad. **4**: 136, 138. 1759.

if not quite certainly synonymous with one of the forms of *Canthium didymum* Gaertn. It may be well to record here that *Plectronia* Linnaeus is a synonym of *Olinia* of the *Oliniaceae* and has been entirely misinterpreted by those authors who considered it synonymous with *Canthium*. In the additions to retained generic names approved by the Cambridge Botanical Congress *Olinia* Thunberg (1799) is conserved over *Plectronia* Linnaeus (1767) for the South African genus typifying the Oliniaceae.

Coffea Linnaeus

Coffea arabica Linn. Sp. Pl. 172. 1753; Lour. Fl. Cochinch. 144. 1790, ed. Willd. 179. 1793, Anamese *cây càphe*.

"Habitat in hortis Cochinchinae, non indigena, sed ex alio loco advecta." The description clearly applies to the Linnaean species, the common coffee plant.

Coffea racemosa Lour. Fl. Cochinch. 145. 1790, ed. Willd. 179. 1793; Hiern. in Oliver Fl. Trop. Afr. **3**: 185. 1877.

Coffea mozambicana DC. Prodr. **4**: 500. 1830 (based on *Coffea racemosa* Lour.).

Coffea ramosa Schult. in Roem. & Schult. Syst. **5**: 198. 1819 (sphalm., *C. racemosa* Lour. intended).

Hexepta racemosa Raf. Sylva Tellur. 164. 1838 (based on *Coffea racemosa* Lour.).

"Habitat agrestis in insula Africana Mozambicco." This species is only known from Loureiro's original description, judging from which, it may not belong in the genus *Coffea*. Hiern (Trans. Linn. Soc. Bot. **1**: 175. 1876), while admitting it as a *Coffea*, states that he had not seen any specimens representing it and (Fl. Trop. Afr. **3**: 185. 1877) repeats that statement, compiling a description from Loureiro's original one. It may prove to be a species of *Canthium*. Loureiro did not see any flowers.

Coffea zanguebariae Lour. Fl. Cochinch. 145. 1790, ed. Willd. 180. 1793; Hiern in Oliver Fl. Trop. Afr. **3**: 182. 1877.

Amajoua africana Spreng. Syst. **2**: 126. 1825 (based on *Coffea zanguebariae* Lour.).

Amazona africana Hiern in Trans. Linn. Soc. Bot. **1**: 172. 1876, in syn., sphalm. (*Amajoua africana* intended).

Hexepta axillaris Raf. Sylva Tellur. 164. 1838 (based on *Coffea zanguebariae* Lour.).

"Habitat in sylvis planis orae Zanguebariae continentis Africanae: indeque a Lusitanis delata, & in hortis culta prope Mozambiccum." Loureiro notes that the odor, taste, and color of the fruit, and the uses of the seed, were the same as for the common coffee, *Coffea arabica* Linn. Hiern cites a specimen from Mozambique, collected by Forbes, as representing the species. Cheney (Coffee, a monograph of the economic species of the genus Coffea 68. *pl. 25.* 1925) gives an amplified description of the species. Loureiro's description typifies the genus *Hexepta* Raf.

Pavetta Linnaeus

Pavetta arenosa Lour. Fl. Cochinch. 73. 1790, ed. Willd. 92. 1793, Chinese *tá sā*.

Pavetta sinica Miq. in Journ. Bot. Néerl. **1**: 107. 1861; Merr. in Sunyatsenia **1**: 39. *pl. 14.* 1930; Brem. in Fedde Repert. **37**: 118. 1934.

Pavetta sinensis Lour. ex Gomes [82] in Mem. Acad. Sci. Lisb. Cl. Sci. Mor. Pol. Bel.-Let. n.s. **4**(1): 28. 1868.

[82] A Loureiro herbarium name here first published by Gomes.

" Habitat in provincia Cantoniensi Sinarum." Loureiro's specimen is preserved in the herbarium of the Paris Museum and has been identified by Pierre as " *Pavetta indica* Linn.? " Loureiro's statement regarding the leaves " tuberculis multis utrinque prominentibus, quasi arenae granulis consita. Inde nomen sini cum *Planta arenosa* " possibly indicates that some of his material had leaves infested with small galls, as is the case with some of the leaves of at least one collection of this *Pavetta* from Hainan, *Tsang 16774*, and a Kwangtung specimen collected by Ford. Loureiro's type is preserved in the herbarium of the Paris Museum, and Doctor Gagnepain kindly examined this for me, reporting on November 4, 1931 as follows: " J'ai vu *Pavetta arenosa* Lour. qui est bien le *P. indica* Linn. mais la var. à feuilles courtement velues en dessous. Aucun reste des ' tuberculis multis utrinque prominentibus.' C'est Desvaux qui a écrit *Pavetta arenosa* Lour. tandis que Loureiro avait écrit sur son étiquette *Pavetta sinensis, dai sa* ou *ta sa.*" This is the pubescent form described by Miquel as *Pavetta sinica* Miq. (Journ. Bot. Néerl. 1: 107. 1861). A photograph of Miquel's type specimen is reproduced in my article on his Kwangtung species (Sunyatsenia 1: 39. *pl. 14.* 1930); it is apparently also the pubescent form illustrated by Ker (Bot. Reg. 3: *pl. 198.* 1817) as *Pavetta indica* Linn. The actual type of *Pavetta indica* Linn. in the Linnaean herbarium is the form with glabrous leaves; one detached leaf on the sheet is slightly pubescent beneath. Bremekamp in his recent monograph of *Pavetta* (Fedde Repert. 37: 1–208. 1934) restricts *Pavetta indica* Linn. to Ceylon and southern India (type from Ceylon) and recognizes a number of distinct species, among them this Kwangtung form, in the collective *Pavetta indica* of modern authors. He retains *Pavetta sinica* Miq. as valid, and regarding *P. arenosa* Lour. states: " species, incertae sedis, probabiliter ad *Tarennam* pertinens." Loureiro's species is the pubescent *Pavetta* characterized as *P. sinica* Miq.

Ixora Linnaeus

Ixora chinensis Lam. Encycl. 3: 344. 1789.

> *Ixora coccinea* (non Linn.) Lour. Fl. Cochinch. 75. 1790, ed. Willd. 95. 1793, Anamese *boung tlang do.*

" Habitat in fruticetis Cochinchinae." I believe Loureiro's species to be the same as *Ixora chinensis* Lam. which is very common in southeastern China. I am unable satisfactorily to separate from Lamarck's species a number of Indo-China specimens placed by Pitard under *Ixora stricta* Roxb. (Lecomte Fl. Gén. Indo-Chine 3: 326. 1924). If Pitard has correctly interpreted Roxburgh's species, it may prove to be a synonym of *Ixora chinensis* Lam., but Lamarck's specific name is much the older, *Ixora stricta* Roxb., dating from 1814. *Flamma sylvarum* (*peregrina*) Rumph. (Herb. Amb. 4: 107. *pl. 47*), described by Rumphius from plants introduced into Amboina from Java, represents *Ixora coccinea* Linn. as I understand that species; it is mentioned by Loureiro as " satis cohaerit." *Flamma sylvarum* Rumph. (Herb. Amb. 4: 105. *pl. 46*), mentioned by Loureiro as *not* representing his species, is the type of *Ixora fulgens* Roxb.

Ixora coccinea Linn. Sp. Pl. 110. 1753; Pitard in Lecomte Fl. Gén. Indo-Chine 3: 323. 1924.

> *Ixora montana* Lour. Fl. Cochinch. 76. 1790, ed. Willd. 96. 1793, Anamese *boung tlang núi.*

" Habitat in locis montanis in Cochinchina." From Loureiro's very short description and especially his description of the leaves as subsessile and cordate, and the flowers as red, I interpret this as a synonym of *Ixora coccinea* Linn. Pitard does not account for Loureiro's species in his consideration of the species of *Ixora* (Lecomte Fl. Gén. Indo-Chine 3: 303–330. 1924). It may be doubted, however, if the true *Ixora coccinea* Linn. occurs in Indo-China as other than a cultivated plant.

Ixora finlaysoniana Wall. List no. 6166. 1832, *nomen nudum;* G. Don Gen. Syst. 3: 572. 1834; Pitard in Lecomte Fl. Gén. Indo-Chine 3: 312. 1924.

Ixora loureiri G. Don Gen. Syst. 3: 573. 1834 (based on *Ixora alba* Lour.).

Ixora alba (non Linn.) Lour. Fl. Cochinch. 76. 1790, ed. Willd. 96. 1793, Anamese *boung tlang tláng.*

" Habitat similiter [i.e., in fruticetis Cochinchinae] in locis planis, sed rarior." In my original manuscript of 1919 I placed this as a doubtful synonym of *Ixora finlaysoniana* Wall., largely on the basis of Loureiro's description of the flowers as white. It may be noted that Pitard cites for Wallich's species the Anamese name *cay bong trang trang,* a cognate form of *boung tlang tláng* as given by Loureiro. The type of *Ixora finlaysoniana* Wall. was probably from Siam or Indo-China.

Psychotria [83] Linnaeus

Psychotria rubra (Lour.) Poir. in Lam. Encycl. Suppl. 4: 597. 1816 (based on *Antherura rubra* Lour.).

Antherura rubra Lour. Fl. Cochinch. 144. 1790, ed. Willd. 178. 1793, Anamese *cây lâu;* Moore in Journ. Bot. 63: 251. 1925.

Psychotria antherura Schult. in Roem. & Schult. Syst. 5: 188. 1819 (based on *Antherura rubra* Lour.).

Polyozus lanceolata Lour. Fl. Cochinch. 75. 1790, ed. Willd. 94. 1793, Chinese *am san cung,* non *Psychotria lanceolata* Nutt.

Psychotria reevesii Wall. in Roxb. Fl. Ind. 2: 164. 1824; Pitard in Lecomte Fl. Gén. Indo-Chine 3: 361. 1924.

Psychotria elliptica Ker. Bot. Reg. 8: *pl. 607.* 1822.

For *Antherura rubra* Loureiro states: " Habitat inculta, passim obvia in Cochinchina." Moore has examined Loureiro's type in the herbarium of the British Museum, and his long critical note (Journ. Bot. 63: 251. 1925) definitely settles the status of Loureiro's genus and species. *Psychotria reevesii* Wall., based on specimens from southeastern China, represents the same species as does *P. elliptica* Ker, which was erroneously localized, in the original description, as having come from Brazil. The species is very common in southeastern China and in Anam. It is represented by *Clemens 3134, 3527, 3632, 4417,* from near the classical locality. Working from Loureiro's description alone in my original manuscript of 1919, I tentatively placed this in the genus *Tarenna;* but Loureiro's description of the seeds as 2, plano-convex, and 5-sulcate, indicates a *Psychotria,* not a *Tarenna.* For *Polyozus lanceolata* Loureiro states: " Habitat Cantone Sinarum, spontanea." The description applies to *Psychotria rubra* Poir., the " flos rubescens " doubtless added from dried specimens due to a change of color in drying. The description of the leaves as lanceolate is better for *P. tutcheri* Dunn. than for *P. rubra* Poir., but the former species has not been found near Canton whereas the latter is common there.

[83] *Psychotria* Linnaeus (1759), conserved name, Vienna Code; older ones are *Myrstiphyllum* P. Browne (1756) and *Psychotrophum* P. Browne (1756).

Lasianthus [84] Jack

Lasianthus verticillatus (Lour.) comb. nov.

Dasus verticillatus Lour. Fl. Cochinch. 142. 1790, ed. Willd. 176. 1793, Anamese *cây caŏng;* Moore in Journ. Bot. **63**: 250. 1925.

Lasianthus andamanicus Pitard in Lecomte Fl. Gén. Indo-Chine **3**: 394. 1924 (non Hook. f. ?).

" Habitat agrestis in Cochinchina." The genus *Dasus* remained of undetermined status until 1925 when Moore (Journ. Bot. **63**: 250. 1925) examined Loureiro's type in the herbarium of the British Museum and verified the suggestion that I made in my original manuscript of 1919, that *Lasianthus* was probably the group represented. I examined the type in 1930 and find there is no doubt as to this generic identification. Moore thought that the species might be *Lasianthus poilanei* Pitard, but I do not think so, because Pitard's species is glabrous. I think that *Dasus verticillatus* Lour., which is represented by *Clemens 3600* from Tourane, is safely the species interpreted by Pitard as *Lasianthus andamanicus*, but as to whether or not Hooker's species is represented, I am not certain. *Lasianthus verticillatus* is very closely related to *L. chinensis* (Champ.) Benth., or at least to certain forms currently referred to the latter. However, Champion's type at Kew has about 13 pairs of lateral nerves, while Loureiro's type of *Lasianthus verticillatus* has only about 8 pairs. Incidentally, *Lasianthus chinensis* as described by Pitard (Lecomte Fl. Gén. Indo-Chine **3**: 391. 1924), keyed out as having *glabrous* fruits, cannot represent Champion's species, as the latter has pubescent fruits.

Paederia [85] Linnaeus

Paederia scandens (Lour.) Merr. in Contr. Arnold. Arb. **8**: 163. 1934 (based on *Gentiana scandens* Lour.).

Gentiana scandens Lour. Fl. Cochinch. 171. 1790, ed. Willd. 213. 1793, Anamese *rau mo,* Chinese *ki sí thân.*

Paederia tomentosa Blume Bijdr. 968. 1826; Pitard in Lecomte Fl. Gén. Indo-Chine **3**: 415. 1924.

Crawfurdia ? loureiri G. Don Gen. Syst. **4**: 200. 1838 (based on *Gentiana scandens* Lour.).

" Habitat, & scandit per arbores, & sepes in Cochinchina, & China." A specimen from Loureiro preserved in the herbarium of the Paris Museum was identified by Pierre as *Paederia foetida* Linn. The Linnaean species as currently interpreted has ovoid or ellipsoid fruits, while Loureiro describes his plant as having a " capsula sub-rotunda." *Gentiana scandens* Lour. is undoubtedly the very common Indo-China species referred by Pitard to *Paederia tomentosa* Blume, which is often confused with *P. foetida* Linn.; the Linnaean species has larger ellipsoid fruits. Pitard failed to account for *Gentiana scandens* Lour., although its type is preserved in the herbarium of the Paris Museum, an oversight readily accounted for, as one would scarcely expect to find the rubiaceous *Paederia* described as a gentianaceous plant. Loureiro states regarding the Chinese form: " In Sinensi folia

---

[84] *Lasianthus* Jack (1823), conserved name, Cambridge Code. *Dasus* Loureiro (1790) is the oldest name for the genus.

[85] *Paederia* Linnaeus (1767), conserved name, Vienna Code; older ones are *Hondbessen* Adanson (1763) = *Hondbession* O. Kuntze 1892, and *Daucontu* Adanson (1763).

saepe videntur cordata levi emarginatura ad basim." This may be the form for which Hance proposed the binomial *Paederia chinensis* which Rehder recognizes as a valid species, as distinct from both *Paederia foetida* Linn. and *P. tomentosa* Blume. I am by no means certain that it is specifically distinct from *Paederia tomentosa* Blume = *P. scandens* (Lour.) Merr.

Serissa Commerson

Serissa foetida (Linn. f.) Comm. in Juss. Gen. 209. 1789.

> *Lycium foetidum* Linn. f. Suppl. 150. 1781.
>
> *Dysoda fasciculata* Lour. Fl. Cochinch. 146. 1790, ed. Willd. 181. 1793, Anamese *man thien huong*, Chinese *mán tsīen yòng*.

" Habitat in Cochinchina, & China, odore ingrata, aspectu pulchra." The description clearly applies to the well-known *Serissa foetida* (Linn. f.) Comm. Loureiro's type is preserved in the herbarium of the British Museum.

Morinda Linnaeus

Morinda citrifolia Linn. Sp. Pl. 176. 1753; Lour. Fl. Cochinch. 140. 1790, ed. Willd. 174. 1793, Anamese *cây nhaù;* Pitard in Lecomte Fl. Gén. Indo-Chine **3**: 423. 1924.

" Habitat culta in hortis Cochinchinensibus." Loureiro's description conforms closely to the characters of the common and widely distributed *Morinda citrifolia* Linn. *Bancudus latifolia* Rumph. (Herb. Amb. **3**: 158. *pl. 99*), cited by Loureiro as a synonym, is correctly placed.

Morinda cochinchinensis DC. Prodr. **4**: 449. 1830; Pitard in Lecomte Fl. Gén. Indo-Chine **3**: 420. 1924 (based on *Morinda umbellata* Lour.).

> *Morinda umbellata* (non Linn.) Lour. Fl. Cochinch. 140. 1790, ed. Willd. 173. 1793, Anamese *cây ngê bà.*
>
> *Morinda vestita* Pierre ex Pitard *l.c.* in syn.
>
> *Morinda trichophylla* Merr. in Philip. Journ. Sci. **23**: 267. 1923.

" Habitat in sylvis Cochinchinae." Pitard has given an amplified description of this species, based on specimens collected in Cambodia by Poilane and by Pierre, although he does not cite Loureiro's description on which de Candolle's binomial is based. He distinguishes it from *M. villosa* Hook. f. by very minor characters. Another very closely allied species is *Morinda jackiana* Korth. Whether or not the specimens cited by Pitard actually represent the species Loureiro described, *M. cochinchinensis* DC. is safely represented by *Clemens 3328, 3746*, from the vicinity of Tourane, near Hue. *Bancudus angustifolius* Rumph. (Herb. Amb. **3**: 157. *pl. 98*), discussed by Loureiro following his description, represents the totally different *Morinda bracteata* Roxb.

Morinda parvifolia Bartl. in DC. Prodr. **4**: 449. 1830; Merr. Enum. Philip. Fl. Pl. **3**: 573. 1923.

> *Morinda royoc* (non Linn.) Lour. Fl. Cochinch. 140. 1790, ed. Willd. 174. 1793, Anamese *cây ngón.*

" Habitat inculta in Cochinchina." Loureiro's description applies closely to Bartling's species, which was based on Luzon material, but which extends from the Philippines to Formosa, southern China, and Indo-China. It is represented by *Robinson 1083, 1552, 1546* from Nhatrang. My interpretation of Loureiro's species as a synonym of *M. parvi-*

folia Bartl. is based on his statement that the plant was a small slender one and that the leaves were small. Pitard (Lecomte Fl. Gén. Indo-Chine **3**: 422. 1924) has manifestly included this small-leaved form in his description of *Morinda umbellata* Linn., giving leaf measurements for that species as small as 2 cm. in length and 1 cm. in width.

Morinda umbellata Linn. Sp. Pl. 176. 1753.

> *Stigmanthus cymosus* Lour. Fl. Cochinch. 146. 1790, ed. Willd. 181. 1793, Anamese *cay buóm rùng;* Moore in Journ. Bot. **63**: 252. 1925.
>
> *Stigmatanthus cymosus* Schult. in Roem. & Schult. Syst. **5**: 225. 1819 (based on *Stigmanthus cymosus* Lour.).
>
> *Cuviera asiatica* Spreng. Syst. **1**: 760. 1825 (based on *Stigmanthus cymosus* Lour.).

"Habitat in sylvis, & montibus Cochinchinae." Loureiro's genus and species remained of entirely doubtful status until 1925, when Moore (Journ. Bot. **63**: 252) solved the problem by the simple expedient of examining the type in the herbarium of the British Museum. The specimen represents the common and widely distributed *Morinda umbellata* Linn.

Borreria G. F. W. Meyer

Borreria articularis (Linn. f.) F. N. Will. in Bull. Herb. Boiss. II **5**: 956. 1905.

> *Spermacoce articularis* Linn. f. Suppl. 119. 1781, excl. syn. Rumph.
>
> *Spermacoce hispida* Linn. Sp. Pl. 102. 1753; Pitard in Lecomte Fl. Gén. Indo-Chine **3**: 439. 1924.
>
> *Spermacoce flexuosa* Lour. Fl. Cochinch. 79. 1790, ed. Willd. 100. 1793, Anamese *deei ruot gà.*
>
> *Borreria hispida* K. Schum. in Engler & Prantl Nat. Pflanzenfam. **4**(4): 144. 1891, non Spruce, 1888.

"Habitat in fruticetis, & sepibus Cochinchinae." Loureiro's description in general applies unmistakably to the common Linnaean species, yet Pitard (Lecomte Fl. Gén. Indo-Chine **3**: 438–442. 1924) fails to account for his binomial. The statement "seminibus 2-cornibus" was probably based on an erroneous observation, probably the calyx-teeth on the capsule-valves being intended. Pitard has erroneously reversed the authorities and citations for *Spermacoce hispida* Linn. and *S. stricta* Linn. f. The oldest valid name for this very common species is that supplied by *Spermacoce articularis* Linn. f. as verified from an examination of the original specimen in the Linnaean herbarium.

Galium Linnaeus

Galium sp.

> *?* *Crucianella angustifolia* (non Linn.) Lour. Fl. Cochinch. 79. 1790, ed. Willd. 100. 1793, Anamese *uy linh tien,* Chinese *uêi lin sĩen.*

"Habitat in China. Florem non vidi, nisi pictum in Herbario sinico." I suspect that this may be *Galium verum* Linn. from the vegetative characters indicated by Loureiro, which in all probability were taken from material secured from an herbalist. The description of the inflorescences was taken from a Chinese illustration, source not indicated, and does not apply to *Galium.* The single illustration cited, *Rubeola angustiore folio* Tournefort (Inst. 130. *pl. 50*), is too imperfect to be of any value in indicating the probable genus represented by Loureiro's inadequate description.

RUBIACEAE OF UNCERTAIN GENERIC STATUS

Ixora novemnervia Lour. Fl. Cochinch. 76. 1790, ed. Willd. 96. 1793, Anamese *buóm rùng tláng*.

"Habitat in desertis locis Cochinchinae." The description is short and imperfect and although it may apply to some rubiaceous plant, certainly no *Ixora* is represented; it is possible that Loureiro had in hand some non-rubiaceous plant. De Pirey's specimen of *buon rung, Chevalier 41253*, is sterile, and I do not recognize it except that it is rubiaceous and is suggestive of *Morinda*.

CAPRIFOLIACEAE

Sambucus (Tournefort) Linnaeus

Sambucus javanica Reinw. in Blume Bijdr. 657. 1826; Danguy in Lecomte Fl. Gén. Indo-Chine 3: 2. *f. 1, 1–2,* 1922.

> *Phyteuma bipinnata* Lour. Fl. Cochinch. 138. 1790, ed. Willd. 172. 1793, Chinese *tcha leàng tsào* (non *Sambucus bipinnata* Cham. & Schlecht.).
> *Phyteuma cochinchinensis* Lour. Fl. Cochinch. 139. 1790, ed. Willd. 172. 1793, Anamese *cây thuóc moi* (non *Sambucus cochinchinensis* Spreng.).
> *Sambucus ebuloides* Desv. ex DC. Prodr. 4: 323. 1830 (based on *Phyteuma bipinnata* Lour.).
> *Sambucus phyteumoides* DC. Prodr. 4: 323. 1830 (based on *Phyteuma cochinchinensis* Lour.).
> *Sambucus chinensis* Lindl. in Trans. Hort. Soc. London 6: 297. 1826.

For *Phyteuma bipinnata* Loureiro states: "Habitat in suburbiis Cantoniensibus"; and for *P. cochinchinensis:* "Habitat in montibus Cochinchinae." Both are erroneously described as having bipinnate leaves and many-seeded fruits. Loureiro's type of the former is preserved in the herbarium of the Paris Museum, and while Danguy cites this Chinese species and Desvaux's binomial based upon it in the synonymy of *S. javanica* Reinw., he curiously does not account for the Indo-Chinese *Phyteuma cochinchinensis* Lour. or its synonym *Sambucus phyteumoides* DC.; Loureiro's description of the corolla-lobes as acute, clearly indicates *Sambucus javanica* Reinw. rather than *S. eberhardtii* Danguy (type from Hue), as the latter has obtuse corolla lobes.

Sambucus cochinchinensis Spreng. Syst. 1: 935. 1825 (based on *Sambucus nigra* Lour.).

> *Sambucus nigra* (non Linn.) Lour. Fl. Cochinch. 181. 1790, ed. Willd. 226. 1793, Anamese *ngô châu duu*, Chinese *û chū yû*.
> *Sambucus ? loureiriana* DC. Prodr. 4: 323. 1830 (based on *Sambucus nigra* Lour.).

"Habitat in montanis Sinensibus." A species of entirely doubtful status, the description in all probability based on fragmentary material secured from an herbalist and also material representing more than one species. The description of the leaflets as entire would indicate some other genus than *Sambucus*. Schwerin (Mitt. Deutsch. Dendr. Ges. **18**: 53. 1909) refers *Sambucus loureiriana* DC. and *S. cochinchinensis* Spreng. to *Turpinia* sp. which was suggested by de Candolle *l.c.* "forte *Turpiniae* species ? aut *Cunoniacea* quaedam? ". But none of the Chinese species of *Turpinia* has entire leaflets, which is true also of all species of *Sambucus*. The species remains one of doubtful status, as the description alone is too indefinite to render identification possible.

Lonicera Linnaeus

Lonicera japonica Thunb. Fl. Jap. 89. 1784.

Lonicera periclymenum (non Linn.) Lour. Fl. Cochinch. 150. 1790, ed. Willd. 185. 1793, Anamese *kim ngân tàu,* Chinese *gín tūm.*

" Habitat in multis locis imperii Sinensis, inculta." Loureiro's species was referred by de Candolle (Prodr. **4**: 333. 1830) with doubt, to *Lonicera confusa* DC., and definitely so by Rehder in his monographic treatment of the genus (Rept. Missouri Bot. Garden **14**: 156. 1903); yet Rehder also cites it (p. 159) as a doubtful synonym of *L. japonica* Thunb. It is probably best placed as a synonym of the latter in view of Loureiro's statement that it occurred in many places in China. *L. confusa* DC. has a relatively restricted distribution, but both species occur in Kwangtung Province whence Loureiro secured most of his Chinese botanical material. His description of the fruit as many-seeded is a manifest error.

Lonicera macrantha (D. Don) DC. Prodr. **4**: 333. 1830; Danguy in Lecomte Fl. Gén. Indo-Chine **3**: 16. 1922.

Caprifolium macranthum D. Don Prodr. Fl. Nepal. 140. 1825.

Lonicera xylosteum (non Linn.) Lour. Fl. Cochinch. 150. 1790, ed. Willd. 186. 1793, Anamese *deei buóm buóm, kim ngân hoa.*

Lonicera cochinchinensis G. Don Gen. Syst. **3**: 447. 1834 (based on *Lonicera xylosteum* Lour.).

" Habitat scandens per sepes, & dumeta in Cochinchina." In my preliminary manuscript of 1919 I placed this as a synonym of *Lonicera affinis* H. & A., but Danguy does not admit this as occurring in Indo-China. On the other hand he cites a number of localities for *Lonicera macrantha,* including Hue, where Loureiro lived. As Loureiro's description conforms to the characters of this species, this reduction may be considered as correct as *Lonicera macrantha* DC. is currently interpreted. It may be noted here that *Lonicera loureiri* DC. (Prodr. **4**: 334. 1830) is typified by *Caprifolium loureiri* Blume (Bijdr. 653. 1826), the type of which was an actual Javan specimen, Blume giving no reference to any species described by Loureiro. De Candolle places *Lonicera xylosteum* Lour. as a doubtful synonym of *L. loureiri* DC., where it certainly does not belong.

CUCURBITACEAE

Melothria Linnaeus

Melothria indica Lour. Fl. Cochinch. 35. 1790, ed. Willd. 43. 1793, Anamese *cung kang deài tlái;* Cogn. in Pflanzenreich **66**(IV–275–1): 98. 1916; Gagnep. in Lecomte Fl. Gén. Indo-Chine **2**: 1064. 1921.

" Habitat in Cochinchina implicata inter sepes." In spite of the fact that Gagnepain seems to have seen no specimens from Indo-China, there is every reason to believe that the current modern interpretation of this species is correct; it is represented by *Clemens 3068* from Tourane. *Cucumis murinus viridis* Rumph. (Herb. Amb. **5**: 463. *pl. 171. f. 2*) cited by Loureiro, is correctly placed. The fresh plant, as Loureiro notes, has the very characteristic odor and flavor of the common cucumber.

Melothria heterophylla (Lour.) Cogn. in DC. Monog. Phan. **3**: 618. 1881; Gagnep. in Lecomte Fl. Gén. Indo-Chine **2**: 1063. 1921.

Solena heterophylla Lour. Fl. Cochinch. 514. 1790, ed. Willd. 629. 1793, Anamese *cu nhang*, Chinese *khū léu, tiēn hōa fuèn*.

Bryonia heterophylla Raeusch. Nomencl. ed. 3, 282. 1797; Steudel Nomencl. ed. 2, 1: 232. 1840 (based on *Solena heterophylla* Lour.).

Bryonia hastata Lour. Fl. Cochinch. 594. 1790, ed. Willd. 731. 1793, Chinese *si toūng quā*.

Juchia hastata M. Roem. Syn. 2: 48. 1846 (based on *Bryonia hastata* Lour.).

Zehneria heterophylla Druce in Rept. Bot. Exch. Club Brit. Isles 4: 653. 1917 (based on *Solena heterophylla* Lour.).

For *Solena heterophylla*, described as a new genus and species, Loureiro states: " Habitat in sylvis Cochinchinae, & Chinae," and for *Bryonia hastata:* " Habitat agrestis circa Cantonem Sinarum." The type of the former is preserved in the herbarium of the British Museum. *Bryonia hastata* Lour. undoubtedly represents the same species, which is an exceedingly variable one in its vegetative characters.

Melothria perpusilla (Blume) Cogn. in DC. Monog. Phan. 3: 607. 1881, Pflanzenreich 66(IV–275–1): 106. 1916; Gagnep. in Lecomte Fl. Gén. Indo-Chine 2: 1061. 1921.

Cucumis maderaspatanus (non Linn.) Lour. Fl. Cochinch. 592. 1790, ed. Willd. 727. 1793, Anamese *cung cang tlòn tlái*.

" Habitat inter sepes Cochinchinae." Loureiro's description does not conform to the characters of the Linnaean species which is *Melochia maderaspatana* (Linn.) Cogn., and which occurs in Indo-China, but in general the characters given by him do conform to those of *Melochia perpusilla* (Blume) Cogn. and I suspect this to be the correct disposition of the form Loureiro described.

Momordica (Tournefort) Linnaeus

Momordica cochinchinensis (Lour.) Spreng. Syst. 3: 14. 1826 (based on *Muricia cochinchinensis* Lour.); Cogn. in DC. Monog. Phan. 3: 444. 1881, Pflanzenreich 88(IV–275–II): 34. 1924; Gagnep. in Lecomte Fl. Gén. Indo-Chine 2: 1068. 1921.

Muricia cochinchinensis Lour. Fl. Cochinch. 596. 1790, ed. Willd. 733. 1793, Anamese *cây góc, mŏuc biet tu*, Chinese *mŏ piĕ sù*.

" Habitat agrestis in Cochinchina, & China." This is a widely distributed, common, and well-known species, extending from India to southern China, Formosa, and Malaysia. Gagnepain gives the Anamese names *gac, mak kao* and *quâ gac*. Loureiro's type is preserved in the herbarium of the British Museum.

Momordica charantia Linn. Sp. Pl. 1009. 1753; Lour. Fl. Cochinch. 589. 1790, ed. Willd. 724. 1793, Anamese *múop dáng*, Chinese *khú quā;* Cogn. in Pflanzenreich 88(IV–275–II): 24. 1924.

" Habitat culta in hortis Cochinchinae, & Chinae." The widely distributed, well-known Linnaean species was correctly interpreted by Loureiro.

Luffa (Tournefort) Linnaeus

Luffa acutangula (Linn.) Roxb. Hort. Beng. 70. 1814, Fl. Ind. ed. 2, 3: 713. 1832; Gagnep. in Lecomte Fl. Gén. Indo-Chine 2: 1075. 1921; Cogn. in Pflanzenreich 88(IV–275–II): 68. 1924.

Cucumis acutangulus Linn. Sp. Pl. 1011. 1753; Lour. Fl. Cochinch. 591. 1790, ed. Willd. 727. 1793, Anamese *muóp khên*.

"Habitat cultus in Cochinchina, & China." The description applies unmistakably to a form of the widely cultivated Linnaean species. *Petola bengalensis* Rumph. (Herb. Amb. **5**: 408. *pl. 149*) is correctly placed as a synonym.

Luffa cylindrica (Linn.) M. Roem. Syn. **2**: 63. 1846 (actually based on *Momordica cylindrica* Lour. which, however, Loureiro correctly credited to Linnaeus); Cogn. in Pflanzen-reich **88**(IV–275–II): 62. 1924.

   Momordica cylindrica Linn. Sp. Pl. 1009. 1753; Lour. Fl. Cochinch. 590. 1790, ed. Willd. 725. 1793, Chinese *sŏy quā*.

   Momordica luffa Linn. Sp. Pl. 1009. 1753; Lour. Fl. Cochinch. 590. 1790, ed. Willd. 724. 1793, Anamese *múop ngot*, Chinese *sū quā*.

   Luffa leucosperma M. Roem. Syn. **2**: 63. 1846 (based on *Momordica luffa* Lour.).

   For *Momordica cylindrica* Loureiro states: "Culta circa Cantonem Sinarum," and for *M. luffa*: "late culta in Cochinchina, & China." Both are manifestly forms of the common sponge gourd, which is widely planted throughout the warmer parts of the Old World. The two species that Linnaeus proposed are merely forms of a single one, and Loureiro's interpretations were essentially correct.

Citrullus Forskål

Citrullus vulgaris Schrad. in Eckl. & Zeyer Enum. Pl. Afr. Austr. 279. 1836, Linnaea 12: 412. 1838; Gagnep. in Lecomte Fl. Gén. Indo-Chine 2: 1056. 1921; Cogn. in Pflanzen-reich **88**(IV–275–II): 103. 1924.

   Cucurbita citrullus Linn. Sp. Pl. 1010. 1753; Lour. Fl. Cochinch. 594. 1790, ed. Willd. 730. 1793, Anamese *dua hâu*, Chinese *sī quā*.

"Habitat frequenter culta in Cochinchina, & China." This is the common water-melon, the Linnaean species being correctly interpreted by Loureiro. *Anguria indica* Rumph. (Herb. Amb. **5**: 400. *pl. 146. f. 1*) is correctly placed as a synonym.

Cucumis (Tournefort) Linnaeus

Cucumis melo Linn. Sp. Pl. 1011. 1753; Lour. Fl. Cochinch. 591. 1790, ed. Willd. 726. 1793, Anamese *dua gang*, Chinese *căn quā;* Gagnep. in Lecomte Fl. Gén. Indo-Chine 2: 1057. 1921.

"Habitat cultus in Cochinchina, & China." The description applies to a form of the common melon.

Cucumis sativus Linn. Sp. Pl. 1012. 1753; Lour. Fl. Cochinch. 591. 1790, ed. Willd. 726. 1793, Anamese *dua chuot*, Chinese *hoâm quā;* Gagnep. in Lecomte Fl. Gén. Indo-Chine 2: 1057. 1921.

"Habitat late culta in Cochinchina, simulque in China." Two varieties of the common cucumber are indicated, one with somewhat insipid fruits about five inches long, the other with fruit having a better odor and flavor and twice as long, for which the Anamese name *dua bà cai* is given.

Benincasa Savi

Benincasa hispida (Thunb.) Cogn. in DC. Monog. Phan. **3**: 513. 1881; Gagnep. in Lecomte
Fl. Gén. Indo-Chine **2**: 1055. 1921; Cogn. in Pflanzenreich **88**(IV–275–II): 164. 1924.
Cucurbita hispida Thunb. Fl. Jap. 322. 1784.
Benincasa cerifera Savi in Bibl. Ital. **9**: 158. 1818.
Cucurbita pepo (non Linn.) Lour. Fl. Cochinch. 593. 1790, ed. Willd. 728. 1793, Ana-
mese *bí dao,* Chinese *tum̄ qūa.*

" Habitat frequentissima in Cochinchina, & China." Loureiro's description clearly
applies to the wax gourd, *Benincasa hispida* (Thunb.) Cogn., not to the pumpkin, *Cucurbita
pepo* Linn. *Camolenga* Rumph. (Herb. Amb. **5**: 395. *pl. 143*), cited by Loureiro as a
synonym, is correctly placed.

Gymnopetalum Arnott

Gymnopetalum cochinchinense (Lour.) Kurz in Journ. Asiat. Soc. Bengal **40**(2): 57. 1871;
Cogn. in DC. Monog. Phan. **3**: 391. 1881; Gagnep. in Lecomte Fl. Gén. Indo-Chine
2: 1049. 1921; Cogn. in Pflanzenreich **88**(IV–275–II): 181. 1924.
Bryonia cochinchinensis Lour. Fl. Cochinch. 595. 1790, ed. Willd. 732. 1793, Anamese
cây qua qua.
Tripodanthera cochinchinensis M. Roem. Syn. **2**: 48. 1846 (based on *Bryonia cochinchinensis* Lour.).
Evonymus chinensis Lour. Fl. Cochinch. 156. 1790, ed. Willd. 194. 1793, Chinese
kām quā.
Gymnopetalum chinense Merr. in Philip. Journ. Sci. **15**: 256. 1919 (based on *Evonymus
chinensis* Lour.).

For *Bryonia cochinchinensis* Loureiro states: " Habitat agrestis in Cochinchina, ad
sepes scandens"; and for *Evonymus chinensis:* " Habitat incultis extra suburbia Cantoniensia in China." It is curious that Loureiro should describe the same species twice, once
in *Bryonia* and once in such a totally unrelated genus as *Evonymus;* yet without question
both descriptions apply to a single species. The type of *Bryonia cochinchinensis* is preserved in the herbarium of the British Museum where it was examined by Moore (Journ.
Bot. **63**: 252. 1925). *Evonymus chinensis* has page priority over *Bryonia cochinchinensis,*
but as noted by Moore " priority of place " is not recognized under the International
Code of Botanical Nomenclature. Recently collected material from Canton bears the
Cantonese names *ye kwa* and *ka shui kwah.*

Gymnopetalum sp.?
Bryonia triloba Lour. Fl. Cochinch. 595. 1790, ed. Willd. 731. 1793, Anamese *deom
ác ba chia.*
Bryonia stipulacea Willd. Sp. Pl. **4**: 620. 1805 (based on *Bryonia triloba* Lour.).
Bryonia agrestis Raeusch. Nomencl. ed. 3, 283. 1797 (based on *Bryonia triloba* Lour.).
" Habitat agrestis in Cochinchina." This has currently been reduced to *Blastania
garcini* (Burm. f.) Cogn. (DC. Monog. Phan. **3**: 629. 1881), but Loureiro's description as
to the leaves and particularly the many-seeded fruit does not apply to *Blastania* and,
moreover, no *Blastania* is known from Indo-China. But for Loureiro's description of the
leaves as " utrinque laevia " I should be inclined to reduce this to *Gymnopetalum monoicum*

Gagnep. which occurs at Hue. Gagnepain makes no attempt to account for Loureiro's species or for the two synonyms based upon it.

Lagenaria Seringe

Lagenaria leucantha (Duch.) Rusby in Mem. Torr. Bot. Club **6**: 43. 1896; Merr. Interpret. Herb. Amb. 493. 1917, Enum. Philip. Fl. Pl. **3**: 584. 1923.

Cucurbita leucantha Duch. in Lam. Encycl. **2**: 150. 1786.

Cucurbita lagenaria Linn. Sp. Pl. 1010. 1753; Lour. Fl. Cochinch. 592. 1790, ed. Willd. 728. 1793, Anamese *cây bâu,* Chinese *hú quā, hô lô.*

Lagenaria vulgaris Seringe in Mém. Soc. Phys. Hist. Nat. Genève **3**: 25. *pl. 2.* 1825, DC. Prodr. **3**: 299. 1828; Cogn. in Pflanzenreich 88(IV–275–II): 201. 1924.

Lagenaria cochinchinensis M. Roem. Syn. **2**: 61. 1846 (based on *Cucurbita lagenaria* Lour.).

" Habitat ubique culta in Cochinchina, & China." The Linnaean species, one of the most commonly cultivated cucurbitaceous plants in the Indo-Malaysian region, was correctly interpreted by Loureiro, who describes three fruit forms of the variable species; Gagnepain omits this genus in his consideration of the Cucurbitaceae of Indo-China (Lecomte Fl. Gén. Indo-Chine **2**: 1030–1095. 1921) possibly because the species occurs there only in cultivation.

Trichosanthes Linnaeus

Trichosanthes anguina Linn. Sp. Pl. 1008. 1753; Lour. Fl. Cochinch. 588. 1790, ed. Willd. 722. 1793, Anamese *muóp saoc;* Gagnep. in Lecomte Fl. Gén. Indo-Chine **2**: 1039. 1921.

" Habitat culta, & esculenta in Cochinchina, & China." Loureiro correctly interpreted the Linnaean species, which is rather widely cultivated in tropical Asia for its edible fruits. *Petola anguina* Rumph. (Herb. Amb. **5**: 407. *pl. 148*), cited by Loureiro as a synonym, is correctly placed.

Trichosanthes cucumerina Linn. Sp. Pl. 1008. 1753; Lour. Fl. Cochinch. 588. 1790, ed. Willd. 722. 1793, Anamese *bát bát tlâu;* Gagnep. in Lecomte Fl. Gén. Indo-Chine **2**: 1040. 1921.

Trichosanthes cochinchinensis M. Roem. Syn. **2**: 96. 1846 (based on *Trichosanthes cucumerina* Lour.).

" Habitat sepes arundinum Bambus in Cochinchina." The description conforms to the characters of the widely distributed Linnaean species.

Trichosanthes pilosa Lour. Fl. Cochinch. 588. 1790, ed. Willd. 723. 1793, Anamese *cây qua;* Gagnep. in Lecomte Fl. Gén. Indo-Chine **2**: 1046. 1921.

Anguina ? pilosa O. Ktz. Rev. Gen. Pl. 254. 1891 (based on *Trichosanthes pilosa* Lour.).

" Habitat agrestis in Cochinchina." Cogniaux (DC. Monog. Phan. **3**: 949. 1881) merely enumerates this as a species of doubtful status, and Gagnepain gives a brief description compiled from Loureiro's short one, with the comment " espèce inconnue." But for Loureiro's description of the leaves as " inferiora palmata, supera triloba," I should be inclined to reduce this to *Trichosanthes villosa* Blume. It is clearly a *Trichosanthes,* if Loureiro's description be correct, but perhaps a species not collected since his time.

Trichosanthes scabra Lour. Fl. Cochinch. 589. 1790, ed. Willd. 723. 1793, Anamese *dua nhà tlòi;* Gagnep. in Lecomte Fl. Gén. Indo-Chine **2**: 1047. 1921.

"Habitat in sylvis Cochinchinae." Cogniaux (DC. Monog. Phan. **3**: 386. 1881) places this as a doubtful synonym of *Trichosanthes integrifolia* (Roxb.) Kurz of Bengal and Burma, where it certainly does not belong. Gagnepain admits it with a short description compiled from Loureiro, with the comment "espèce inconnue." It may ultimately be proven not to be a *Trichosanthes*, if Loureiro's description is correct, for the "Pomum . . . 12-lobum" is scarcely a *Trichosanthes* character, but suggests the ribbed fruits of some species of *Gymnopetalum*.

Trichosanthes tricuspidata Lour. Fl. Cochinch. 589. 1790, ed. Willd. 723. 1793, Anamese *bát bát rùng.*

Anguina tricuspidata O. Ktz. Rev. Gen. Pl. 254. 1891 (based on *Trichosanthes tricuspidata* Lour.).

"Habitat agrestis in Cochinchina." If Loureiro's description be correct in the "stipulis sub-orbiculatis, crenatis, crassis" and in its "poma . . . bilocularia, bisperma," apparently no *Trichosanthes* is represented; he does not describe the flowers, other than to state that they are white, monoecious and in axillary spikes. I cannot, however, suggest any reduction of it. If we may judge from the description, it is clearly not the form described as *T. tricuspidata* Lour. by Cogniaux (DC. Monog. Phan. **3**: 374. 1881) nor the one described by Gagnepain as *T. tricuspidata* Blume (Lecomte Fl. Gén. Indo-Chine **2**: 1042. 1921).

Cucurbita (Tournefort) Linnaeus

Cucurbita maxima Duch. in Lam. Encycl. **2**: 151. 1786.

Cucurbita melopepo (non Linn.) Lour. Fl. Cochinch. 593. 1790, ed. Willd. 729. 1793, Anamese *bí ngô*, Chinese *năn quā, fán quā.*

"Habitat late culta in Cochinchina, & China." The description apparently applies to a form of the common squash which is widely cultivated in the warmer parts of the Old World, rather than to the pumpkin, *Cucurbita pepo* Linn. *Pepo indicus* Rumph. (Herb. Amb. **5**: 399. *pl. 145*), cited by Loureiro as representing the species, is *Cucurbita maxima* Duch. Gagnepain does not admit any species of *Cucurbita* in his treatment of the family (Lecomte Fl. Gén. Indo-Chine **2**: 1030–1095. 1921) possibly because the plants are found there only in cultivation.

Coccinia Wight & Arnott

Coccinia cordifolia (Linn.) Cogn. in DC. Monog. Phan. **3**: 529. 1881; Gagnep. in Lecomte Fl. Gén. Indo-Chine **2**: 1054. 1921.

Bryonia cordifolia Linn. Sp. Pl. 1012. 1753.

Bryonia grandis Linn. Mant. **1**: 126. 1767; Lour. Fl. Cochinch. 595. 1790, ed. Willd. 731. 1793, Anamese *deom ác ngu chia.*

Coccinia grandis Voigt Hort. Suburb. Calcut. 59. 1845; M. Roem. Syn. **2**: 93. 1846.

Coccinia loureiriana M. Roem. Syn. **2**: 93. 1846 (based on *Bryonia grandis* Lour.).

"Habitat in sylvis Cochinchinae." The Linnaean species was, I believe, correctly interpreted by Loureiro. *Vitis alba indica* Rumph. (Herb. Amb. **5**: 448. *pl. 166. f. 1*), cited by Loureiro as a synonym, is correctly placed.

CAMPANULACEAE

Sphenoclea [86] Gaertner

Sphenoclea zeylanica Gaertn. Fruct. 1: 113. *pl. 24.* 1788; Danguy in Lecomte Fl. Gén. Indo-Chine 3: 692. 1930.

> *Rapinia herbacea* Lour. Fl. Cochinch. 127. 1790, ed. Willd. 157. 1793, Anamese *bôn bôn tlòn lá.*

"Habitat inculta in hortis Cochinchinae." This is clearly the correct disposition of Loureiro's genus and species in spite of his description of the calyx as 8-partite; he apparently included as sepals not only the 5 calyx segments but also the bract and two bracteoles subtending each flower. It is a common and widely distributed rice-paddy weed. Loureiro's type is preserved in the herbarium of the British Museum.

Lobelia (Plumier) Linnaeus

Lobelia chinensis Lour. Fl. Cochinch. 514. 1790, ed. Willd. 628. 1793, Chinese *puón fuên liên.*

> *Lobelia erinus* Thunb. Fl. Jap. 325. 1784, non Linn.
> *Lobelia radicans* Thunb. in Trans. Linn. Soc. 2: 330. 1794.

"Habitat inculta prope Cantone Sinarum." This species is not uncommon in open grassy places near Canton. I can find no constant characters by which the Japanese form can be distinguished from the Kwangtung type. Loureiro's specific name should be retained.

COMPOSITAE

Vernonia [87] Schreber

Vernonia cinerea (Linn.) Less. in Linnaea 4: 291. 1829; Gagnep. in Lecomte Fl. Gén. Indo-Chine 3: 484. 1924.

> *Conyza cinerea* Linn. Sp. Pl. 862. 1753.
> *Conyza chinensis* Linn. Sp. Pl. 862. 1753 et herb. Linn.!
> *Pteronia tomentosa* Lour. Fl. Cochinch. 489. 1790, ed. Willd. 597. 1793, Chinese *chú hōa mú.*
> *Calea cordata* Lour. Fl. Cochinch. 488. 1790, ed. Willd. 595. 1793, Anamese *cây bac dâu.*
> *Conyza candida* (non Linn.) Lour. Fl. Cochinch. 495. 1790, ed. Willd. 605. 1793, Anamese *bac dâu com.*
> *Conyza odorata* (non Linn.) Lour. Fl. Cochinch. 495. 1790, ed. Willd. 605. 1793.
> *Eupatorium sinuatum* Lour. Fl. Cochinch. 487. 1790, ed. Willd. 595. 1793.
> *Eupatorium hispidum* Pers. Syn. 2: 402. 1807 (based on *E. sinuatum* Lour.).
> *Gnaphalium indicum* (non Linn.) Lour. Fl. Cochinch. 497. 1790, ed. Willd. 608. 1793, Anamese *cây son.*

For *Pteronia tomentosa* Loureiro states: "Habitat spontanea prope Cantonem Sinarum"; and for *Calea cordata:* "Habitat spontanea in Cochinchina." The former is manifestly *Conyza cinerea* Less. *Calea cordata* Lour. is in all probability a form of this

---

[86] *Sphenoclea* Gaertner (1788), conserved name Vienna Code; an older one is *Pongati* Adanson (1756, 1759 = *Pongatium* Jussieu 1789).

[87] *Vernonia* Schreber (1791), conserved name, Vienna Code; an older one is *Behen* Hill (1762).

very common, widely distributed and variable species, in spite of Loureiro's description of the leaves as cordate. It is interesting to note that Gagnepain (Lecomte Fl. Gén. Indo-Chine **3**: 484. 1924) who does not attempt to identify *Calea cordata* Lour., quotes practically the same Anamese name for *Vernonia cinerea* Less., *cây bat dau*, as Loureiro's *cây bac dâu* for *Calea cordata*. *Conyza chinensis* Linn. was based on a specimen collected in China by Toren, and this specimen in the Linnaean herbarium is *Vernonia cinerea* (Linn.) Less. For *Conyza odorata* Loureiro states: " Habitat agrestis in Cochinchina." The description seems to apply to a luxurious form of *Vernonia cinerea* Less. For *Conyza candida* Loureiro states: " Habitat inculta in Cochinchina." The rather meager description indicates a form of the ubiquitous *Vernonia cinerea* Less. As partial confirmation of the correctness of this reduction, Gagnepain, as noted above, cites the Anamese name *cô bac dâu* as one of those recorded for Lessing's species. For *Eupatorium sinuatum* Loureiro states: " Habitat agreste in insula Africana Mozambicco." The description is short and apparently applies to a form of *Vernonia cinerea* Less. Oliver & Hiern (Fl. Trop. Afr. **3**: 301. 1877) merely state that Loureiro's species was unknown to them, quoting de Candolle's suggestion that maybe a *Vernonia* was represented. For *Gnaphalium indicum* Loureiro states: " Habitat agreste in Cochinchina." The description applies definitely to *Vernonia cinerea*, not to *Gnaphalium*. The reference to Plukenet (Almag. 172. *pl. 187. f. 5*) apparently copied from Linnaeus, must be excluded; it represents the Linnaean species but not the form Loureiro described.

Elephantopus Linnaeus

Elephantopus scaber Linn. Sp. Pl. 814. 1753.

Scabiosa cochinchinensis Lour. Fl. Cochinch. 68. 1790, ed. Willd. 85. 1793, Anamese *co luoi mèo*, Chinese *ti tan tsào*.

Asterocephalus cochinchinensis Spreng. Syst. **1**: 380. 1825 (based on *Scabiosa cochinchinensis* Lour.).

" Habitat inculta in Cochinchina: etiam in China." Loureiro's description clearly applies to the common and widely distributed *Elephantopus scaber* Linn. The Cantonese name on recently collected material appears as *ta tom cho*. Bretschneider, as noted by Hemsley, first indicated that the Chinese name cited by Loureiro appertains to *Elephantopus scaber* Linn. Neither Arènes, in his treatment of the *Dipsacaceae*, nor Gagnepain, in his treatment of the Compositae (Lecomte Fl. Gén. Indo-Chine **3**: 448–663. 1924) accounts for Loureiro's binomial or the synonyms based upon it.

Adenostemma J. R. & G. Forster

Adenostemma lavenia (Linn.) O. Ktz. Rev. Gen. Pl. 304. 1891.

Verbesina lavenia Linn. Sp. Pl. 902. 1753.

Adenostemma viscosum Forst. Char. Gen. 90. 1776.

Spilanthes tinctorius Lour. Fl. Cochinch. 484. 1790, ed. Willd. 590. 1793, Anamese *chàm lón lá*.

Adenostemma tinctorium Cass. Dict. Sci. Nat. **25**: 364. 1822 (based on *Spilanthes tinctorius* Lour.).

" Habitat cultus in Cochinchina, & China." Loureiro's description applies in essentials to the very common *Adenostemma lavenia* O. Ktz. I do not know whether this species is ever cultivated and have no information concerning its use as a dye which Loureiro discusses. *Abecedaria* Rumph. (Herb. Amb. **6**: 145. *pl. 65*), discussed by Loureiro following his description, is *Spilanthes acmella* Murr., a species totally different from the one Loureiro described.

Ageratum Linnaeus

Ageratum conyzoides Linn. Sp. Pl. 839. 1753.

> *Ageratum ciliare* (non Linn.) Lour. Fl. Cochinch. 484. 1790, ed. Willd. 591. 1793.

"Habitat incultum prope Cantonem Sinarum." The description conforms rather closely to the characters of the ubiquitous *Ageratum conyzoides* Linn. The rather poor figure in Plukenet cited by Loureiro may represent *Ageratum conyzoides* Linn.

Eupatorium (Tournefort) Linnaeus

Eupatorium quaternum DC. Prodr. **5**: 183. 1836; Gagnep. in Lecomte Fl. Gén. Indo-Chine **3**: 509. 1924 (based on *Eupatorium purpureum* Lour.).

> *Eupatorium purpureum* (non Linn.) Lour. Fl. Cochinch. 487. 1790, ed. Willd. 594. 1793, Anamese *cây bach son.*

"Habitat spontaneum in agris Cochinchinae." This is known only from Loureiro's description, and from its 8-flowered heads I believe it to be a true *Eupatorium*. The description conforms rather closely to the characters of *Eupatorium lindleyanum* DC., but this species is unknown from Indo-China, and even if it does occur there it would probably not be a low-altitude form. Gagnepain's description is compiled from that of Loureiro.

Solidago (Vaillant) Linnaeus

Solidago virgaurea Linn. Sp. Pl. 880. 1753.

> *Solidago decurrens* Lour. Fl. Cochinch. 501. 1790, ed. Willd. 612. 1793, Chinese *hoâng kăm siong.*
>
> *Dectis decurrens* Raf. Fl. Tellur. **2**: 43. 1837 (based on *Solidago decurrens* Lour.).
>
> *Solidago cantoniensis* Lour. Fl. Cochinch. 501. 1790, ed. Willd. 612. 1793, Chinese *kăm siŏng hŏa.*

For *Solidago decurrens* Loureiro states: "Habitat inculta prope Cantonem Sinarum"; and for *S. cantoniensis* he states: "Habitat inculta prope Cantonem Sinarum"; both descriptions apparently apply to a single species and this the form of *Solidago virgaurea* Linn. not uncommon on low barren hills in the vicinity of Canton. This southern Chinese form is the one described as *Amphirhapis leiocarpa* Benth. in Hook. Lond. Journ. Bot. **1**: 488. 1842 = *Solidago virgaurea* Linn. var. *leiocarpa* Miq. in Néerl. Journ. Bot. **1**: 101. 1861.

Grangea Forskål

Grangea maderaspatana (Linn.) Poir. in Lam. Encycl. Suppl. **2**: 825. 1812; Gagnep. in Lecomte Fl. Gén. Indo-Chine **3**: 574. *f. 62, 2-3.* 1924.

> *Artemisia maderaspatana* Linn. Sp. Pl. 849. 1753.
>
> *Cotula anthemoides* (non Linn.) Lour. Fl. Cochinch. 493. 1790, ed. Willd. 602. 1793, Anamese *rau kaóc.*

"Habitat agrestis in agris, & hortis Cochinchinae." Gagnepain (Lecomte Fl. Gén. Indo-Chine **3**: 578. 1924) admits *Cotula anthemoides* as occurring in Indo-China, citing several collections and crediting this binomial to Loureiro. In my manuscript of 1919 I reduced *Cotula anthemoides* Lour. to *Grangea maderaspatana* Poir. on the basis of Loureiro's description of the leaves as pubescent, the heads as long-peduncled, the flower-heads as globose, the flowers as yellow, and the plant as branched and a foot high. These are *Grangea* characters, not those of *Cotula anthemoides* Linn. It is suspected that Gagnepain erred in crediting *Cotula anthemoides* to Loureiro but that Linnaeus was intended, as *Cotula anthemoides* Linn. is widely distributed in Asia.

Boltonia L'Héritier

Boltonia indica (Linn.) Benth. Fl. Hongk. 174. 1861; Gagnep. in Fl. Gén. Indo-Chine 3: 621. 1924.

Aster indicus Linn. Sp. Pl. 876. 1753.

Matricaria cantoniensis Lour. Fl. Cochinch. 498. 1790, ed. Willd. 609. 1793, Chinese *hí su tsu.*

Hisutsua cantoniensis DC. Prodr. 6: 44. 1837 (based on *Matricaria cantoniensis* Lour.).

" Habitat spontanea Cantone Sinarum." Loureiro's description applies in all respects to the species commonly known as *Aster indicus* Linn., the type of which was also from China and probably from Canton. Loureiro's type is preserved in the herbarium of the Paris Museum. The species is common in the vicinity of Canton. *Hisutsua,* a new genus proposed by de Candolle, was based on Loureiro's species, the name being derived from its Chinese one cited by Loureiro.

Callistephus [88] Cassini

Callistephus chinensis (Linn.) Nees Gen. Sp. Aster. 222. 1832.

Aster chinensis Linn. Sp. Pl. 877. 1753.

Callistemma hortense Cass. Dict. Sci. Nat. 6: Suppl. 46. 1817.

Callistephus hortensis Cass. ex Nees Gen. Sp. Aster. 222. 1832, *in syn.*

Aster indicus (non Linn.) Lour. Fl. Cochinch. 503. 1790, ed. Willd. 615. 1793, Chinese *mà lân hōa.*

" Habitat spontaneus, cultusque apud Sinas." Loureiro's description conforms sufficiently well with the characters of the China aster, *Callistephus chinensis* (Linn.) Nees, to warrant this reduction. He probably observed it in cultivation in Canton, for the China aster is widely planted for ornamental purposes; it is wild in Hopei province and in other parts of northern China. Cassini does not publish the binomial in the Dict. Sci. Nat. 37: 491. 1825, merely mentioning the genus *Callistephus.*

Aster (Tournefort) Linnaeus

Aster sampsoni (Hance) Hemsl. in Journ. Linn. Soc. Bot. 23: 415. 1888.

Heterocarpus sampsoni Hance in Journ. Bot. 5: 370. 1867.

Erigeron hirsutum Lour. Fl. Cochinch. 500. 1790, ed. Willd. 611. 1793, Chinese *ha sĭ koŭc,* non *Aster hirsutus* Harv.

" Habitat agreste circa Cantonem Sinarum." The type of Hance's species was from the West River, Kwangtung Province. I have what I take to be the same species from

---

[88] *Callistephus* Cassini (1825), conserved name, Vienna Code; an older one is *Callistemma* Cassini (1817).

the White Cloud Hills at Canton, *Levine 1797*, and Loureiro's description conforms to its characters. Hemsley (Journ. Linn. Soc. Bot. **23**: 418. 1888) mentions Loureiro's species, following *Erigeron chinense* Jacq., as an obscure one; the description of the ray-flowers as blue, points unmistakably to *Aster*.

Blumea [59] de Candolle

Blumea balsamifera (Linn.) DC. Prodr. **5**: 447. 1836; Gagnep. in Lecomte Fl. Gén. Indo-Chine **3**: 547. 1924.

Conyza balsamifera Linn. Sp. Pl. ed. 2, 1208. 1763.

Baccharis salvia Lour. Fl. Cochinch. 494. 1790, ed. Willd. 603. 1793, Anamese *cây dai bi.*

Loureiro correctly interpreted the Linnaean species, a very characteristic, common, and widely distributed one in the warmer parts of the Old World; in fact, essentially, *Baccharis salvia* is merely a new binomial for *Conyza balsamifera* Linn., which Loureiro cites as a synonym. *Conyza odorata* Rumph. (Herb. Amb. **6**: 55. *pl. 24. f. 1*), cited as a synonym, is correctly placed. Loureiro does not cite any locality, but the Anamese name indicates an Indo-China plant.

Blumea hieracifolia (D. Don) DC. in Wight Contrib. Bot. Ind. 15. 1834, Prodr. **5**: 442. 1836; Gagnep. in Lecomte Fl. Gén. Indo-Chine **3**: 534. 1924.

Erigeron hieracifolium D. Don Prodr. Fl. Nepal. 172. 1825.

Conyza hieracifolia Spreng. Syst. **3**: 514. 1826.

Conyza hirsuta (non Linn.) Lour. Fl. Cochinch. 496. 1790, ed. Willd. 606. 1793, Anamese *co duôi hùm*, Chinese *hô m̄i tsâo.*

"Habitat inculta in Cochinchina, & in China." The form Loureiro described is clearly referable to *Blumea hieracifolia* DC. as interpreted by Gagnepain. The species is a common and widely distributed one in the Indo-Malaysian region and Loureiro's description conforms to its characters. It is represented by *Clemens 3519* from Tourane.

Blumea laciniata (Roxb.) DC. Prodr. **5**: 436. 1836.

Conyza laciniata Roxb. Hort. Beng. 61. 1814, *nomen nudum*, Fl. Ind. ed. 2, **3**: 427. 1832.

Serratula multiflora (non Linn.) Lour. Fl. Cochinch. 483. 1790, ed. Willd. 589. 1793, Chinese *mu mîn sŏ.*

"Habitat inculta Cantone Sinarum." The description rather clearly appertains to *Blumea* and doubtless to a form of the common and widely distributed *Blumea laciniata* (Roxb.) DC.

Blumea laevis (Lour.) comb. nov.

Placus laevis Lour. Fl. Cochinchina 497. 1790, ed. Willd. 607. 1793, Anamese *cúc bánh ít, hoa vàng.*

Baccharis laevis Spreng. Syst. **3**: 466. 1826 (based on *Placus laevis* Lour.).

Blumea virens DC. in Wight Contrib. Bot. Ind. 14. 1834, Prodr. **5**: 439. 1836; Gagnep. in Lecomte Fl. Gén. Indo-Chine **3**: 536. 1924.

Conyza virens Wall. List no. 3037. 1831, *nomen nudum.*

? Baccharis dioscoridis (non Linn.) Lour. Fl. Cochinch. 494. 1790, ed. Willd. 603. 1793, Anamese *cây tu bi*, Chinese *laóng fú su.*

[59] *Blumea* De Candolle (1833), conserved name, Vienna Code; an older one is *Placus* Loureiro (1790).

For *Placus laevis* Loureiro states: " Habitat agrestis in Cochinchina." The description is sufficiently definite and in my opinion applies unmistakably to the widely distributed Indo-Malaysian species currently known as *Blumea virens* DC. Gagnepain does not mention Loureiro's species or Sprengel's synonym based upon it in his treatment of the Compositae of Indo-China (Lecomte Fl. Gén. Indo-Chine 3: 448–663. 1924). For *Baccharis dioscoridis* Loureiro states: " Habitat tam agrestis, quam culta in Cochinchina, & China." The species that he described is most certainly a *Blumea;* it may be a form of *B. laevis* (Lour.) Merr.

Blumea lanceolaria (Roxb.) Druce in Rept. Bot. Exch. Club Brit. Isles **4**: 609. 1917.
 Conyza lanceolaria Roxb. Fl. Ind. ed. 2, **3**: 432. 1832.
 Gorteria setosa (non Linn.) Lour. Fl. Cochinch. 507. 1790, ed. Willd. 620. 1793.
 Gorteria loureiriana DC. Prodr. **6**: 501. 1837 (based on *Gorteria setosa* Lour.).
 Conyza squarrosa Wall. List no. 3025. 1830, *nomen nudum*, non Linn.
 Blumea myriocephala DC. Prodr. **5**: 445. 1836; Gagnep. in Lecomte Fl. Gén. Indo-Chine **3**: 546. 1924.
 Blumea spectabilis DC. Prodr. **5**: 445. 1836.
 Conyza chinensis (non Linn.) Lour. Fl. Cochinch. 496. 1790, ed. Willd. 606. 1793, Anamese *rau soúng ăn goi.*

For *Gorteria setosa* Loureiro states: " Habitat agrestis prope Cantonem Sinarum," and his description conforms closely with the general characters of the Kwangtung form currently referred to *Blumea myriocephala* DC. For *Conyza chinensis* Loureiro states: " Habitat agrestis in Cochinchina." The Linnaean species is *Vernonia cinerea* Less. as is *Senecio amboinicus* Rumph. (Herb. Amb. **6**: 36. *pl. 14. f. 2*) erroneously cited by Loureiro as representing his species; Loureiro's description applies to a totally different plant and conforms closely with the characters of *Blumea myriocephala* DC. I interpret de Candolle's species to be the same as the older *Conyza lanceolaria* Roxb. = *Blumea lanceolaria* (Roxb.) Druce.

Blumea mollis (D. Don) Merr. in Philip. Journ. Sci. **5**: Bot. 395. 1910, Enum. Philip. Fl. Pl. **3**: 603. 1923.
 Erigeron molle D. Don Prodr. Fl. Nepal. 172. 1825.
 Blumea trichophora DC. Prodr. **5**: 436. 1836.
 Blumea wightiana DC. in Wight Contrib. Bot. Ind. 14. 1834, Prodr. **5**: 435. 1836; Gagnep. in Lecomte Fl. Gén. Indo-Chine **3**: 541. 1924.
 Placus tomentosus Lour. Fl. Cochinch. 497. 1790, ed. Willd. 607. 1793, Anamese *cúc bánh ít, hoa tím*, non *Blumea tomentosa* A. Rich.
 Baccharis cochinchinensis Spreng. Syst. **3**: 466. 1826 (based on *Placus tomentosus* Lour.).

" Habitat agrestis, cultusque in Cochinchina." Loureiro's description applies to the very common and widely distributed Indo-Malaysian species currently known as *Blumea wightiana* DC. This is the type of the genus *Placus* Loureiro, which antedates *Blumea* de Candolle by many years, but the latter is conserved. Bentham (Benth. & Hook. f. Gen. Pl. **2**: 290. 1873) states: " *Placus* Lour. . . . quoad *P. tomentosus*, est fide speciminis auctoris *Blumeae* species." Loureiro's type is preserved in the herbarium of the British Museum.

Blumea pubigera (Linn.) Merr. in Philip. Journ. Sci. **14**: 250. 1919.

Conyza pubigera Linn. Mant. **1**: 113. 1767; Lour. Fl. Cochinch. 495. 1790, ed. Willd. 604. 1793, Anamese *dầu xuong rừng*.

Blumea chinensis DC. Prodr. **5**: 444. 1836; Gagnep. in Lecomte Fl. Gén. Indo-Chine **3**: 527. *f. 54. 1–8.* 1924, non *Conyza chinensis* Linn.[90]

" Habitat agrestis in Cochinchina." *Sonchus volubilis* Rumph. (Herb. Amb. **5**: 299. *pl. 103. f. 2*), cited by Loureiro, after Linnaeus, as a synonym, may represent *Blumea pubigera* (Linn.) Merr. as here interpreted, or it may represent the closely allied *Blumea riparia* (Blume) DC. which Gagnepain retains as a valid species. Loureiro's description applies to the Linnaean species.

Blumea sinuata (Lour.) comb. nov.

Gnaphalium sinuatum Lour. Fl. Cochinch. 497. 1790, ed. Willd. 608. 1793.

Conyza laciniata Roxb. Hort. Beng. 61. 1814, *nomen nudum*, Fl. Ind. ed. 2, **3**: 427. 1832.

Blumea laciniata DC. Prodr. **5**: 436. 1836; Gagnep. in Lecomte Fl. Gén. Indo-Chine **3**: 532. 1924.

" Habitat agreste in Cochinchina." Loureiro's concise description applies unmistakably to the common and widely distributed species of *Blumea* currently known as *B. laciniata* DC. which is apparently fairly common in Indo-China and which occurs at Hue (*P. Couderc*), Loureiro's classical locality.

Blumea sp.

? Xeranthemum chinense Lour. Fl. Cochinch. 498. 1790, ed. Willd. 608. 1793, Chinese *siaô louc ngi*.

" Habitat incultum prope Cantonem Sinarum." The species might be a *Blumea*, an *Erigeron*, a *Solidago*, or a representative of some other genus. The description is very indefinite.

Blumea sp.

Xeranthemum retortum (non Linn.) Lour. Fl. Cochinch. 498. 1790, ed. Willd. 609. 1793, Anamese *cúc ràng kua*.

Helichrysium cochinchinense Spreng. Syst. **3**: 482. 1826 (based on *Xeranthemum retortum* Lour.).

" Habitat spontaneum in agris Cochinchinae." De Candolle (Prodr. **6**: 210. 1837) under *Helichrysium cochinchinense* Spreng. queries: " An Conyzae seu Blumeae sp.? ". The description impresses me as appertaining to a species of *Blumea*, but beyond this I am unable to carry the identification.

Blumea sp.

Erigeron philadelphicum (non Linn.) Lour. Fl. Cochinch. 500. 1790, ed. Willd. 611. 1793, Anamese *cây con hát*.

Erigeron cochinchinense Lour. Fl. Cochinch. 500. 1790, ed. Willd. 611. 1793, *in nota*.

" Habitat agreste in Cochinchina." Loureiro noted that his specimen did not agree fully with the published description of the Linnaean species and at the end of his discussion

---

[90] The actual type of *Conyza chinensis* Linn. in the Linnaean herbarium is a form of *Vernonia cinerea* (Linn.) Less. This is the name-bringing synonym of *Blumea chinensis* DC., but de Candolle actually described a *Blumea*, not the *Vernonia*.

published the specific name *cochinchinense* as follows: " si praeterea valde differre observetur, Cochinchinense vocabitur." The description apparently applies to some species of *Blumea*.

Sphaeranthus (Vaillant) Linnaeus

Sphaeranthus africanus Linn. Sp. Pl. ed. 2, 1314. 1763; Gagnep. in Lecomte Fl. Gén. Indo-Chine **3**: 566. 1924.

Sphaeranthus cochinchinensis Lour. Fl. Cochinch. 510. 1790, ed. Willd. 623. 1793, Anamese *co bò xít*.

" Habitat spontaneus inter segetes, & in hortis Cochinchinae, & Chinae." The description clearly applies to the common and widely distributed Linnaean species.

Gnaphalium Linnaeus

Gnaphalium luteo-album Linn. Sp. Pl. 851. 1753.

Chrysocoma villosa (non Linn.) Lour. Fl. Cochinch. 486. 1790, ed. Willd. 594. 1753, Anamese *rau cúc*.

" Habitat agrestis in Cochinchina." Loureiro's description clearly applies to the Linnaean species. As confirmation of the correctness of this reduction, at least as to genus, Gagnepain (Lecomte Fl. Gén. Indo-Chine **3**: 558. 1924) cites *cây rau khuc* as one of the local names of *Gnaphalium indicum* Linn. Since Loureiro describes his plant as one and one-half feet high, *G. luteo-album* Linn. rather than *G. indicum* Linn. is indicated.

Inula Linnaeus

Inula cappa (Ham.) DC. Prodr. **5**: 469. 1836.

Conyza cappa Ham. in DC. Prodr. Fl. Nepal. 176. 1825.

Baccharis chinensis Lour. Fl. Cochinch. 494. 1790, ed. Willd. 604. 1793, Chinese *xān po leng* (non *Inula chinensis* Rupr.).

" Habitat inculta prope Cantonem Sinarum." Loureiro's description is definite and applies unmistakably to *Inula cappa* DC., which is abundant in open grassy places in Kwangtung Province. Loureiro's specific name is much older than Hamilton's, but is invalidated in *Inula* by *I. chinensis* Rupr. (1859).

Xanthium (Tournefort) Linnaeus

Xanthium strumarium Linn. Sp. Pl. 987. 1753; Lour. Fl. Cochinch. 563. 1790, ed. Willd. 689. 1793, Anamese *cây ké, nguu bàng;* Gagnep. in Lecomte Fl. Gén. Indo-Chine **3**: 588. 1924.

" Habitat agreste in agris, & sepibus Cochinchinae, & Chinae, passim obvium." The description applies to the Linnaean species, which is common and widely distributed in China and Indo-China.

Siegesbeckia Linnaeus

Siegesbeckia orientalis Linn. Sp. Pl. 900. 1753; Lour. Fl. Cochinch. 504. 1790, ed. Willd. 616. 1793, Anamese *nu ao ria;* Gagnep. in Lecomte Fl. Gén. Indo-Chine **3**: 600. 1924.

" Habitat ubique inculta in hortis Cochinchinae; puto, quod etiam in China, sed ibi non obvia." The description applies unmistakably to the common and widely distributed Linnaean species.

Enhydra (*Enydra*) Loureiro

Enhydra (Enydra) fluctuans Lour. Fl. Cochinch. 511. 1790, ed. Willd. 625. 1793, Anamese *rau ngu oúng;* Gagnep. in Lecomte Fl. Gén. Indo-Chine **3**: 563. 1924.

Meyera fluctuans Spreng. Syst. **3**: 602. 1826 (based on *Enhydra fluctuans* Lour.).

"Habitat in paludibus Cochinchinae." A well-known species of wide distribution in tropical Asia, the type of the genus. The spelling of the generic name in both editions is *Enydra;* it is corrected to *Enhydra.* Gagnepain does not account for Sprengel's synonym, which was based on Loureiro's binomial. Sprengel also erroneously adds *Sobreyra* (*Sobrya*) *sessilifolia* Ruíz & Pav., as a synonym. Loureiro's specimen listed as being among his plants in the herbarium of the British Museum has not been located.

Eclipta [91] Linnaeus

Eclipta alba (Linn.) Hassk. Pl. Jav. Rar. 528. 1848.

Verbesina alba Linn. Sp. Pl. 902. 1753.

Eclipta erecta Linn. Mant. **2**: 286. 1771; Lour. Fl. Cochinch. 505. 1790, ed. Willd. 617. 1793, Anamese *co muc.*

"Habitat inculta in hortis Cochinchinae." Loureiro's description applies to this very common and widely distributed weed. *Ecliptica* Rumph. (Herb. Amb. **6**: 43. *pl. 18. f. 1*) is correctly placed as a synonym.

Wedelia Jacquin

Wedelia chinensis (Osbeck) Merr. in Philip. Journ. Sci. **12**: Bot. 111. 1917.

Solidago chinensis Osbeck Dagbok Ostind Resa 241. 1757.

Verbesina calendulacea Linn. Sp. Pl. 902. 1753; Lour. Fl. Cochinch. 506. 1790, ed. Willd. 619. 1793, Chinese *fan khi kŏúc.*

Wedelia calendulacea Less. Syn. Compos. 222. 1832, non Pers. 1807.

"Habitat inculta prope Cantonem Sinarum." The Linnaean species was correctly interpreted by Loureiro, but the specific name is invalid in *Wedelia,* hence the adoption of Osbeck's name. The species is very common in the vicinity of Canton. *Wedelia calendulacea* Pers. was based on an actual Australian specimen and has nothing to do with *Wedelia calendulacea* Less.

Helianthus Linnaeus

Helianthus annuus Linn. Sp. Pl. 904. 1753; Lour. Fl. Cochinch. 509. 1790, ed. Willd. 622. 1793.

Helianthus giganteus (non Linn.) Lour. Fl. Cochinch. 509. 1790, ed. Willd. 623. 1793, Anamese *hoa kùy,* Chinese *hoâm qūei hōa.*

For *Helianthus annuus* Loureiro states: "Habitat cultus in insula Africana Mozambico"; and for *H. giganteus* "Habitat cultus in Cochinchina, & China." Both descriptions I believe to refer to forms of the variable and widely distributed *Helianthus annuus* Linn.

Spilanthes Jacquin

Spilanthes acmella (Linn.) Murr. Syst. 610. 1774; Gagnep. in Lecomte Fl. Gén. Indo-Chine **3**: 598. 1924.

[91] *Eclipta* Linnaeus (1771), conserved name, Brussels Code; an older one is *Eupatoriophalacron* Adanson (1763).

Verbesina acmella Linn. Sp. Pl. 901. 1753.

Eclipta prostrata (non Linn.) Lour. Fl. Cochinch. 505. 1790, ed. Willd. 618. 1793, Anamese *cây nu ao tlon*.

"Habitat inculta in hortis Cochinchinae." This reduction was made from Loureiro's description in my original manuscript of 1919 and is accepted by Gagnepain. Loureiro's description conforms to the characters of *Spilanthes acmella* Murr.

Bidens (Tournefort) Linnaeus

Bidens biternata (Lour.) Merr. & Sherff in Bot. Gaz. **88**: 293. 1921 (based on *Coreopsis biternata* Lour.).

Coreopsis biternata Lour. Fl. Cochinch. 508. 1790, ed. Willd. 622. 1793, Chinese *cā ap chiŏc*.

Actinea biternata Spreng. Syst. **3**: 574. 1826 (based on *Coreopsis biternata* Lour.).

Bidens chinensis Willd. Sp. Pl. **3**: 1719. 1803; Schulz in Bot. Jahrb. **50**: Suppl. 178. 1914.

"Habitat agrestis prope Cantonem Sinarum." For a detailed discussion of this case, with the citation of many specimens illustrating Loureiro's species, see Sherff's paper referred to above. As noted there, O. E. Schulz in his detailed study of *Bidens chinensis* Willd., apparently overlooked Loureiro's earlier description of this species. According to Schulz (Urban Symb. Antill. **7**: 135. 1911) *Bidens chinensis* Willd. (Sp. Pl. **3**: 1719. 1803, and herb. Willd.) is the same as *Bidens pilosa* Linn. var. *dubia* O. E. Schulz, but Sherff places it as a synonym of *Bidens biternata*.

Bidens bipinnata Linn. Sp. Pl. 832. 1753; Lour. Fl. Cochinch. 488. 1790, ed. Willd. 596. 1793 (err. *bipinmata*), Anamese *cây loùng dèn;* Gagnep. in Lecomte Fl. Gén. Indo-Chine **3**: 608. 1924.

"Habitat spontanea in Cochinchina, & China." Doctor Sherff thinks that the plant Loureiro had represents one of the forms of the Linnaean species approaching *B. biternata* (Lour.) Merr. & Sherff, but the 3-awned achenes, indicated by Loureiro, indicate *B. bipinnata* Linn. rather than *B. biternata* (Lour.) Merr. & Sherff. The Plukenet reference should be excluded, as it represents *Bidens biternata* (Lour.) Merr. & Sherff.

Bidens pilosa Linn. Sp. Pl. 832. 1753; Lour. Fl. Cochinch. 488. 1790, ed. Willd. 596. 1793, Gagnep. in Lecomte Fl. Gén. Indo-Chine **3**: 606. 1924.

"Habitat agrestis in Cochinchina." Loureiro seems to have had material representing the Linnaean species, and Doctor Sherff, who has made an intensive study of the genus, confirms this interpretation of it. *Agrimonia molucca* Rumph. (Herb. Amb. **6**: 38. *pl. 15. f. 2*), cited by Loureiro as representing the species, may or may not represent the Linnaean species.

Bidens pilosa Linn. var. **minor** (Blume) Sherff in Bot. Gaz. **80**: 387. 1925, **86**: 443. 1928.

Bidens sundaica Blume var. *minor* Blume Bijdr. 914. 1826.

Bidens ? leucorrhiza (*leucorhiza*) DC. Prodr. **5**: 605. 1836 (based on *Coreopsis leucorrhiza* Lour.).

Coreopsis leucorrhiza Lour. Fl. Cochinch. 508. 1790, ed. Willd. 622. 1793 (*leucorhiza*), Anamese *phaong phung, fâm fūm*.

Kerneria dubia Cass. in Dict. Sci. Nat. **24**: 398. 1822, pro parte.

Bidens pilosa Linn. var. *dubia* O. E. Schulz in Urban Symb. Antill. **7**: 135. 1911.
" Habitat agrestis prope Cantonem Sinarum." Judging from Loureiro's description and from specimens available for study from the vicinity of Canton, I believe Sherff to be correct in his disposition of *Coreopsis leucorrhiza* Lour.

Chrysanthemum (Tournefort) Linnaeus

Chrysanthemum coronarium Linn. Sp. Pl. 890. 1753; L. H. Bailey Gent. Herb. **1**: 48. *f. 17*. 1920.

> *Buphthalmum oleraceum* Lour. Fl. Cochinch. 506. 1790, ed. Willd. 618. 1793, Anamese *cúc tăng ô.*

" Habitat cultum in hortis Cochinchinae, & Chinae." The plant that Loureiro described is undoubtedly the Chinese form referred by Hemsley (Journ. Linn. Soc. Bot. **23**: 438. 1888) to *Chrysanthemum segetum* Linn. This form, with somewhat fleshy leaves, is cultivated by the Chinese for food; a Cantonese name on recently collected material is *tong ko;* in Manila, where it is cultivated by the Chinese, it is known as *tung hao.* Hemsley gives the local name as *t'ung-hao-ts'ai.* I believe it to be referable to *C. coronarium* Linn.

Chrysanthemum indicum Linn. Sp. Pl. 889. 1753.

> *Chrysanthemum procumbens* Lour. Fl. Cochinch. 499. 1790, ed. Willd. 610. 1793, Chinese *kim cúc,* Anamese *siaô kiŏ hōa.*
>
> *Pyrethrum procumbens* Kostel. Allgem. Med.-Pharm. Fl. **2**: 691. 1831 (based on *Chrysanthemum procumbens* Lour.).

" Habitat spontaneum, cultumque in Cochinchina, & China." Loureiro's description applies to the wild form that is common about Canton. He inadvertently reversed the Anamese and Chinese names; on recently collected material the Cantonese names are *pin kuk* and *wong kook.* Loureiro included various floral forms in his description, as apparently did Linnaeus in his original publication of the binomial, if we may judge by the synonyms cited, that really belong with the next species.

Chrysanthemum morifolium Ramat. in Journ. Nat. Hist. **2**: 240. 1792; L. H. Bailey Gent. Herb. **1**: 131. 1923.

> *Chrysanthemum indicum* (non Linn.) Lour. Fl. Cochinch. 499. 1790, ed. Willd. 610. 1793, Anamese *daì cúc,* Chinese *tá kiŏ hōa.*
>
> *Chrysanthemum sinense* Sabine in Trans. Linn. Soc. **14**: 145. 1825.

" Habitat in hortis Cochinchinae, Chinae, late cultum ob pulchritudinem floris, cujus diameter 3 pollices, & amplius aequat." The description includes various cultivated forms of the garden *Chrysanthemum,* with white, pink, reddish, purple, violet, and yellow flowers. L. H. Bailey has shown that *C. morifolium* Ram. is the oldest valid binomial for this cultigen.

Centipeda Loureiro

Centipeda minima (Linn.) A. Br. & Aschers. Ind. Sem. Hort. Berol. App. 6. 1867.

> *Artemisia minima* Linn. Sp. Pl. 849. 1753.
>
> *Centipeda orbicularis* Lour. Fl. Cochinch. 493. 1790, ed. Willd. 602. 1793, Anamese *co the;* Gagnep. in Lecomte Fl. Gén. Indo-Chine **3**: 587. 1924.

" Habitat inculta in agris Cochinchinae." This is identical with the older *Artemisia minima* Linn., a common, very widely distributed, and well-known species. The oldest specific name is here adopted.   Loureiro's type is preserved in the herbarium of the British Museum.

Tagetes Linnaeus

Tagetes patula Linn. Sp. Pl. 887. 1753; Lour. Fl. Cochinch. 504. 1790, ed. Willd. 616. 1793, Anamese *cúc van tho.*

"Habitat ubique culta in hortis Cochinchinae: etiam in China, & in multis Indiae locis." Loureiro was doubtless correct in his interpretation of the Linnaean species, one of American origin, now widely cultivated in the warmer parts of the Old World and occasionally spontaneous.

Artemisia Linnaeus

Artemisia annua Linn. Sp. Pl. 847. 1753; Lour. Fl. Cochinch. 490. 1790, ed. Willd. 599 1793, Chinese *tsaò cāo.*

? Artemisia abrotanum (non Linn.) Lour. Fl. Cochinch. 490. 1790, ed. Willd. 598 1793, Anamese *thanh hao,* Chinese *ȳn chiñ hāo.*

Artemisia annua Linn. was apparently interpreted correctly by Loureiro; his specimens were from "prope Pekinum Sinarum." I place here *A. abrotanum* "incultum, cultumque in Cochinchina, & China," as described by Loureiro, although Gagnepain (Lecomte Fl. Gén. Indo-Chine **3**: 584. 1924) thinks it may perhaps belong with *A. carvifolia* Wall.

Artemisia vulgaris Linn. Sp. Pl. 848. 1753; Lour. Fl. Cochinch. 491. 1790, ed. Willd. 600. 1793, Anamese *thuóc kúu,* Chinese *ngái yĕ;* Gagnep. in Lecomte Fl. Gén. Indo-Chine **3**: 584. 1924.

"Habitat inculta, cultaque in Cochinchina, & China." Loureiro's description applies to the Linnaean species as interpreted by most authors, including Gagnepain. *Artemisia latifolia* Rumph. (Herb. Amb. **5**: 261. *pl. 91. f. 2*), cited by Loureiro as a synonym, is correctly placed in accordance with this interpretation of *Artemisia vulgaris* Linn. As Pampanini has segregated the species in this group, this would probably be referable to *Artemisia indica* Willd. as interpreted by him (Nuov. Giorn. Bot. Ital. n.s. **33**: 457. 1926).

Artemisia sp.

Verbesina spicata Lour. Fl. Cochinch. 507. 1790, ed. Willd. 620. 1793, Anamese *cúc ăn rau,* Chinese *thīen caì tsái.*

Eclipta spicata Spreng. Syst. **3**: 603. 1826 (based on *Verbesina spicata* Lour.).

"Habitat culta in hortis Cochinchinae, & Chinae: et cum acetariis mensis apponitur." The description apparently appertains to *Artemisia,* and possibly to a form of *A. vulgaris* Linn.

Crossostephium Lessing

Crossostephium chinense (Linn.) Makino in Bot. Mag. Tokyo **20**: 33. 1906; Merr. in Philip. Journ. Sci. **15**: 260. 1919.

Artemisia chinensis Linn. Sp. Pl. 849. 1753, excl. syn. Gmelin; Lour. Fl. Cochinch. 492. 1790, ed. Willd. 600. 1793, Chinese *khí ngái.*

Artemisia judaica (non Linn.) Lour. Fl. Cochinch. 489. 1790, ed. Willd. 597. 1793, Anamese *ngaoe phu duong,* Chinese *ngaoc fù yong.*

Crossostephium artemisioides Less. ex Cham. & Schlecht. in Linnaea **6**: 220. 1831; Gagnep. in Lecomte Fl. Gén. Indo-Chine **3**: 576. *f. 62. 4–6.* 1924.

Artemisia loureiri Kostel. Allgem. Med.-Pharm. Fl. 2: 699. 1831 (based on *A. judaica* Lour.).

For *Artemisia chinensis* Loureiro states: " Habitat Cantone Sinarum," and it is clear that he correctly interpreted the Linnaean species, the type of which was a specimen collected by Lagerstroem in China. For *A. judaica* he states: " Habitat in Cochinchina, & China," and I believe the description also appertains to *Crossostephium chinense* (Linn.) Makino. The species is commonly cultivated in Canton, is also grown in pots in Manila, and is cultivated in Indo-China. This is Gagnepain's reduction of *A. judaica* Lour.

Petasites (Tournefort) Linnaeus

Petasites japonicus (Sieb. & Zucc.) F. Schmidt, Reise. Amurland. 145. 1868; Miq. ex Franch. & Sav. Enum. Pl. Jap. 1: 220. 1875.

Nardosmia japonica Sieb. & Zucc. Fl. Jap. Fam. Nat. 2: 181. 1846.

Tussilago farfara (non Linn.) Lour. Fl. Cochinch. 502. 1790, ed. Willd. 614. 1793, Anamese *khoan doung hoa*, Chinese *koàn tūm hōa*.

" Habitat inculta in locis Borealibus imperii Sinensis." Loureiro's material was doubtless secured from an herbalist. His description applies to the northern form currently referred to *Petasites japonica* F. Schmidt. Bretschneider notes that the Chinese name cited by Loureiro is used in Japan for the latter species.

Gynura [92] Cassini

Gynura divaricata (Linn.) DC. Prodr. 6: 301. 1837.

Senecio divaricatus Linn. Sp. Pl. 866. 1873; Lour. Fl. Cochinch. 502. 1790, ed. Willd. 613. 1793, Chinese *kām siūn lin*.

Gynura ovalis DC. Prodr. 6: 300. 1837.

Cacalia ovalis Ker Bot. Reg. 2: *pl. 101.* 1816.

" Habitat incultus prope Cantonem Sinarum." Although *Gynura divaricata* DC. and *G. ovalis* DC. are currently retained as distinct species, I believe them to be identical. The species, as I interpret it, is a common and conspicuous one in the vicinity of Canton. The Linnaean type, of which I have a photograph through the courtesy of the secretary of the Linnaean Society, is a specimen collected by Osbeck near Canton. It is clearly the same as *Gynura ovalis* DC.

Gynura procumbens (Lour.) Merr. Enum. Philip. Fl. Pl. 3: 618. 1923 (based on *Cacalia procumbens* Lour.).

Cacalia procumbens Lour. Fl. Cochinch. 485. 1790, ed. Willd. 592. 1793, Anamese *rau lui*.

Cacalia sarmentosa Blume Bijdr. 907. 1826.

Gynura sarmentosa DC. Prodr. 6: 298. 1837; Gagnep. in Lecomte Fl. Gén. Indo-Chine 3: 510. 1924.

Gynura affinis Turcz. in Bull. Soc. Nat. Mosc. 24(1): 201. 1851.

Gynura scabra Turcz. *l.c.*

" Habitat in Cochinchina, & China tam culta quam agrestis." Loureiro's description applies unmistakably to the common, widely distributed, and well-known species currently

---

[92] *Gynura* Cassini (1825), conserved name, Vienna Code; an older one is *Crassocephalum* Moench (1794).

known as *Gynura sarmentosa* DC. *Sonchus volubilis* Rumph. (Herb. Amb. **5**: 299. *pl. 103. f. 2*), cited by Loureiro as representing his species, must be excluded as it represents *Blumea pubigera* (Linn.) Merr. (*B. chinensis* auctt., non *Conyza chinensis* Linn. which is *Vernonia cinerea* Less.).

Gynura pseudo-china (Linn.) DC. Prodr. **6**: 299. 1837; Gagnep. in Lecomte Fl. Gén. Indo-Chine **3**: 511. 1924.

Senecio pseudo-china Linn. Sp. Pl. 867. 1753.

Cacalia bulbosa Lour. Fl. Cochinch. 485. 1790, ed. Willd. 592. 1793, Anamese *cây tam thăt*, Chinese *sān săt*.

Senecio biflorus Burm. f. Fl. Ind. 181. 1768.

Gynura biflora Merr. in Philip. Journ. Sci. **19**: 386. 1921.

Gynura bulbosa Hook. & Arn. Bot. Beechey's Voy. 194. 1836 (based on *Cacalia bulbosa* Lour.).

"Habitat culta, incultaque in China, & Cochinchina." I interpret Loureiro's species as being identical with *G. pseudo-china* (Linn.) DC. on the basis of his description of the leaves as "folia radicalia, . . . lyrata. . . ." The form that Hooker and Arnott actually had when they transferred Loureiro's specific name to *Gynura* was unquestionably *Gynura divaricata* (Linn.) DC. which, as I interpret it, has characteristically lineolate leaves. *Senecio pseudo-china* Linn. was based on *Senecio madraspatanus rapi folio* Dill. (Hort. Elth. 345. *pl. 258. f. 335.* 1732) which Gagnepain cites, but at the same time he omitted the Linnaean binomial which is the name-bringing synonym of the species.

Gynura segetum (Lour.) Merr. in Philip. Journ. Sci. **15**: 260. 1919 (based on *Cacalia segetum* Lour., *C. pinnatifida* Lour.).

Cacalia pinnatifida (non Linn.) Lour. Fl. Cochinch. 486. 1790, ed. Willd. 593. 1793, Chinese *cién sān săt*.

Gynura pinnatifida DC. Prodr. **6**: 301. 1837 (based on *Cacalia pinnatifida* Lour., non Linn.).

Cacalia segetum Lour. Fl. Cochinch. 486. 1790, ed. Willd. 593. 1793, in nota.

"Habitat prope Cantonem Sinarum inter Oryzae segetes: unde vernaculum nomen Sinense *Cacalia Segetum*." It is to be noted that Loureiro described this as a new species independent of *Cacalia pinnatifida* Linn. Hemsley (Journ. Linn. Soc. Bot. **23**: 448. 1888) was apparently correct in his interpretation of Loureiro's species, which, however, needs a new specific name and I have supplied it from Loureiro's casually published Latin translation of the Chinese vernacular name, quoted above. *Cacalia pinnatifida* Linn. is a totally different species.

Senecio (Tournefort) Linnaeus

Senecio scandens Ham. in Don. Prodr. Fl. Nepal. 178. 1825.

Cineraria repanda Lour. Fl. Cochinch. 501. 1790, ed. Willd. 613. 1793, Chinese *cau li măn* (non *Senecio repandus* Thunb.).

Cineraria chinensis Spreng. Syst. **3**: 549. 1826 (based on *Cineraria repanda* Lour.).

Senecio chinensis DC. Prodr. **6**: 363. 1837 (based on *Cineraria repanda* Lour.).

"Habitat inculta prope Cantonem Sinarum." This is abundant in thickets near Canton, recently collected specimens bearing the local name *kout li ming*, a cognate form of the one Loureiro gives.

Emilia Cassini

Emilia sonchifolia (Linn.) DC. Prodr. **6**: 302. 1837.

Cacalia sonchifolia Linn. Sp. Pl. 835. 1753; Lour. Fl. Cochinch. 486. 1790, ed. Willd. 593. 1793, Anamese *cây mắt tlang*.

" Habitat spontanea in agris, & hortis minus cultis in Cochinchina." The description conforms very well with the characters of the common, variable, and widely distributed *Emilia sonchifolia* DC. *Sonchus amboinicus* Rumph. (Herb. Amb. **5**: 297. *pl. 103. f. 1*) is correctly cited by Loureiro, after Linnaeus, as a synonym.

Carthamus Linnaeus

Carthamus tinctorius Linn. Sp. Pl. 830. 1753; Lour. Fl. Cochinch. 481. 1790, ed. Willd. 587. 1793, Anamese *cây rum, dieu kanh*, Chinese *hûm lân hôa*.

" Habitat abundanter cultus in Cochinchina, & China." The safflower was correctly interpreted by Loureiro. *Cnicus indicus* Rumph. (Herb. Amb. **5**: 215. *pl. 79. f. 2*), cited, after Linnaeus, as a synonym, is correctly placed. Formerly this species was commonly planted in the Orient, but is now only occasionally found in cultivation.

Cnicus [93] (Linnaeus) Gaertner

Cnicus chinensis (Gardn. & Champ.) Benth. ex Maxim. in Bull. Acad. Sci. St. Pétersb. **19**: 510. 1874; Mél. Biol. **9**: 331. 1874.

Cirsium chinense Gardn. & Champ. in Hook. Journ. Bot. Kew Gard. Miscel. **1**: 323. 1849.

Carduus chinensis DC. Prodr. **6**: 629. 1837 (based on *Carduus lanceolatus* Lour.).

Cirsium oreithales Hance in Walp. Ann. **2**: 944. 1852.

Carduus lanceolatus (non Linn.) Lour. Fl. Cochinch. 482. 1790, ed. Willd. 588. 1793, Chinese *la di tsào, sião kŷ*.

" Habitat incultus prope Cantonem Sinarum." Loureiro's description applies closely to *Cnicus chinensis* Benth., which, however, was based on *Cirsium chinense* Gardn. & Champ., not on the somewhat older *Carduus chinensis* DC., the latter being merely a new name for *Carduus lanceolatus* Lour. Hemsley (Journ. Linn. Soc. Bot. **23**: 461. 1888) notes that *Carduus linearis* Thunb. (Fl. Jap. 305. 1784) is probably identical with *Cnicus chinensis* Benth. If this proves to be the case the accepted name would be *Cnicus linearis* (Thunb.) Benth. & Hook. f. ex Franch. & Savat. Enum. Pl. Jap. **1**: 261. 1875.

Cnicus japonicus (DC.) Maxim. in Bull. Acad. Sci. St. Pétersb. **19**: 503. 1874; Mél. Biol. **9**: 322. 1874.

Cirsium japonicum DC. Prodr. **6**: 640. 1837.

Carduus tuberosus (non Linn.) Lour. Fl. Cochinch. 482. 1790, ed. Willd. 589. 1793, Chinese *thù gîn sên*.

" Habitat incultus prope Cantonem Sinarum." Loureiro's description manifestly applies to the Kwangtung form that is currently referred to *Cnicus japonicus* (DC.) Maxim. Only two species of the genus are known from the vicinity of Canton.

[93] *Cnicus* Gaertner (1791), conserved name, Vienna Code; an older one is *Carbenia* Adanson (1763).

Gerbera [94] (Gronovius) Cassini

Gerbera anandria (Linn.) Schultz-Bip. in Flora 27: 782. 1844.

Tussilago anandria Linn. Sp. Pl. 865. 1753; Lour. Fl. Cochinch. 503. 1790, ed. Willd. 614. 1793, Anamese *khoan doung hoa*, Chinese *lū chāu, koàn tūm hōa*.

" Habitat agrestis loca borealia apud Sinas." Loureiro's description applies to the Linnaean species, which is common and widely distributed in China; he doubtless secured his material from an herbalist.

Taraxacum [95] (Linnaeus) Wiggers

Taraxacum officinale [Web. in] Wigg. Prim. Fl. Holsat. 56. 1780.

Leontodon taraxacum Linn. Sp. Pl. 798. 1753.

Leontodon sinense Lour. Fl. Cochinch. 479. 1790, ed. Willd. 584, 1793, Anamese *bô coung anh*, Chinese *pû cūm tsào*.

" Habitat incultum apud Sinas." In spite of Loureiro's description of his species as an annual, I believe that he had one of the numerous forms of the common dandelion.

Lactuca (Tournefort) Linnaeus

Lactuca debilis (Thunb.) Benth. ex Maxim. in Bull. Acad. Sci. St. Pétersb. 19: 523. 1874, Mél. Biol. 9: 365. 1874.

Prenanthes debilis Thunb. Fl. Jap. 300. 1784.

Youngia debilis DC. Prodr. 7: 194. 1838.

Ixeris debilis A. Gray in Mem. Amer. Acad. II. 6: 397. 1859.

Picris repens Lour. Fl. Cochinch. 478. 1790, ed. Willd. 583. 1793, Anamese *hô hoàng lien*, Chinese *hû hôam liên*.

Borkhausia repens Spreng. Syst. 3: 652. 1826 (based on *Picris repens* Lour.).

" Habitat spontanea Cantone Sinarum." Loureiro's description clearly applies to the form not uncommon in the vicinity of Canton that is currently, and apparently correctly, referred to *Lactuca debilis* (Thunb.) Benth. The species is not known to occur in Indo-China and the Anamese name cited by Loureiro was undoubtedly derived from an herbalist, as the species has its place in Chinese materia medica.

Lactuca pinnatifida (Lour.) comb. nov.

Scorzonera pinnatifida Lour. Fl. Cochinch. 479. 1790, ed. Willd. 584. 1793.

Sonchus goraeensis Lam. Encycl. 3: 397. 1791.

Scorzonera africana Poir. in Lam. Encycl. Suppl. 5: 114. 1817 (based on *Scorzonera pinnatifida* Lour.).

Lactuca goraeensis Schultz-Bip. in Flora 25: 422. 1842; Oliv. & Hiern in Oliv. Fl. Trop. Afr. 3: 452. 1877.

" Habitat spontanea in continenti Africae Orientalis, prope Mozambicum." Judging from a comparison of descriptions, Loureiro's species is a synonym of *Lactuca goraeensis* Schultz-Bip., which occurs in the Mozambique District (Zanzibar). No true *Scorzonera* is known from tropical Africa. Oliver & Hiern (Fl. Trop. Afr. 3: 461. 1877) merely state that Loureiro's species was unknown to them. Pages 361 to 755 of Lamarck's Encyclo-

[94] *Gerbera* Cassini (1817), conserved name, Brussels Code; an older one is *Aphyllocaulon* Lagasca (1811).

[95] *Taraxacum* Wiggers (1780), conserved name, Vienna Code; an older one is *Hedypnois* Scopoli (1772).

pédie were not published until 1791, Loureiro's specific name thus being one year older than Lamarck's.

Lactuca indica Linn. Mant. 2: 278. 1771; Gagnep. in Lecomte Fl. Gén. Indo-Chine 3: 654. 1924.

> *Lactuca saligna* (non Linn.) Lour. Fl. Cochinch. 480. 1790, ed. Willd. 585. 1793, Anamese *rau diep hoang.*
>
> *Sonchus floridanus* (non Linn.) Lour. Fl. Cochinch. 480. 1790, ed. Willd. 586. 1793, Chinese *nieù li soi.*
>
> *Sonchus sibiricus* (non Linn.) Lour. Fl. Cochinch. 481. 1790, ed. Willd. 586. 1793, Chinese *xān tû.*
>
> *Prenanthes laciniata* Houtt. Nat. Hist. II 10: 381. *pl. 66. f. 1.* 1779.
>
> *Prenanthes squarrosa* Thunb. Fl. Jap. 303. 1784.
>
> *Lactuca brevirostris* Champ. in Hook. Journ. Bot. Kew Gard. Miscel. 4: 237. 1852.
>
> *Lactuca laciniata* Makino in Bot. Mag. Tokyo 17: 88. 1903.

All three of Loureiro's descriptions apply unmistakably to this common, widely distributed, variable, and much named species. For the first he states: " Habitat agrestis in Cochinchina "; the second and third: " Habitat incultus prope Cantonem Sinarum."

Lactuca sativa Linn. Sp. Pl. 795. 1753; Lour. Fl. Cochinch. 479. 1790, ed. Willd. 585. 1793, Anamese *rau diep taù,* Chinese *yĕ tsái kiú.*

> *Lactuca indica* (non Linn.) Lour. Fl. Cochinch. 480. 1790, ed. Willd. 585. 1793, Anamese *rau diep nhà.*

The first is indicated as: " Habitat culta in Cochinchina, & Macai Sinarum ex semine ab Europa oriundo," and the second: " Colitur in Cochinchina, sapore multo inferior sativa." Manifestly both descriptions appertain to forms of the common lettuce.

Cichorium (Tournefort) Linnaeus

Cichorium endivia Linn. Sp. Pl. 813. 1753; Lour. Fl. Cochinch. 478. 1790, ed. Willd. 583. 1793, Anamese *khô thao,* Chinese *khú tsái.*

" Habitat in locis Borealibus imperii Sinensis." The endive is cultivated for food in China and Loureiro was doubtless correct in his interpretation of the Linnaean species. He probably secured his material from some dealer in drug plants.

Crepis (Vaillant) Linnaeus

Crepis japonica (Linn.) Benth. Fl. Hongk. 194. 1861.

> *Prenanthes japonica* Linn. Mant. 1: 107. 1767.
>
> *Lapsana rhagadiolus* (non Linn.) Lour. Fl. Cochinch. 481. 1790, ed. Willd. 587. 1793, Anamese *cai nhà tlòi.*

" Habitat inculta in hortis, & agris Cochinchinae." The characters given by Loureiro conform very closely with those of this wide-spread, common, and well-known weed.

COMPOSITAE OF UNCERTAIN GENERIC STATUS

Artemisia aquatica Lour. Fl. Cochinch. 490. 1790, ed. Willd. 598. 1793, Anamese *cây thuy tung,* Chinese *hài tūm.*

" Habitat culta in Cochinchina, & China, puto, quod etiam agrestis. Aquam amat, indeque nomen vernaculum. Per multos annos servatur in vase aqua pleno crescens, & florens, a terra prosus remota." Hemsley (Journ. Linn. Soc. Bot. **23**: 441. 1888) considers that this is probably not an *Artemisia*, while Gagnepain (Lecomte Fl. Gén. Indo-Chine **3**: 584. 1924) definitely excludes it from the genus; neither suggests any identification.

Centaurea cochinchinensis DC. Prodr. **6**: 601. 1837 (based on *Centaurea phrygia* Lour.).
> *Centaurea phrygia* (non Linn.) Lour. Fl. Cochinch. 508. 1790, ed. Willd. 621. 1793, Anamese *bac dâu loung*.

" Habitat inculta in Cochinchina." Certainly no *Centaurea* is represented, but I am unable to suggest the proper reduction of this species. It is not accounted for by Gagnepain in his treatment of the Compositae of Indo-China (Lecomte Fl. Gén. Indo-Chine **3**: 448–663. 1924). If Loureiro's description be correct in the involucral bracts being " subulatis, imbricatis, in senectute recurvatis, *plumosis, argenteis*," the species ought to be determinable, yet I am unable to refer it with any degree of satisfaction to any of the genera admitted by Gagnepain as occurring in Indo-China.

Serratula scordium Lour. Fl. Cochinch. 483. 1790, ed. Willd. 590. 1793, Anamese *cây muoi túoi, trach lan*, Chinese *tsĕ lân*.

" Habitat spontanea, cultaque in Cochinchina, & China: hic vero duae aliae plantae, habitu prorsus diversae eodem nomine indicantur." It is suspected that Loureiro's description was based on more or less fragmentary material, perhaps secured from an herbalist. I do not know any eastern Asiatic composite having the combination of characters indicated in the description.

Genera and Species of Wholly Uncertain Status

Abutua africana Lour. Fl. Cochinch. 631. 1790, ed. Willd. 776. 1793.

" Habitat agrestis in ora Orientali Africae." The type of the genus *Abutua* is *A. indica* Lour. = *Gnetum indicum* Merr. *Abutua africana* was described from a sterile specimen. From the description of the leaves as " ternatis " and again as " 3-nata " (i.e., 3-foliolate) *Abutua africana* cannot possibly represent the genus *Gnetum*. Probably some species of the Leguminosae is indicated. The description is very inadequate.

Buddleia ternata Lour. Fl. Cochinch. 72. 1790, ed. Willd. 91. 1793, Anamese *cây lao linh*.

" Habitat agrestis in Cochinchina." I can suggest no reduction for this imperfectly described species. It is manifestly not a *Buddleia*. The description is doubtless incorrect in some details.

Callicarpa umbellata Lour. Fl. Cochinch. 70. 1790, ed. Willd. 88. 1793, Anamese *cây ma ca*.
> *Agonon umbellata* Raf. Sylva Tellur. 161. 1838 (based on *Callicarpa umbellata* Lour.).

" Habitat in sylvis Cochinchinae." If Loureiro's description be correct this is not a *Callicarpa* because of the alternate leaves, 5-flowered umbels and 4-fid calyces, yet I can suggest no reduction at this time; Schauer merely queries " An *Premnae* species? " *Ehretia* of the Boraginaceae is also suggested by the description, but the 4-merous flowers, and sessile stamens (anthers) and stigmas eliminate this group as a possibility. In some respects Myrsinaceae is also suggested, but here as with Verbenaceae and Boraginaceae there are also discrepancies in Loureiro's description. It is highly probable that the fruit characters were deliberately added by Loureiro to make his description conform to the generic characters of *Callicarpa*. The native name cited throws no light on the situation; Gagnepain records the same form as one of the Anamese names of the totally different *Antidesma ghaesembilla* Gaertn. The genus *Agonon* Rafinesque was based on Loureiro's description.

Canarina zanguebar Lour. Fl. Cochinch. 195. 1790, ed. Willd. 240. 1793.

" Habitat inculta in ora Africae Zanguebar." A. de Candolle (Prodr. **7**: 422. 1839) repeats Loureiro's description, with the comment: " Verisimiliter generis diversi propter folia alterna et capsulam basi dehiscentam." This is apparently neither a *Canarina* nor a representative of the Campanulaceae. One familiar with the flora of Zanzibar could doubtless place it from Loureiro's description through the process of elimination.

Cephalanthus procumbens Lour. Fl. Cochinch. 67. 1790, ed. Willd. 84. 1793, Anamese *deei trôp*.

> *Silamnus procumbens* Raf. Sylva Tellur. 61. 1838 (based on *Cephalanthus procumbens* Lour.).
>
> *Stilbe procumbens* Spreng. Syst. **1**: 418. 1825 (based on *Cephalanthus procumbens* Lour.).
>
> *Cephalanthus dioicus* Lour.[96] ex Gomes in Mem. Acad. Sci. Lisb. Cl. Sci. Mor. Pol. Bel.-Let. n.s. **4**(1): 26. 1868.

" In Cochinchina." The description, if correct, indicates a very characteristic plant, but not a *Cephalanthus*, nor even a rubiaceous species. I can suggest no reduction. The description is doubtless erroneous in important characters. The genus *Silamnus* Raf. was based wholly on Loureiro's description of this species.

Diosma asiatica Lour. Fl. Cochinch. 161. 1790, ed. Willd. 200. 1793.

> *Pseudiosma asiatica* Juss. in Mém. Mus. Hist. Nat. [Paris] **12**: 519. 1825 (based on *Diosma asiatica* Lour.).

" Habitat in monte *Hòn chén* ex adverso urbis Huaeae Cochinchinae metropolis." *Pseudiosma* was proposed by Jussieu as a new generic name for the species Loureiro erroneously ascribed to the Linnaean genus *Diosma*, although Jussieu saw no material representing Loureiro's species. In Index Kewensis *Pseudiosma* is reduced to *Zanthoxylum* or *Euodia*. The alternate leaves eliminate the latter, and the assumed simple leaves and unarmed characters of the plant eliminate the former, while the androecium characters eliminate both. Loureiro's description, if correct, indicates a distinctly characteristic species, but I do not recall any known Indo-China plant that presents the combination of characters given for this species. It is suspected that the description was based on a mixture of material, or that it is erroneous in essential characters.

Dorstenia chinensis Lour. Fl. Cochinch. 90. 1790, ed. Willd. 114. 1793, Anamese *bach chi*, Chinese *pě chí*.

> *Procris chinensis* Spreng. Syst. **3**: 846. 1826 (based on *Dorstenia chinensis* Lour.).

" Habitat in provinciis borealibus imperii Sinensis." The description is very imperfect, manifestly based on fragmentary material secured from some herbalist and on an illustration which Loureiro saw in some Chinese publication. Weddell was apparently correct in excluding the species from the Urticaceae. Hemsley (Journ. Linn. Soc. Bot. **26**: 456. 1894) notes that Bretschneider was unable to locate the plant by the Chinese name cited

---

[96] A Loureiro herbarium name here first published by Gomes.

by Loureiro and states that it was " possibly an *Elatostema*." But *Elatostema* does not grow in northern China, and Loureiro's description, poor as it is, manifestly does not apply to this genus.

Euclea herbacea Lour. Fl. Cochinch. 629. 1790, ed. Willd. 773. 1793, Chinese *xĕ lin tsù*.

" Habitat inculta prope Cantonem Sinarum." The description is unusually short, an herbaceous dioecious plant with 5 sepals, 5 petals and 15 stamens. The leaves are not described. Hiern (Trans. Cambr. Philos. Soc. **12**: 106. 1873) suggests that this may be some euphorbiaceous plant. I know of no Kwangtung species that presents the combination of floral characters indicated by Loureiro for *Euclea herbacea*.

Galium tuberosum Lour. Fl. Cochinch. 79. 1790, ed. Willd. 99. 1793, Anamese *hùynh tinh*, Chinese *hoâm cĭm*.

" Habitat cultum in agris Cochinchinae, & Chinae." Hemsley (Journ. Linn. Soc. Bot. **23**: 395. 1888) quotes Bretschneider to the effect that the Chinese name cited by Loureiro appertains to various species of *Polygonatum* of the Liliaceae. It is probable that Loureiro had fragmentary material, probably secured from an herbalist, on which he based his description. The root characters do not appertain to *Galium*. The *Galium* characters given by him may have been added to make his description conform to the characters of the genus.

Glabraria tersa (non Linn.) Lour. Fl. Cochinch. 471. 1790, ed. Willd. 576. 1793, Anamese *cây pháo luói*.

" Habitat in altis sylvis Cochinchinae." Loureiro definitely states that he did not examine the flowers and his description is otherwise so imperfect that I have been unable to refer this species to its proper group. It is probable that he had specimens of some lauraceous plant, perhaps a *Litsea*, as his conception of what he mistook to be the Linnaean species must have been based on *Lignum leve angustifolium* (*minus*) Rumph. (Herb. Amb. **3**: 71. *pl. 44*) which he cites as a synonym, and which represents some species of *Litsea*. Linnaeus (Mant. **2**: 276. 1771) seriously erred in referring the Rumphian illustration to his *Glabraria tersa*. The actual type in the Linnaean herbarium, to which the Linnaean description conforms, is a species of *Boschia* of the Bombacaceae; see Merrill Interpret. Herb. Amb. 235. 1917, sub *Litsea* sp.

Ixora violacea Lour. Fl. Cochinch. 76. 1790, ed. Willd. 97. 1793, Anamese *buóm rùng tiá*.

" Habitat loca inculta, ubi arborum agrestium ramis innititur in Cochinchina." The description is inadequate. It is probable that no rubiaceous plant is represented; in some respects the description suggests *Loranthus*.

Lycium cochinchinense Lour. Fl. Cochinch. 134. 1790, ed. Willd. 165. 1793, Anamese *cây son lút*.

" Habitat in sylvis Cochinchinae." I can make no suggestion as to what this may be; it is clearly not a *Lycium* and is not a solanaceous plant. It is not mentioned by Bonati in his treatment of the Solanaceae of Indo-China (Lecomte Fl. Gén. Indo-Chine **4**: 313–341. 1915–27). Dunal (DC. Prodr. **13**(1): 511. 1852) compiled a short description from Loureiro's original and queried: " An *L. subglobosum?* ".

Penaea scandens Lour. Fl. Cochinch. 73. 1790, ed. Willd. 92. 1793, Anamese *deei dinh dang.*

Loureiro's material was from Indo-China " nostrae Asiaticae, & Cochinchinae in-diginae sunt," and he notes that the characters of his two species did not conform entirely with those of the Linnaean genus *Penaea; Penaea nitida* Lour. is *Gluta nitida* (Lour.) Merr. I cannot make a definite reduction of *P. scandens* Lour. from the data at present available. Except for the many-seeded fruit, the species might be a convolvulaceous one, but this is unlikely in view of the identity of Loureiro's other species of *Penaea.*

Primula sinensis Lour. Fl. Cochinch. 105. 1790, ed. Willd. 128. 1793, Anamese *ngaoc trâm hoa,* Chinese *yù tsuan hõa.*

" Habitat in imperio Sinensi." The description is so indefinite that I cannot suggest a reduction for the species. It is clearly neither a *Primula* nor a primulaceous plant. One would suspect, from the habitat cited, and from the Anamese as well as a Chinese name, that Loureiro secured his material from an herbalist, yet he indicates no medicinal uses for the plant. The other species he described, *Primula mutabilis* Lour., is *Hydrangea opuloides* (Lam.) K. Koch. Pax and Knuth (Pflanzenreich 22(IV–237): 160. 1905) in correctly excluding the species from *Primula,* make no suggestion as to what genus might be represented.

Ptelea ovata Lour. Fl. Cochinch. 82. 1790, ed. Willd. 104. 1793, Anamese *cây hôt man.*

" Habitat agrestis in Cochinchina." I cannot suggest any definite reduction from Loureiro's description. Clearly no *Ptelea* is represented and probably no rutaceous plant. Loureiro described a shrub with simple, entire, glabrous leaves, small 4-merous flowers, and coriaceous petals. He saw only staminate flowers and thought that the species was a dioecious one. Possibly some genus of Euphorbiaceae is represented. D. Don (Gen. Syst. 2: 12. 1832) erred in reducing Loureiro's species to *Ptelidium ovatum* Poir. (*Seringia ovata* Spreng.), but probably followed Sprengel (Syst. 1: 441. 1825) as the latter placed Loureiro's species as a doubtful synonym of *Seringia ovata* Spreng. *Ptelidium ovatum* Poir. was based on Madagascar material.

Salsola didyma Lour. Fl. Cochinch. 173. 1790, ed. Willd. 216. 1793.

Isgarum didymum Raf. Fl. Tellur. 3: 46. 1837 (based on *Salsola didyma* Lour.).

" Habitat inculta in insula Mozambicco, in Africa." Rafinesque proposed the new generic name *Isgarum* based on *Salsola didyma* Lour. because of the 2-lobed, 2-celled, 1-seeded capsules. Moquin (DC. Prodr. 13(2): 192. 1849) states: " Ob capsulam bilocu-larem et monospermam (vel unilocularem et dispermam) certe non *Salsolacea.*" De Dalla Torre and Harms (Gen. Siphon. 145. 1900), apparently in error, place *Isagarum* Raf. as a definite synonym of *Salsola.* Neither of the binomials given above is accounted for in the Flora of Tropical Africa, and I am unable to suggest a reduction of *Salsola didyma* Lour. (*Isgarum didymum* Raf.) from the description alone.

Sarothra gentianoides (non Linn.) Lour. Fl. Cochinch. 182. 1790, ed. Willd. 227. 1793, Anamese *cây cu gà.*

Sarothra loureiriana Schult. in Roem. & Schult. Syst. 6: 679. 1820 (based on *Sarothra gentianoides* Lour.).

" Habitat in locis arenosis Cochinchinae, prope Metropolim [Hue]." The description, if correct, is definite and from it the plant should be identifiable, yet I cannot make any satisfactory suggestion as to what it may be.

Sideroxylon cantoniense Lour. Fl. Cochinch. 122. 1790, ed. Willd. 151. 1793, Chinese
sān cŏt.

"Habitat in suburbis Cantoniensibus Sinarum." I cannot make any suggestion as
to what Loureiro intended to characterize; certainly nothing allied to *Sideroxylon* is represented. It is suspected that there was either a mixture of material on Loureiro's part,
or serious misinterpretations of morphological characters.

Thalictrum sinense Lour. Fl. Cochinch. 346. 1790, ed. Willd. 423. 1793, Anamese *bôi mâu*,
Chinese *pói mù.*

"Habitat agreste in China." An altogether doubtful species, certainly not a *Thalictrum*, and apparently not a ranunculaceous plant. De Candolle (Syst. 1: 187. 1818, Prodr.
1: 16. 1824) states: "An Thalictrum? an Ranunculaceae?" Loureiro's imperfect description was apparently based on material secured from an herbalist. G. Don (Gen.
Syst. 1: 15. 1831) suggests that perhaps a species of *Ranunculus* is represented.

INDEX TO NATIVE NAMES (MOSTLY ANAMESE AND CHINESE)

INDEX TO LATIN NAMES

413

www.ingramcontent.com/pod-product-compliance
Lightning Source LLC
Chambersburg PA
CBHW081339190326
41458CB00018B/6052